EXTRAGALACTIC RADIO SOURCES

INTERNATIONAL ASTRONOMICAL UNION

UNION ASTRONOMIQUE INTERNATIONALE

EXTRAGALACTIC RADIO SOURCES

PROCEEDINGS OF THE 175TH SYMPOSIUM OF THE
INTERNATIONAL ASTRONOMICAL UNION,
HELD IN BOLOGNA, ITALY
10–14 OCTOBER 1995

EDITED BY

R. EKERS

Australia Telescope, CSIRO,
Epping, Australia

C. FANTI

Istituto di Radioastronomia,
CNR, Bologna, Italy

and

L. PADRIELLI

Istituto di Radioastronomia,
CNR, Bologna, Italy

KLUWER ACADEMIC PUBLISHERS

DORDRECHT / BOSTON / LONDON

Library of Congress Cataloging-in-Publication Data

Extragalactic radio sources / edited by R. Ekers and C. Fanti and L.
  Padrielli.
      p.   cm.
  Proceedings of the 175th IAU Symposium held in Bologna, October
10-14, 1995.
  Includes indexes.
  ISBN-13: 978-0-7923-4122-2
  1. Radio sources (Astronomy)--Congresses. 2. Galaxies-
-Congresses.   I. Ekers, R. D. (Ron D.) II. Fanti, C.
III. Padrielli, L. IV. International Astronomical Union. Symposium
(175th : 1995 : Bologna, Italy)
  QB857.E938 1996
  523.8'2--dc20   96-9022

ISBN-13: 978-0-7923-4122-2   e-ISBN-13: 978-94-009-0295-4
DOI: 10.1007/978-94-009-0295-4

Published on behalf of
the International Astronomical Union
by
Kluwer Academic Publishers, P.O. Box 17, 3300 AA Dordrecht, The Netherlands.

Kluwer Academic Publishers incorporates
the publishing programmes of
D. Reidel, Martinus Nijhoff, Dr W. Junk and MTP Press.

Sold and distributed in the U.S.A. and Canada
by Kluwer Academic Publishers,
101 Philip Drive, Norwell, MA 02061, U.S.A.

In all other countries, sold and distributed
by Kluwer Academic Publishers Group,
P.O. Box 322, 3300 AH Dordrecht, The Netherlands.

TABLE OF CONTENTS

Relation between Radio and Other Wavelenghts

Unification Issues

Radio Source Modelling and Emission Mechanisms

Cosmological Implications

PREFACE

IAU 175 is a Symposium on extragalactic radio sources held in Bologna, October 10–14, 1995 to celebrate the 100th anniversary of Marconi's successful experiment on radio broadcasting. This is most appropriate, not only because Bologna is Marconi's home town, but because radio astronomy is a science made possible by the technical innovation resulting from the commercial exploitation of radio communications.

Of all the areas of astronomy influenced by radio astronomy surely the most far–reaching has been the discovery of radio galaxies by John Bolton and his colleagues nearly 50 years ago. In these 50 years much has been learned, about both the radio galaxies and the Universe they probe, but many questions remain unanswered and many more have been raised since the last meeting on extragalactic radio sources held in Albuquerque more than a decade ago. Bologna, with its rich tradition of scientific discussion and hospitality, and with a long background in radio astronomy provided the perfect environment for this Symposium on radio sources.

There were 249 active participants from 24 different countries at the Symposium. 18% were women and they contributed a similar fraction of the talks. While this is a little better than the average for IAU membership if you look at the distribution over countries you will see that Italy alone provided almost half of the female astronomers present! Furthermore the Bologna women included the co-chair and 100% of the LOC!

The main objective of this Symposium was to update our knowledge of the physics and statistical properties of extragalactic radio sources since the Albuquerque Symposium.

This past decade has seen major advances in resolution, sensitivity and imaging techniques. High quality images are now being routinely produced with resolution down to better than a milliarcsecond.

While radio astronomy was the unifying theme of this Symposium, cross fertilization was achieved by inviting experts from a range of other fields to review results from studies at γ, X–ray, Optical and IR wavelengths. There was no emphasis on AGN's or starbursts galaxies at this meeting because these objects have been covered in many other meetings. Likewise the discussions on unification were treated in less detail here than at other

AGN meetings. We also emphasized the powerful radio galaxies and did not cover radio emission from normal galaxies like our own.

Part of the session on theoretical modeling was held as a parallel session. Although this provided participants more opportunities to present results, the concept of parallel sessions in a Symposium on one topic of astronomy was not well accepted by the participants.

A special panel style session was held to discuss the new big surveys of radio sources. A noteworthy outcome was the strong group endorsement of an open policy for the dissemination of information from the surveys.

There are some small changes in the organization of the proceedings compared to the presentations in the meeting which have been made to give a more coherent structure and to incorporate the posters in the appropriate places. This may result in an occasional non-sequiter in the discussion but the proceedings are now more coherently organized from small to large size, from low to high power and from low to high frequency.

I would like to thank Lucia Padrielli and her LOC team who have had to sacrifice their active participation in the scientific discussions in order to provide the organizational effort needed to run the meeting so smoothly. This is always a difficult undertaking and in this occasion especially so as they had to deal with the first use of a brand new lecture theater. Also thanks to the many postdocs and students who manned the microphones, distributed questionnaires and performed all the other miscellaneous duties required to make a large meeting run smoothly.

Finally many thanks to the SOC members who each took responsibility for a segment of the program, and especially to my co-chair Carla Fanti for taking on much more than an equal share in the scientific organization of this meeting.

The editors of these Proceedings warmly thank Tasso Tzioumis for his help in deciphering the (sometimes illegible) handwriting of the about 150 questions and corresponding answers and for carefully checking them after they were typed.

Ron Ekers

SCIENTIFIC ORGANIZING COMMITTEE

- G. Bicknell
- R. Ekers (co–chair)
- C. Fanti (co–chair)
- M. Inoue
- V. Kapahi
- K. Kellermann
- G. Miley
- B. Partridge
- R. Perley
- P. Scheuer

LOCAL ORGANIZING COMMITTEE

- C. Fanti
- L. Feretti
- L. Gregorini
- L. Padrielli (chair)
- P. Parma
- T. Venturi

LIST OF PARTICIPANTS

Zulema Abrahm
IAG/USP BRAZIL
CP 9638
01065-970 Sao Paulo, Brazil
zulema@iagusp.ansp.br

Chidi E. Akujor
IMO State University
P.M.B. 2000
Owerri, Nigeria
chidi@mpifr-bonn.mpg.de

Walter Alef
Max Planck Inst. Radioastronomie
Auf dem Hugel 69,
D-53121 Bonn, Germany
walef@mpifr-bonn.mpg.de

Margo F. Aller
University of Michigan
Department of Astronomy,
817 Dennison Building,
Ann Arbor, MI 48109-1090, U.S.A.
margo@astro.lsa.umich.edu

Hugh D. Aller
University of Michigan
Department of Astronomy,
830 Dennison Building,
Ann Arbor, Michigan,
48109-1090, U.S.A.
hugh@astro.lsa.umich.edu

M. Ramana Athreya
NCRA - TIFR
P.O. Bag 3
Pune University Campus
Pune - 411007, India
ramana@gmrt.ernet.in

Joanne C. Baker
MRAO, Cavendish Laboratory,
Madingley Road,
Cambridge, CB3 0HE, U.K.
jcb@mrao.cam.ac.uk

Sandro Bardelli
Istituto di Radioastronomia
Via P. Gobetti, 101
40133 Bologna, Italy
bardelli@astbo1.bo.cnr.it

Tony Beasley
NRAO
P.O. Box O
Socorro, NM 87801, U.S.A.
tbeasley@aoc.nrao.edu

Robert Becker
UC - Davis/LLNL
Physics Dept.
Univ. of California
Davis CA 95616, U.S.A.
bob@igpp.llnl.gov

Mitchell C. Begelman
JILA, University of Colorado
Box 440,
Boulder, CO 80309-0440, U.S.A.
mitch@jila.colorado.edu

Giuseppe Bertin
Scuola Normale Superiore
P.za Cavalieri 7,
I-56126 Pisa, Italy
bertin@vaxsns.sns.it

Geoffrey V. Bicknell
ANU Astroph. Theory Centre
School Mathematical Sc.
Australian National University
Canberra, ACT 0200 Australia
geoff@mso.anu.edu.au

Luc Binette
Observatoire de Lyon
9, Avenue Charles-Andre
F-69561 Saint-Genis-Laval
Cedex, France
binette@obs.univ-lyon1.fr

Katherine Blundell
Oxford University Astrophysics
Keble Road
Oxford, OXI 3RH, U.K.
k.blundell1@physics.oxford.ac.uk

Gianluigi Bodo
Osservatorio Astronomico-Torino
Strada dell'Osservatorio, 20
10025 Pino Torinese (TO), Italy
bodo@to.astro.it

Hans Boeheringer
MPI fuer Extraterr. Physik
Giessenbachstrasse
85740 Garching, Germany
hxb@rosat.mpe-garching.mpg.de

Marco Bondi
Istituto di Radioastronomia
Via Gobetti, 101
40129 Bologna, Italy
bondi@astbo1.bo.cnr.it

Roy Booth
Onsala Space Observatory
Onsala, S-43992, Sweden
roy@oso.chalmers.se

Silvia Bosio
Osservatorio Astronomico-Torino
Strada Osservatorio, 20
10025 Pino Torinese (TO), Italy
bosio@to.astro.it

Geoffrey Bower
UC Berkeley Astronomy Dept.
601 Campbell
Berkeley, CA 94720, U.S.A.
gcbower@astron.berkeley.edu

Alessandro Braccesi
Dipartimento di Astronomia
Via Zamboni, 33 - Bologna, Italy
braccesi@astbo3.bo.astro.it

Wolfgang Brinkmann
MPI fuer Extraterr. Physik
Giessenbachstrasse
85740 Garching, Germany
wpb@mpe-garching.mpg.de

Gianfranco Brunetti
Istituto di Radioastronomia
Via Gobetti, 101
40129 Bologna, Italy
gbrunetti@astbo1.bo.cnr.it

Bernard Burke
M.I.T
Research Labor.Electronics
Room n.26-331
Cambridge,02139-4307
Massachusetts U.S.A

Alessandro Caccianiga
Osservatorio Astronomico-Brera
Via Brera, 28
20121 Milano, Italy
caccia@bach.mi.astro.it

Massimo Calvani
Osservatorio Astronomico
Vicolo Osservatorio
35122 Padova, Italy
calvani@astrpd.pd.astro.it

Alessandro Capetti
Space Telescope Science Inst.
3700, San Martin Drive,
Baltimore, MD 21209, U.S.A.
capetti@stsci.edu

Chris Carilli
SAO
Mail Stop 6
60 Garden St.
Cambridge, MA 02138, U.S.A.
ccarilli@cfa.harvard.edu

Rene' Carrillo
Inst. Astronomia, UNAM
Ag. Postal 70-264
Cd. Universitaria
C.P. 04510, Mexico D.F., Mexico
rene@astroscu.unam.mx

Tim Cawthorne
Univ. Central Lancashire
Dept. of Physics and Astronomy
Preston, Lancashire
PRI 2HE, U.K.
tvc@star.uclan.ac.uk

Patrick Charlot
Observatoire de Paris
61 Avenue de l'Observatoire
75014 Paris, France
charlot@mesioa.obspm.fr

K. Tadeusz Chyzy
Astronom. Obs. Jagellonian Univ
ul. Orla 171
30-244 Krakow, Poland
uochyzy@kinga.cyf-kr.edu.pl

Paolo Ciliegi
Osservatorio Astronomico - Brera
Via Brera, 28
20121 Milano, Italy
ciliegi@bach.mi.astro.it.

Andrea Cimatti
IGPP-LLNL
7000 East Ave.
P.O. Box 808
L-413 Livermore CA, 94550 U.S.A.
cimatti@igpp.llnl.gov

Marshall Cohen
Caltech
Caltech 105-24
Pasadena, CA 91125 U.S.A.
mhc@astro.caltech.edu

Andrea Comastri
Osservatorio Astronomico
Via Zamboni, 33
40126 Bologna, Italy
comastri@astbo1.bo.cnr.it

James J. Condon
NRAO
520 Edgemont Road
Charlottesville,
VA 22903, U.S.A.
jcondon@nrao.edu

John Conway
Onsala Space Observatory
Onsala, S-43992
Sweden
Jconway@oso.chalmers.se

William Cotton
NRAO
520 Edgemont Road
Charlottesville,
VA 22903-2475 U.S.A
bcotton@nrao.edu

Philippe Crane
European Southern Observatory
Karl-Schwarzschild-Strasse2,
Garching-bei-Munchen
D-85748 Germany
crane@eso.org

Rustam D. Dagkesamanskii
Lebedev Physical Institute
Astro Space Center
Radioastronomy Department
Leninsky Prosp. 53
117924 Moscow, Russia
rdd@rasfian.serpukhov.su

Daniele Dallacasa
Istituto di Radioastronomia
Via P. Gobetti, 101
40129 Bologna, Italy
dallacasa@astbo1.bo.cnr.it

Ruth A. Daly
Princeton University
Dept of Physics
Princeton, NJ, 08544, U.S.A.
Daly@pupgg.princeton.edu

Ger de Bruyn
NFRA & Kapteyn Institute
Postbus 2
7990 AA Dwingeloo,
The Netherlands
ger@nfra.nl

Jane Dennett-Thorpe
MRAO
Cavendish Labs.
Madingley Road,
Cambridge, CB3 0HE, U.K.
jdt@mrao.cam.ac.uk

Hans De Ruiter
Osservatorio Astronomico
Via Zamboni, 33
40126 Bologna, Italy
deruiter@astbo1.bo.cnr.it

Vincent Despringre
Laboratoire d'astrophysique
14 Avenue Edouard Belin
31400 Toulouse, France
despring@obs-mip.fr

David De Young
Kitt Peak National Observatory
950 N. Cherry Ave.
P.O. Box 26732
Tucson, Az 85726, U.S.A.
deyoung@noao.edu

Phil Diamond
NRAO - SOC
P.O. Box O,
Socorro, NM 87801, U.S.A.
pdiamond@nrao.edu

Tristano di Girolamo
Istituto di Radioastronomia
Via Gobetti, 101
40129 Bologna, Italy
tdigirolamo@astbo1.bo.cnr.it

S. di Serego Alighieri
Osserv. Astrofisico di Arcetri
Largo E. Fermi,
I-50125 Firenze, Italy
sperello@arcetri.astro.it

Vladimir A. Dogiel
P.N.Lebedev Physical Institute
117924 Moscow,
Leninski pr.53, Russia
vad@mpe-garching.mpg.de

Oliver Dreissigacker
LSW Heidelberg
Konigstuhl
D-69117 Heidelberg, Germany
odreissi@mail.lsw.uni-heidelberg.de

James S. Dunlop
University of Edinburgh
Institute for Astronomy
Department Physics and Astronomy
Royal Observatory Blackford Hill
Edinburg EH9 3HJ, U.K.
jsd@roe.ac.uk

Philip G. Edwards
ISAS
3-1-1 Yoshinodai,
Sagamihara,
Kanagawa 229, Japan
pge@orihime.isaslan1.isas.ac.jp

Jean Eilek
Physics Dept. New Mexico Tech
Socorro, NM 87801 U.S.A.
jeilek@aoc.nrao.edu

Ron Ekers
Australia Telescope
P.O. Box 76
Epping 2121
Australia
rekers@atnf.csiro.au

Silviarosa Facondi
Dipartimento di Astronomia
Via Zamboni, 33 - Bologna, Italy
sfacondi@astbo1.bo.cnr.it

Heino Falcke
Astronomy Dept.
University of Maryland
College Prk
MD 20742, U.S.A.
hfalcke@mpifr-bonn.mpg.de

Carla Fanti
Istituto di Radioastronomia
Via Gobetti, 101
40129 Bologna - Italy
cfanti@astbo1.bo.cnr.it

Roberto Fanti
Istituto di Radioastronomia
Via Gobetti, 101
40129 Bologna - Italy
rfanti@astbo1.bo.cnr.it

Luigina Feretti
Istituto di Radioastronomia
Via Gobetti, 101
40126 Bologna, Italy
lferetti@astbo1.bo.cnr.it

Attilio Ferrari
Universita' di Torino
Istituto Fisica Generale
Via P. Giuria, 1
10125 Torino, Italy
ferrari@astto2.to.astro.it

Pierre Ferruit
CRAL
9 av.Charles Andre
F-69561 St. Genis Laval Cedex,
France
ferruit@obs.univ-lyon1.fr

Edward Fomalont
NRAO
520 Edgemont Road
Charlottesville,
VA 22903, U.S.A.
efomalont@nrao.edu

Jose A. Font Roda
Valencia University
Departamento de Fisica Teorica
Dr. Moliner, 50
12005 Burjassot, Valencia
Spain
font@kerr.ific.uv.es

Didier Fraix-Burnet
Lab. Astrophysique Grenoble
BP 53X F-38041,
Grenoble Cedex, France
fraix@gag.observ-gr.fr

Denise C. Gabuzda
Astro Space Center
Radioastronomy Dept.
Lebedev Physical Institute
53 Leninsky Prospekt
117924 Moscow, Russia
gabuzda@sci.lpi.ac.ru

Simon Garrington
NRAL Jodrell Bank
University of Manchester
Macclesfield, Cheshire,
SK11 9DL, U.K.
stg@jb.man.ac.uk

G. Gavazzi
Osservatorio Astronomico-Brera
Via Brera, 28
20121 Milano, Italy
gavazzi@bach.mi.astro.it

Richard Gelderman
NRC, NASA/GSFC
code 681
Lab. for Astronomy
and Solar Physics,
Greenbelt, MD 20771, U.S.A.
gelderman@gsfc.nasa.gov

Gabriele Ghisellini
Osservatorio Astronomico-Torino
Strada Osservatorio, 20
10025 Pino Torinese (TO)
Italy
ghisellini@to.astro.it

Isabella Gioia
Istituto di Radioastronomia
Via Gobetti, 101
40129 Bologna, Italy
igioia@astbo1.bo.cnr.it

Gabriele Giovannini
Istituto di adioastronomia
Via Gobetti, 101
40129 Bologna, Italy
ggiovannini@astbo1.bo.cnr.it

Nectaria Gizani
Manchester / Jodrell Bank
Nuffield Radio Astronomy
Laboratories
Macclesfield
Cheshire SK11 9DL, U.K.
ng@jb.man.ac.uk

Ignacio Gonzalez-Serrano
Instituto de Fisica de Cantabria
Facultad de Ciencias,
Universidad de Cantabria
39005 Santander, Spain
gserrano@astro.unican.es

Miller Goss
NRAO
P.O. Box O
Socorro, NM 87801, U.S.A.
mgoss@nrao.edu

Ann Gower
University of Victoria
Dept. of Physics and Astronomy
University of Victoria
P.O. Box 3055 MS 7700
Victoria, BC, Canada
agower@oyster.phys.uvic.ca

Loretta Gregorini
Istituto di Radioastronomia
Via Gobetti, 101
40129 Bologna, Italy
gregorini@astbo1.bo.cnr.it

Gavril Grueff
Univ. Bologna, Dip. Astronomia
Via Zamboni, 33
40126 Bologna, Italy
ggrueff@astbo1.bo.astro.it

Carlotta Gruppioni
Istituto di Radioastronomia
Via Gobetti, 101
40129 Bologna, Italy
cgruppioni@astbo1.bo.cnr.it

Leonid Gurvits
Joint Inst. for VLBI in Europe
P.O. Box 2
7990 AA Dwingeloo,
The Netherlands
lgurvits@nfra.nl

Francois Hammer
DAEC - Obs. Paris - Meudon
92195 Meudon Principal,
France
hammer@gin.obspm.fr

Martin J. Hardcastle
Mullard Radio Astronomy Obs.
Cavendish Laboratory,
Madingley Road,
Cambridge CB3 OHE, U.K.
mjh22@mrao.cam.ac.uk

Philip E. Hardee
University of Alabama
Dept. of Physics & Astronomy
Box 870324
Tuscaloosa,
AL 35487-0324, U.S.A.
hardee@venus.astr.ua.edu

Daniel E. Harris
Centre for Astrophysics
MS-3, 60 Garden St.
Cambridge, MA 02138, U.S.A.
harris@cfa.harvard.edu

Jochen Heidt
Landessternwarte Heidelberg
Koenigstuhl,
69117 Heidelberg, Germany
jheidt@hp2.lsw.uni-heidelberg.de

Jackie Hewitt
Massachusetts Inst of Technology
Department of Physics
Room 26-327
Cambridge, MA 02139, U.S.A.
jhewitt@mit.edu

Paul Hickson
University of British Columbia
c/o Oss. Astronomico di Brera
Via Brera 28
20121 Milano, Italy
hickson@bach.mi.astro.it

Stephen Higgins
John Moores University
Astrophysics Group,
School of Computing and
Mathematical Sciences
Byrom Street
Liverpool L3 3AF, U.K.
swh@staru1.livjm.ac.uk

Hisashi Hirabayashi
ISAS
3-1-1 Yoshinodai,
Sagamihara, Kanagawa, Japan
hirax@vsop.isas.ac.jp

Dick Hunstead
University of Sidney
School of Physics
NSW 2006, Australia
rwh@astrop.physics.usyd.edu.au

John B. Hutchings
DAO, Canada
Dominion Astrophysical Obs.
5071 West Saanich Rd,
Victoria, B. C. V8X 4M6, Canada
Hutchings@dao.nrc.ca

Chris Impey
University of Arizona
Steward Observatory
Tucson, AZ 85721, U.S.A.
cimpey@as.arizona.edu

Hajime Inoue
ISAS
3-1-1, Yoshinodai,
Sagamihara, Kanagawa 229, Japan
inoue@astro.isas.ac.jp

Makoto Inoue
Nobeyama Radio Observatory
Osawa, Mitaka, Tokio 181
JAPAN
inoue@nao.ac.jp

Carole Jackson
IoA, Cambridge
Royal Greenwich Observatory
Madingley Road, Cambridge
CB1 4YJ, U.K.
c.jackson@ast.cam.ac.uk

Dave Jauncey
Australia Telescope, CSIRO
c/o Cossa
GPO Box 3023
Canberra, ACT 2601, Australia
djauncey@thor.cbr.cossa.csiro.au

Dayton Jones
Jet Propulsion Laboratory
Mail Code 238-332
4800 Oak Grove Drive,
Pasadena, CA. 91109 U.S.A.
dj@bllac.jpl.nasa.gov

Paul A. Jones
Physics Dept.
Univ. of Sydney Nepean,
P.O. Box 10,
Kingswood, NSW, 2747 Australia
p.jones@st.nepean.uws.edu.au

Junji Inatani
Nobeyama Radio Observatory
N.A.O. of Japan
Nobeyama, Nagano
384-13, Japan
inatani@nro.nao.ac.jp

Christian Kaiser
MRAO
Cavendish Laboratories
Madingley Road
Cambridge, CB3 OHE, U.K.
ckaiser@mrao.cam.ac.uk

Vijay K. Kapahi
NCRA
T.I.F.R., Poona Univ. Campus
Pune - 411007, India
vijay@gmrt.ernet.in

Ken Kellermann
NRAO
Edgemont Road
Charlottesville VA 22903 U.S.A.
kkellerm@nrao.edu

Edward King
University of Tasmania/ATNF
Department of Physics
GPO Box 252C,
Hobart 7001, Australia
edward.king@phys.utas.edu.au

John Kirk
MPI Kernphysik
Postfach 10 39 80
D 69027 Heidelberg, Germany
kirk@boris.mpi-hd.mpg.de

Uli Klein
Radioastronomisches Institut
Bonn University
Auf dem Hugel 71
53121 Bonn - Germany
uklein@astro.uni-bonn.de

Pavel Yu. Kochanev
Sternberg Astronomical Inst.
Universitetsky Prospekt 13
119899 Moscow, Russia
gabuzda@sci.lpi.ac.ru

Anton Koekemoer
Inst. d'Astrophysique Paris
98 Bis, Bd. Arago,
75014 Paris, France
koekemoe@iap.fr

S. Komissarov
The University of Leeds
Dep. Applied Mathematical Stud.
Leeds, LS2 9JT, U.K.
serguei@amsta.leeds.ac.uk

Merkurii Konjukov
Lebedev Physical Institute
Radioastronomy Department
Leninsky Prosp., 53
117924, Moscow, Russia
konjukov@rasfian.serpukhov.su

Yuri Y. Kovalev
Moscow State University
Sternberg Astronomical Institute
Universitetskij pr. 13,
119899 Moscow, Russia
yyk@sai.msk.su

Thomas P. Krichbaum
MPifR, Bonn
Auf dem Hugel 69
53121 Bonn - Germany
p459kri@mpifr-bonn.mpg.de

Gopal Krishna
NCRA - TIFR
P.O. Ganeshkhind
Pune University Campus
Pune - 411007, India
krishna@gmrt.ernet.in

Marek J. Kukula
University of Edinburgh
Institute of Astronomy
Royal Observatory
Edinburgh EH9 3HJ, U.K.
mjk@staru1.livjm.ac.uk

Mark Lacy
Astrophysics, Oxford
Keble Road
Oxford, OXI 3Rh, U.K.
m.lacy1@physics.ox.ac.uk

Robert Laing
Royal Greenwich Observatory
Madingley Road
Cambridge CB3 OEZ, U.K.
rl@ast.cam.ac.uk

Georg Lamer
IAA Tuebingen
Waldhaeuserstr. 64
d-72076-Tuebingen, Germany
lamer@ait.physik.uni-tuebingen.de

Lucas Lara
IRA (CNR)/IAA (CSIC)
Via P. Gobetti, 101
I-40129 Bologna, Italy
lucas@astbo1.bo.cnr.it

Carlo Lari
Istituto di Radioastronomia
Via Gobetti, 101
40126 Bologna, Italy
lari@astbo1.bo.cnr.it

Gianni Latini
Torino Observatory
Strada Osservatorio, 20
10025 Pino Torinese, (TO)
Italy
latini@to.astro.it

J. Duncan Law-Green
NRAL Jodrell Bank
Lower Withington
Macclesfield
Cheshire SK11 9DL, U.K.
dlg@jb.man.ac.uk

J. Patrick Leahy
Manchester / Jodrell Bank
Lower Withington
Macclesfield
Cheshire SK11 9DL, U.K.
jpl@jb.man.ac.uk

Michael Ledlow
New Mexico State University
Dept. of Astronomy
Box 30001/dept. 4500
Las Cruces, New Mexico
88003-0001, U.S.A.
mledlow@nmsu.edu

Jim Lovell
University of Tasmania
Physics Department,
GPO Box 252C, Hobart 7001,
Australia
jim.lovell@phys.utas.edu.au

Tommaso Maccacaro
Osservatorio Astronomico-Brera
Via Brera, 28
20121 Milano, Italy
tommaso@bach.mi.astro.it

Duccio Macchetto
ST Science Institute
3700 San Martin Drive
Baltimore, MD 21218, U.S.A.
macchetto@stsci.edu

Jerzy Machalski
OAUJ Cracow
Astron. Observatory
Jagellonian Univ.
ul. Orla 171, Cracow, Pl-30244,
Poland
machalsk@oa.uj.edu.pl

Karl-Heinz Mack
Radioastronomisches Institut
Universitat
Auf Dem Hugel 71
53121 Bonn, Germany
kmack@astro.uni-bonn.de

Franco Mantovani
Istituto di Radioastronomia
Via Gobetti, 101
40129 Bologna, Italy
fmantovani@astbo1.bo.cnr.it

Laura Maraschi
Dipartimento di Fisica
Universita' di Milano
via Celoria 16 20133 Milano
Italy
32366::maraschi

Maria Marcha
Univ. of Lisbon
DICE
Av. Prof. Gama Pinto, 2
1699 Lisboa Codex, Portugal
mmarcha@milkyway.cc.fc.ul.pt

Silvano Massaglia
Osservatorio Astronomico
Strada Osservatorio, 20
10025 Pino Torinese (TO)
Italy
massaglia@astto2.to.astro.it

Leonid Matveenko
Space Research Institute
Profsojuznaja 84/32
Moscow, 117810 Russia
lmvlbi@iki3.iki.rssi.ru

Brian R. McNamara
Harvard-Smithsonian CfA
Center for Astrophysics
60 Garden St.
Cambridge, MA, U.S.A.
brm@cfa241.harvard.edu

David L. Meier
Caltech/JPL
Jet Propulsion Laboratory,
238-332, Pasadena,
CA 91109, U.S.A.
dlm@cena.jpl.nasa.gov

Klaus Meisenheimer
Royal Observatory
Blackford Hill
Edinburgh EH9 3Hj, Scoltland,
U.K.
meise@mpia-hd.mpg.de

Don Melrose
Sydney University
School of Physics
Sidney, NSW 2006, Australia
melrose@physics.usyd.edu.au

T. K. Menon
University of British Columbia
Dept. of Geophysics & Astronomy
2219 Main Mall,
Vancouver, British Columbia
V6T 1 Z4 Canada
menon@astro.ubc.ca

Evert Meurs
Dunsink Observatory
Castleknock,
Dublin 15, Ireland
ejam@dunsink.dias.ie

Peter F. Michelson
Stanford University
Dept. of Physics,
Stanford, CA 94305 U.S.A.
e7.f69@forsythe.stanford.edu

George Miley
Sterrewacht Leiden
Leiden University
P.O. Box 9513,
NielsBohrweg 2, 2300 RA
Leiden, The Netherlands
miley@strw.leidenuniv.nl

Joseph Miller
Lick Observatory
University of California
Santa Cruz, California
95064, U.S.A.
miller@helios.ucsc.edu

James M. Moran
Harvard-Smithsonian Center Astr.
MS 42, 60 Garden St.
Cambridge, MA 02138, U.S.A.
moran@cfa.harvard.edu

Raffaella Morganti
IRA (Bologna) / ATNF
CSIRO - ATNF
P.O. Box 76
Epping - NSW 2121, Australia
rmorganti@atnf.csiro.au

David Murphy
Jet Propulsion Laboratory
MS 238-332
4800 Oak Grove Drive
Pasadena CA 91109-8099 U.S.A.
dwm@casa.jpl.nasa.gov

Anita Mucke
Max-Planck-Inst.Extraterr.Physik
Giessenbachstrasse
85740 Garching, Germany
afm@mpe-garching.mpg.de

Robert Mutel
Dept. of Physics and Astronomy
Univ. of Iowa
Iowa City IA 52242, U.S.A.
rlm@astro.physics.uiowa.edu

George Nicolson
HartRAO
Hartebeesthoek Radioastronomy
Observatory
P O Box 443
Krugersdorp 1740, SUD AFRICA
george@bootes.hartrao.ac.za

Laura Norci
Dunsink Observatory
Castleknock,
Dublin 15, Ireland
ln@dunsink.dias.ie

Colin Norman
JHU/STScI
3700 San Martin Drive
Baltimore, MD 21218 U.S.A.
norman@stsci.edu

Ray Norris
Australia Telescope
Radiophysics Lab.,
P.O. Box, 76
Epping, NSW 2121, Australia
rnorris@atnf.csiro.au

Aileen O'Donoghue
St. Lawrence University
Physics Dept.
Canton NY 13617, U.S.A.
aodonogh@vm.stlawu.edu

Frazer N. Owen
NRAO
P.O. Box O
Socorro, NM 87801, U.S.A.
fowen@aoc.nrao.edu

Franco Pacini
Oss. Astrofisico Arcetri
Largo E. Fermi 5
50125, Firenze, Italy
pacini@arcetri.astro.it

Paolo Padovani
Physics Dept., II Univ. of Rome
Via della Ricerca Scientifica 1,
00133 Roma, Italy
padovani@roma2.infn.it

Giorgio G.C. Palumbo
Dipartimento di Astronomia
Via Zamboni, 33
40126 Bologna, Italy
ggcpalumbo@astbo3.bo.cnr.it

Lucia Padrielli
Istituro di Radioastronomia
Via Gobetti, 101
40129 Bologna - Italy
padrielli@astbo1.bo.cnr.it

Yuri Parijskij
Spacial Astrophysical Obs.
357147 Karachay Cherkessia,
Nizhny Arkhyz, Russia
par@sao.stavropol.su

Paola Parma
Istituto di Radioastronomia
Via P. Gobetti, 101
40129 Bologna, Italy
parma@astbo1.bo.cnr.it

R. Bruce Partridge
Haverford College
Haverford,
PA 19041, U.S.A.
bpartrid@haverford.edu

Alok Patnaik
MPIfR, Bonn
Auf dem Huegel 69
D 53121 Bonn, Germany
apatnaik@mpifr-bonn.mpg.de

Ivan Pauliny-Toth
Max Planck Inst. Radioastronomie
Auf Dem Huegel 69
D-53121 Bonn, Germany
p033ptt@sun12.mpifr-bonn.mpg.de

Tim Pearson
Caltech
Mail Code 105-24
Pasadena, CA 91125, U.S.A.
tjp@astro.caltech.edu

Richard Perley
NRAO
P.O. Box O
Socorro NM 87801 U.S.A.
rperley@nrao.edu

Eric Perlman
LHEA/GSFC
Mail Code 668
Goddard Space Flight Center
Greenbelt, MD 20771, U.S.A.
perlman@rosserv.gsfc.nasa.gov

Joseph E. Pesce
STScI
3700 San Martin Dr.
Baltimore, MD 212128, U.S.A.
pesce@stsci.edu

Martin Pohl
MPE
Postfach 1603
85740 Garching - Germany
mkp@mpe-garching.mpg.de

Richard W. Porcas
MPIfR, Bonn
Auf dem Huegel 69,
D 53121 Bonn, Germany
porcas@mpifr-bonn.mpg.de

Isabella Prandoni
Istituto di Radioastronomia
Via P. Gobetti, 101
40129 Bologna, Italy
prandoni@astbo1.bo.cnr.it

Bob Preston
JPL
238-332 Jet Propulsion Lab.
4800 Oak Grove Drive
Pasadena, California 91109,
U.S.A.
rap@sgra.jpl.nasa.gov

Eugen Preuss
MPIfR
Auf dem Huegel 69
53121 Bonn, Germany
epreuss@mpifr-bonn.mpg.de

V. Radhakrishna
Raman Research Institute
C V Raman Avenue
Sadashivanagar, Bangalore
560080 INDIA
rad@rri.ernet.in

Claudia M. Raiteri
Osservatorio Astronomico-Torino
Strada Osservatorio, 20
10025 Pino Torinese (TO), Italy
raiteri@to.astro.it

Frederik T. Rantakyro
Istituto di Radioastronomia
Via P. Gobetti, 101
40129 Bologna, Italy
fredrik@astbo1.bo.cnr.it

Tony Readhead
Caltech
1200 East California Boulevard
Pasadena, 91125
California U.S.A
acr@deimos.caltech.edu

Travis A. Rector
CASA - University of Colorado
CU Campus Box 389
Boulder, CO 80309-0389 U.S.A.
rector@casa.colorado.edu

Andrew Reid
The University of Sydney
The Astrophysics Department,
The School of Physics,
NSW 2006, Australia
areid@physics.usyd.edu.au

Roeland Rengelink
Leiden University
P.O. Box 9513
2300 RA Leiden,
The Netherlands
rengelin@strw.leidenuniv.nl

Eric A. Richards
University of Virginia
Dept. of Astronomy
P.O. Box 3818
Charlottesville,
VA 22903, U.S.A.
er4n@virginia.edu

Maria Rioja
JIVE
Postbus, 2
7990 AA Dwingeloo,
The Netherlands
rioja@nfra.nl

B. Rocca-Volmerange
Inst. d'Astrophysique de Paris
98bis Bd. Arago
75014 Paris, France
rocca@iap.fr

Marina M. Romanova
Space Research Institute,
Academy of Sciences of Russia
Profsoyuznaya 84/32
Moscow 117810, Russia
mromanov@esoc1.iki.rssi.ru

Jonathan D. Romney
NRAO
P.O. Box O
Socorro, NM 87801, U.S.A.
jromney@nrao.edu

Eduardo Ros
Dept. Astronomy y Astrophysics
University of Valencia
E-46100 Burjassot, Valencia
Spain
ros@vlbi.matapl.uv.es

Paola Rossi
Osservatorio Astronomico
Strada Osservatorio, 20
10025 Pino Torinese (TO), Italy
rossi@to.astro.it

Huub Rottgering
Sterrewacht Leiden
P.O. Box 9513
2300 RA Leiden, The Netherlands
rottgeri@strw.leidenuniv.nl

Lawrence Rudnick
University of Minnesota
Dept. of Astronomy
116 Church St SE,
Minneapolis MN 55455, U.S.A.
larry@mazel.spa.umn.edu

Corrado Ruscica
Osservatorio Astronomico-Brera
Via Brera, 28
I20121 Milano, Italy
ruscica@bach.mi.astro.it

Stanislaw Rys
Astronom. Obs. Jagellonian Univ
ul. Rozrywka 22/44
31-419 Krakow, Poland
strys@oa.uj.edu.pl

Marco Salvati
Osservatorio di Arcetri
Largo E.Fermi 5
Firenze 50125, Italy
salvati@arcetri.astro.it

Renzo Sancisi
Kapteyn Institute
Postbus 800
9700 AV Groningen
The Netherlands
sancisi@astro.rug.nl

Lakshmi Saripalli
Indian Institute of Astrophysics
Koramangala
Bangalore - 560034, India
lakshmi@iiap.ernet.in

Richard Schilizzi
Joint Inst. for VLBI Europe
Postbox 2,
7990 AA Dwingeloo,
The Netherlands
rts@nfra.nl

Arno P. Schoenmakers
Sterrewacht Leiden
P. O. Box 9500
2300 RA Leiden, The Netherlands
schoenma@fys.ruu.nl

Rolf Schwartz
Max-Plank-Inst.Radioastronomie
Auf dem Hugel 69
53121 Bonn, Germany
p589evn@mpifr-bonn.mpg.de

Giancarlo Setti
Istituto di Radioastronomia
Via Gobetti, 101
40129 Bologna - Italy
setti@astbo1.bo.cnr.it

David B. Shaffer
Radiometrics Inc.
1742 Saddleback Court
Henderson, Nevada 89014, U.S.A.
dbs@bootes.gsfc.nasa.gov

Peter Shaver
ESO
Karl-Schwarzschild-Strasse 2
D-85748 Garching bei Munchen,
Germany
pshaver@eso.org

Joachim Siebert
MPE fuer Extraterr. Physik
Giessenbachstrasse
85740 Garching, Germany
jos@mpe-garching.mpg.de

Aimo Sillanpaa
Tuorla Observatory
Fin-21500 Piikkio,
Finland
aimosill@sara.utu.fi

Susan Simkin
Michigan State University
Dept. Physics-Astronomy
Michigan State Univ.
East Lansing, MI 48824-1116
U.S.A.
simkin@grus.pa.msu.edu

Ashok K. Singal
Physical Research Laboratory
Astronomy and Astrophysic Div.
Navrangpur,
Ahmedabad-389999, India
asingal@prl.ernet.in

Joel G. Smith
Jet Propulsion Laboratory
4800 Oak Grove Dr, (MS 238-332)
Pasadena, CA 91109, U.S.A.
barbara.a.swanson@jpl.nasa.gov

Ignas Snellen
Leiden Observatory
P.O. Box 9513
2300 RA Leiden, The Netherlands
snellen@strw.leidenuniv.nl

Konstantin P. Sokolov
Institute of Radio Astronomy
Nat. Academy Sciences Ukraine
4, Krasnoznamennaya Street,
Kharkov 310002,Ukraine
rai@ira.kharkov.ua

Helene Sol
Observatoire de Paris - Meudon
Place J. Janssen
F-92195 Meudon, France
sol@obspm.fr

Carlo Stanghellini
Ist. di Radioastronomia-Noto
Contrada Renna bassa
Localita' Case di mezzo, C.P.141
96017 Noto (SR), Italy
carlo@eloro.ira.noto.cnr.it

W. Steffen
Univ. of Manchester
Oxford Rd.
Manchester M13 9PL, U.K.
wsteffen@ast.man.ac.uk

John Stocke
Univ. of Colorado at Boulder
CASA
Campus Box 389
Boulder, CO 80309-0389, U.S.A.
stocke@hyades.colorado.edu

R. Syunyaev
Astro Space Center
Profsoyuznaya 84/32
Moscow 119899, Russia
sunyaev@hea.iki.rssi.ru

Clive Tadhunter
University of Sheffield
Department of Physics
Sheffield S3 7RH, U.K.
c.tadhunter@sheffield.ac.uk

Leo O. Takalo
Tuorla Observatory
Vaisalantie 20,
21500 Piikkio, Finland
takalo@sara.utu.fi

Greg Taylor
NRAO
P.O. Box 0
Socorro, NM 87801, U.S.A.
gbt@astro.caltech.edu

Harri Terasranta
Metsahovi
02540 Kylmala
Finland
hte@vipunen.hut.fi

Steven J. Tingay
Mount Stromlo Observ.
Private Bag
Weston Creek P.O.
A.C.T., 2611, Australia
tingay@mso.anu.edu.au

Thomas Toniazzo
Scuola Normale Superiore
P.za Cavalieri N.7,
I-56126 Pisa
Italy
ttoniazz@astro.sns.it

Vincenza Tornatore
Ist. di Radioastronomia-Noto
Contrada Renna bassa
Localita' Case di mezzo
C.P. 141 96017 Noto (SR),
Italy
tornatore@ira.noto.cnr.it

Marja Tornikoski
Metsahovi Radio Research St.
Metsahovintie,
FIN-02540 Kylmala, Finland
merja.tornikoski@hut.fi

Edoardo Trussoni
Osservatorio Astronomico-Torino
Strada dell'Osservatorio, 20
10025 Pino Torinese (TO), Italy
trussoni@to.astro.it

Zlatan Tsvetanov
Johns Hopkins University
Dep. of Physics and Astronomy
34th & North Charles Street
Baltimore, MD 21218, U.S.A.
zlatan@pha.jhu.edu

Tasso Tzioumis
ATNF, CSIRO
P. O. Box 76,
Epping, NSW 2121, Australia
atzioumi@atnf.csiro.au

Marie-Helene Ulrich
ESO
Karl-Schwarzschild-Str. 2
D-85748 Garching bei Munche,
Germany
mhulrich@eso.org

Meg Urry
Space Telescope Science Inst.
3700 San Martin Dr.
Baltimore, MD 21227, U.S.A.
cmu@stsci.edu

Fausto Vagnetti
Tor Vergata Astrophysics
Dipartimento di Fisica,
Universita' di Roma Tor Vergata
Via della Ricerca Scientifica
00133 Roma, Italy
vagnetti@roma2.infn.it

Wil van Breugel
IGPP/LLNL
LLNL L-413, 7000 East Av.
P.O.Box 808,
Livermore, CA 94550, U.S.A.
wil@igpp.llnl.gov

Tiziana Venturi
Istituto di Radioastronomia
Via Gobetti, 101
40133 Bologna, Italy
tventuri@astbo1.bo.cnr.it

Rene Vermeulen
Caltech
Radio Astronomy 105-24
Pasadena, CA91125, U.S.A.
rcv@astro.caltech.edu

Gianpaolo Vettolani
Istituto di Radioastronimia
Via P. Gobetti, 101
40129 Bologna, Italy
gvettolani@astbo1.bo.cnr.it

Lourdes Vicente
Observatoire de Meudon (DARC)
5 Place Jules Janssen
92195 Meudon
Cedex, France
vicente@obspm.fr

Mario Vietri
Osservatorio Astronomico di Roma
00040 Monte Porzio Catone (Roma)
Italy
vietri@coma.mporzio.astro.it

Mario Vigotti
Istituto di Radioastronomia
Via P. Gobetti 101
40129, Bologna, Italy
vigotti@astbo1.bo.cnr.it

Daniela Villani
Dip. Astronomia - Univ. Firenze
Largo E. Fermi, 5
50125 Firenze, Italy
villani@arcetri.astro.it

Massimo Villata
Osservatorio Astronomico-Torino
Strada dell'Osservatorio, 20
10025 Pino Torinese (TO), Italy
villata@to.astro.it

Stefan Wagner
Inst.Landessternwarte
Koenigstuhl
69117 Heidelberg, Germany
swagner@mail.lsw.uni-heidelberg.de

Craig Walker
NRAO - Socorro
P.O. Box 0
Socorro, NM 87801, U.S.A.
cwalker@nrao.edu

Mark Walker
RCfTA
School of Physics A28
University of Sidney,
NSW 2006, Australia
m.walker@physics.usyd.edu.au

Jasper V. Wall
Royal Greenwich Observatory
Madingley Rod,
Cambridge, CB3 OEZ, U.K.
jvw@ast.cam.ac.uk

Andrew S. Wilson
University of Maryland
Astronomy Department
College Park,
MD 20742, U.S.A.
wilson@astro.umd.edu

Anna Wolter
Osservatorio Astronomico-Brera
Via Brera 28
20121 Milano, Italy
anna@bach.mi.astro.it

L. Woltjer
Observ. Haute Provence
04870 St. Michel l'Observatoire
France

Alan Wright
Parkes Observatory, ATNF
P.O. Box 276,
Parkes NSW 2870, Australia
awright@atnf.csiro.au

Joan M. Wrobel
NRAO
P.O. Box 0,
Socorro, New Mexico
87801, U.S.A.
jwrobel@nrao.edu

Koujun Yamashita
Nagoya University
Furo-cho, Chikusa-ku
Nagoya 464-01, Japan
yamasita@satio.phys.nagoya-u.ac.jp

Yu Zhi-yao
Shangai Astronomical Observatory
80 Nandan Road,
Shangai 200030, China
xytan@fudan.ihep.ac.cn

Alessandra Zanichelli
Istituto di Radioastronomia
Via P. Gobetti, 101
40129 Bologna, Italy
azanichelli@astbo1.bo.cnr.it

Anton Zensus
NRAO
520 Edgemont Road
Charlottesville, VA 22902,
U.S.A.
azensus@nrao.edu

Elena Zucca
Istituto di Radioastronomia
Via P. Gobetti, 101
40133 Bologna, Italy
zucca@astbo1.bo.cnr.it

GUGLIELMO MARCONI AND RADIOASTRONOMY

GIANCARLO SETTI
Dipartimento di Astronomia, University of Bologna
Istituto di Radioastronomia CNR
via P. Gobetti 101, 40129 Bologna, Italy

The decision to hold this IAU Symposium at Bologna on the occasion of the 100th anniversary of the discovery of the wireless telegraphy represents in itself a recognition of the widespread feeling of a historical link between the great discovery of Guglielmo Marconi and the birth of radioastronomy. Obviously, it is not a direct link. We all well know that the birth of radioastronomy must be traced back to the year 1932 when Karl Jansky recognized for the first time the existence of a radio signal probably associated with a celestial source. This a classical example of a 'serendipitous' discovery made while Jansky was investigating for the Bell Telephone Laboratories the sources of radio interferences with a rotating antenna array operating at about 14 m wavelength. The study of local disturbances was of primary importance in the rapid development of radio communications which had been geared by the Marconi's discovery.

Marconi was born in Bologna on April 25th, 1874. Almost precisely 100 years ago with respect to the time of this Symposium (September 1895) he sent a radio signal from Villa Griffone (Fig.1) which was successfully received over a distance of about 1600 m on the far side of an interposed hill, the "Celestini's hill". It is said that the successful result of the experiment was signaled to him by a gun shot fired by a farmer who stood on the side of the receiving apparatus with Guglielmo's brother. Villa Griffone, located on a hill site at about 10 Km south-west of Bologna, was the country mansion of the Marconi's family. It is now the seat of the "Guglielmo Marconi Foundation" which was created in 1938 with the aim of promoting research in radio communications and of perpetuating the memory and work of Marconi. On the back wall of the Villa there is a memorial tablet precisely placed under the window from where Marconi supposedly sent the signal by switching on his spark emitter connected with a parabolic antenna. This is one of the windows of the 'silk worm' room in the Villa

1

R. Ekers et al. (eds.), Extragalactic Radio Sources, 1–4.
© *1996 IAU.*

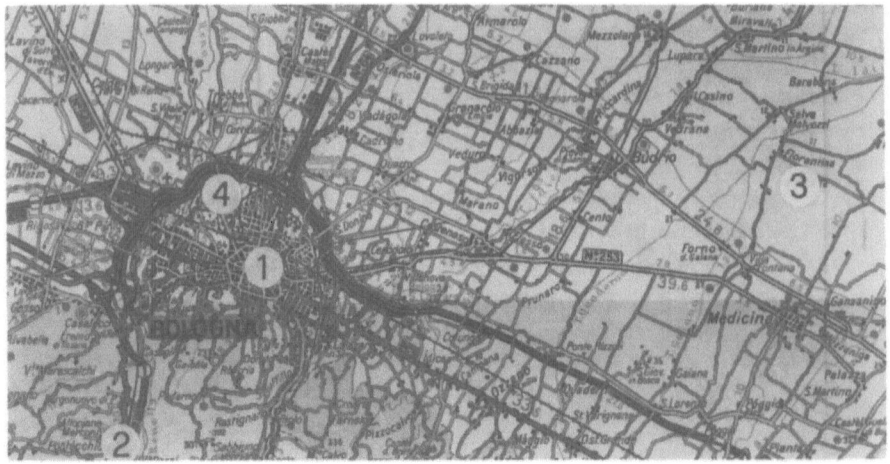

Figure 1.   (1) Institute of Physics 'A. Righi'; (2) Villa Grifone; (3) Radioastronomy Station; (4) Radioastronomy Institute CNR.

attic where the father, Giuseppe, had confined the young Guglielmo while he was doing those strange and, he might have thought, perhaps useless experiments. At that time Guglielmo was only 20 years old! On the other hand the mother, Annie Jameson, immediately captured the potential of the discovery made by the son and sent him to relatives in London where, she thought, it would have been easier to find the financial support required for the practical developments of the invention. The "Wireless Telegraph Trading Signal Co.Ltd.", which later acquired the name of "Marconi's Wireless Telegraph Co.", was founded in 1897 with a capital of 100.000 pounds. It owned worldwide rights on Marconi's patents with the exception of Italy.

In 1901 a signal sent from the Marconi's station in Poldhu (Cornwall) was successfully received at St.John's (Newfoundland). In the following year Marconi crossed the Atlantic on board the "Carlo Alberto" ship leaving Plymouth for Sydney (New Scotland). Strong signals launched from Poldhu were continuously received until the ship finally cast anchor in the bay of Sydney (October 31st) at a distance of 4000 Km, thus definitely establishing the possibility of a transcontinental radio link. A 2500–fold increase in the distance covered by wireless communications in only six years! This gives an idea of the rapidity with which Marconi was progressing in the revolution of the world communication system and, with it, of all socio-economical and political aspects on planet Earth. In 1909 he was awarded the Nobel Prize in Physics. In 1918, toward the end of World War I, he succeeded in sending wireless telegraphy messages from England to Australia.

This was followed twelve years later by a spectacular experiment in which he switched on the lights of Sydney's Town Hall by means of pulses sent from the yacht "Elettra", his navigating laboratory, at anchor in the harbor of Genoa. It is interesting to note that in 1935, the same year in which Jansky conclusively demonstrated the radio emission from the Galaxy, Marconi performed several sighting experiments by means of radio waves on the old Roman road 'Aurelia' in the presence of chief Governmental Authorities, thus anticipating, albeit on a minor scale, the application of radar techniques. The same year he became professor of Electromagnetic Waves at the University of Rome, where he prematurely died on July 20, 1937.

It would take much more time than that available to this opening speech to dwell on Marconi's gigantic personality, his frantic activity, achievements and battles to defend himself against powerful economical interests set in motion by his invention. What I want to do, instead, is to present some less known aspects which see the young Guglielmo regularly attending the lectures delivered by Augusto Righi, a famed physicist and director of the Institute of Physics of Bologna University. At that time Professor Righi was conducting a series of experiments aimed at proving that the properties of the hertzian waves strictly conformed to Maxwell theory. According to a testimony reported in a recent study (Dragoni, 1995) "...It was from the experiments carried out by Righi that Marconi started out on his brilliant discovery. Let us mention, among the experiments concerning the transmission at a distance, that made by Righi in 1889 in Marconi's presence over a distance of 20 m, that is from one end to the other of a corridor facing a lecture hall at the Faculty of Arts...". At that time Marconi was only 15 years old! The above mentioned study reports quotations from Degna, the daughter of Marconi's first marriage, in her father's biography "...Guglielmo, lacking of regular learning, could not enroll at the University of Bologna. He had to adapt to follow Righi's lectures as a simple hearer. Professor Righi did not encourage Marconi very much. However, he enabled him to work in his laboratory..." and from a recently found letter of Guglielmo to his brother written in 1892 "...I am studying very hard to get a Technical High School degree as Professor Righi wishes...". All this, and other testimonies not quoted here, clearly indicate the deep influence that must have been exerted by Righi on the young Guglielmo.

On the other hand the nature of this relationship between the famous scientists and the young inventor was openly stated in a meeting held at the Archiginnasio, the old seat of the University of Bologna, on September 21st, 1902, less than one year after the first transatlantic transmission. Marconi is quoted to have said "It gives me particular pleasure to see Professor Righi present here. He has undertaken major studies on electromagnetic waves and the results of his profound researches have been very useful for

my own discoveries". To which Righi replied "I am grateful for the opportunity to offer him (Marconi) my sincere congratulations. Probably no one
was better placed than I was to appreciate, while they were developing, his
exceptional creative flair and rare intellectual gifts". On the one hand the
physicist who was fighting to produce precise and efficient electrical apparatus working at short wavelengths, which would enable him to conduct
experiments on a laboratory size, on the other the young inventor who constructed efficient means of producing and controlling longer wavelengths to
connect the world.

Thus one may say that Bologna was at the heart of the scientific and
technological events which have dramatically shaped the development of
our society at the turn of the last century. It is perhaps not a chance
that this 'love affair' with radio waves was reaffirmed in 1959 when it was
decided that the Institute of Physics 'A. Righi' of the University of Bologna
would have embarked on a major project in radioastronomy. Six years later
the "Northern Cross" radiotelescope was opened at a location close to the
little town of Medicina, about 26 Km from Bologna toward east. In the year
1970 the Italian National Research Council (CNR) established the Institute
of Radioastronomy which took over from the University the running of
the "Northern Cross" and became the leading Italian institution in this
field of research. It was hosted in the building of the Institute of Physics
'A. Righi' and moved two years ago in the newly constructed premises of
the CNR in Bologna where this Symposium is now being held. It can be
noted that Marconi himself was President of the CNR, taking over this
prestigious office in the course of a solemn ceremony held on the Capitol
Hill on February 2nd, 1928.

Similarly to what happened in the world of communications, the development of radioastronomical techniques has brought about a revolution in
our understanding of the universe. There is practically no field of astronomy that has not been affected by the analysis of the information carried by
the the radio waves, from the study of the planets to the discovery of the
cosmic microwave background radiation, a milestone in the cosmological
inquiry. And, of course, there is still the hope that soon or later one might
be able to identify radio signals which will tell us about the existence of
intelligent life established somewhere else in the Galaxy, thus widening the
scope and profound implications of Marconi's invention.

References

Dragoni, G. (1995), The Dawns of Wireless telegraphy: The Relationship between Au-
    gusto Righi and Guglielmo Marconi, in *International Congress on the First Centenary
    of the Applied Electromagnetic Waves*, Russian Society 'A.S.Popov', 4–6 May 1995,
    Moscow, in press.

PARSEC-SCALE STRUCTURE OF COMPACT RADIO SOURCES

J. A. ZENSUS
National Radio Astronomy Observatory
520 Edgemont Road, Charlottesville, VA 22903, USA

Abstract. High-dynamic range imaging and monitoring with Very Long Baseline Interferometry have considerably increased our knowledge of the parsec-scale properties of compact radio sources. I review some of the properties of individual sources in areas where particular progress has been made in the last few years.

1. Imaging of Luminous Objects

The study of prominent flat-spectrum sources with Very Long Baseline Interferometry (VLBI) provides arguably the best opportunity for testing physical models of parsec-scale radio jets. Routine high-quality observations are possible with the Very Long Baseline Array (VLBA), the global VLBI campaigns with various other networks, and the "World Array" campaigns. At millimeter wavelengths, ad hoc arrays have been used to make images with highest resolution, albeit modest overall fidelity. Space VLBI missions in preparation bear the potential of imaging at yet higher resolution.

For a review of the various VLBI surveys and for morphological classifications see Wilkinson (1995), and Vermeulen (these Proceedings) for a statistical analysis of superluminal motions. Here, I will focus on several recent results from various ongoing studies of individual objects. I will then review some of the results from detailed VLBI monitoring studies of specific sources.

The high-luminosity VLBI sources typically have well collimated core-jet structure at centimeter and millimeter wavelengths. The cores are identified based on compactness, flat radio spectra, and strong flux density variability. Stronger misalignment between the parsec-scale and the kiloparsec-

5

scale jets have been found for BL Lac objects compared to quasars (Conway & Murphy, 1993). No clear counter-jets have been found in core-dominated, i.e., presumably strongly boosted sources, but counter-jets are seen in some lobe-dominated objects, e.g., Cygnus A and 3C 84. Structural variability and in particular apparent superluminal motion are frequently observed, and together with the one-sidedness and total flux variability this has been taken as direct evidence for the presence of bulk relativistic motion in such objects. Many of the observed jets are curved, and in some cases semi-oscillating trajectories or ridge lines have been observed.

High-quality images that show a great detail of detail in the jet structure are available for a growing number of sources. Good examples include 3C 120 (see Craig Walker for a tour de force in imaging), 3C 273 (Unwin *et al.*, 1994), 3C 345 (Zensus *et al.*, 1996), and 0836+71 (Hummel *et al.*, 1992). Such luminous sources typically show continuous jets with rich substructures, markedly different from the simple structures often seen in maps from the era of the last IAU Symposium on "Extragalactic Radio Sources". The main features in these images correspond to the "distinct components" seen in older maps, but we can now study weaker features and often directly the underlying continuous jet emission. The evidence for the apparent superluminal motions has remained strong, but it is now clear that infrequent sampling in time is bound to cause misidentifications and confusion.

2. Curved Trajectories and Variable Speeds

Statistical studies (e.g., Vermeulen, these proceedings) assume that the motion measured for a particular superluminal source component is occurring along ballistic trajectories at constant speed. We now know that in a given source not only can different components have different speeds, but accelerations and decelerations are well established. Rarely are the apparent motions along straight ballistic trajectories: kinks and bends are frequently seen and in a small number of cases, complicated curved trajectories have been found. Note that such curvature sometimes is measured not for a given component, but indirectly by tracing several components, e.g., along a jet that might be characterized by a well-defined ridge line or time-averaged mean jet axis. The curvature seen so often is perhaps the strongest evidence against the paradigm of truly moving plasmons as the physical nature of superluminally moving components.

Figure 1 shows the apparent curved trajectories for superluminal features of 3C 345 (Zensus *et al.*, 1996), within 1 mas from the stationary core D. At distances from the core larger than about 2 mas the components appear to roughly follow the same curved trajectory in the north-west direction towards the arcsecond structure near the core the trajectories differ

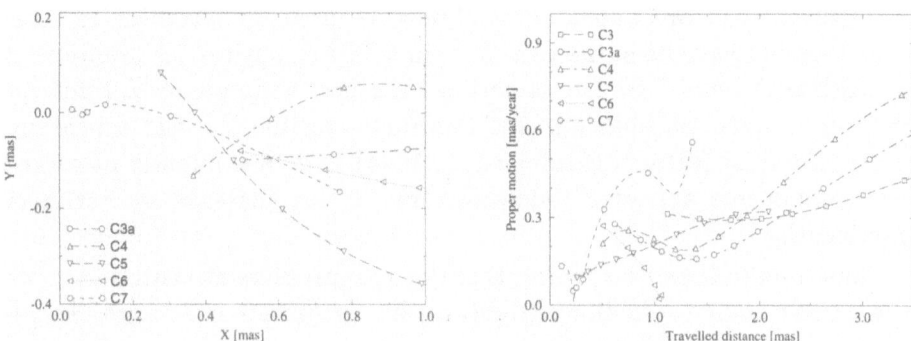

Figure 1. *Left:* Trajectories of superluminal features in 3C 345 within 1 mas from the core (Zensus *et al.*, 1995b; Zensus *et al.*, 1996), measured with respect to the core D. *Right:* Proper motions in 3C 345 (Zensus *et al.*, 1995b; Zensus *et al.*, 1996).

substantially. In this and other sources the curvature appears most pronounced near the core. Frequency-dependencies of component shapes and positions and of the observed trajectories—indicative of opacity effects—have been measured, e.g., in 3C 345 and in the quasar 4C 39.25 (Alberdi *et al.*, 1996). Similar curved trajectories or jet ridge lines are reported in a growing number of cases, especially from high-resolution observations at millimeter wavelengths (Krichbaum *et al.*, 1994).

The velocity changes that occur in 3C 345 are complex (see Fig. 1): to first order, the component speeds increase with separation from the core, but there is evidence for a systematic minimum speed at 1–1.5 mas from the core, indicating a possible transition between physically different (the inner and outer) regions of the jet (Lobanov, 1996). The kinematics and luminosity changes suggest that the acceleration measured are intrinsic (Zensus *et al.*, 1996; Lobanov & Zensus, 1995).

Of the many other examples for variable speeds, I mention here only the quasar 4C 39.25—previously thought first to be a stationary double source, and later a candidate for superluminal contraction. It shows a component moving between two stationary features. Recent observations at 22 and 43 GHz demonstrate that the apparent speed slows down as the moving feature approaches the eastern stationary feature (Alberdi *et al.*, 1993). (There is also some evidence that there are apparent differences in the curvature in the 22 and 43 GHz images (Alberdi *et al.*, 1996).) This has been explained in terms of repeated bending of the jet, where the stationary regions correspond to the Doppler enhanced underlying jet flow where it is oriented almost along the line of sight.

Most observers assume that in the sources observed with VLBI there is *no difference between pattern and fluid speed;* see Vermeulen (these Proceedings) for a discussion of the potential pitfalls.

In several sources, appearances of new components have been reported to be related to events in the total flux, in γ, X-ray, optical, IR-, millimeter, or centimeter bands. BL Lac is perhaps the best example for a tight correlation between variability in the centimeter regime and the appearance of superluminal features (Mutel *et al.*, 1994), but in a number of sources new components appeared following a major flux increase at centimeter wavelengths.

Superluminal motion requires that the jet must be relativistic and viewed at a narrow angle to the line of sight, so that projection effects are likely to be significant. Assuming a value of the motion's Lorentz factor, it is possible to reconstruct the three-dimensional trajectory of a moving feature (cf. Zensus, Cohen, and Unwin 1995). For example, the derived intrinsic jet curvature for 3C 345 is small, but it is greatly amplified by projection effects, and the angle of the jet to the line of sight increases smoothly with radius from the core from about 1 deg to about 4 deg. Scenarios to explain the curved trajectories include precession and orbital motion, and a number of interesting models have been proposed based on helical motion to explain the curved, quasi-periodic trajectories seen in 3C 345, 1803+78, 4C39.25, and similar sources.

We still do not know the true physical nature of the components we observe in parsec-scale radio jets. The properties at least of the core-dominated sources are thought by many to result from bulk-relativistic motion in two oppositely directed jets of plasma originating in the nucleus of the source (Blandford & Königl, 1979). Thus the optically thick cores in the VLBI images represent the base of the jet (located near the apex of the jet cone), and the superluminal features are regions of enhanced emission moving along the jet. The general properties of this "standard model" are well established (Pearson & Zensus, 1987), but in-depth comparisons between observations and models have generally been unsatisfactory.

Until recently, most monitoring studies were confined to images of the total intensity of a given source in one particular observing band. This suffices to determine basic morphology and component kinematics but it lacks important physical information that can only be obtained from measuring the jet spectra (from multi-frequency work) and the magnetic field distributions (from polarization imaging). Several sources are now being studied at multiple frequencies to determine spectral properties for comparison with models. For 3C 345, crude spectral information from the long-term monitoring can be used to measure the basic parameters of the component synchrotron spectra: turnover flux density and frequency, and integrated flux in the range 4–25 GHz (Lobanov & Zensus, 1995). At least for component C4, the observed luminosity variations of the jet components seem to require a variable pattern Lorentz factor, and the same may be true for the other components. Spectral properties are important for testing the basic idea that the jet components are caused by relativistic shocks, but the evidence in favor of the shock model is inconclusive so far. The total flux variations in some sources have been adequately explained with shocks (Hughes *et al.*, 1989), and in 3C 345 the total flux changes are well correlated with variations within about 5 pc from the core, suggesting shocks may dominate in the jet at least near the core (Lobanov & Zensus, 1995). alternative models that have been proposed involves interaction with the ambient plasma (Rose *et al.*, 1987). Marcaide et al. (1994) apply a detailed model of shock components moving on twisted trajectories to the quasar 4C39.25. BL Lac is perhaps the best understood case for the interpretation of VLBI components by shocks. In this source, the position angles of the axes of the curved trajectories of subsequent components shifting, suggesting evidence for precession caused by a binary black hole system (Mutel *et al.*, 1994).

Linear polarization sensitive VLBI observations of compact sources yield direct information on the structure and order of the underlying magnetic fields, the presence and nature of thermal material, the energy of relativistic electrons, and the geometry of emission on sub-milliarcsecond scales. There is evidence for a difference in the polarization properties of quasars and BL Lac objects, and the latter show significantly lower speeds, which argues for a physical and not merely an orientation difference between the two classes of source (Gabuzda *et al.*, 1994). Wardle et al. have studied the polarization structure at 5 GHz of 3C 345 and interpreted their results in terms of a comprehensive shock model. First polarization images of 3C 345 at 22 GHz are basically in agreement with the interpretation by Wardle et al. (Leppänen *et al.*, 1995).

3. Conclusion

VLBI monitoring studies at centimeter and millimeter wavelengths, enhanced by spectral and polarization imaging, can discriminate detailed physical models. Combined with broad-band total flux density and polarization observations they can be used to determine the overall physical conditions in parsec-scale radio jets. Some of the intriguing aspects where new insights may soon become feasible include the formation of the jets, possible jet acceleration mechanisms, the nature of any non-relativistic matter involved (e.g., the ambient medium and thermal outflows), and the influence of the central region on the jet dynamics.

I thank Craig Walker and Antxon Alberdi who provided results prior to publication. This paper was prepared in part during a visit at the MPIfR in Bonn, made possible through a Humboldt Award of the Alexander v. Humboldt Stiftung. The National Radio Astronomy Observatory is a facility of the National Science Foundation, operated under cooperative agreement by Associated Universities, Inc.

References

Alberdi, A., Krichbaum, T. P., Marcaide, J. M., Witzel, A., Graham, D. A., Inoue, M., Morimoto, M., Booth, R. S., Rönnäng, B. O., Colomer, F., Rogers, A. E. E., Zensus, J. A., Readhead, A. C. S., Lawrence, C. R., Bartel, N., Shapiro, I. I., Burke, & F., B. 1993. *Astron. Astrophys.*, **271**, 93–100.

Alberdi, A., Krichbaum, T. P., Witzel, A., Zensus, J. A., *et al.* 1996. *Astron. Astrophys.* in press.

Blandford, R. D., & Königl, A. 1979. *Astrophys. J.*, **232**, 34–48.

Conway, J. E., & Murphy, D. W. 1993. *Astrophys. J.*, **411**, 89–102.

Gabuzda, D. C., Mullan, C. M., Cawthorne, T. V., Wardle, J. F. C., & Roberts, D. H. 1994. *Astrophys. J.*, **435**, 140–161.

Hughes, P. A., Aller, H. D., & Aller, M. F. 1989. *Astrophys. J.*, **341**, 68–79.

Hummel, C. A., Muxlow, T. W. B., Krichbaum, T. P., Quirrenbach, A., Schalinski, C. J., Witzel, A., & Johnston, K. J. 1992. *Astron. Astrophys.*, **266**, 93–100.

Krichbaum, T. P., Witzel, A., Standke, K. J., Graham, D. A., Schalinski, C. J., & Zensus, J. A. 1994.

Leppänen, K. J., Zensus, J. A., & Diamond, P. D. 1995. *Astron. J.* In press.

Lobanov, A. P. 1996. Ph.D. thesis, New Mexico Institute of Mining and Technology.

Lobanov, A. P., & Zensus, J. A. 1995. *Astrophys. J.* submitted.

Marcaide, J. M., Alberdi, A., Gómez, J. L., Guirado, J. C., Marscher, A. P., & Zhang, Y. F.

Mutel, R. L., Denn, G. R., & Dryer, M. J. 1994.

Pearson, T. J., & Zensus, J. A. 1987.

Rose, W. K., Beall, J. H., Guillory, J., & Kainer, S. 1987. *Astrophys. J.*, **314**, 95–102.

Unwin, S. C., Davis, R. J., & Muxlow, T. W. B. 1994. In *Compact Extragalactic Radio Sources*, eds. K. Kellermann and J. A. Zensus (Green Bank: NRAO), 81–86.

Wardle, J. F. C., Cawthorne, T. V., Roberts, D. H., & Brown, L. F. 1994. *Astrophys. J.*, **437**, 122–135.

Zensus, J. A., Lobanov, A. P., Leppänen, K. J., Unwin, S. C., & Wehrle, A. E. 1995a. *Astron. J.* In preparation.

Zensus, J. A., Cohen, M. H., & Unwin, S. C. 1995b. *Astrophys. J.*, **443**, 35–53. In preparation.

THE SUB-PARSEC SCALE JETS OF AGN

T.P. KRICHBAUM, W. ALEF, A. WITZEL

Max-Planck-Institut für Radioastronomie, Bonn, Germany

With an angular resolutions of 0.05 – 0.2 mas, millimeter-VLBI[1] observations (at 22, 43, and 86 GHz) allow to investigate the very central – sub-parsec scale – regions of active galactic nuclei (AGN), which are self-absorbed at lower frequencies. Here we briefly present preliminary results from recent observations of Cygnus A at 22 & 43 GHz, which reveal evidence for subluminal motion in jet and counter-jet, and 86 GHz VLBI observations of two extreme γ-blazars, suggesting a tight correlation between their γ-ray activity and the generation of jets.

Cygnus A: Its proximity ($z = 0.057$) and large radio luminosity makes the archetypical FR II radio galaxy Cygnus A particularly interesting for high spatial resolution imaging. The detection of motion in jet and counter-jet will allow – at least in principle – to determine the orientation of the jet, its intrinsic velocity and the cosmological distance of Cygnus A. After initial detection with VLBI at 43 GHz (Krichbaum *et al.*, 1993, *AA 275, 375*), we obtained an improved second epoch 43 GHz VLBI image. From this and 22 GHz observations performed in parallel, a two-sided core-jet structure is revealed (Fig.1). We tentatively determined the position of the VLBI-core using its compactness and inverted spectrum. With this identification, we find subluminal motion with velocities in the range of $v/c \simeq 0.1 - 0.3$ ($H_0 = 100$ km s^{-1} Mpc^{-1}, $q_0 = 0.5$) for at least 4 distinct jet components located at $r \leq 3 - 4$ mas core separation (Krichbaum *et al.*, 1996, *Greenbank workshop, eds. Carilli & Harris, p.92*). At 5 GHz slightly higher velocities in the range $0.2 \leq v/c \leq 0.5$ for components located at larger core separations ($r > 5$ mas, Carilli *et al.*, 1994, *AJ 108, 64*) are found and indicate either apparent acceleration along the jet or the presence of pattern speeds. From the 22 GHz maps (Fig.1) and from an unpublished map at 18 cm (Fig.2) we obtained a jet to counter-jet ratio in the range $R \simeq 2 - 8$.

0528+134: Two 3 mm VLBI-maps (1993.26 & 1994.0) reveal strong evidence for the emergence of a new feature east of the main component (Fig.2). Since at larger core separations ($r > 1$ mas) the jet is oriented to the north (eg. Zhang *et al.*, 1994, *ApJ 432, 91*), motion along a strongly bent path ($\Delta p.a. \simeq 90°$ for $r \leq 1$ mas) is suggested. Back-extrapolation of the motion ($\beta_{app} \simeq 5$) of the new component yields strong evidence that it was

11

R. Ekers et al. (eds.), Extragalactic Radio Sources, 11–13.

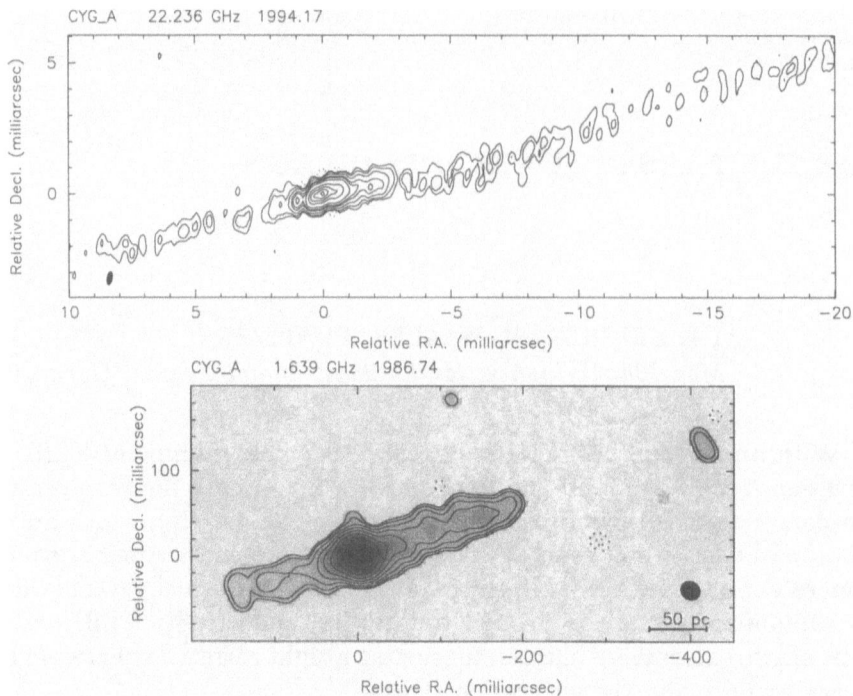

Figure 1. *Top:* Jet- and counter-jet of Cygnus A at 22 GHz (1994.17). Only the central part of a larger map is shown. Contour levels are -0.1, 0.1, 0.3, 1, 3, 5, 10, 30, 50, 70, 90 % of the peak. The restoring beam size is 0.5 × 0.2 mas, *p.a.* = −11°, 0.2 mas correspond to 0.15 pc. *Bottom:* EVN-VLBI map of Cygnus A at 1.6 GHz. Contour levels are -0.5, -0.3, 0.3, 0.5, 1, 2, 4, 8, 16, 32, 64 % of the peak, the beam size is 23.6 × 20.7 mas, *p.a.* = 54°. The jet is oriented along P.A. ≃ 287°, well aligned with the kpc-jet. From this map a jet-to-counterjet ratio of $R \simeq 2$ is obtained. The bar indicates a linear scale of 50 pc.

ejected during the correlated mm-/γ−ray outburst of 1993.5 (Krichbaum, *et al.*, 1996, *Proc. Nat. Acad. Sci. USA, in press*). This supports the idea that the γ−radiation originates from highly relativistic jets (eg. Sikora *et al.*, 1994, *ApJ 421, 153*).

3C 454.3: From two 3 mm VLBI-maps (1993.26 & 1994.0) an evolving complex core-jet structure is revealed (Fig.3), showing superluminal motion with $\beta_{app} \simeq 6 - 7$ at $r \leq 1$ mas. This is considerably slower than the motion reported by Pauliny-Toth et al. (*in preparation*) at larger core separations, who find component acceleration from $\beta_{app} \simeq 8$ at $r = 2$ mas to $\beta_{app} \simeq 21$ at $r = 5$ mas. If interpreted geometrically, an ultra-relativistic flow ($\gamma \simeq 20$) along a spatially bent path and differential Doppler-boosting (eg. like in 3C 345, Zensus, *this conference*) must be considered.

VLBI imaging at 86 GHz therefore can provide important tests of such geometrical interpretation schemes and will reveal – owing to its high angular and spatial resolution – early detections of new jet components, allowing to study the relation between jet production and broad-band flux density activity (eg. Aller *et al.* & Romanova *et al.*, *this conference*).

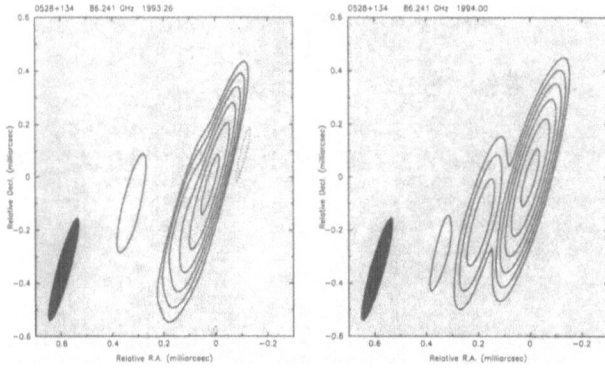

Figure 2. The evolution of the sub-mas structure of 0528+134 at 86 GHz between 1993.26 (left) and 1994.00 (right). The restoring beam size is $50 \times 400\ \mu as$, $p.a. = -16°$. Contour levels in both maps are -2.5, 2.5, 5, 10, 20, 40, 80 % of the peak flux density of 0.68 (left) respectively 0.65 (right) Jy/beam. Note that for unknown reasons the flux densities seen in both maps are by a factor ~ 3 lower than the total flux density. The maps and Gaussian modelfits to the data reveal the ejection of a new component from the compact selfabsorbed core.

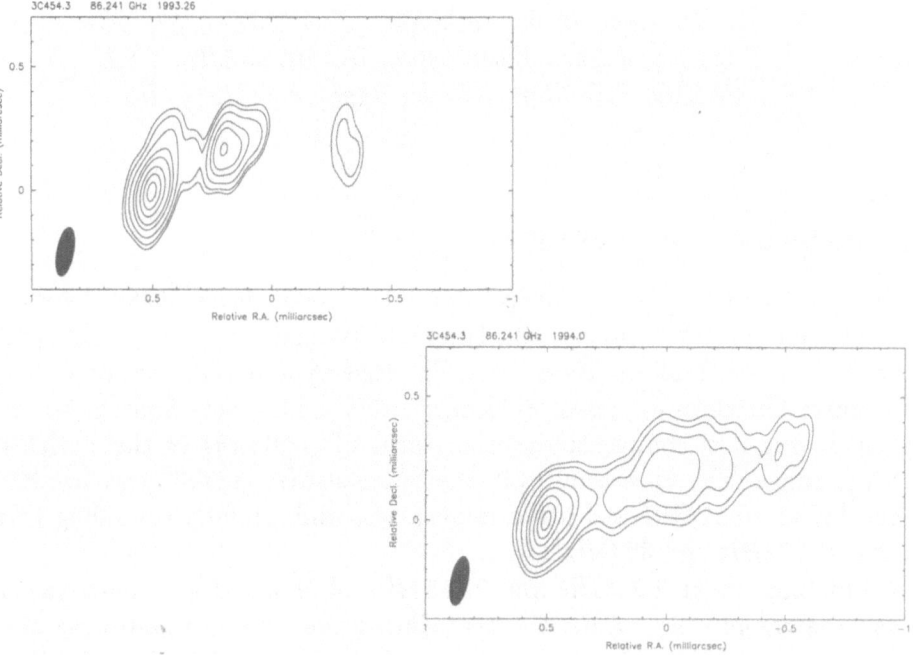

Figure 3. Clean-maps of 3C454.3 at 86 GHz obtained in 1993.26 (top) and 1994.0 (bottom). Both maps are convolved with a beam of size $70 \times 200\mu as$, $p.a. = -12°$. Contour levels are -3, 3, 5, 10, 20, 30, 50, 70, 90 % of the peak flux density of 0.94 Jy/beam (top) and 1.35 Jy/beam (bottom). Next to the stationary assumed bright and compact core component located at the map center, two features located at $r \simeq 0.3$ mas and $r \simeq 0.8$ mas (in 1993.26) seem to separate from the core with $\beta_{app} \simeq 6 - 7$.

[1]*mm-VLBI is a joint effort of the observatories at Effelsberg, Onsala & SEST, Pico Veleta, Haystack, Quabbin, Kitt Peak, Owens Valley, Hat Creek and the VLBA. T.P.K. appreciates support of the BMFT-Verbundfoschung.*

RADIO OBSERVATIONS OF THE γ-RAY BLAZAR 0528+134

M. POHL[1], W. REICH[2], T. P. KRICHBAUM[2], K. STANDKE[2,3],
S. BRITZEN[2], H. P. REUTER[2], P. REICH[2], R. SCHLICKEISER[2],
H. UNGERECHTS[4], R. L. FIEDLER[5], E. B. WALTMAN[5],
K. J. JOHNSTON[6] AND F. D. GHIGO[7]

[1] *MPE, Postfach 1603, 85740 Garching, Germany*
[2] *MPIfR, Postfach 2024, 53010 Bonn, Germany*
[3] *University Bonn, Nussallee 17, 53121 Bonn, Germany*
[4] *IRAM, Ave. Divina Pastora 7, 18012 Granada, Spain*
[5] *NRL, Code 7210, Washington, DC 20375-5351, USA*
[6] *US Naval Obs., Washington, DC 20392-5420, USA*
[7] *NRAO, P.O. Box 2, Green Bank, WV 24944, USA*

1. Radio monitoring of 0528+134

We report multifrequency observations of the γ-ray blazar 0528+134 with the Effelsberg 100-m telescope, the IRAM 30-m telescope at Pico Veleta and the NRL Green Bank Interferometer. The observing methods are described elsewhere (Reich *et al.*, 1993; Pohl *et al.*, 1995). The radio lightcurves are given in Fig.1 in comparison to the status of 0528+134 in the EGRET energy range. The uncertainties in the flux densities quoted there are less than 5% at 10.55 GHz and lower frequencies, while slightly exceeding this value at 32 GHz and 86 GHz.

Fluctuations at 2.25 GHz and 2.695 GHz of about 20% – much larger than the errors – are rather frequent during the time of monitoring and these fluctuations are significantly larger than those at 4.75 GHz, 8.3 GHz and 10.55 GHz. These fluctuations may be explained as stochastic inter-stellar scattering and are probably not intrinsic. The unusual depression in July 1993 is most likely an extreme scattering event (Pohl *et al.*, 1995).

0528+134 underwent a major radio and mm outburst in 1993 a few month after a very strong outburst in high energy γ-rays. The evolution of the source during the outburst follows the canonical behaviour of appearing first at high frequencies, and then with decreasing optical depth and energy

14

R. Ekers et al. (eds.), Extragalactic Radio Sources, 14–16.

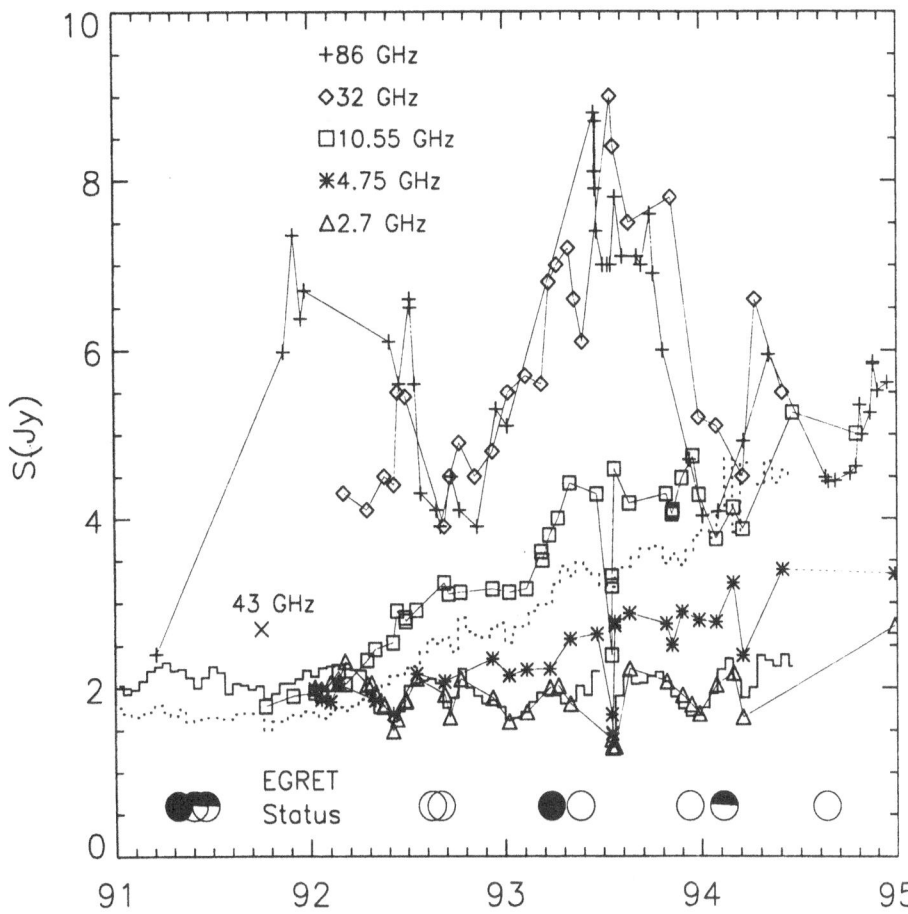

Figure 1. Radio lightcurves of 0528+134 from the Effelsberg 100-m, the IRAM 30-m and the NRL-GBI telescopes. The Green Bank data are given as averages over two weeks in histogram mode, where the full line denotes data at 2.25 GHz and the dotted lines give data at 8.3 GHz. The source activity in the EGRET range is indicated by open circles for low level, half-filled circles for medium activity, and filled circles for a high γ-ray flux.

loss of the radiating particles being detected at lower and lower frequencies. A similar behaviour is indicated for the weaker γ-ray outburst in May 1991 and the mm activity in the beginning of 1992. This result fits nicely to the general finding that blazars are bright in γ-rays preferentially at the beginning of a radio outburst (Valtaoja and Teräsranta, 1995; Mücke *et al.*, 1996).

2. VLBI observations of 0528+134

0528+134 was observed with a global VLBI array at 22 GHz in November 1992. Within the IRIS-S and EUROPE geodetic VLBI observing campaigns the source was also observed regularly at 8.4 GHz. The VLBI structure of 0528+134 at 22 GHz (epoch 1992.85) shows a one-sided core jet structure of ∼ 5 mas length, which is also typical for other epochs.

One of the components exhibits superluminal motion with an apparent velocity of $\beta_{app} = 4.4 \pm 1.7$ (for $H_0 = 100$ km/s/Mpc, $q_0 = 0.5$). Superluminal motion in 0528+134 was expected since Doppler boosting is required to satisfy the compactness limit at MeV photon energies (McNaron-Brown et al., 1995). Due to the strong reduction of the pair production cross section in the Klein-Nishina limit there is no similar requirement for γ-rays in the EGRET range (Pohl et al., 1995).

Since only few EGRET sources have been observed with sufficient resolution in VLBI, the high percentage of superluminal sources in the sample of EGRET detected quasars yields strong evidence for the assertion that high energy γ-ray emission from AGN originates in relativistic jets and is strongly boosted around the direction of the jet.

A new VLBI component which is not present in data at 22 GHz from 1991 (Zhang et al., 1994) also exhibits superluminal motion. Backextrapolating in time we find that the new component was expelled from the core in the first half of 1991, near the first γ-ray outburst in May 1991. Newest VLBI data indicate the appearance of another new component in 1994 (Krichbaum et al., 1995). Although this new component has yet been found only at two epochs, it appears to move superluminally and may have been expelled from the core between fall 1992 and summer 1993, i.e. near the time of the second gamma ray outburst and at the beginning of a strong radio outburst.

If the evidence for such a correlation can be solidified in future measurements, it would nicely complement our view that activity in the central source of AGN results in the expulsion of plasma blobs (or shocks) at relativistic velocity, in which relativistic particles, most probably electrons, produce high energy gamma ray emission when the blob is still near to the core, and which after some cooling and expansion become visible at mm and cm wavelengths.

References

Krichbaum T.P., Britzen S., Standke K.J. et al.: 1995, in *Quasars and AGN: High Resolution Radio Imaging*, Eds. M.Cohen and K.Kellermann, Proc. Nat. Acad. Sci. USA, in press

McNaron-Brown K., Johnson W.N., Jung G.V. et al., 1995, ApJ **451**,575

Mücke A., Pohl M., Reich P. et al., 1996, *Proceedings of the 3rd Compton Symposium*, A&AS submitted

Pohl M., Reich W., Krichbaum T.P. et al., 1995, A&A **303**, 383

Reich W., Steppe H., Schlickeiser R. et al., 1993, A&A **273**, 65

Valtaoja E., Teräsranta H., 1995, A&A **297**, L13

Zhang Y.F., Marscher A.P., Aller H.D. et al., 1995, ApJ **432**, 91

A DRAMATIC (AND INVISIBLE!) FLARE IN NRAO 530

3 mm λ VLBI Observations

GEOFFREY C. BOWER, DON BACKER, RICK FORSTER AND
MELVIN WRIGHT

UC Berkeley Radio Astronomy Lab
601 Campbell, Berkeley CA 94720 USA

3 mm λ VLBI observations of the extremely flat spectrum QSO NRAO 530 at the peak of its brightest recorded flare show a resolved core only slightly changed from its pre-outburst state. Observations with Haystack, Kitt Peak and Owens Valley observatories in April 1994, prior to the onset of the flare found a single component with $T_b = 8 \times 10^{10}$ K. One year later, within one week of the peak of flare at 3 mm λ, observations with Haystack, Kitt Peak and Hat Creek observatories revealed a component comparable in size and flux density with $T_b = 7 \times 10^{10}$ K. Table 1 details the source models for the two experiments.

The 3 mm λ light curve taken at Hat Creek Radio Observatory (Figure 1) shows the sudden rise in flux density between the two experiments. The peak in April 1995 is the highest in 10 years of monitoring at HCRO. Furthermore, the UMRAO database shows that this is the brightest flare at centimeter wavelengths in almost 30 years (Aller, H. & Aller, M. 1995, private communication). The flare may be associated with a gamma ray flare observed with EGRET in Spring and Summer of 1994 (Mattox, J.R. 1995, private communication).

Despite the sparse *uv*-coverage and the difficulty of calibration at millimeter wavelengths, we can place limits on the structure external to the core. In the April 1994 data, the closure phases are consistent with no structure while the short-baseline amplitudes account for most of the zero-baseline flux. In the April 1995 data, the amplitudes reach a maximum of 6 ± 1 Jy on the shortest baselines, about half of the the zero-baseline flux of 12 ± 1 Jy. Much of the flux is, therefore, resolved or hidden by destructive interference. Unfortunately, technical problems in the determination of the closure phase in this experiment currently prevent us from addressing this issue more specifically. We hope to settle this issue in the near future.

If the flux is resolved, it may be in the form of either isolated components

17

R. Ekers et al. (eds.), Extragalactic Radio Sources, 17–18.

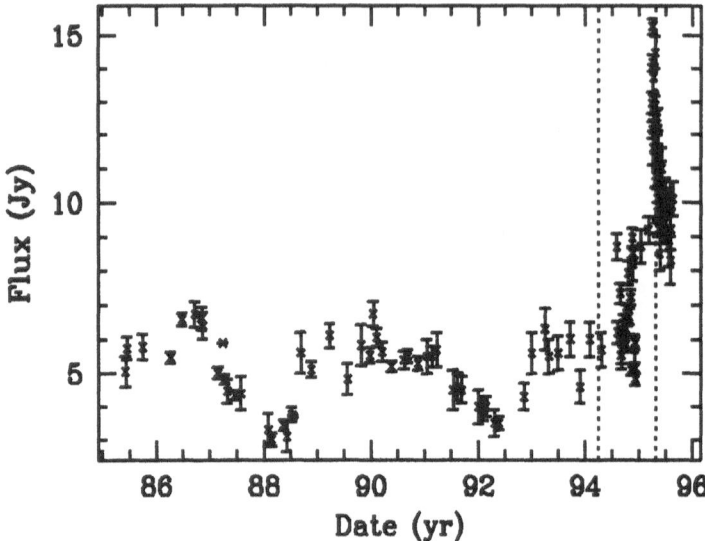

Figure 1. 3 mm λ fluxes of NRAO 530 from Hat Creek Radio Observatory. The vertical bars mark the dates of the VLBI experiments.

or a halo but it must be separated from the core by greater than 0.5 milli-arcsec. Associating the creation of the component with the onset of the flare in Fall 1994 requires an apparent velocity $\beta_{app} > 9h^{-1}$ for $q_0 = 0.5$ and $H_0 = 100h$ km/s/Mpc. This velocity falls near the upper limit of the distribution presented by Vermeulen in these proceedings, making NRAO 530 one of the most beamed sources known. The detection of NRAO 530 by EGRET further confirms the connection between superluminal blazars and gamma ray sources (von Montigny, C. *et al.* 1995, *ApJ*, **440**, 525).

The simple structure apparent at this and other wavelengths suggests that the spectrum, flat from 80 MHz to > 230 GHz, cannot be accounted for by the superposition of distinct components. Instead, a self-similar model for the emission is favored (*e.g.*, Blandford and Konigl 1979, *ApJ*, **232**, 34).

We thank the staff of Haystack, Kitt Peak, Owens Valley and Hat Creek observatories for their assistance in these observations.

TABLE 1. NRAO 530 Model Parameters

Experiment Date	Size	Flux	Total Flux	T_b
April 1994	85 μarcsec	4 ± 0.5 Jy	6 ± 1 Jy	8×10^{10} K
April 1995	120 μarcsec	7 ± 1 Jy	12 ± 1 Jy	7×10^{10} K

VLBI OBSERVATIONS OF SOUTHERN EGRET IDENTIFICATIONS

J.E.J. LOVELL[1], S.J. TINGAY[2], P.G. EDWARDS[3],
D.L. JAUNCEY[4] AND R.A. PRESTON[5]

[1] *Department of Physics, University of Tasmania - Hobart, 7001, Australia*
[2] *Mount Stromlo and Siding Spring Observatories - ACT 2611, Australia*
[3] *Institute of Space and Astronautical Science - Sagamihara, Kanagawa 229, Japan*
[4] *Australia Telescope National Facility - Epping, 2121, Australia*
[5] *Jet Propulsion Laboratory - Pasadena, 91109, USA*

We present high resolution VLBI images of three southern radio sources: PKS 0208−512, PKS 0521−365 and PKS 0537−441. These sources have been identified as > 100 MeV gamma-ray sources with the Energetic Gamma-Ray Telescope (EGRET) on board the *Compton Gamma-Ray Observatory* (Thompson *et al.* 1995). These are the first results in a continuing program of VLBI observations of southern EGRET identifications with the Southern Hemisphere VLBI Experiment (SHEVE) array of telescopes (Jauncey *et al.*, 1994).

Our observations of **PKS 0208−512** at 4.8 GHz (figure 1(left)) show that the radio source consists of a bright, unresolved core and a jet-like extension at a position angle of 233±5°. We estimate that the compact core has a brightness temperature of $> 1.2 \times 10^{12}$ K. This is at the inverse Compton limit for synchrotron radiation. Since we have only one epoch of VLBI data we cannot estimate any apparent motion of the jet-like feature relative to the core.

We observed **PKS 0521−365** at 4.8 GHz in November 1992, February 1993 and May 1993 and at 8.4 GHz in October 1993. The data from these four epochs have been used to estimate any apparent motion of the jet component relative to the core. The data are consistent with no motion over this 0.9 year period. The brightness temperature of the core, measured from our highest resolution image (figure 1(centre)) is 1.1×10^{11} K, well below the inverse Compton limit.

R. Ekers et al. (eds.), Extragalactic Radio Sources, 19–20.

Our VLBI images of **PKS 0537−441** at 4.8 and 8.4 GHz show that the source is dominated by a slightly resolved component, which we identify as the core together with a jet-like component extending towards the north. We infer a brightness temperature of 6.5×10^{11} K from the slightly resolved core component in our 4.8 GHz image (figure 1(right)).

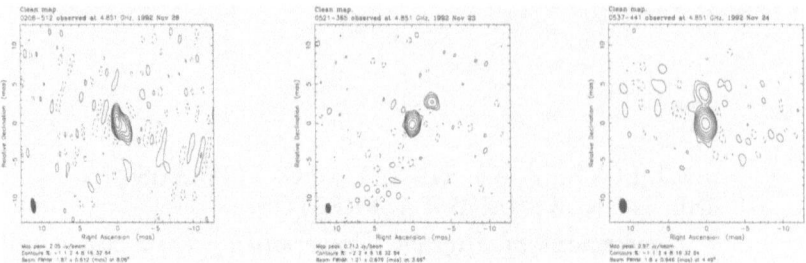

Figure 1. 4.8 GHz SHEVE images of PKS 0208−512 (left), PKS 0521−365 (centre) and PKS 0537−441 (right).

The images of PKS 0208−512, PKS 0521−365 and PKS 0537−441 show that these compact radio sources are dominated by > 1 Jy cores and resolved into core-jet morphologies. In each case the position angle of the milliarcsecond-scale structure seen with VLBI is aligned with the position angle of the arcsecond-scale radio structure.

The brightness temperatures of the cores suggest that PKS 0208−512 and PKS 0537−441 may be more highly beamed than PKS 0521−365. This difference in brightness temperature correlates with the strong statistical EGRET detection of PKS 0208−512 and PKS 0537−441 (> 5σ significance) and the weak statistical detection of PKS 0521−365 (4σ to 5σ significance), even though PKS 0521−365 is much closer to us. Unfortunately, on the basis of the brightness temperatures for these three sources we cannot make a strong statement concerning the importance of relativistic beaming for EGRET detectability. Observations at later epochs will place more constraints on the properties of these sources.

VLBI observations of sources over the full range of gamma-ray strength will be required before we can find possible differences or similarities in VLBI properties between EGRET detections and non detections. The need for southern hemisphere observations of the weak EGRET detections is especially apparent since 7/11 lie south of $\delta = 0°$ and only 1/11 lies north of $\delta = +20°$. Our continuing VLBI observations are aimed at addressing this need.

References

Jauncey, D.L. et al.: 1994 *"Very High Angular Resolution Imaging"*, J.G. Roberston and
    W.J. Tango (eds), p 131-134
Thompson, D.J. et al.: 1995, *Astrophys. J., Suppl. Ser.*, in press

MONITORING THE JET IN CENTAURUS A
AT 0.1 PARSEC RESOLUTION

R.A. PRESTON[1], S.J. TINGAY[2], D.L. JAUNCEY[3]
J.E. REYNOLDS[3], J.E.J. LOVELL[4], P.M. MCCULLOCH[4]
A.K. TZIOUMIS[3], M.E. COSTA[5], D.W. MURPHY[1], D.L. MEIER[1]
D.L. JONES[1], R.W. CLAY[6], P.G. EDWARDS[7], S.P. ELLINGSEN[4]
R.H. FERRIS[3], R.G. GOUGH[3], P. HARBISON[8], P.A. JONES[9]
E.A. KING[4], A.J. KEMBALL[10], V. MIGENES[1], G.D. NICOLSON[10]
M.W. SINCLAIR[3], T.D. VAN OMMEN[1], R.M. WARK[3]
AND G.L. WHITE[9]

[1] *Jet Propulsion Laboratory, Caltech, Pasadena, CA, USA*
[2] *Mt. Stromlo & Siding Spring Obs., Canberra, ACT, Australia*
[3] *Australia Telescope National Facility, Epping, NSW, Australia*
[4] *University of Tasmania, Hobart, Tasmania, Australia*
[5] *University of Western Australia, Nedlands, WA, Australia*
[6] *University of Adelaide, Adelaide, SA, Australia*
[7] *Institute of Space & Astronautical Science, Sagamihara, Japan*
[8] *British Aerospace Australia, Canberra, ACT, Australia*
[9] *University of Western Sydney, Kingswood, NSW, Australia*
[10] *Hartebeesthoek Radio Astr. Obs., Krugersdorp, South Africa*

Abstract. Centaurus A is the closest active extragalactic radio source, at a distance of approximately 3.5 Mpc, and is identified with the peculiar elliptical galaxy NGC 5128. As such it is a very important target for observations of the small-scale (sub-parsec) and large-scale (kpc) structures in extragalactic jets. Here we present Mk-II VLBI observations made at 8.4 GHz over a 4.3 year period from early 1991 until mid-1995, as well as a 4.8 GHz observation that was co-eval with one of the 8.4 GHz observations. All of the observations were made with the SHEVE array except for the last observation which was made with the VLBA. The dual-frequency observations identify the core of the radio source, while the multi-epoch observations show the complex structural evolution at a resolution of 0.1 pc. Subluminal motion of $\approx 0.15c$ is evident. Structural changes are observed on time scales shorter than four months.

R. Ekers et al. (eds.), Extragalactic Radio Sources, 21–22.

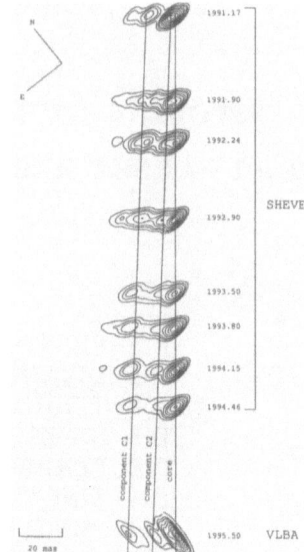

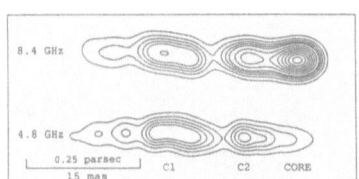

Figure 1.

In November 1992 two observations of Centaurus A were made approximately three days apart, one at 4.8 GHz and the other at 8.4 GHz. Figure 1a shows the two images resulting from these observations. The brightest feature in the 8.4 GHz image is almost completely absent in the 4.8 GHz image, whereas there is a good coincidence between other features in the two images. On the basis of this result the brightest feature at 8.4 GHz is a highly inverted spectrum component and the other main features seen in both images (C1 and C2) are steeper spectrum components. These identifications naturally point to the inverted spectrum component as the core of the radio source, with C1 and C2 being components within a sub-parsec-scale jet. The highly inverted spectrum of the core may be due to synchrotron self-absorption, although free-free absorption may also be involved (see paper by Jones *et al.* in these proceedings).

Having identified the core of the radio source we can now examine the evolution of the sub-parsec-scale structure with multi-epoch monitoring observations. Figure 1b presents the results of such a monitoring campaign, giving us a quantitative description of how the various components in the radio source changed their relative positions, dimensions, and flux densities over a 4.3 year period. The images are vertically aligned so the core remains stationary, with slanted lines superposed onto components C1 and C2 to indicate a motion of 0.15c. Note that C1 appears to undergo significant internal evolution during the series of observations, particularly between 1991.17 and 1992.24. This rapid fading and brightening of a component ≈ 4.5 mas (0.25 ly) in extent could be explained if the apparent jet flow speed is actually greater than 0.74c (true jet flow speed >0.59c).

A COUNTERJET IN THE NUCLEUS OF CENTAURUS A

D.L. JONES[1], S.J. TINGAY[2], R.A. PRESTON[1], D.L. JAUNCEY[3],
J.E. REYNOLDS[3], J.E.J. LOVELL[4], P.M. MCCULLOCH[4]
A.K. TZIOUMIS[3],M.E. COSTA[5], D.W. MURPHY[1], D.L. MEIER[1]
R.W. CLAY[6], P.G. EDWARDS[7], S.P. ELLINGSEN[4], R.H. FERRIS[3]
R.G. GOUGH[3], P. HARBISON[8], P.A. JONES[9], E.A. KING[4]
A.J. KEMBALL[10], V. MIGENES[1], G.D. NICOLSON[10]
M.W. SINCLAIR[3], T.D. VAN OMMEN[1], R.M. WARK[3]
AND G.L. WHITE[9]

[1] *Jet Propulsion Laboratory, Caltech, Pasadena, CA, USA*
[2] *Mt. Stromlo & Siding Spring Obs., Canberra, ACT, Australia*
[3] *Australia Telescope National Facility, Epping, NSW, Australia*
[4] *University of Tasmania, Hobart, Tasmania, Australia*
[5] *University of Western Australia, Nedlands, WA, Australia*
[6] *University of Adelaide, Adelaide, SA, Australia*
[7] *Institute of Space & Astronautical Science, Sagamihara, Japan*
[8] *British Aerospace Australia, Canberra, ACT, Australia*
[9] *University of Western Sydney, Kingswood, NSW, Australia*
[10] *Hartebeesthoek Radio Astr. Obs., Krugersdorp, South Africa*

Abstract. Centaurus A (NGC 5128) is the nearest giant radio galaxy. It is
a Fanaroff-Riley type 1 (low luminosity) radio source, but the compact radio
source in the nucleus is strong enough that VLBI imaging has been possible
with both the SHEVE array and the VLBA at several frequencies. These
observations have detected a sub-parsec scale counterjet. This shows that
jet formation in at least some FR I sources is intrinsically two-sided over
very small distances and the radio jets in Centaurus A are probably only
moderately relativistic. We also find evidence that the center of activity in
Centaurus A is partially obscured by a disk or torus of dense plasma.

We observed Centaurus A at 8.4 GHz with a global (SHEVE+VLBA)
array in October 1993. The resulting image is shown in figure 1a. The
brightest peak corresponds to the inverted-spectrum core (as determined
from nearly-simultaneous 4.8 and 8.4 GHz SHEVE images; Jauncey *et al.*

R. Ekers et al. (eds.), Extragalactic Radio Sources, 23–24.

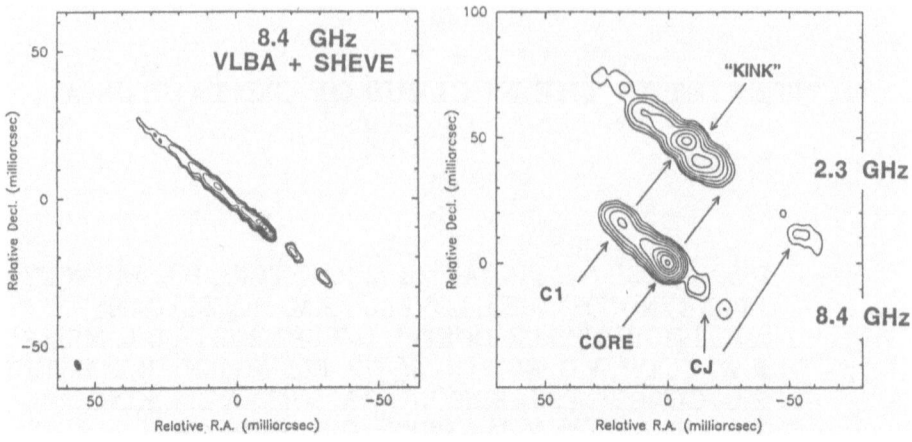

Figure 1. (a) 8.4 GHz global image; (b) Simultaneous 2.3 & 8.4 GHz VLBA images.

1995, Tingay *et al.* 1995, Preston *et al.* 1996). The two counterjet peaks are seen in both Mk-II and Mk-IIIA data sets from this experiment. Figure 1b shows images from a dual-frequency (2.3 and 8.4 GHz) VLBA experiment in July 1995. Both images have been convolved with the same 6 mas restoring beam. The core component in Centaurus A appears to have a remarkably inverted spectrum between 2.3 and 8.4 GHz ($\alpha \sim 4$), while the rest of the main jet has a flat or steep spectrum ($\alpha \leq 0$). In addition, the separation between the core and the first detectable peak in the counterjet is much greater at 2.3 GHz than at 8.4 GHz. Both of these effects can be explained if the central 0.4-0.8 pc of Centaurus A are seen through a nearly edge-on disk or torus of ionized gas. Vermeulen *et al.* (1994) and Walker *et al.* (1994) found a similar situation in 3C84. A 2-3 pc path through 10^4K gas with a mean electron density of $10^4 - 10^5$ cm^{-3} will give a spectral turnover frequency > 10 GHz due to free-free absorption. The jet/counterjet brightness ratio is quite small ($\sim$4-8) and the proper motions seen in the main jet are sub-luminal (Tingay *et al.* 1994, 1995). This supports the belief that the jets in FR I radio galaxies are not highly relativistic. Continuing VLBI observations of Centaurus A will allow proper motions in the jet and counterjet to be compared, which will set new constraints on the geometry.

References

Jauncey, D.L., *et al.* 1995, in *Proc. Nat. Acad. Sci.*, in press
Preston, R.A., *et al.* 1996, these proceedings
Tingay, S.L., *et al.* 1994, *Aust. J. Phys.*, **447**, 619
Tingay, S.L., *et al.* 1995, in *Jets from Stars and AGN*, in press
Vermeulen, R.C., Readhead, A.C.S. & Backer, D.C., 1994, *Ap.J.*, **430**, L41
Walker, R.C., Romney, J.D. & Benson, J.M., 1994, *Ap.J.*, **430**, L45

HIGH FREQUENCY RADIO OBSERVATIONS
OF THE NUCLEUS OF CEN A

ZULEMA ABRAHAM

Instituto Astronômico e Geofísico, Universidade de São Paulo
CP 9638, 01065-970, São Paulo, SP, Brazil

1. Introduction

Cen A (NGC5128) is a giant radio galaxy that in several ways behaves like a weak active galactic nucleus. It presents three pairs of radio lobes and a compact core with a one sided jet (Jauncey et al. 1995). The compact core is highly variable in wavelengths ranging from the radio region to X and γ rays (Kinzer et al. 1995). In the radio domain, the emission from the core dominates at high frequencies and a considerable effort was made to study variability at 22 and 43 GHz for more than ten years (Botti & Abraham 1995).

2. Observation and Results

The observations of the nucleus of NGC5128, reported here, were made with the Itapetinga radiotelescope at the frequency of 43 GHz, with a HPBW of 2'.2, during the period 1988-1994. The observational procedure consisted of scans across the galaxy, passing through the core and the two inner lobes. Virgo A was used as a primary calibrator and the northern lobe, located at a distance of 4' from the core was used as a secondary calibrator. The flux density of the central source was obtained after subtraction of the inner lobes, represented by two gaussian fits. Notice that this procedure is different from what it was used by Botti & Abraham (1993), where scans were made perpendicular to the plane of the galaxy. In that case, the contribution of the southern lobe was not completely eliminated and the calibration was not instantaneous, as in the data described here.

The results are presented in Fig. 1. We can see that the flux density of the core changed by a factor of two in time scales of the order of months,

R. Ekers et al. (eds.), Extragalactic Radio Sources, 25–26.
© 1996 *IAU.*

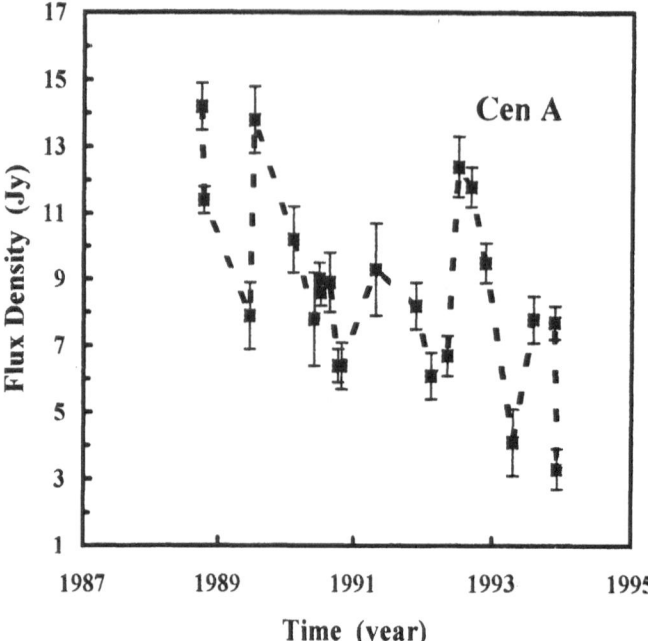

Figure 1. Flux density as a function of time of the nucleus of Cen A, at 43 GHz

with very sharp increases and slower decreases in intensity. X-ray observations (40 - 500 keV) with the CGRO (Kinzer et al. 1995) did not show
any correlation with our radio results, except for the 1991 observations,
when the source presented strong intensity at both wavelengths. VLBI observations at 8.4 GHz between 1991 and 1994 (Jauncey et al. 1995) showed
a core-jet structure, with the jet formed by several components separating from the core with constant velocity. The variability observed by us at
higher frequencies could be correlated to variability in the intensity of the
individual components, (e.g. 1991) or to the ejection of a new component
from the core (e.g. 1988).

This work was partially supported by the Brazilian Agency CNPq. Itapetinga
radiotelescope is operated by the University of Mackenzie, USP, UNICAMP and
INPE.

References

Botti, L.C.L., Abraham Z. (1993) Long-term radio observation of the nucleus of NGC5128
    (Centaurus A), *MNRAS*, **264**, 807-812
Jauncey D.L. et al. (1995) Sub-Parsec Scale Structure and Evolution of Centaurus A
    (NGC5128) *Quasars and AGN: High Resolution Radio Imaging.* National Academy
    of Sciences Pub., in press.
Kinzer R.L. et al. (1995) OSSE observations of gamma-ray emission from Centaurus A,
    ApJ, **449**, 105-118

SIMULTANEOUS MULTI-FREQUENCY IMAGING OF THE NUCLEUS OF NGC 1275

J.D. ROMNEY[1], R.C. WALKER[1], K.I. KELLERMANN[2],
R.C. VERMEULEN[3] AND V. DHAWAN[1]

[1] *NRAO - Socorro, NM, USA*
[2] *NRAO - Charlottesville, VA, USA*
[3] *Caltech - Pasadena, CA, USA*

1. Introduction

An unusual counterjet feature was discovered in 3C 84, the compact radio nucleus of NGC 1275, in the "First Science" observations on the VLBA at 8.4 GHz (Walker *et al.*, 1994), and simultaneously in Global VLBI observations at 22 GHz (Vermeulen *et al.*, 1994). Comparison of these images indicated a strongly inverted spectrum in this feature, but the interpretation was clouded by the two-year difference between the epochs of observation. To resolve this ambiguity, and to study the spectrum of the counterjet, we exploited the capabilities of the VLBA to make nearly simultaneous observations of 3C 84 at 2.3, 5.0, 8.4, 15.4, 22, and 43 GHz, in four apparitions over a 16-day period in January 1995. These observations also served to continue structural monitoring programs at 15 and 22 GHz. This paper presents preliminary images from those observations. A companion contribution by Walker *et al.* (1995) discusses the interpretation of the images.

2. Observations

Individual observing sessions were scheduled at 15.4, 22, and 43 GHz, while simultaneous observations at 2.3 and 8.4 GHz alternated with 5.0 GHz in a fourth session. All ten VLBA stations participated, except that high winds disabled the Mauna Kea station in the 15 GHz session, and an instrumental failure occurred at Fort Davis in the multi-frequency session. The observa-

R. Ekers et al. (eds.), Extragalactic Radio Sources, 27–29.
© *1996 IAU.*

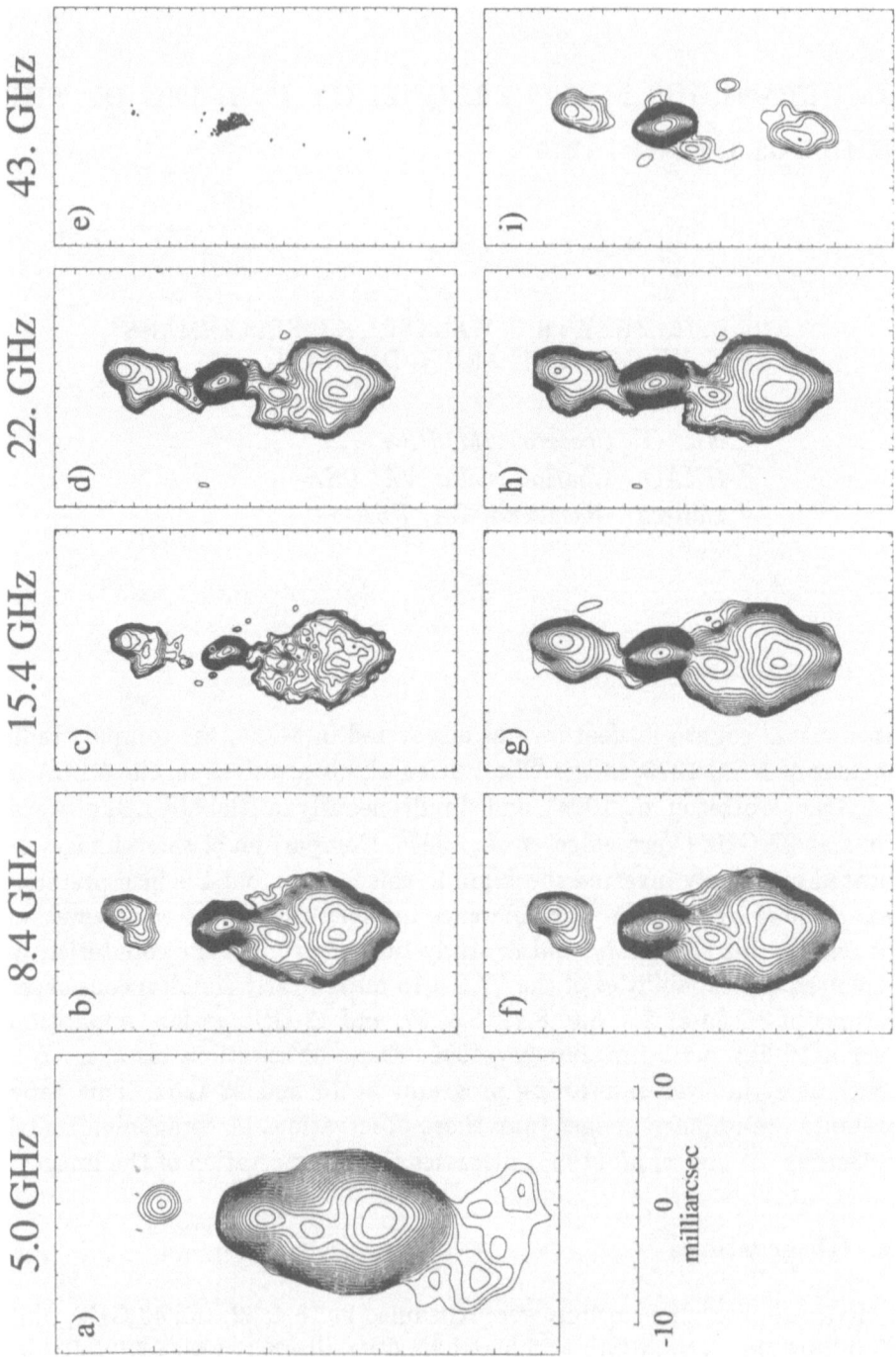

Figure 1.   Images of 3C 84 at five frequencies. The image pairs at the highest four frequencies are explained in the text. The common angular scale is shown in the lower left. A common set of contour levels are used, with the lowest contour at 5 mJy/beam except in (a) 1.86-mJy/beam and (e) 13.3 mJy/beam. Ascending contours are logarithmic with seven steps per decade (1., 1.39, 1.93, 2.68, 3.73, 5.18, 7.20, 10., . . .).

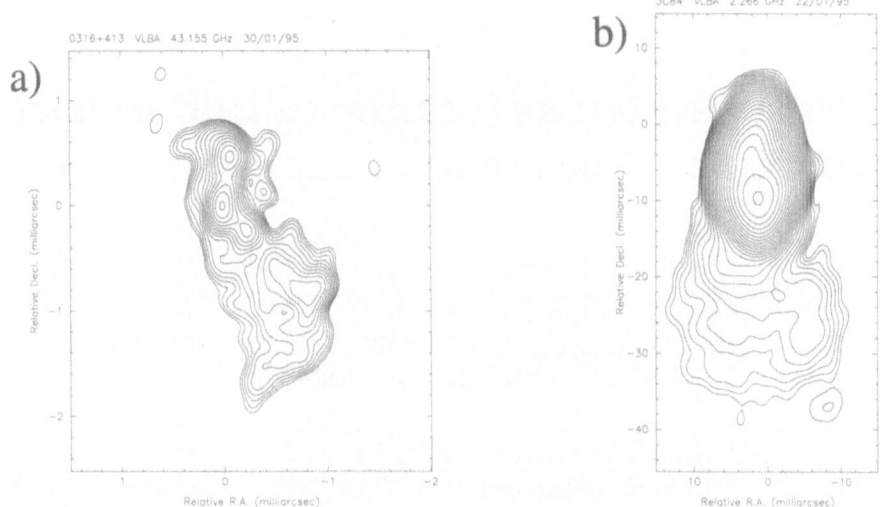

Figure 2. (a) Fine-scale image at 43 GHz. (b) Extended image at 2.3 GHz. The lowest contour in both images is 2.5 mJy/beam, with the same logarithmic ascending contours used in Figure 1.

tions were correlated within 4 weeks, and imaged using primarily the AIPS system, as well as the DIFMAP and VISAMAT programs.

Figure 1 presents two different views of the images at the five highest frequencies. The upper set of images have been restored with the conventional approximation to the synthesized beam. Those in the lower row have been restored with the beam appropriate to the 5.0-GHz observations, to facilitate comparisons and to bring out the jet and counterjet structures at 43 GHz. The 43-GHz image also required a substantial (u, v) taper in the imaging process to reveal the counterjet. The images are aligned relative to the central "core" component which is brightest at most frequencies.

Two images which can not be conformed to this scheme are shown separately. Figure 2a presents the core region at 43 GHz, at a scale more suitable to the extremely compact structures which exist at that frequency. Figure 2b is the lowest-frequency image, at 2.3 GHz, which is sufficiently extended that it could not be combined with those at the higher frequencies.

We gratefully acknowledge the contributions of W. Alef, D. C. Backer, J. M. Benson, and A. C. S. Readhead to the 3C 84 observational program of which this paper is part.

References

Vermeulen, R.C., Readhead, A.C.S., and Backer, D.C. (1994), *Ap. J.*, **430**, L41.
Walker, R.C., Romney, J.D., and Benson, J.M. (1994) *Ap. J.*, **430**, L45.
Walker, R.C. *et al.* (1995) *This Volume.*

CONSTRAINTS ON THE ACCRETION REGION IN NGC 1275 FROM VLBA OBSERVATIONS OF THE COUNTERJET

R.C. WALKER[1], J.D. ROMNEY[1], R.C. VERMEULEN[2],
V. DHAWAN[1] AND K.I. KELLERMANN[3]

[1] *N.R.A.O. - Socorro, NM, USA*
[2] *Caltech - Pasadena, CA, USA*
[3] *N.R.A.O. - Charlottesville, VA, USA*

1. Introduction

A northern feature, that is most likely related to a counterjet, was found on parsec scales in 3C 84 (which is in NGC 1275) in "First Science" observations on the VLBA[1] at 8.4 GHz (Walker *et al.*, 1994) and in Global VLBI observations at 22 GHz (Vermeulen *et al.*, 1994). The jet/counterjet length ratio, brightness ratio at 22 GHz, and speeds, as measured over many years with VLBI, fit a simple beaming model with symmetric jets oriented at 30–50 degrees to the line-of-sight and traveling at a speed of 0.3–0.5 times the speed of light. These ranges allow for Hubble constants of between 50 and 100 km s^{-1} Mpc^{-1}. The brightness of the counterjet at 8.4 GHz was very much lower than expected with this model, assuming that the spectral indices in the near and far side jets are similar. This was interpreted as the result of free-free absorption in an ionized medium that lies in front of the counterjet but not in front of the near-side jet. An accretion disk or torus has an appropriate geometry to show this effect (for an analysis, see Levinson *et al.*, 1995).

If the free-free absorption concept is correct, this opens the exciting possibility of constraining the physical parameters in the accretion region by using VLBI to measure $L\langle n^2 g/T^{3/2}\rangle$ where L is the path length, n is the density, T is the temperature, and g is the Gaunt factor. Both the value

---

[1]The National Radio Astronomy Observatory is operated by Associated Universities, Inc., under cooperative agreement with the National Science Foundation

R. Ekers et al. (eds.), Extragalactic Radio Sources, 30–32.
© 1996 *IAU.*

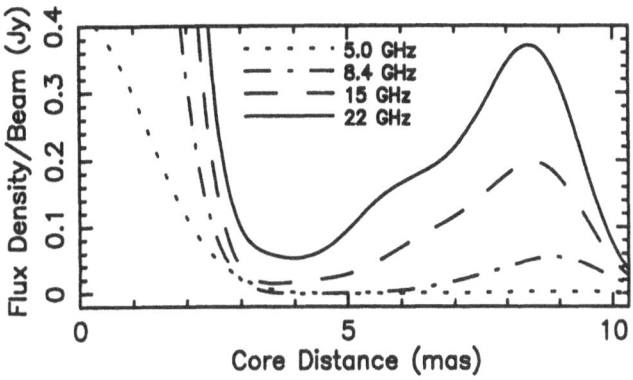

Figure 1. Slices along the ridge of the counterjet in 3C 84. The flux densities have been scaled so that they would match if the spectral index were −0.7.

and the radial dependence of this parameter can be measured on sub-parsec scales (at $z = 0.018$, 1 pc = 4 h mas, which is easily resolved).

In January 1995, in order to image the counterjet simultaneously at many frequencies, we made VLBA observations of 3C 84 at 2.3, 5.0, 8.4, 15, 22, and 43 GHz. The preliminary images are presented in the companion contribution by Romney *et al.* (1995) in this volume. The best images to use to study the absorption are those at 5.0, 8.4, 15, and 22 GHz. The counterjet is not seen at all at 2.3 GHz and, at 43 GHz, there were inadequate short spacings to image it. The images show clearly that the counterjet is cut off strongly at low frequencies.

2. The Data and Model

Figure 1 shows slices through the counterjet at the four good frequencies with the flux densities adjusted so that they would be the same if the spectral index were −0.70 as it is in the southern jet. To compare with absorption models, the effects of source structure need to be removed. This can be done by assuming a constant spectral index and by dividing the slices by the 22 GHz slice. A set of such ratios is shown in Figure 2. The region ≤ 4.5 mas from the core should be ignored: it is dominated by the core. The strong cutoff with frequency and the radial dependence of that cutoff in the counterjet can clearly be seen in the ratios beyond about 4.5 mas.

Figure 3 shows a very simple model for the absorption. In this model, $Ln^2/T^{3/2} = 3286R^{-2.5}$ (L and R in pc, n in cm^{-3}, and T in K) and $T = 10^4$K in the Gaunt factor. The constant and the power law index have been chosen to match the data curves for 8.4 GHz and to give the correct separation between 8.4 GHz and 5 GHz curves. The match to the data is reasonably good. Varying the index by about 0.5, while adjusting the

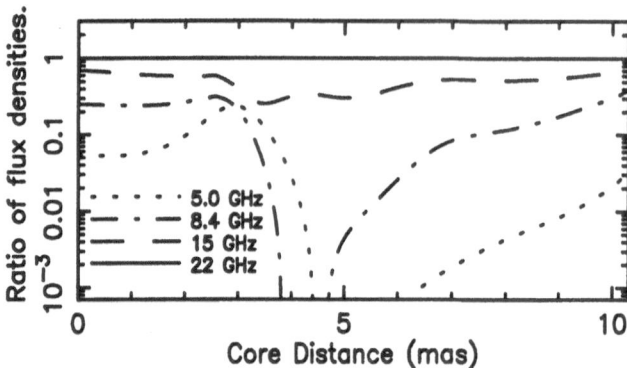

Figure 2.   Ratio of scaled slices to the 22 GHz slice. This is a good observable for comparison with absorption models. Recall 1 pc = 4 h mas.

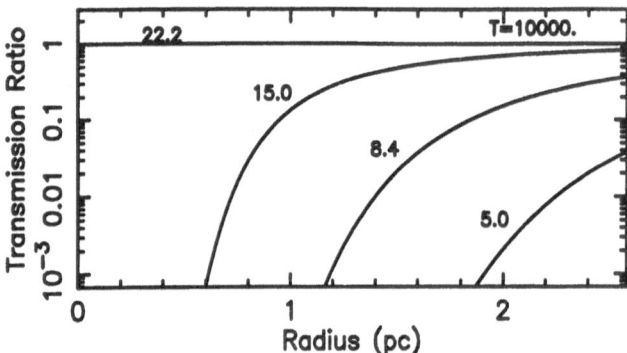

Figure 3.   Model ratios of slices to the 22 GHz slice along the spine of the 3C 84 jet for the simple power law model described in the text.

constant to match the middle of the 8.4 GHz curve, significantly degrades the fit. This is a preliminary result both because the effect of the finite resolution of the observations has not yet been taken into account and because the 22 GHz image is preliminary. However, it is clear that these data can provide a good constraint on the parameters of the ionized gas near this active galactic nucleus and on its radial dependence.

We gratefully acknowledge the contributions of A.C.S. Readhead, D.C. Backer, W. Alef, and J.M. Benson to the 3C 84 observational program.

References

Levinson, A., Laor, A., and Vermeulen, R. C. (1995) *Ap. J.*, **448**, 589.
Romney, J.D. *et al.* (1995) *This Volume.*
Vermeulen, R.C., Readhead, A.C.S., and Backer, D.C. (1994), *Ap. J.*, **430**, L41.
Walker, R.C., Romney, J.D., and Benson, J.M. (1994) *Ap. J.*, **430**, L45.

CONTINUUM VLBI POLARIMETRY OF 3C454.3 AT 43 GHZ

A. J. KEMBALL AND P. J. DIAMOND
National Radio Astronomy Observatory
Socorro, NM 87801, USA

Polarization VLBI calibration at high frequencies has traditionally been difficult due to poor sensitivity and high antenna instrumental polarization across inhomogeneous networks. The higher observing frequency and increased spatial resolution diminishes the chances of finding ideal VLBI polarization calibrators. The advent of the Very Long Baseline Array (VLBA), which has standardized feeds with low instrumental polarization, has minimized these observational difficulties. Recent work in polarization calibration has suggested that somewhat resolved sources may be used in an iterative polarization calibration scheme (Cotton 1993). A full generalization of this method has been developed by Leppanen, Zensus and Diamond (1995) in calibrating 22 GHz polarization observations with the VLBA.

We have mapped the quasar 3C454.3 in full polarization at 7mm wavelength using data obtained with the VLBA. This strong radio source at a redshift $z = 0.859$, is an optically violent variable (OVV) and is classified as a high optical polarization ($p > 3\%$) quasar (HPQ). It has been the subject of extensive VLBI observations (Pauliny-Toth *et al.* 1987; and references therein).

The observations were conducted on December 1, 1994 using the full VLBA (except FD). The source 3C454.3 was observed for a total of four hours, and 0420-014, a candidate polarization calibrator, for approximately two hours. The data were reduced within the Astronomical Image Processing System (AIPS) maintained by NRAO, using standard polarization calibration techniques (Kemball, Diamond and Cotton 1995), and including a correction for atmospheric opacity. The source 0420-014 was used as an iterative polarization calibrator under the similarity approximation developed by Cotton (1993) discussed above.

3C454.3 has a known core-jet morphology at a dominant position angle of -65 degrees (Padrielli *et al.* 1986). There is an apparent break in the jet direction within a few milliarcseconds of the core, where the dominant

R. Ekers et al. (eds.), Extragalactic Radio Sources, 33-34.

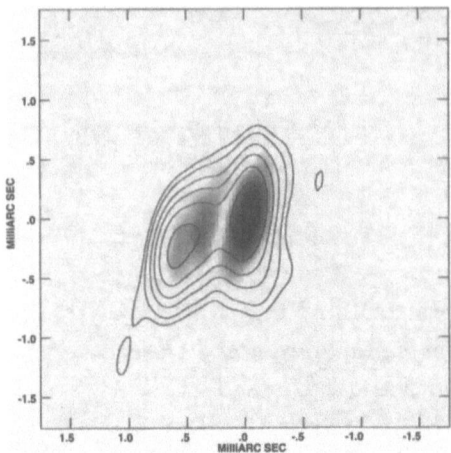

Figure 1.   Fig. 1. A map of the linearly polarised intensity for 3C454.3 at 7mm, with a contour plot of total intensity superimposed. Stokes I contour levels are at (-30, -15, -7.5, -3, -1.5, 1.5, 3, 7.5, 15, 30, 60, 90) % of the peak.

projected jet position angle is ~ -95 degrees, as shown in 8 GHz and 22 GHz VLBI observations (Pauliny-Toth *et al.* 1987; Charlot 1990). The 43 GHz polarization map is presented in Fig. 1.

Integrated polarization measurements at 33.5 GHz (Flett and Henderson 1983) show variability about a mean degree of polarization $\bar{p} \sim 4.5\%$. The 43 GHz data indicate a core polarization comparable to this with roughly twice that polarization for the inner component. Previous VLBI polarimetry at 5 GHz has found that quasar cores are rarely significantly polarized above $p = 2\%$ (Cawthorne *et al.* 1993). This may be explained by spatial blending of fine-scale polarization structure in the core.

The polarization calibration method evaluated here appears to work adequately for a source of this flux density when observed with a network with low instrumental polarization, such as the VLBA. Feed D-terms were found to be of the order of a few percent uniformly across the array.

References

Cawthorne, T.V., Wardle, J.F.C., Roberts, D.H., and Gabusda, D.C. (1993), *Ap.J*, **416**, 519.

Charlot, P. (1990), *AA*, **229**, 51.

Cotton, W.D. (1993), *AJ*, **106**, 1241.

Flett, A.M., and Henderson, C. (1983), *MNRAS*, **204**, 1285.

Kemball, A.J., Diamond, P.J., and Cotton, W.D. (1995), *A&AS*, **110**, 383.

Leppanen, K.J., Zensus, J.A., and Diamond, P.J. (1995), *in press*.

Padrielli, L., Romney, J.D., Bartel, N., Fanti, R., Ficarra, A., Mantovani, F., Matveyenko, L., Nicolson, G.D., and Weiler, K.W. (1986), *AA*, **165**, 53.

Pauliny-Toth, I.I.K., Porcas, R.W., Zensus, J.A., Kellermann, K.I., Wu, S.Y., Nicolson, G.D., and Mantovani, F. (1987), *Nature*, **328**, 778.

MULTI-FREQUENCY VLBI OBSERVATIONS OF 3C390.3

E. PREUSS[1], W. ALEF[1], K. I. KELLERMANN[2]

[1] Max-Planck-Institut für Radioastronomie, Auf dem Hügel 69, 53121 Bonn, Germany
[2] National Radio Astronomy Observatory, 520 Edgemont Road, Charlottesville, VA 22903, USA

Abstract. We have mapped the broad line radio galaxy 3C390.3 at 1.3, 6, and 18 cm wavelengths with resolutions ranging from 0.2 to 2.7 mas. Our new 6 cm image is consistent with a stationary bright region located 4 to 5 mas from the core and motion of the other features with apparent velocities of 0.4 to 0.7 mas/yr. Our high resolution 1.3 cm image indicates that the 'jet' breaks up into 5 or more distinct features.

The N galaxy 3C 390.3 (1845+797) is one of the closest radio galaxies with observed component motion and is also one of the nearest objects of FR type II (z=0.057; 1 mas = 1.5 pc; H_0 = 50 km/s/Mpc, q_0 = 0.5). VLA observations with arcsec resolution show two lobes, each with compact structure, and a weak one-sided thin jet joining the NW lobe to a central core (Alef et al. 1995, Leahy and Perley 1995).

Our previous 6 cm VLBI observations made between 1978 and 1989 suggested the ejection of multiple radio components at a rate of about once every four years with an apparent component motion of 0.7 mas/year (v/c = 3.5), but we also found evidence for a stationary region of enhanced emission located between 4 and 5 mas from the 'core' (Alef et al. 1995, Alef et al. 1988).

In order to clarify the morphology and apparent kinematics of the central region of 3C 390.3 we have made new observations at 1.3, 6, and 18 cm which we report here.

The 6 cm observations, made with the VLBA in 1995.29, extends the range of the previous sequence of observations by an additional 6 years. The map (Fig. 1) shows the same general features seen previously. As before we

R. Ekers et al. (eds.), Extragalactic Radio Sources, 35–37.

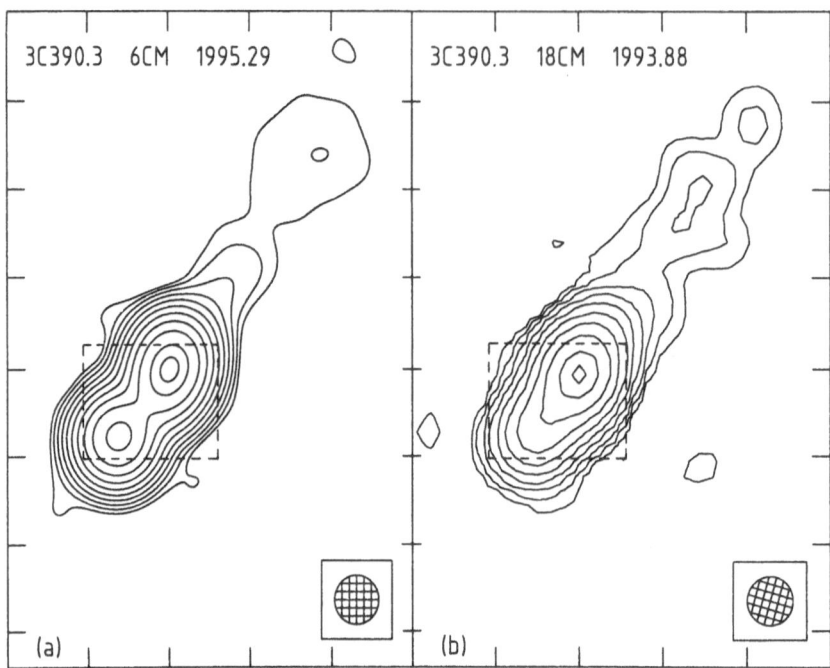

Figure 1. VLBI maps of 3C390.3 at a resolution of 2.7 mas. a) λ =6 cm: contour intervals are at 0.125, 0.25, 0.5, 1, 2, 4, 8,16, 32, 64, and 90 % of the peak brightness of 350 mJy/ beam. b) λ =18 cm: Contour intervals are the same as in (a) except the lowest contour level is at 0.25 % of the peak brightness of 268 mJy/ beam. The spacing between tick marks is 5 mas. The dashed rectangles mark the area covered by Fig. 2

see that the brightest jet feature is located ~5.3 mas away from the core, but it has now become as strong as the core component, whereas in 1989 and earlier it was much weaker.

Alef et al. (1995) noted that in the 1989 image, the jet component at 4-5 mas from the core appears to bend through an angle of ~ 3°. The new data show a sharper bend of ~ 12° which obviously mirrors the even sharper bend in the ridge line of the 1.3 cm image (Fig. 2). The faint features seen at about 12 and 20 mas from the core may be identified with features seen in the earlier observations if they have moved at a rate of 0.5 to 0.6 mas/yr; but the association with the features present six or more years ago is not unambiguous.

The 18 cm image, shown in Fig.1b, was obtained from a 15 station observation using the VLBA plus antennas at Green Bank (140 ft), Jodrell Bank (Mk I), Effelsberg, Medicina and Onsala, and was intended to trace intermediate scale structure whose surface brightness is too low to be seen in the 6 cm images. The 18cm map traces the jet feature out to a distance of 20 mas (30 pc) from the core. Comparison with the 6 cm image with the same resolution shows the outer jet to be much brighter at the longer

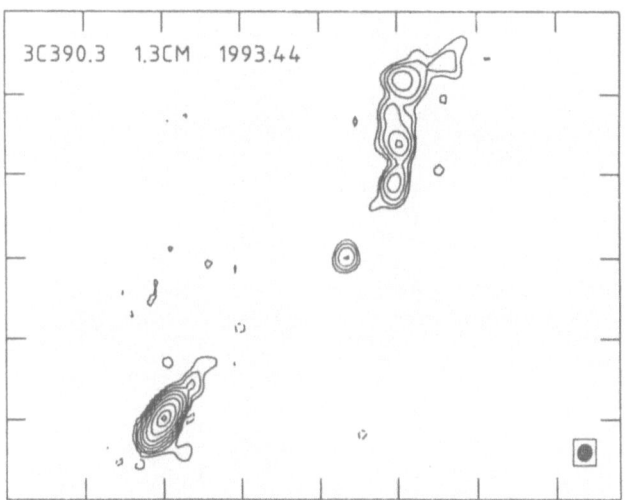

Figure 2. VLBI map of 3C390.3 at λ =1.3 cm with a resolution of 0.2 mas. Contour levels are at -0.5, 0.5, 1, 2, 4, 8, 16, 32, 64, and 90 % of the peak brightness of 143 mJy/ beam. The spacing between tick marks is 1 mas.

wavelength.

The 1.3 cm observations were made in 1993.44 and used all antennas of the VLBA (except Mauna Kea, Owens Valley and Saint Croix) plus the Effelsberg, Medicina, Noto and Onsala (20 m) antennas. This global array gives a resolution of ~0.2 mas so that features as small as 0.3 pc are well resolved. The 1.3 cm image (Fig. 2) has a very different appearance from the lower resolution 6 cm image. At 1.3 cm, the strong SE component, which we identify with the core, is resolved. The inner part of the jet is clearly visible pointing toward a jet feature 2.5 mas away. Apparently the jet is formed on a scale ≤ 1 pc, and remains remarkably well collimated over more than 100 kpc. The NW jet feature, which we identify with the bright 6 cm feature located at the same position, shows remarkable detail and appears to break up into at least five discrete components which lie along a curved locus. At ~5 mas from the core, the jet shows a sharp bend and then points toward the extended jet (PA~ −35°).

The NRAO is a facility of the National Science Foundation and is operated by Associated Universities Inc. under a cooperative agreement.

References

Alef, W., Götz, M.M.A., Preuss, E., Kellermann, K.I. (1988) Structural changes in the nucleus of the double radio galaxy 3C390.3, *Astron. Astrophysics*, **192**, pp. 53–56

Alef, W., Wu, S.Y., Preuss, E., Kellermann, K.I. and Qiu, Y.H. (1995) 3C390.3: a Lobe-Dominated Radio Galaxy with a Possible Superluminal Nucleus, *Astron. Astrophysics*, in press

Leahy, J.P. and Perley, R.A. (1995) The Jets and Hotspots of 3C390.3, *MNRAS*, in Press

STRUCTURAL CHANGES IN CTA102

F.T. RANTAKYRÖ[1,2] AND L.B.BÅÅTH[3]

[1] *Istituto di Radioastronomia-CNR – Bologna – Italy*
[2] *JIVE*
[3] *CBD- Halmstad University - Halmstad - Sweden*

1. Introduction

A study over several epochs and frequencies has demonstrated that the variability in CTA 102 (QSO 2230+114, z = 1.037) at $\lambda < 32$ cm is mainly due to intrinsic processes (Rantakyrö et al. 1995, *A&A*, in press, hereafter R1995). Observations of CTA 102 with the EGRET telescope at the high-energy γ-ray waveband have shown that the source exhibits a strong γ-ray luminosity, $L_\gamma = 5 \times 10^{47}$ ergs s^{-1} (Nolan, P.L., et al. 1993, *ApJ*, **414**, 82). This luminosity dominates the emission seen at all other wavebands, a common feature among BLAZARs. The common explanation for the high L_γ is that the emission is the result of a beamed jet with a high Lorentz factor.

2. Observations and results

The maps presented here (Figure 1) are epochs 1992.45 (R1995) and 1994.18. Both maps are made with identical contourlevels. The position of the two central components in the source has been determined by fitting Gaussian point sources to the final images. Identification of which component that has moved where is problematical due to the long time between the epochs. Thus we give two alternatives for the movement of the outer component, **case 1:** where component E has moved to the position of G, and **case 2:** where component F has moved to the position of G. The results of the fits are presented in table 1. The $\lambda 1.3$ cm observations suggests a very high proper motion, either a contraction or expansion. One should view this value with strong caution as we have only two sessions with somewhat

R. Ekers et al. (eds.), Extragalactic Radio Sources, 39–40.

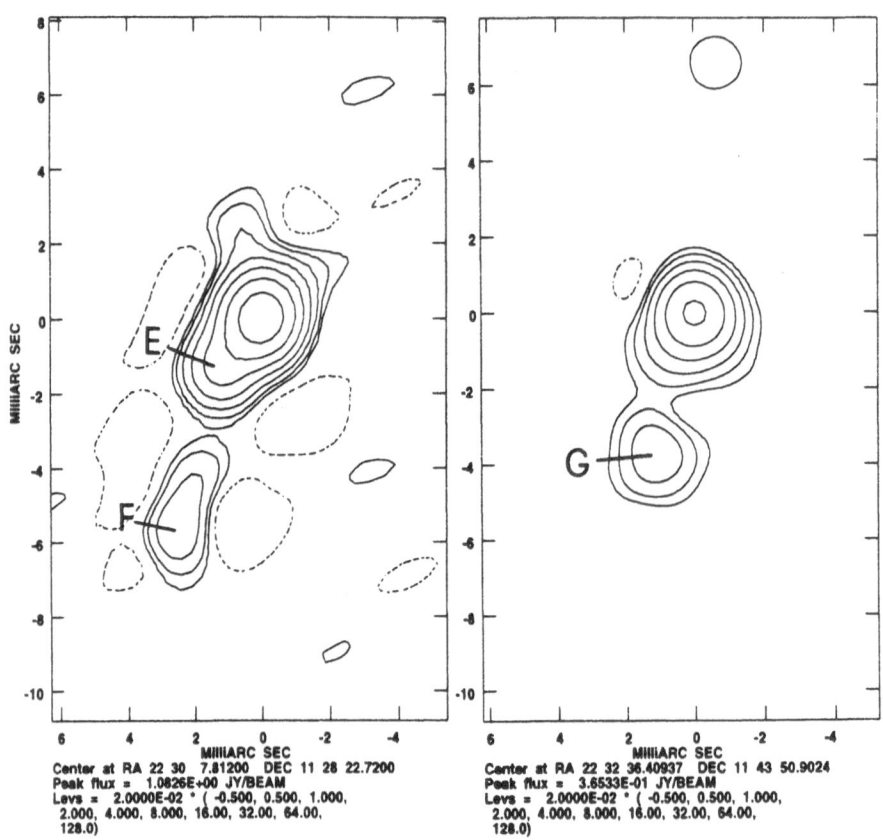

Figure 1. **Left:** EVN $\lambda 1.3$cm 1992.45. **Right:**EVN $\lambda 1.3$cm 1994.18

TABLE 1. Proper motion and distances from core for components. r_0 is the distance from the core.

Epoch	ID	λ [cm]	r_0 [mas]	μ_{app} [mas/yr]	β_{app} [v/c]
1992.45	E	1.3	1.7±0.1		
1992.45	F	1.3	6.2±0.5		
1994.18	G (case 1)	1.3	4.0±0.1	1.3±0.2	36±6
1994.18	G (case 2)	1.3	4.0±0.1	-1.3±0.3	-36±8

sparse uv-coverage at $\lambda 1.3$ cm (4 and 5 antennas respectively), and the relative time between epochs is large. We aim to investigate this source further, both at 22 GHz and at 3mm, and with both groundbased and satellite based telescopes.

VLBA STUDIES OF BL LAC

G. DENN AND R. MUTEL

University of Iowa
Dept. Physics and Astronomy
Iowa City, IA 52240
U.S.A.

1. Introduction

The radio core of BL Lac has been monitored with VLBI techniques since 1980. At least seven superluminal events have been seen, usually associated with flux outbursts. The components have been interpreted as weak relativistic shocks associated with opposed jets. Recent dramatic changes in the nucleus of BL Lac have been seen: optical spectra (Lawrence *et al.* 1995) show the presence of broad-line Hα and [NII] emission lines. The single-dish radio flux has also increased by 3 times in the past year. Often in BL Lac's past, a flux outburst immediately precedes the ejection of a new superluminal component.

2. Observations

We are currently engaged in a multifrequency monitoring program of BL Lac with the VLBA at several epochs separated by two month intervals. Table 1 lists the current status of the observations. Stokes-I maps have been produced for epochs 1995.5 and 1995.7 at all three frequencies. The 3.6 cm maps show one central component, slightly elongated along p.a. = 180°. The 2.0 cm and 1.3 cm maps show an emerging component at p.a. = 180°, approximately 1.25 mas from the core. This is in marked contrast to maps made as late as 1994.8, in which the jet p.a. was nearly 200°. Convolving the 1.3 cm model with the 2.0 cm and 3.6 cm beams produced maps nearly identical to the images at those frequencies.

R. Ekers et al. (eds.), Extragalactic Radio Sources, 41–42.

TABLE 1. Observational summary (1995)

	Jun 05	Aug 02	Oct 09	Dec
λ(cm)	3.6,2.0,1.3	3.6,2.0,1.3	2.0,1.3	TBD
Status	Correlated	Correlated	Observed	Scheduled
I-maps	yes	yes	not yet	-
P-maps	no	no	-	-

3. Interpretation

Simple conical jet models (e.g. König1 1981) of a shocked-emission region suggests that there should be a frequency dependent separation between the core position and a shock component along the jet. For typical physical values, the ratio of distances between core and shocked component should be $D_{1.3}/D_2 =\sim 0.75$. One-dimensional slices indicate a ratio of $D_{1.3}/D_2 = 1.04\pm0.2$, which does not agree with the simple conical model. More realistic models (*e.g.* Hughes *et al.*) will be tested after the polarization mapping is completed.

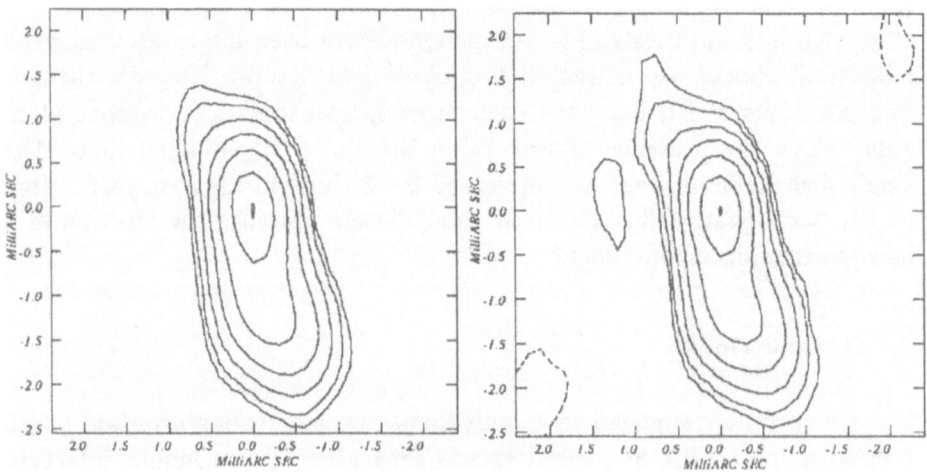

Figure 1.   15 GHz maps of the 1995.5 and 1995.7 epochs. The lowest contour is 2%.

References

Hughes, P.A., Aller, H. D., and Aller, M. F. (1989), *Ap. J.*, **341**, 68.
König1, A., (1981), *Ap. J.*, **243**, 700.
Lawrence, *et al.* , (1995), preprint.

HELICAL MOTION IN THE BL LAC OBJECT OJ287 ?

L. VICENTE[1], P. CHARLOT[2], H. SOL[1]

[1] Observatoire de Paris-Meudon, CNRS/UPR 176,
5 Place Jules Janssen, 92195 Meudon Principal Cedex, France
[2] Observatoire de Paris, CNRS/URA 1125,
61 Avenue de l'Observatoire, 75014 Paris, France

The structural evolution of the BL Lac object OJ287 has been studied with milliarcsecond resolution by using 8.4 GHz geodetic VLBI data from the Crustal Dynamics Program. Such data provide valuable maps which are useful to track superluminal components on short time scales (Charlot 1992). We have mapped and model-fitted OJ287 at ten epochs separated by intervals of a few months between 1985.4 and 1988.1 (Fig. 1a). The structure of OJ287 over this time span consists of two components, one of which is the source core, while the other one is identified with the knot K3, ejected during the 1983–1984 optical outburst. Our models indicate that the knot K3 moves along a non-radial path, with evidence for an apparent deceleration by mid-1986 and a reacceleration afterwards (Fig. 1b). We explain this peculiar motion by a projection effect due to a helical morphology of the jet. Our proposed helical model, estimated by considering the sky positions of K3 together with those of the earlier knots K1 and K2 (Roberts et al. 1987, Gabuzda et al. 1989), corresponds to a helix with a pitch angle of 20° on a narrow cone of half opening angle 3.4°, and central axis inclined at 17° relative to the line of sight. Ejection of VLBI components appears to occur simultaneously with optical outbursts and at mid-time between them. This finding is consistent with the supermassive binary black hole model previously proposed to explain those periodical outbursts (Sillanpää et al. 1988). Comparison of our helical fit with hydrodynamical models (Hardee et al. 1994) provides estimates of the radius ($R \simeq 0.015$ pc at a distance of 5 pc from the core), opening angle ($\psi \simeq 0.1°$), and Mach number ($M \simeq 60$) of the jet, and shows that the orbital 9.0 yr period of the binary black hole system is adequate to drive the helical perturbation if the jet propagates in a hot ambient medium with an external sound speed of 5000 km/s.

R. Ekers et al. (eds.), Extragalactic Radio Sources, 43–44.

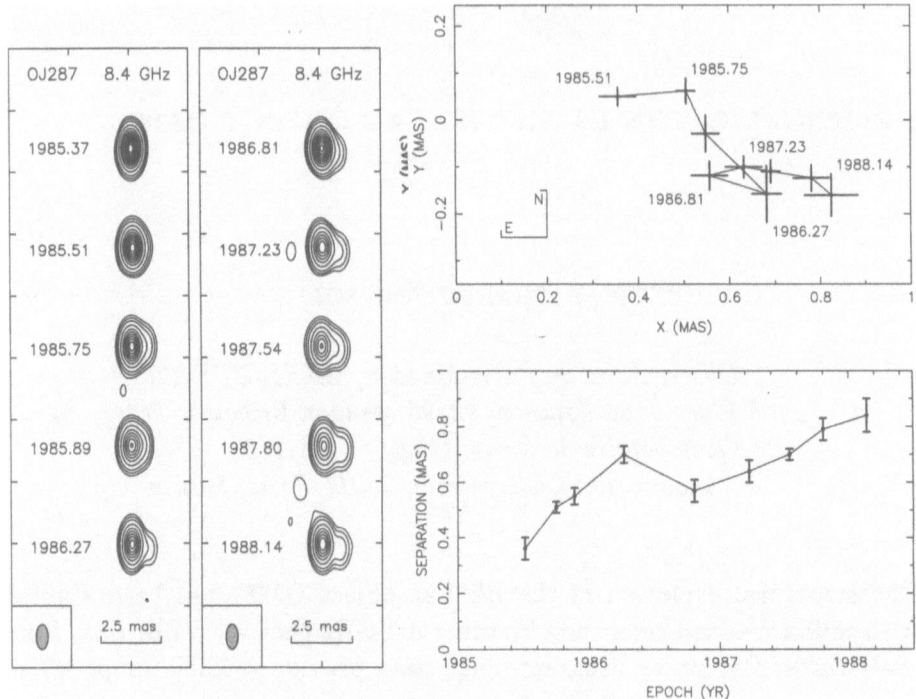

Figure 1. Left (a): 8.4 GHz VLBI maps of OJ287 at ten epochs. The main component is the core, while the western extension represents the knot K3. Contour levels are ± 0.05, 0.1, 0.2, 0.5, 1, 1.5, 2, 2.5, 3, 3.5, 4, 4.5, 5 and 5.5 Jy/beam (1×0.5 mas^2 at PA $= 0°$). Right (b): sky positions (top) and separations (bottom) of K3 from 1985.51 to 1988.14. The successive positions are connected with straight lines to indicate the time evolution.

Acknowledgements

We are grateful to the GSFC VLBI group for providing the data analyzed in this report. The analysis was carried out with the Caltech VLBI package.

References

Charlot P. (1992), Monitoring of extragalactic radio source structures with astrometric/geodetic VLBI experiments - the case of 3C273, *Extragalactic radio sources - from beams to jets*, Eds J. Roland, H. Sol and G. Pelletier, Cambridge University Press, pp. 112–118

Gabuzda D. C., Wardle J. F. C., Roberts D. H. (1989), Superluminal motion in the BL Lacertae object OJ287, *ApJ*, Vol. no. **336**, pp. L59–L62

Hardee P. E., Cooper M. A., Clarke, D. A. (1994), On jet response to a driving frequency and the jets in 3C 449, *ApJ*, Vol. no. **424**, pp. 126–137

Roberts D. H., Gabuzda D. C., Wardle J. F. C. (1987), Linear polarization structure of the BL Lacertae object OJ287 at milliarcsecond resolution, *ApJ*, Vol. no. **323**, pp. 536–542

Sillanpää A., Haarala S., Valtonen M. J., Sundelius B., Byrd G. G. (1988), OJ287: binary pair of supermassive black holes, *ApJ*, Vol. no. **325**, pp. 628–634

VARIABILITY CHARACTERISTICS OF BLAZAR OJ 287

L.O. TAKALO[1], A. SILLANPÄÄ[1], T. PURSIMO[1], H.J. LEHTO[1],
K. NILSSON[1], P. TEERIKORPI[1], P. HEINÄMÄKI[1], M. KIDGER[2],
J.A. DE DIEGO[2], T. MAHONEY[2], J.-M. RODRíGUEZ-ESPINOSA[2],
J.N. GONZÁLEZ-PÉREZ[2], P. BOLTWOOD[3], D. DULTZIN-HACYAN[4],
E. BENITEZ[4], G. TURNER[5], J. ROBERTSON[5], R. HONEYCUT[5],
YU.S. EFIMOV[6], N. SHAKHOVSKOY[6], P. A. CHARLES[7], D. KÜHL[8],
K.J. SCHRAMM[8], U. BORGEEST[8], J.V. LINDE[8], W. WENEIT[8],
T. SCHRAMM[8], A. SADUN[9], R. GRASHUIS[10], J. HEIDT[11], H. BOCK[11],
S. WAGNER[11], M. KÜMMEL[11], A. HEINES[11], M. FIORUCCI [12],
G. TOSTI[12], C. RAITERI[13], M. VILLATA[13], G. LATINI[13], S. BOSIO[13],
G. GHISELLINI[13] AND G. DE FRANCESCO[13]

[1] *Tuorla Observatory, FIN-21500 Piikkiö, Finland*
[2] *Instituto de Astrofísica de Canarias, La Laguna, Tenerife, Spain*
[3] *1655 Main St. Stittsville, Ont. K2S 1N6, Canada*
[4] *Instituto de Astronomía-UNAM, Apdo. Postal 70-264, 04510, Mexico, D.F. Mexico*
[5] *Department of Astronomy, Indiana University, Swain West 319, Bloomington, IN 47405, USA*
[6] *Crimean Astrophysical Observatory, P/O Nauchny, 334413 Crimea, Ukraine*
[7] *University of Oxford, Dept. of Astrophysics, Keble Road Oxford, OX1 3RH, England*
[8] *Hamburger Sternwarte, Hamburg Universität, Gojensbergweg 112, D-21029 Hamburg 80, Germany*
[9] *Bradley Observatory, Agness Scott College, Decantur, GA 30030, USA*
[10] *Capilla Peak Observatory, Albuquerque, NM 87131, USA*
[11] *Landessternwarte Königstuhl, D-69117 Heidelberg, Germany*
[12] *Osservatorio Astronomico, Universita di Perugia, I-601123 Perugia, Italy*
[13] *Osservatorio Astronomico di Torino, Strada Osservatorio 20, I-10025, Pino Torinese, Italy*

R. Ekers et al. (eds.), Extragalactic Radio Sources, 45–46.

Blazar OJ 287 is one of the best observed extragalactic objects. It's historical light curve goes back to 1890's. Based on the historical behaviour Sillanpää et al. (1988) showed that OJ 287 displays large periodic outbursts, with a period of 11.7 years. We have monitored OJ 287 intensively for two years, during the OJ-94 project. This project was created for monitoring OJ 287 during its predicted new outburst in 1994. In the data archive we have over 7000 observations on OJ 287, in the radio, infrared and optical bands. This data archive contains the best ever obtained light curves for any extragalactic object. The optical light curve shows continuous variability down to time scales of tens of minutes. The variability observed in OJ 287 can be broken down to (at least) four different categories:

1. The large outbursts that occur every 11.7 years. The last one of these happened during November 1994, almost at the predicted time. These outbursts can be due to the binary black hole model proposed by Sillanpää et al. (1988).

2. There are quite frequent, but randomly occurring flares that last from a few days to two weeks. During the project we have observed ten such flares, with the flare amplitudes ranging from half to over one magnitudes. These flares seem to have a synchrotron origin, being probably caused by shocks in the jet.

3. Small amplitude "flickering", that is present all the time. Time scales for this ranges from tens of minutes to hours, and the amplitude from 0.2 to 0.5 magnitudes. This flickering can be due to some instabilities in the jet or in the accretion disk.

4. Microvariability, that is seen occasionally. There is no clear correlation between it and the other variations. This microvariability can be due to the same events that cause the flickering.

References

Sillanpää, A., et al. (1988), *ApJ*, Vol. **325.**, pp. 628-634

THE ENVIRONMENT OF OJ 287:
NEARBY GALAXIES AND A LONG OPTICAL JET?

A. SILLANPÄÄ, L. TAKALO, K. NILSSON, T. PURSIMO
P. TEERIKORPI
Tuorla Observatory
FIN-21500 Piikkiö, Finland

J. HEIDT
Landessternwarte Königstuhl
D-69117 Heidelberg, Germany

AND

D. DULTZIN-HACYAN AND E. BENÍTEZ
Instituto de Astronomía-UNAM
Apdo. postal 70-264, 04510 Mexico, D.F. Mexico

1. Introduction

A widely accepted model for BL Lac objects is that they are radio galaxies with a relativistic jet pointing almost directly towards us. But we need a clear trigger mechanism for these jets. One possibility is the close interaction between the BL Lac host and the closeby galaxies (e.g. Heckman et al. 1986). This interaction has been seen many times in the case of quasars (Hutchings et al. 1989) but not so much is known about the close surroundings of the BL Lac objects although there has been some pioneer work like Stickel et al. (1993). The problem has usually been that the images are not deep enough and that the seeing has not been so good. To clarify the situation we have started an observing program to get very deep images in the subarcsecond seeing conditions from the whole 1 Jy sample (Stickel et al. 1991) of BL Lac objects. The aims of this study are: 1. to search for very close companions to the BL Lacs, 2. to study the large scale galaxy clustering around the BL Lacs and 3. to study the BL Lac hosts themselves.

47

R. Ekers et al. (eds.), Extragalactic Radio Sources, 47–48.

2. Observations and Results

We have taken deep CCD images of OJ 287 in the very wide frequency range from B to K. The V-band images were taken with the 2.1m telescope on San Pedro Mártir, Mexico; B and I-images with the 2.56m NOT-telescope on La Palma, Canary Islands; R and K-images with the 2.2m telescope on Calar Alto, Spain. The main results of our OJ 287 study can be summarized as follows:

1. We have found at least six objects closer than 10 arcseconds from OJ 287. If these objects are really connected to OJ 287 their projected distances are smaller than 60 Kpc. The closest companion is only 3.4 arcsecs from OJ 287.

2. There is a "jet"-like feature to the south-west from the object. We can see at least six "knots" in almost perfect chain starting from the object. The projected length of this feature is about 25 arcsecs.

3. There is also another "jet"-like feature to the west of the object. The length of this feature is only ten arcsecs and it coincides with the beginning of the radio jet observed with the VLA (Perlman and Stocke 1994).

4. There are quite many galaxies close to OJ 287 which are possibly seen also in the VLA image (Kollgaard et al. 1992). If this is true it means that there are many active galaxies around OJ 287.

5. From our R- and K-band images we have calculated that in the 2'x2' area around OJ 287 there is a galaxy overdensity by a factor of two (Metcalfe et al. 1991 and Cowie et al. 1990). The R-K colours for the brightest of these galaxies are about 3.6. These values are typical for galaxies at z=0.3 which means that they may be associated with OJ 287.

References

Cowie, L.L., et al. 1990, *ApJ*, Vol. no. **360**, pp. L1-L5
Heckman, T.M., et al. 1986, *ApJ*, Vol. no. **311**, pp. 526-547
Hutchings, J.B., et al. 1989, *ApJ*, Vol. no. **342**, pp. 660-665
Kollgaard, R.I., et al. 1992, *AJ*, Vol. no. **104**, pp. 1687-1705
Metcalfe, N., et al. 1991, *MNRAS*, Vol. no. **249**, pp. 498-522
Perlman, E.S., Stocke, J.T. 1994, *AJ*, Vol. no. **108**, pp. 56-63
Stickel, M., et al. 1991 *ApJ*, Vol. no. **374**, pp. 431-439
Stickel, M., et al. 1993 *A&AS*, Vol. no. **98**, pp. 393-442

NEW RESULTS FROM VLBI POLARIZATION OBSERVATIONS OF BL LACERTAE OBJECTS

D.C. GABUZDA
Lebedev Physical Institute
Moscow, RUSSIA

1. A Systematic Study of the VLBI Polarization Characteristics of BL Lacertae Objects

The major distinguishing features of BL Lacertae Objects are weak or absent line emission and strong and variable optical, infrared, and radio polarization (Angel and Stockman 1980; Kollgaard 1994). The radio emission and much of the optical emission is believed to be synchrotron radiation. In nearly every BL Lacertae object in which polarization structure has been detected, the polarization position angles χ in knots in the jets are nearly parallel to the VLBI structural axis. Assuming the jet components to be optically thin, the inferred magnetic fields **B** are nearly perpendicular to the jet direction θ ; perhaps the most natural interpretation of this is that the knots are associated with shocks that compress an initially tangled **B** field as they propagate down the VLBI jet, enhancing **B** transverse to the compression (Laing 1980; Hughes, Aller, & Aller 1989). The superluminal speeds observed in the jets of BL Lacertae objects are on average lower than those observed in a comparably core-dominated population of quasars (Gabuzda *et al.* 1994). The distribution of core polarization position angles χ_{core} at $\lambda = 6$ cm is bi-modal, suggesting that χ_{core} is roughly perpendicular to θ when the cores are quiescent, and aligns with θ at epochs when polarization from newly emerging shock components is blended with the core polarization (Gabuzda *et al.* 1994).

In order to understand these clear systematic trends, it is useful to examine the polarization properties of a complete sample of BL Lacertae objects. A sample convenient for such studies is that defined by Kühr and Schmidt (1990). $\lambda = 6$ cm VLBI polarization images have already been published for 15 sources in this sample, for 10 sources at more than one epoch (Gabuzda *et al.* 1994 and references therein). $\lambda = 6$ cm observations of 18

R. Ekers et al. (eds.), Extragalactic Radio Sources, 49–50.

additional BL Lacertae objects are now being reduced and analyzed; these complete first epoch imaging of all sources in the Kühr and Schmidt sample. Second epoch $\lambda = 6$ cm polarization observations for this sample were completed in May of this year. These observations are being complemented by observations of a smaller number of sources at shorter wavelengths (see, e.g., Gabuzda, Pushkarev, & Cawthorne, these proceedings).

2. Recent Results

Recently analyzed $\lambda = 3.6$ cm images, as well as $\lambda = 6$ cm images currently being analyzed, support the trends noted above, but are also providing some surprises. The u-v coverage for comparatively short baselines for these data is somewhat better than at the earlier epochs, and as a result, the extended VLBI structure is better reproduced in the resulting images. These are revealing more sources with complex jet structure, sometimes with evidence that the jet χ vectors maintain their alignment with the local θ when the direction of motion for a given component changes substantially. These new images also support the idea that the $\lambda = 6$ cm distribution of χ_{core} is bimodal, with χ_{core} preferably either aligned with the small scale jet direction or perpendicular to it.

Perhaps most surprising is that these new images are revealing more sources in which the χ vectors in jet components indicate longitudinal **B** fields. In some sources, some jet components have χ aligned with θ (transverse **B**) and some have χ transverse to θ (longitudinal **B**). This indicates that some jet components in BL Lacertae objects are either not shocks, or are rather weak shocks imposed on a comparatively strong underlying longitudinal **B** field. There is some indication that the jet components in which **B** is longitudinal tend to be more extended than those in which **B** is transverse, but this is still uncertain. Completion of the analysis for the first epoch observations for the Kühr and Schmidt BL Lacertae object sample will give more information about how common jet components with longitudinal **B** field are, and how these components differ from those in which **B** is transverse. These studies will as well provide information about the relationship between BL Lacertae objects and quasars, in which longitudinal jet **B** fields are much more common.

References

Angel & Stockman (1980) *Annual Reviews of Astronomy and Astrophysics.* 8, 321.
Gabuzda *et al.* (1994) *Astrophysical Journal.* 435, 140.
Hughes, Aller, & Aller (1989) *Astrophysical Journal.* 341, 68.
Kollgaard (1994) *Vistas in Astronomy,* 38, 29.
Kühr and Schmidt 1990, *Astronomical Journal,* 99, 1.
Laing (1980) *Monthly Notices of the Royal Astronimcal Society,* 193, 439.

3.6 CM VLBI TOTAL INTENSITY AND POLARIZATION IMAGES OF BL LACERTAE OBJECTS

D.C. GABUZDA AND A.B. PUSHKAREV

Lebedev Physical Institute
Moscow, RUSSIA

AND

T.V. CAWTHORNE

University of Central Lancashire
Preston, Lancashire, UNITED KINGDOM

1. Introduction

The major distinguishing features of BL Lacertae Objects are weak or absent line emission and strong and variable optical, infrared, and radio polarization (Angel and Stockman 1980; Kollgaard 1994). The radio emission and much of the optical emission is believed to be synchrotron radiation. There are now some 20 BL Lacertae objects for which VLBI polarization (VLBP) images have been made at $\lambda = 6$ cm (Gabuzda *et al.* 1994 and references therein). In nearly every BL Lacertae object in which polarization structure has been detected, the polarization position angles in knots in the jets are nearly parallel to the VLBI structural axis. Assuming the jet components to be optically thin, the magnetic fields inferred by this orientation are nearly perpendicular to the direction of the jet; perhaps the most natural interpretation of this is that the knots are associated with shocks that compress an initially tangled magnetic field as they propagate down the VLBI jet, enhancing the magnetic field transverse to the compression (Laing 1980; Hughes, Aller, & Aller 1989).

At $\lambda = 6$ cm, the degrees of polarization of the cores of BL Lacertae objects ($\sim 2 - 5\%$) are typically higher than that of quasars ($\leq 2\%$); one possible origin for this is that the total observed polarizations for most of the cores in BL Lacertae objects include a substantial contribution from newly emerging knots (Gabuzda *et al.* 1994).

R. Ekers et al. (eds.), Extragalactic Radio Sources, 51–52.
© 1996 *IAU.*

2. Results of 3.6 cm VLBI Polarization Observations of BL Lacertae Objects

3.6 cm polarization images of BL Lacertae objects clearly show the tendency for χ to be aligned with the local jet direction, although there is evidence that the dominant magnetic field in components can *occasionally* be longitudinal. A striking example of this is offered by a knot in 1219+285, which is polarized with χ clearly perpendicular to the jet direction. The considerable degree of polarization of this component (10%) argues that it is optically thin, so that the inferred dominant magnetic field is longitudinal. This clearly suggests that at least occasionally, the discrete components observed in the jets of BL Lacertae objects are either not transverse shocks, or are weak shocks in which the compression is insufficient to dominate an underlying longitudinal magnetic field.

In 0735+178 and OJ 287 (Gabuzda & Cawthorne, in preparation) there is evidence for the detection of jet polarization in places where there is no clear evidence for an I component. In both cases, this polarization is close to transverse to the jet direction, implying an associated longitudinal magnetic field, if the emission is optically thin. If this is indeed the case, this may be interknot polarization, indicating the presence of a dominant longitudinal magnetic field in the underlying (unshocked) flow.

The 3.6 cm cores are on average somewhat more weakly polarized than at 6 cm. Only 2 of 22 measurements of 6 cm core polarizations (or limits to core polarization) indicated the cores to be polarized less than 2%, whereas the 3.6 cm measurements indicate 5 of 9 cores to be polarized less than 2%. Both of the sources for which polarization was detected in jet components but not in the cores (1219+285 and BL Lac) have comparatively low redshifts, so that the linear scales which are observed are somewhat smaller than those for the other sources. These tendencies are all consistent with a picture in which the observed "core" polarizations are typically a superposition of a quite weakly polarized core and newly emerging jet components (shocks).

These results are discussed more completely in a paper by Gabuzda & Cawthorne, to be submitted to *MNRAS* in November, 1995.

References

Angel & Stockman (1980) *Annual Reviews of Astronomy and Astrophysics.* 8, 321.
Gabuzda et al. (1994) *Astrophysical Journal.* 435, 140.
Hughes, Aller, & Aller (1989) *Astrophysical Journal.* 341, 68.
Kollgaard (1994) *Vistas in Astronomy,* 38, 29.
Laing (1980) *Monthly Notices of the Royal Astronimcal Society,* 193, 439.

EXTENDED EMISSION IN BL LAC OBJECTS

M. BONDI, D. DALLACASA AND C. STANGHELLINI
IRA, Bologna, Italy

AND

R. DELLA CECA
JHU, Baltimore, USA

The study of the extended emission and polarization properties of BL Lacs is an important step for the identification of their parent population. FRI radio sources, the supposed parent population of BL Lacs, have weaker extended radio luminosity and a dominant inferred magnetic field perpendicular to the jet, while FRII radio sources, the supposed parent population of quasars, have stronger extended radio power and an inferred magnetic field parallel to the jet. The only complete sample of radio selected BL Lacs (1 Jy sample, Stickel et al. 1991, *ApJ*, **374**, 431) contains 34 objects. Unfortunately, about half of 1 Jy BL Lacs do not have very high dynamic range images, necessary to detect the low emissivity radio emission surrounding the bright compact source, either because the object was never observed, or because the observation was carried out at the beginning of 1980s with low sensitivity. In 1994 we started a programme using the VLA (A, B, and D configuration, see Table 1) and the WSRT (W in Table 1) to complete the high sensitivity radio imaging of the 1 Jy sample. We aim to investigate morphology and polarization properties, as well as the luminosity of the extended emission. This contribution presents the L band observations. The results are very preliminary, some of the data reduction is still in progress as well as the statistical analysis. The sources in Table 1 have been roughly classified as extended (E), or point-like (P) if no extended feature was detected. Among the 15 sources observed at the highest resolution 13 were classified as extended. In many sources we detect significantly much more extended flux than previously reported from earlier observations. Almost all the BL Lac objects we observed at the highest resolution show some extended features; furthermore, in a few cases, we detected emission on the arcminute scale. The power of the extended luminosity covers 3 orders of

R. Ekers et al. (eds.), Extragalactic Radio Sources, 53–54.

TABLE 1. For the sources with extended features, the Table lists the redshift, the largest angular sizes in arcsecond and kpc, the flux density and the power of the extended emission ($H_0 = 100$ kms^{-1}Mpc^{-1}). When a redshift was not available we used a value of 0.5, close to the median value for the redshift distribution of the 1 Jy sample.

IAU Name	Obs.	Morph.	z	LAS arcsec.	LAS kpc	S_{ext} mJy	$P_{1.4}$ $\times 10^{24}$ W/Hz
0048−097	A,B,D	E		28	100	130	34.8
0118−272	A,B,D	E	>0.557	35	130	105	32.6
0138−097	A,B,D	E	>0.501	20	71	27	7.3
0426−380	A,B,D	E	>1.030	5	21	16	18.2
0454+844	W	P					
0537−441	A,B,D	E	0.896	17	71	187	160.6
0716+714	W	E		41	146	180	48.2
0814+425	W	E	0.258	24	60	15	1.1
0820+225	D,W	E	0.951	25	107	124	120.0
0823+033	D	P					
0828+493	W	P					
0851+202	W	P					
0954+658	A,B,D,W	E	0.367	7	22	16	2.3
1144−379	A,D	P					
1147+245	A,B,D,W	E		38	135	48	12.8
1308+326	D,W	P					
1418+546	D,W	E	0.152	227	393	86	2.1
1514−241	A,B	E	0.049	57	38	134	0.3
1519−273	A,B	P					
1652+398	A,B	E	0.033	35	16	74	0.1
1749+701	W	P					
1803+784	A,B,D,W	E	0.684	95	376	35	17.5
1807+698	A,B,D	E	0.051	341	260	961	2.7
1823+568	W	E	0.664	32	126	114	53.8
2007+777	W	E	0.342	52	154	41	5.1
2131−021	A,B,D	E	0.557	70	260	92	30.5
2240−260	A,B,D	E	0.774	24	98	344	220.5
2254+074	D	P					

magnitude, 3 objects (0537−441, 0820+225, and 2240−260) have values typical of a FRII radio source. These new data will be used for an updated statistical analysis of the properties of the extended emission in the 1 Jy sample of BL Lac objects.

Acknowledgements: MB and DD acknowledge for financial support the European Union EU Fellows under the contract CHBGCT920212.

ARCSECOND SCALE POLARIZATION
OF BL LAC OBJECTS

C. STANGHELLINI[1], P. CASSARO[2], M. BONDI[1,5]

AND

D. DALLACASA[1,3], R. DELLA CECA[4], R. A. ZAPPALÀ [2]

[1] *Istituto di Radioastronomia CNR - Italy*
[2] *University of Catania - Italy*
[3] *JIVE, Dwingeloo - The Netherlands*
[4] *JHU, Baltimore - USA*
[5] *NRAL, Jodrell Bank - UK*

1. Introduction

BL Lac objects are an enigmatic class of active galactic nuclei. They are characterized by high luminosity, a flat radio spectrum that steepens at higher energies, relatively high optical and radio polarization, rapid variability and an optical continuum with weak or absent emission lines (see Urry and Padovani, 1995 for a recent review).

These properties have been interpreted in terms of a relativistic jet closely aligned to the line of sight (Blandford and Rees, 1978, Ghisellini et al., 1993). This model, known as the beaming model, implies that there must be a so called "parent population" of radio sources intrinsically identical to BL Lac objects, but with the jets oriented at large angles to the line of sight. Browne (1983) was the first to propose the low luminosity FR I radio galaxies as the most likely candidates for the "parent population" of the core dominated BL Lac objects. An outcome of the beaming model is that all the properties not depending on orientation should be shared by the BL Lac objects and the FR I radio galaxies.

Prompted by this problem we observed several BL Lac objects from the 1 Jy sample (Stickel at al. 1991) with the VLA and WSRT in order to

R. Ekers et al. (eds.), Extragalactic Radio Sources, 55–56.
© 1996 *IAU.*

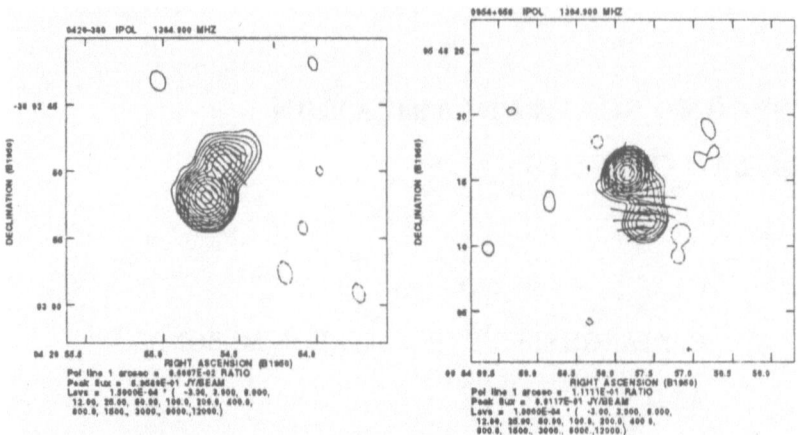

investigate the presence of arcsecond extended emission (see contribution by Bondi et al.) and to study their polarization properties. This aimed to complete the radio information on the complete sample.

2. Preliminary Results

The data reduction is still in progress but the results obtained so far are very promising. We detect polarized emission from extended structures reaching and exceeding 40%, while the cores show a fractional polarization between 1% and 5%. We find several objects where polarized emission is detected along the jet. These can be observed in more detail to study the magnetic field orientation in relation to bends in the jets. This will help in the interpretation of the observed properties in terms of projection effects, jet-ambient interaction and unification models.

An interesting preliminary result is that in some objects (see figures) the E field of the polarized emission is oriented perpendicular to the jet direction, i.e. the magnetic field structure is basically longitudinal to the jet. This behaviour is commonly seen in the jets of FR II radio sources. The completion of the analysis is necessary to infer statistically significant conclusions in the comparison between the polarization properties of the observed BL Lac objects and the polarization properties of the proposed parent population.

References

Blandford, R., and Rees, M.J., 1978, in Proc. Pittsburgh Conf. on BL Lac Objects, ed. A.N. Wolfe, p/328.
Browne I.W.A., 1983, *MNRAS*, **204**, 23
Ghisellini, G., et al., 1993, *Ap.J.*, **407**, 65.
Stickel M, Padovani P., Urry C.M., Fried J.W., Kühr H, 1991, *ApJ*, **374**, 431
Urry, C.M., and Padovani, P., 1995, *PASP*, **107**, 803.

SUPERLUMINAL MOTIONS: THE FIRST 100 SOURCES

R.C. VERMEULEN
California Institute of Technology,
Astronomy 105-24, Pasadena, CA 91125, USA

Abstract. First results from a large homogeneous superluminal motion survey are presented. The data do not show compelling evidence for the existence of intrinsically different populations of galaxies, BL Lac objects, or quasars. β_{app} in the range 1–$5\,h^{-1}$ occur with roughly equal frequency; higher values, up to $\beta_{app} = 10\,h^{-1}$, are rather more scarce than appeared to be the case from earlier work, which evidently concentrated on sources which are not representative of the general population. The β_{app} distribution suggests that there might be a skewed distribution of Lorentz factors over the sample, with a peak at $\gamma_b \approx 2\,h^{-1}$ and a tail up to at least $\gamma_b \approx 10\,h^{-1}$. There appears to be a clearly rising upper envelope to the β_{app} distribution when plotted as a function of observed 5 GHz luminosity; a combination of source counts and the apparent velocity statistics in a larger sample could provide much insight into the properties of radio jet sources.

1. Introduction

This review has been adapted from a presentation first given at the conference on "High Resolution Radio Imaging of Quasars and AGN" (Vermeulen 1995). It emphasizes statistics. Some other contributions to this Symposium deal with the wealth of data available for a select few sources. For example, in 3C 345 (Zensus 1996) prolonged, intensive monitoring suggests that in the inner parsecs the jet has a constant Lorentz-factor ($\gamma\sim10$), but bends away from the line-of-sight by a few degrees (e.g. Zensus et al. 1995).

In contrast, the measurements presented here are mostly rather sparse. But, as long as the source selection criteria are well known, a statistical approach has the virtue that population models can be constructed, involving

R. Ekers et al. (eds.), Extragalactic Radio Sources, 57–62.

a distribution of Lorentz factors, jet bending, pattern motions, acceleration, or whatever other complexity is thought to be indicated by the data. These models can be compared to the observed distribution of (superluminal) apparent velocities with the help of Monte Carlo simulations. Thus, the apparent velocity statistics of the sample reveal for the population as a whole complexities which are missed in individual objects.

Friedmann cosmology with $H_0 = 100h$ km s^{-1} Mpc^{-1} is used throughout this review, and where appropriate the dependence of the results on h and q_0 is shown. Indeed, since superluminal radio sources can be observed over a wide range of redshifts, their statistics can contribute to a discrimination between different cosmological models (e.g. Cohen et al. 1988, Vermeulen & Cohen 1994). Furthermore, if observable, individual two-sided relativistic jets near the plane of the sky would have a "standard velocity" of $2c$, so that h follows directly from the observed angular separation rate. This promise might be fulfilled by a recently discovered CSO, 1946+708 (Taylor, Vermeulen, & Pearson 1995).

2. Randomly Oriented and Beamed Samples; Unification Tests

The model predictions in this review, based mostly on Monte Carlo simulations, follow the formalism and assumptions outlined earlier (Vermeulen & Cohen 1994). If a sample has randomly oriented jets, then many will be pointed near the plane of the sky, and have $\beta_{app} = v_{app}/c \approx 1$. There will be a modest fraction of superluminals, from jets at angles of $\sin \theta \sim 1/\gamma$, and a small percentage of knots with $\beta_{app} < 1$, in jets pointed almost straight towards us. If, on the other hand, a flux-limited sample is selected on beamed radiation (such as flat-spectrum centimeter radio emission), then most of the observed motions should be close to the maximum possible velocity, $\beta_{app} = \beta\gamma$, because the width of the beaming cone is similar to the angle at which the largest apparent motion occurs.

The distribution of velocities in a sample is therefore potentially an indicator of the range of angles to the line-of-sight, useful to test unification models, subject to some caveats (e.g. Vermeulen & Cohen 1994). In quasars selected at low frequency (to minimize orientation bias, see Hough 1994, and Zensus & Porcas 1987), the apparent velocities tend decrease with decreasing core dominance (R, which is thought to be another orientation indicator, e.g. Orr & Browne 1982). However, there is significant scatter (e.g. Vermeulen et al. 1993).

Taking all data from the literature (an inhomogeneous collection, Vermeulen & Cohen 1994), it seems that at the very largest R, which often occur in BL Lacs, the apparent velocities may decrease again, as expected for jets viewed within the $1/\gamma$ beaming cone. The β_{app} statistics presently

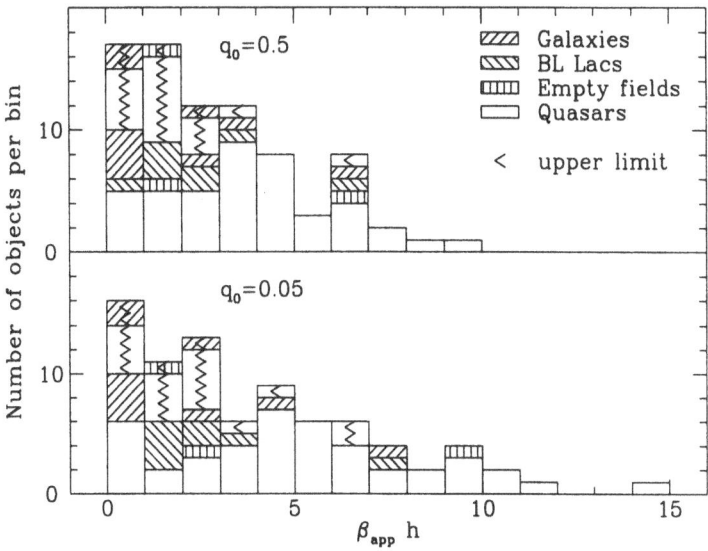

Figure 1. The observed apparent velocity distribution for 81 objects in the homogeneous PR+CJ flat-spectrum sample, showing the scarcity of higher values, compared to earlier work and to predictions for beamed samples. There is no firm evidence for differences related to optical identification.

available do not require the BL Lacs to be intrinsically different objects, as is sometimes suggested (e.g. Gabuzda et al. 1994, Urry & Padovani 1995), but more work on homogeneous samples is needed.

3. First Results from the CJ Survey

The combination of the Pearson-Readhead (Pearson & Readhead 1988) and Caltech-Jodrell Bank VLBI surveys (Polatidis et al. 1995, Taylor et al. 1994) yields a complete flux-limited sample of 293 flat-spectrum sources brighter than 0.35 Jy at 5 GHz. Second epoch observations, with a time interval of 2–3 years, are in progress, and will hopefully be completed in 1995. Apparent velocities or useful upper limits are now available for 81 of these sources; they are shown in Figure 1.

Ignoring upper limits and empty fields, the mean apparent velocity is slightly smaller for the galaxies ($N = 7$, $\langle \beta_{app} \rangle = 2.1$) and the BL Lacs ($N = 8$, $\langle \beta_{app} \rangle = 2.3$) than for the quasars ($N = 44$, $\langle \beta_{app} \rangle = 3.2$); all for $q_0 = 0.5$. This is in the sense expected for unification models. However, the KS test yields a probability of 23% that the galaxies and the quasars have the same β_{app} distribution, and a probability of 30% for the BL Lac objects and the quasars. With the full sample, stronger tests of unification models will be possible.

There *is* a *substantial* fraction (~25%) of stationary features or upper limits. Then, if $q_0 = 0.5$ it seems that the majority of β_{app} are evenly spread over the range $1-5\,h^{-1}$, with a tail (17%) of higher values, up to $10h^{-1}$. If $q_0 = 0.05$, the β_{app} distribution tapers off even more gently; most values are still below $10h^{-1}$. The fraction of sources with $\beta_{app} = 5-10\,h^{-1}$ in the older literature (Vermeulen & Cohen 1994) was more than twice as high, 36%. The scarcity in CJ of these somewhat faster motions is not caused by undersampling in time. It seems that there used to a bias towards studying and publishing fast superluminals. A lot of the sources previously studied are highly variable (many appear in the Variable Source Sample defined by Wehrle et al. 1992, for example), so this is further evidence for a correlation between variability at high frequency and a large apparent velocity (e.g. Teräsranta & Valtaoja 1994).

Setting aside the upper limits, and adopting $q_0 = 0.5$ (the least extreme case), the β_{app} distribution can be reproduced if there is a wide range of Lorentz factors in the sample, with in particular a long tail to high values. As a numerical example for illustrative purposes only, if $h = 0.55$ then $\gamma_b \approx 4$ could be the peak of a skewed distribution spanning the range $\gamma_b \approx 2-18$. Interestingly, such a distribution is akin to that derived from largely independent data, involving radio source counts (Padovani & Urry 1992).

Well-fitting models can also be found without such a broad range of bulk Lorentz factors (γ_b), but assuming that the observed features are patterns with a $\gamma_p \neq \gamma_b$. This breaks the match between the (bulk) beaming cone and the angle at which the maximum (pattern) velocity occurs, thus yielding a larger fraction of relatively slower motions. The predicted velocity statistics are degenerate between slow and fast patterns, but this can be resolved by taking into account other measurements of the bulk Doppler factor, such as provided by the inverse-Compton X-ray deficit (Ghisellini et al. 1993) or the equipartition Doppler factor (Readhead 1994, Güijosa & Daly 1996).

To fit the data with $q_0 = 0.5$ one would need either $r = \gamma_p/\gamma_b \approx 0.25$ or $r \approx 10.0$; intermediate values are ruled out, unless there is also a considerable range of γ_b values. Such low or high r are un-appealing; $r = 0.25$ because, since β_{app} up to $10\,h^{-1}$ occur, it requires a rather high $\gamma_b \geq 40\,h^{-1}$ in all objects; $r = 10$ because, conversely, it requires that almost all objects have $\gamma_b \leq 1\,h^{-1}$, in contradiction with other evidence for substantial Doppler beaming. It seems that this high r case is akin to an incarnation of the light-echo models (Ekers & Liang 1990), in which relativistic motion is admitted, but Doppler beaming is not. While it is entirely plausible that pattern velocities do play a role, for example in causing apparently stationary patterns in relativistic jets, further evidence that Doppler beaming is in fact important is given by the observed luminosity dependence of the β_{app} distribution.

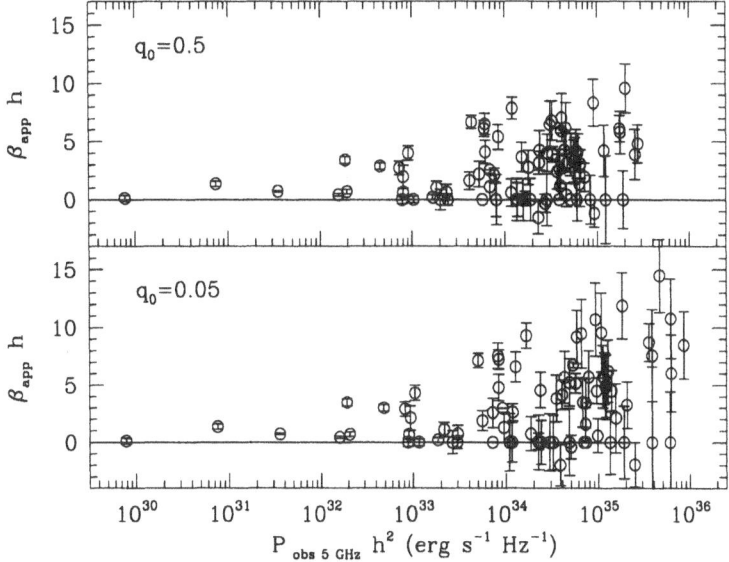

Figure 2. The observed apparent velocity distribution for 81 objects in the homogeneous PR+CJ flat-spectrum sample, illustrating that the upper envelope rises as a function of the observed 5 GHz monochromatic luminosity. Upper limits are plotted as error bars on $\beta_{app} = 0$.

Figure 2 shows the dependence of the observed β_{app} on observed monochromatic luminosity $P_{5,obs}$, calculated assuming isotropic emission. While low β_{app} can be found at any observed luminosity, there seems to be a striking correlation of the largest β_{app} with observed luminosity: the upper envelope rises considerably. Most of our flat-spectrum sources at the high observed luminosity end ($P_{5,obs} \sim 10^{35}\ h^{-2}\ \mathrm{erg\,s^{-1}\,Hz^{-1}}$) seem to conform roughly to the simple model: they have highly relativistic jets, and are in our sample because their observed luminosity is considerably enhanced by Doppler beaming; most of them are from a parent population 2 or 3 orders of magnitude down in intrinsic (isotropic) luminosity. But furthermore, the absence of similarly high β_{app} at $P_{5,obs} \leq 10^{33}\ h^{-2}\ \mathrm{erg\,s^{-1}\,Hz^{-1}}$ suggests that those objects are rather less beamed, which in turn implies that there is no substantial population down another 2 or 3 orders of magnitude in intrinsic luminosity, from which members can get Doppler beamed up. Thus, it would seem that highly relativistic jets may occur only in objects with a restricted range of 5 GHz intrinsic luminosities, perhaps predominantly near $10^{32}\ h^{-2}\ \mathrm{erg\,s^{-1}\,Hz^{-1}}$, and that there is a correlation between intrinsic 5 GHz radio luminosity and Lorentz factor, akin to that often postulated for the low frequency radio luminosity (*e.g.* the FR-I – FR-II division). Clearly, further analysis of the effects seen in Figure 2 is needed, and a combination of source counts and the apparent velocity statistics of a larger sample could provide much insight into the properties of radio jet sources.

Acknowledgements. The material in this paper was drawn from previous reviews (Vermeulen 1995, 1996), and updates earlier work (Vermeulen & Cohen 1994), by including many new measurements of Caltech-Jodrell survey sources (Polatidis et al. 1995, Taylor et al. 1994), which will be published separately after further analysis. I thank all of my collaborators for their indispensable contributions, and Matthew Lister for pointing out an error in the original manuscript. This survey has been supported in part by the National Science Foundation under grants AST 88–14554, AST 91–17100, and AST 94–20018. Support to attend IAU Symposium 175 was provided by the American Astronomical Society and the International Astronomical Union.

References

Cohen, M. H., Barthel, P. D., Pearson, T. J., & Zensus, J. A. 1988 *Ap. J.*, **329**, 1.

Ekers, R. D., & Liang, H. 1990, in *Parsec-Scale Radio Jets*, eds. Zensus, J. A., & Pearson, T. J. (Cambridge University Press: Cambridge), p. 333.

Gabuzda, D. C., Mullan, C. M., Cawthorne, T. V., Wardle, J. F. C., & Roberts, D. H. 1994 *Ap. J.*, **435**, 140.

Ghisellini, G., Padovani, P., Celotti, A., & Maraschi, L. 1993 *Ap. J.*, **407**, 65.

Güijosa, A., & Daly, R. A. 1996, *Ap. J.*, in the press.

Hough, D. H. 1994, in *Compact Extragalactic Radio Sources*, eds. Zensus, J. A., & Kellermann, K. I. (NRAO: Charlottesville), p. 169.

Orr, M. J. L., & Browne, I. W. A. 1982 *Mon. Not. Royal Astron. Soc.*, **200**, 1067.

Padovani, P., & Urry, C. M. 1992 *Ap. J.*, **387**, 449.

Pearson, T. J., & Readhead, A. C. S. 1988 *Ap. J.*, **328**, 114.

Polatidis, A. G., Wilkinson, P. N., Xu, W., Readhead, A. C. S., Pearson, T. J., Taylor, G. B., & Vermeulen, R. C. 1995 ApJS 98, 1

Readhead, A. C. S. 1994 *Ap. J.*, **426**, 51.

Taylor, G. B., Vermeulen, R. C., Pearson, T. J., Readhead, A. C. S., Henstock, D. R., Browne, I. W. A., & Wilkinson, P. N. 1994 *Ap. J. Suppl.*, **95**, 345.

Taylor, G. B., Vermeulen, R. C., & Pearson, T. J. 1995 *Proc. Natl. Acad. Sci. USA*, **92**, 11381.

Teräsranta, H., & Valtaoja, E. 1994 *Astron. Astrophys.*, **283**, 51.

Urry, C. M, & Padovani, P. 1995 *Publ. Astron. Soc. Pac.*, **107**, 803.

Vermeulen, R. C. 1995 *Proc. Natl. Acad. Sci. USA*, **92**, 11385.

Vermeulen, R. C. 1996, in *Energy Transport in Radio Galaxies and Quasars*, eds. P. Hardee, A. Bridle, & A. Zensus (Astronomical Socity of the Pacific: San Francisco), in the press.

Vermeulen, R. C. & Cohen, M. H. 1994 *Ap. J.*, **430**, 467.

Vermeulen, R. C., Bernstein, R. A., Hough, D. H., & Readhead, A. C. S. 1993 *Ap. J.*, **417**, 541.

Wehrle, A. E., Cohen, M. H., Unwin, S. C., Aller, H. D., Aller, M. F., & Nicolson, G. 1992 *Ap. J.*, **391**, 589.

Zensus, J. A. 1996, in *Extragalactic Radio Sources*, proc. IAU Symp. 175, ed. C. Fanti (Kluwer: Dordrecht), in the press (this volume).

Zensus, J. A., & Porcas, R. W. 1987, in *Superluminal Radio Sources*, eds. Zensus, J. A., & Pearson, T. J. (Cambridge University Press: Cambridge), p. 126.

Zensus, J. A., Cohen, M. H., & Unwin S. C. 1995 *Ap. J.*, **443**, 35.

CSS VERSUS LARGE SIZE SOURCES

R. FANTI[1] AND R. E. SPENCER [2]

[1] *Istituto di Radioastronomia del CNR, Bologna*
[2] *Nuffield Radio Astronomy Laboratories,*
University of Manchester, Jodrell Bank

1. Introduction

A large fraction of the sources in flux density limited radio samples have angular sizes < 2 arcsec (and hence projected linear sizes $\leq$ 10–15 kpc for H_0 = 100 Km/(sec Mpc), and steep (α > 0.5, S$\propto \nu^{-\alpha}$) high frequency spectra (Kapahi, 1981; Peacock and Wall 1982). The proportion of these *Compact Steep-spectrum Sources* (CSSs) is high (15–30% depending on the selection frequency) amongst distant (z > 0.2) radio sources of high power, both galaxies and quasars. We include in this class the *GHz Peaked Spectrum Radio Sources* (GPS), sub–kpc objects whose radio spectra are peaked at GHz frequencies (see, e.g., O'Dea et al, 1991).

Statistical analyses have shown that the majority of CSSs cannot be larger sized sources foreshortened by projection effects (Fanti et al., 1990) so that their radio emission actually originates on sub–galactic scales.

What are the connections between CSSs and larger sized radio sources? Are CSSs part of an evolutionary sequence, in which they represent the early stage (*youth scenario*; e.g. Phillips and Mutel, 1982; Carvahlo, 1985)? Or are they instead a separate class of objects, where unusual conditions in the interstellar medium, such as higher density and/or turbulence, inhibit the radio source from growing to large dimensions (*frustration scenario*; e.g. van Breugel et al., 1984)?

Well studied samples of CSSs are those from the 3CR catalogue, selected at 178 MHz, and from the Peacock and Wall catalogue, selected at 2.7 GHZ (Fanti et al., 1995 and references therein; Sanghera et al., 1995). A new sample of CSS has been produced as a result of VLBI surveys on sources selected

R. Ekers et al. (eds.), Extragalactic Radio Sources, 63–66.

at 5 GHz (hereafter called CJ, see Readhead 1995a, and references therein).
The frequency of selection is important since high frequency samples tend
to include more sources with self-absorption low frequency turnovers. For
example the CJ sample contains objects whose peak frequencies are larger
than those in the PW sample and with smaller angular sizes.

2. Radio Morphology

The core fractional luminosities of CSSs are similar to those of large size
radio sources of similar power, for both radio galaxies and quasars (median
value $\approx 3\%$ for quasars and $\leq 0.4\%$ for radio galaxies). Boosting is therefore
similar in the CSSs and larger radio sources and hence the CSSs are believed
to have similar orientations in the sky to the large size double sources.

Double–lobed structure in CSSs is common in both quasars ($\geq 50\%$) and
galaxies ($\approx 90\%$). By double–lobed structure we mean that extended radio
emission, excluding jet(s), is seen on both sides of the centre of activity,
identified by the radio core. In fact radio cores have been detected in only
70% of quasars and 40% of radio galaxies. Nevertheless we are confident, as
indicated by the similarity of morphologies, that we can correctly classify
those with truly symmetric structure even when the core is undetected

In order to emphasize the symmetry of the radio structure, we follow
Readhead (1995a), using the name CSO's (Compact Symmetric Objects)
for double lobed CSSs of small linear size (≤ 0.5 Kpc, mostly from the CJ
sample), and MSO's (Medium-size Symmetric Objects) for the two lobed
CSS's with sizes ≥ 0.5 Kpc(mostly from the PW and 3CR samples).

These sources seem to be down–scaled versions of the large size doubles.
However, asymmetries in lobe flux density ratio, or in arm ratio, or both,
are more pronounced (Fanti et al., 1990; Sanghera et al., 1995).

A second minor group of CSSs is composed of objects with either complex morphology or strongly asymmetric emission with respect to the core.
The majority of them are quasars with most of the luminosity being contributed by the jet (Spencer 1994). These distortions may be due to amplification by projection of small intrinsic bending, but the weakness of the
core requires the line of sight to be larger for the core than the jet.

Bright jets are found also among the MSOs, mainly in quasars and less
frequently in radio galaxies. They are certainly more prominent than in the
large size doubles. The jet luminosity could be due not only to boosting,
but also to interactions of the jet with the external medium. Interactions
can convert the kinetic energy carried by the jet to internal energy and
then to radiation, causing the jet to brighten.

In conclusion, CSSs seem to contain two classes of objects. One class
is characterized by a symmetric structure and mimics the characteristics

of the large size doubles. It is tempting to identify these sources with the progenitors of the large doubles.

The other class, contains objects with very distorted structure and often with very bright jets. These could be cases of "frustrated" radio sources. The class is dominated by quasars but it is unclear if this is because of an intrinsic property of quasars.

3. Physical parameters and source evolution

For a source with average luminosity in the sample, the equipartition pressure in the lobes can be parameterized as $p_{eq} \approx 5 \times 10^{-8} D_{kpc}^{-2}$ dyne cm^{-2}, where D_{kpc} is the half size of the source. We ask if, or under what conditions, a source with MSO properties remains confined or grows progressively larger with time.

Using simple one dimensional ram–pressure arguments, the advance speed of the lobes is determined by the equilibrium between the internal pressure and the ram–pressure of the external medium, $p_i \approx n_{ext} m_p v_l^2$ (where n_{ext} is the particle density of the external medium, m_p is the proton mass and v_l the lobe advance speed).

We assume an external density profile of the King type with central density n_o, core radius r_c and asymptotic behaviour $n_{ext} \propto$ distance$^{-\beta}$, with $1.5 \leq \beta \leq 2$. If we take $n_0 \approx 0.1$ cm^{-3} and $r_c = 2$ kpc, then the time required by an average MSO to attain a size of ≈ 10 kpc is $\approx 1.5 \times 10^6$ years. In order to keep the source small (≤ 15 Kpc over 2×10^7 years) combinations of central density and core radius ranging from $n_o \approx 10$ cm^{-3} and $r_c = 1$ Kpc to $n_o \approx 1$ cm^{-3} and $r_c \approx 5$ Kpc are required.

If the medium is either hot (10^7 K) or tepid (10^4 K), we would expect high luminosities in X-rays or in emission lines, which have not been observed (Fanti et al., 1995). So we think it unlikely that the bulk of the CSOs and MSOs are frustrated radio sources (see also De Young, 1993).

4. CSOs and MSOs as young radio sources

If the CSOs and MSOs are young objects, with ages $\leq 10^6$ years, and the typical time scale of activity of the central engine is $\approx 2 \times 10^7$ years, then these objects should be a small fraction of the population of which they are the young phase. Since the MSOs comprise ≈ 30 - 40 % of the high luminosity source population, there seems to be a contradiction. This difficulty is avoided if, during their evolution, radio sources either increase their velocity of expansion, or decrease in radio luminosity. These alternative possibilities have been discussed recently by Fanti et al. (1995), and Readhead et al. (1995 a,b).

For the first scenario the "ram pressure" model requires an external density $n_e \propto r^{-3.3}$, and a rather high central density ($n_o \approx 70 cm^{-3}$). These figures seem quite unrealistic.

The second scenario requires a luminosity $\propto size^h$ where $h \approx 0.4 - 0.5$ to be consistent with the observed size distribution. This also fits with the prediction of the continuous energy supply model derived from Scheuer (1974) and discussed by Baldwin (1982). In this model the radio luminosity decreases as the source grows, due to lobe expansion in the external medium. The decrease in luminosity required for CSOs/MSOs being young sources is obtained for $\beta \approx 1.5 - 2$.

5. Conclusions

We have distinguished between double lobed CSSs (for which we adopt the names CSOs/MSOs, following Readhead 1995a) and complex morphology or asymmetric jet dominated CSSs. We suggest that the first class contains the precursors of the large size double lobed radio sources. The required condition is that while they increase in size they undergo either an increase in their expansion velocity or a decrease in their luminosity. We think that the second possibility is the most realistic.

References

Baldwin, J., in *Extragalactic Radio Sources*, IAU Symp. 97, D.S. Heeschen and C.M. Wade, Reidel, 1982, 21

van Breugel, W.J.M., Miley, G.K., Heckman, T.A.,1984, *Astron. J.* **89**, 5

Carvahlo J.C., 1985, *Monthly Notices Roy. Astron. Soc.* **215**, 463

de Young, D.S., 1993, *Astrophys.J.* **402**, 95

Fanti, R., Fanti, C., Schilizzi, R.T., Spencer, R.E., Nan Rendong, Parma, P., van Breugel, W.J.M., Venturi, T., 1990, *Astron. Astrophys.* **231**, 333

Fanti, C., Fanti, R., Dallacasa D., Schilizzi, R.T., Spencer, R.E., Stanghellini C., 1995, *Astron. Astrophys.*, **302**, 317

Kapahi V.K., 1981, *Astron. Astrophys. Suppl.* **43**, 381

Mutel, R.L. and Phillips, R.B., in *The impact of VLBI on Astrophysics and Geophysics*, IAU Symp. 129, M.J. Reid and J.M. Moran, Reidel, 1988, 73

O'Dea, C.P., Baum, S.A., Stanghellini, C., 1991, *Astrophys.J.* **380**, 66

Peacock J.A., Wall J.V., 1982, *Monthly Notices Roy. Astron. Soc.* **198**, 843

Phillips T.J., Mutel R.L., 1982, *Astron. Astrophys.* **106**, 21

Readhead A.C.S., 1995a, *Quasars and AGN: High Resolution Imaging*, Irvine 1995, eds. Cohen and Kellermann, Publications National Academy of Sciences, in press

Readhead A.C.S., Taylor G.B., Pearson T.J., Wilkinson P.N., 1995b, *Astrophys. J*, in press

Sanghera H.S., Saikia D.J., Ludke E., Spencer R.E., Foulsham P.A., Akujor C.E., Tzioumis A.K., 1994, *Astron. Astrophys.* in press

Scheuer, P.A.G., 1974, *Monthly Notices Roy. Astron. Soc.* **166**,513

Spencer R.E., 1994, in *Compact Extragalactic Radio Sources*, Proceedings of a workshop held at Socorro, eds. Zensus J.A. and Kellermann K.I., p. 35

A VLBI STUDY OF
GHZ-PEAKED-SPECTRUM RADIO SOURCES

C. STANGHELLINI[1], C.P. O'DEA[2], S.A. BAUM[2]
D. DALLACASA[1,3], R. FANTI [4] AND C. FANTI[4]

[1] *Istituto di Radioastronomia CNR - Italy*
[2] *STScI, Baltimore - USA*
[3] *JIVE, Dwingeloo - The Netherlands*
[4] *University of Bologna - Italy*

1. Introduction

The GHz-Peaked-Spectrum (GPS) radio sources are powerful ($L_{radio} \approx$ 10^{45} erg sec^{-1}) and compact (10 – 100 mas, 10 – 1000 parsecs) sources characterized by a simple convex spectrum which peaks near 1 GHz (O'Dea et al. 1991 and references therein).

Optical and radio observations lead to the conclusion that GPS sources are formed when the radio source is confined to the narrow line region (or an even smaller scale) by a dense and clumpy medium. This could lead to 2 different evolutionary scenarios: as first suggested by Phillips and Mutel (1982) GPS radio sources could be classical double radio sources at the very first stage of their life, or alternatively they will never become as large as the classical doubles since the dense and turbulent environment is able to confine and trap the radio emitting region on the scale of the NLR (Baum et al. 1990).

Early VLBI observations revealed that GPS quasars had asymmetric or complex radio morphologies and GPS galaxies had compact double (CD) or triple sources. Deeper and more sensitive VLBI observations of CSS (a similar class with moderately larger sizes) and GPS radio sources show that double structures are common also in quasars, accounting for (in this morphological classification) around half of the observed quasars, while complex morphologies are often seen also among GPS or CSS radio galaxies (Dalla-

R. Ekers et al. (eds.), Extragalactic Radio Sources, 67–68.

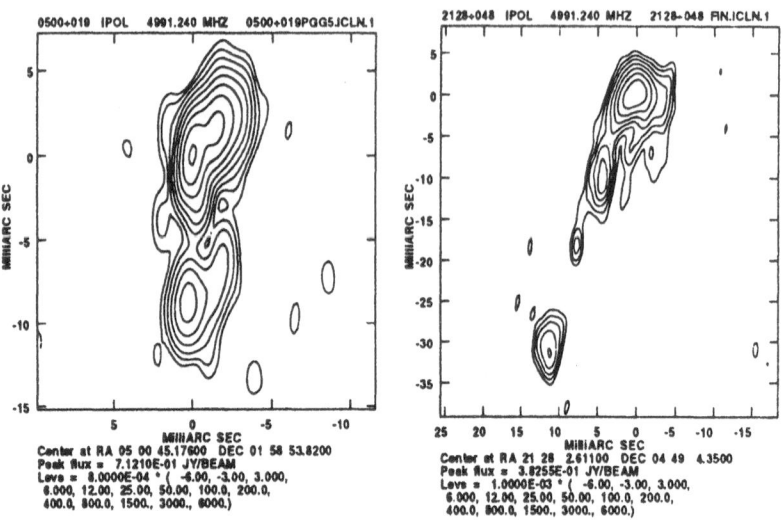

casa et al. 1995, Fanti et al. 1995, Stanghellini et al. in prep.) . Most of the GPS radio sources studied so far show low variability and no superluminal motion suggesting that in general these objects are not strongly Doppler boosted.

2. Observations and First Results

We observed 10 GPS radio sources in February 1993 at 4.992 GHz using the global VLBI network with the MK II recording system.

These GPS radio sources show a variety of morphologies and have angular sizes ranging from a few to a hundred mas corresponding to linear sizes from a few to hundreds parsecs.

Two objects (0500+019, 2128+048, see figures) might be classified as Compact Symmetric Objects (CSO) (Readhead et al 1994). Two others (2210+016, 1345+125) may belong to the CSO class but need further observations. The other objects are marginally resolved or show complex morphologies.

References

Dallacasa D., et al. 1995 *Astron. Astrophys.*, **295**, 27.

Fanti, C., Fanti, R., Dallacasa, D., Schilissi, R. T., Spencer, R. E., and Stanghellini, C., 1995, *Astron. Astrophys.*, **302**, 317.

O'Dea C.P. ,Baum S.A., Stanghellini C., 1991, *Astrophys. J.*, **380**, 66

Readhead A. C. S., et al. 1994, CERS meeting

2.23 AND 8.29 GHZ OBSERVATIONS OF CSS-GPS CANDIDATES

D. DALLACASA, M. BONDI, F. MANTOVANI

Istituto di Radioastronomia - CNR, Bologna, Italy

AND

W. ALEF

Max Planck Institut für Radioastronomie, Bonn

Compact Steep-Spectrum (CSS) and GHz-Peaked Spectrum (GPS) radio sources are intrinsically small objects (Fanti et al. 1990) with high frequency steep spectra ($\alpha > 0.5$ with $S \propto \nu^{-\alpha}$), found at moderate or high redshifts ($z > 0.2$ but many objects have $z > 1.5$). Their characteristics have been interpreted in terms of youth (e.g. Fanti et al. 1995) or "frustration" (e.g. van Breugel et al. 1984). Their radio spectra turn over at tens or a few hundreds of MHz (CSS) or at higher frequencies (GPS), interpreted as due to synchrotron self-absorption. They represent a significant fraction of flux limited catalogues (15–30 %, depending on the frequency).

The samples studied so far (Spencer et al. 1989, Fanti et al. 1990, Dallacasa et al. 1995) were limited to the most powerful objects. Sources with lower power will help in understanding the relation between the CSS-GPS objects and other classes of radio sources.

We present the results of S/X (2.23/8.29 GHz) VLBI snapshot observations of 29 sources drawn from a list of CSS and GPS candidates. Data were recorded in Mk3 mode A during two separate VLBI sessions using the antennas in Effelsberg, Medicina, Noto, Onsala, Matera and Wettzell. Typical resolutions were about 30×12 mas and 9×3.5 mas at S and X bands respectively. Each source was observed for a number (3–5) of 13-min scans in a range of hour angle and due to the limited uv-coverage only simple structures could be imaged. Only two sources were not detected. The r.m.s. noise level in the images is typically of 1-2 mJy and the dynamic range achieved is between a few tens and a few hundreds. A summary of the parameters from the images is presented in Table 1.

69

R. Ekers et al. (eds.), Extragalactic Radio Sources, 69–70.

TABLE 1. Summary of the observational results: In column 3, P = pointlike; S = single component, resolved either at S or at X band (or both); D = double; C–J = core–jet; T = triple; CPLX = complex. ND= Not Detected. The flux densities reported in column 4 (S_{CFD}) and column 6 (X_{CFD}) refer to the integrated values measured on the images.

IAU Name	LAS mas	Morph.	S_{CFD} Jy	HPBW$_S$ mas (p.a)	X_{CFD} Jy	HPBW$_X$ mas (p.a)
0201+113		P	0.56	25.1x10.5 (60)	0.67	9.7x3.3 (53)
0237−027		S	0.26	20.9x13.7 (58)	0.80	7.5x4.3 (64)
0237−233		S?	3.27	49.7x16.9 (37)	1.93	21.0x4.5 (33)
0500+019	10	S	1.56	26.2x11.4 (70)	1.36	9.8x3.4 (68)
0511−220		S?–CJ?	0.33	38.2x19.2 (42)	0.86	14.7x5.4 (49)
0743−006		P	0.70	26.7x13.0 (-78)	1.71	8.4x4.5 (84)
0922+005		S?–CJ?	0.45	24.4x12.5 (-73)	0.41	10.0x4.5 (-87)
0941−080	50	D	0.92	37.9x18.0 (-57)	0.20	8.2x5.4 (-81)
0941+261	–	–	ND	——	ND	——
1143−245		S	1.11	40.7x20.3 (41)	0.92	16.8x6.0 (39)
1237−101		S	0.82	23.4x14.3 (51)	0.67	7.9x4.2 (47)
1317−005		P	0.18	20.9x13.0 (-54)	0.11	7.1x3.8 (50)
1402−012		P	0.49	19.1x11.8 (-83)	0.29	5.2x3.3 (-84)
1502+036		S	0.51	28.6x11.3 (56)	0.61	13.5x3.2 (46)
1518+047	150	D	1.69	22.1x10.2 (53)	0.40	10.5x3.6 (44)
1602+576	120	D?	0.57	17.8x10.6 (89)	0.34	4.1x2.4 (90)
1607+268	60	D	2.39	18.7x 9.9 (66)	0.90	9.1x3.7 (48)
1629+120	1150	D	0.29	31.9x10.0 (53)	0.11	10.0x4.1 (44)
1629+680	120	C–J	0.79	20.6x11.0 (84)	0.20	5.2x3.5 (70)
1801+010		P	1.64	21.6x13.6 (68)	1.05	5.5x3.2 (80)
1848+283		P	0.24	22.1x10.6 (54)	1.44	8.3x3.4 (51)
2044−027	40	S	0.29	28.0x17.9 (62)	0.03	10.0x4.3 (50)
2053−201	–	–	ND	——	ND	——
2126−158		P	0.56	33.6x14.7 (42)	0.86	13.0x3.9 (41)
2128+048	60	D?	2.25	24.1x12.1 (61)	1.28	8.1x3.7 (53)
2137+209	50	S	0.63	24.6x 9.8 (52)	0.23	9.6x3.3 (46)
2210+016	100	T–CPLX	1.20	23.6x12.3 (60)	0.40	9.7x3.7 (54)
2223+210		S	0.79	22.3x10.0 (57)	1.23	8.8x2.9 (50)
2351−006		S	0.29	24.2x12.6 (60)	0.27	8.7x3.9 (65)

References

van Breugel, W.J.M., Miley, G.K., Heckman, T.M. 1984, *AJ*, **89**, 5.
Dallacasa, D., et al. 1995, *A&A*, **295**, 27.
Fanti, C., et al. 1995, *A&A*, **302**, 317.
Fanti, R., et al. 1990, *A&A*, **231**, 333.
Spencer, R.E., et al. 1989, *MNRAS*, **240**, 657.

LARGE BENT JETS IN THE INNER REGION OF CSS

F. MANTOVANI [1], W. JUNOR[2], M. BONDI[1], L. PADRIELLI[1],
W. COTTON[3] AND E. SALERNO[1]

[1] *Istituto di Radioastronomia, Bologna, Italy*
[2] *Dept. of Astrophysycs, University of New Mexico, USA*
[3] *National Radio Astronomy Observatory, Charlottesville, USA*

1. Introduction

Recently we focussed our attention on a sample of Compact Steep-spectrum Sources (CSSs) selected because of the large bent radio jets seen in the inner region of emission. The largest distortions are often seen in sources dominated by jets, and there are suggestions that this might to some extent be due to projection effects. However, superluminal motion is rare in CSSs. The only case we know of so far is 3C147 (Alef at al. 1990) with a mildly superluminal speed of $\sim 1.3v/c$. Moreover, the core fractional luminosity in CSSs is $\sim 3\%$ and $\leq 0.4\%$ for quasars and radio galaxies respectively. Similar values are found for large size radio sources *i.e.* both boosting and orientations in the sky are similar for the two classes of objects. An alternative possibility is that these bent-jet sources might also be brightened by interactions with the ambient media. There are clear indications that intrinsic distortions due to interactions with a dense inhomogeneous gaseous environment play an important role. Observational support comes from the large RMs found in CSSs (Taylor et al. 1992; Mantovani et al. 1994; Junor et al. these proc.) and often associated with strong depolarization (Garrington & Akujor, t.p.). The CSSs also have very luminous Narrow Line Regions emission, with exceptional velocity structure (Gelderman, t.p.).

2. Our investigation

We are investigating a sample of sources showing large bent jets on the *mas* scale. Table 1 lists the selected sources. It is possible that these are extreme

R. Ekers et al. (eds.), Extragalactic Radio Sources, 71–72.

TABLE 1. Source parameters.

Source	ref	dist mas	dist pc	ΔPA deg	z	O.I.	RM×$(1+z)^2$ rad m^{-2}
0127+233 3C43	a	225	780	93	1.459	Q	−1088
0358+004 3C99	b	15	46	60	0.425	G	40
0429+415 3C119	c	39	142	55	1.023	Q	3400
0538+498 3C147	d	200	664	90	0.545	Q	−3222/621
0548+165	e	80	253	90	0.474	Q	1934
1328+254 3C287	f			58	1.055	Q	–
1442+101 OQ172	g	15	39	90	3.531	Q	22400
1629+680 4C68.18	h	76	226	66	2.475	Q	–
1741+279 B2	e	300	862	84	0.372	Q	−219/293
2033+187	e	40	129	80	(0.5)	–	n
2147+145	e	23	75	90	(0.5)	–	n

Note: In column 8, the mark - means no measurements available, the mark n means polarization not detected. References: a - Spencer at al. (1991); b - Mantovani et al. (1992); c - Nan Ren-dong et al. (1991); d - Alef et al. (1990); e - this investigation; f - Fanti et al. (1989); g - Dallacasa et al. (1995); h - Dallacasa et al. (in preparation)

cases. Features common to these sources are: (a) - a core-jet structure; (b) - a change in the jet major axis Position Angle $\Delta PA > 50°$; (c) - a bend that occur at a linear separation which is < 1 kpc from the core.

Their *mas scale* structures lend support for the view of a strong interaction between the jet flow and dense gas clouds. Norman & Balsara (1993) have investigated the physics of a jet/cloud collision with 3-D hydrodynamical simulations. Their work reveals interesting features that further support this interpretation. For example, the reflected jet inherits the stability properties of the original jet. Nevertheless, the images of sources like 3C119, 0548+165 and 3C287 represent a puzzle since they would require more than one collision to produce structures that are spiral-like or bent by about 180°.

References

Alef, W. *et al.* (1990) *Proceedings CSS & GPS Radio Sources* Workshop held in Dwingeloo, ed. by C. Fanti, R.Fanti, C.P. O'Dea and R.T. Schilizzi

Dallacasa, D. *et al.* (1995) *A&A* **295**, pp. 27

Fanti, C. *et al.* (1989) *A&A* **217**, pp. 44

Mantovani, F. *et al.* (1994) *A&A* **292**, pp. 59

Nan Ren-dong *et al.* (1991) *A&A* **245**, pp. 449

Norman, M.L. & Balsara D.S. (1993) *Jet in Extragalactic Radio Sources* ed. by K. Meisenheimer & H.J. Roeser, Berlin, Springer Verlag

Spencer R. S. *et al.* (1991) *MNRAS* **250**, pp. 225

Taylor, G. B., Inoue, M. and Tabara, H. (1992) *A&A* **264**, pp. 421

HIGH-RESOLUTION STRUCTURE OF SOUTHERN COMPACT STEEP SPECTRUM SOURCES

A. TZIOUMIS[1], R. MORGANTI[1,2], C. TADHUNTER[3], R. DICKSON[3]
C. FANTI[2], D. DALLACASA[2], J. REYNOLDS[1], D. JAUNCEY[1],
R. PRESTON[4], P. MCCULLOCH[5], E. KING[5,1], S. TINGAY[6],
P. EDWARDS[7], M. COSTA[5], D. JONES[4], J. LOVELL[5], R. CLAY[7],
D. MEIER[4], D. MURPHY[4], R. GOUGH[1], R. FERRIS[1], G. WHITE[8]
AND P. JONES[8]

[1] *ATNF – CSIRO, Epping, Australia*
[2] *Istituto di Radioastronomia, Bologna, Italy*
[3] *Department of Physics, University of Sheffield, UK*
[4] *Jet Propulsion Laboratory, Caltech, Pasadena, CA, USA*
[5] *University of Tasmania, Hobart, Tasmania, Australia*
[6] *Mt. Stromlo & Siding Spring Obs., Canberra, ACT, Australia*
[7] *University of Adelaide, Adelaide, SA, Australia*
[8] *University of Western Sydney, Kingswood, NSW, Australia*

Two important factors for understanding the physical nature of compact steep spectrum (CSS) radio sources are determining the correct radio morphological classification of these objects together with their characteristics in wavebands different from the radio (Fanti et al. 1995, A&A, 302, 317). Seven CSS sources (linear dimensions < 30kpc for $H_o = 50$ $km\,s^{-1}Mpc^{-1}$ and $\alpha > 0.5$, $S \sim \nu^{-\alpha}$) have been found in a complete sample of strong southern radio sources. This group of CSS sources is particularly interesting because some optical and X-ray information is already available as part of a more general study of southern radio sources (Morganti et al. & Siebert et al. these Proceedings). The spectra of all the sources were presented in Tadhunter et al. (1993, MNRAS, 263, 999.) Here we present VLBI observations for three of these sources (0252-71, 1306-09 and 1814-63). The remaining four have already been imaged with VLBI (King et al. these Proceedings).

Results and Discussion: The images obtained by observations at 2.3 GHz from the Southern Hemisphere MKII VLBI network (SHEVE) are shown in Figure 1. For **1306-095** and **1814-637** less than half of the total flux

R. Ekers et al. (eds.), Extragalactic Radio Sources, 73–74.
© 1996 *IAU.*

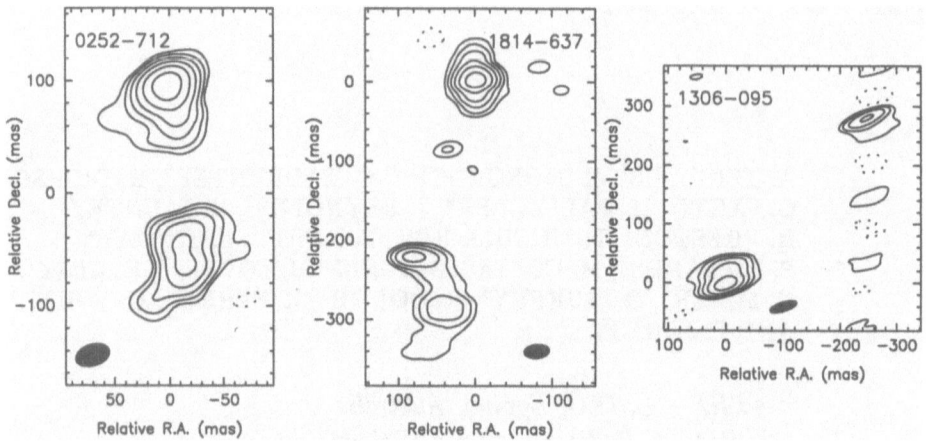

Figure 1.   VLBI images at 2.3 GHz from the SHEVE network, observed on 20-22 Feb.
1993. Contour levels are at -2,2,4,8,16,32,64 % of peak flux density, except for 1306-095
where the contour levels start at 4%. The noise level in the images is 6-8 mJy. The
peak flux densities detected are 1.14 Jy for 0252-712, 1.7 Jy for 1814-637 and 0.3 Jy for
1306-095.

density has been detected in the VLBI observations, implying the existence
of substantial diffuse structures around these objects. In contrast, in the
0252-712 VLBI image the total flux density is detected and there is no
diffuse structure.

All seven sources in our complete sample show the double-lobed radio
structure typical of the majority of known CSS objects. However, radio
spectral index information is still necessary for some of the objects, to
investigate the possible presence of a flat-spectrum core.

The optical spectra are characterized by strong [O III]λ5007 and [O II]
λ3727 lines, as observed in extended radio sources of similar power. In
the cases where it could be measured (0023-26 and 1934-63), we found an
electron temperature higher than in other extended radio galaxies. The
optical polarization is similar to the extended objects. In some cases we
found significant polarization and for 1934-63 the optical polarization axis
is perpendicular to the radio axis, as predicted by the beaming/scattering
model (Tadhunter et al. 1994, MNRAS, 271, 807). For the broad-line galaxy
2135-20 we have only an upper limit to the polarization, as for most of the
extended broad-line radio galaxies observed (Shaw et al. 1995, MNRAS,
275, 703).

A SINGLE BASELINE VLBI SURVEY OF
SOUTHERN PEAKED SPECTRUM SOURCES

E.A. KING[1], P.M. MCCULLOCH[2], D.L. JAUNCEY[1],
J.E. REYNOLDS[1], R.A. PRESTON[3], D.L. MEIER[3], D.W. MURPHY[3],
A.K. TZIOUMIS[1], J.E.J. LOVELL[2], T.D. VAN OMMEN[3] D.L. JONES[3]

[1] *ATNF, CSIRO, Epping, NSW, Australia*
[2] *University of Tasmania - Hobart, Tas., Australia*
[3] *Jet Propulsion Laboratory, Pasadena, CA, USA.*

A sample of 38 southern peaked-spectrum radio sources from the Parkes Catalogue have been observed using single-baseline VLBI. Thirty three objects were successfully detected on baselines of $> 30 M\lambda$ at 2.3 GHz. For 21 of these sources, the flux density in the compact components contributes more than half the total flux density of the radio emission. Twenty sources showed structure more complex than a point-source.

The compact morphology of each source was determined by fitting simple image-plane components to the visibility data. This procedure used a toolbox of simple components and was controlled by a stringent set of rules intended to minimize the possibility of introducing false structure into the models. A detailed comparison with the full images from multi-baseline VLBI observations of a subset of the sources has confirmed the general validity of the resulting models. In particular, this comparison shows that single baseline VLBI observations of $\sim 6 - 8$ h duration are capable of providing reliable estimates of component separations and relative position angles, as well as position angles and lower bounds on the flux densities of individual components. However, actual flux densities and component sizes can not normally be measured reliably with only a single baseline.

A companion flux density monitoring survey conducted over a three-year period with the University of Tasmania's 26 m antenna at both 2.3 and 8.4 GHz showed that the incidence of variability in these objects is systematically low, despite the significant contribution made by compact components. Clearly, despite the morphological similarities, the compact structure in these objects is different from that seen in the more variable flat-spectrum sources.

R. Ekers et al. (eds.), Extragalactic Radio Sources, 75–76.

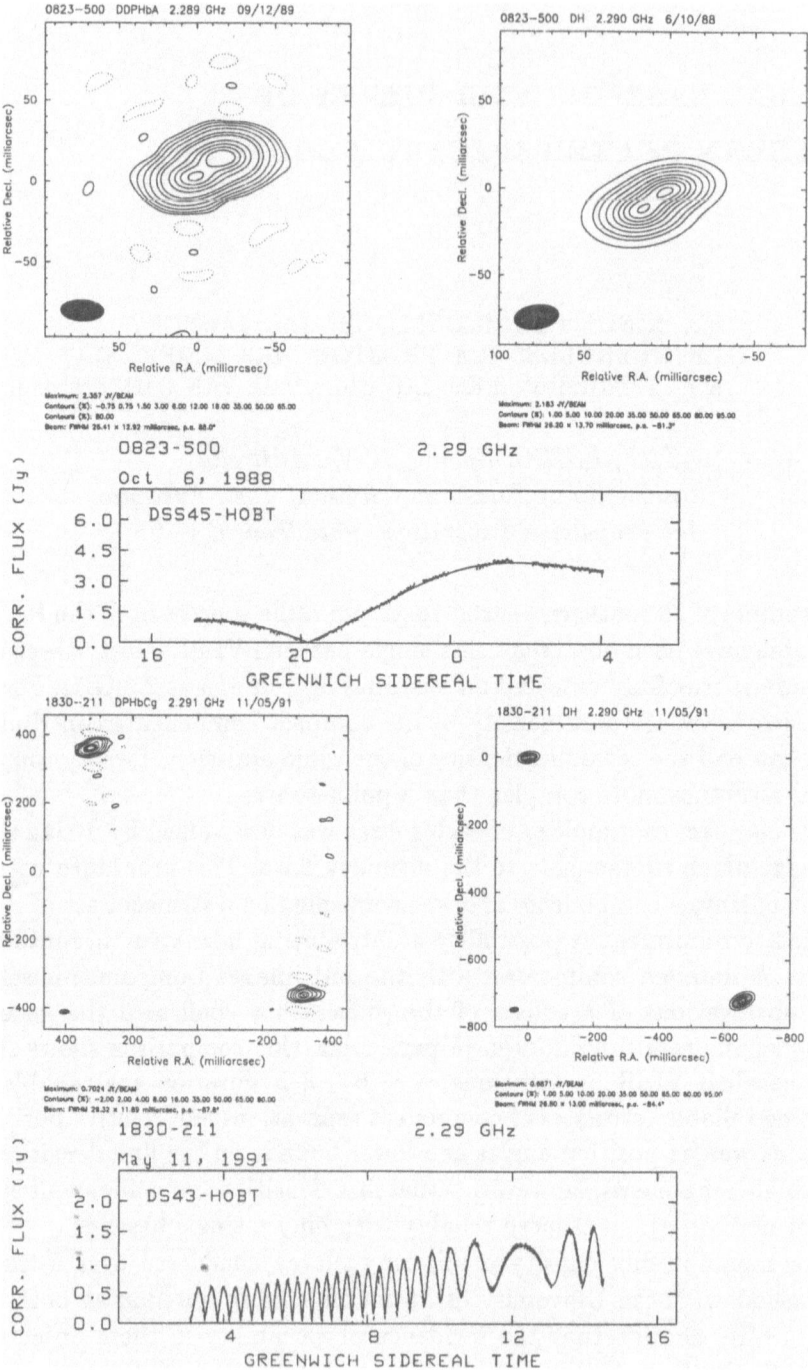

Figure 1.   Images, models and modelfits for PKS0823−500 & PKS1830−211. For each object, the multi-baseline VLBI image is on the left, the single-baseline model on the right. The solid line in the lower plots indicates the model visibility.

EXTREME DEPOLARIZATION IN
COMPACT STEEP SPECTRUM SOURCES

S.T. GARRINGTON
The University of Manchester, NRAL Jodrell Bank, UK
AND
C.E. AKUJOR
Imo State University, Owerri, Nigeria.

Compact steep spectrum (CSS) sources show all the features of the powerful extended radio galaxies and quasars but on a scale at least ten times smaller. The debate continues over whether this is primarily due to a difference in age or the result of denser gas surrounding CSS sources (Fanti et al. 1990); although two of the most recent investigations favour the view that CSS sources are young (Readhead et al. and Fanti et al., these proceedings)

Radio polarization observations offer a powerful probe of the environments of radio sources and there have been several suggestions that Faraday effects in CSS sources may be particularly strong (eg van Breugel et al. 1984, Kato et al. 1987). We have observed almost all of the CSS sources in the 3C sample with the VLA at 8.4 GHz (Akujor and Garrington 1995). We find that the degree of polarization of the CSS components is very low (median value 2 – 3%) compared with more extended sources (eg Garrington et al 1991). By combining new and published observations at 15 GHz (eg van Breugel et al. 1992) we find that about half of the components show significant Faraday depolarization between 8 and 15 GHz. About one quarter of the components are unpolarized at both 8.4 and 15 GHz. New observations at 22 GHz reveal higher polarization in most cases, confirming that these are indeed cases of extreme depolarization.

CSS sources show some of the most striking examples of depolarization asymmetry. In the radio galaxy 3C67, the northern lobe depolarizes slowly between 8.4 and 1.5 GHz, while the southern lobe is completely depolarized at all frequencies below 22 GHz (Figure 1). The CSS sources follow the trends first established in more extended sources for the depolarization to be stronger in the lobe which is closer to the radio nucleus and/or opposite a

R. Ekers et al. (eds.), Extragalactic Radio Sources, 77–78.

one-sided radio jet (see Laing, these proceedings, for a review). This implies that the CSS sources are not deeply embedded in the depolarizing gas.

In the CSS sources the Faraday dispersion $\Delta \sim 2500$ cm$^{-3}\mu$Gpc, roughly 30 times larger than typical extended sources (Garrington and Conway 1991) From the estimates of Δ we estimate the product of the thermal electron density and magnetic field strength $nB \sim 1$ cm$^{-3}\mu$G surrounding the CSS sources. The CSS sources lie on the established trend for smaller sources to show stronger depolarization. If CSS sources are simply young then we would expect all sources to have intrinsically similar environments. In this case Figure 2 may simply represent the radial decline of the density of this environment; we estimate $n(r) \propto r^{(-1.1\pm0.1)}$ following Garrington and Conway (1991). This is similar to the radial density decline required in the models of Readhead et al. to explain the high incidence of CSS sources by their increased luminosities while they are expanding through denser gas.

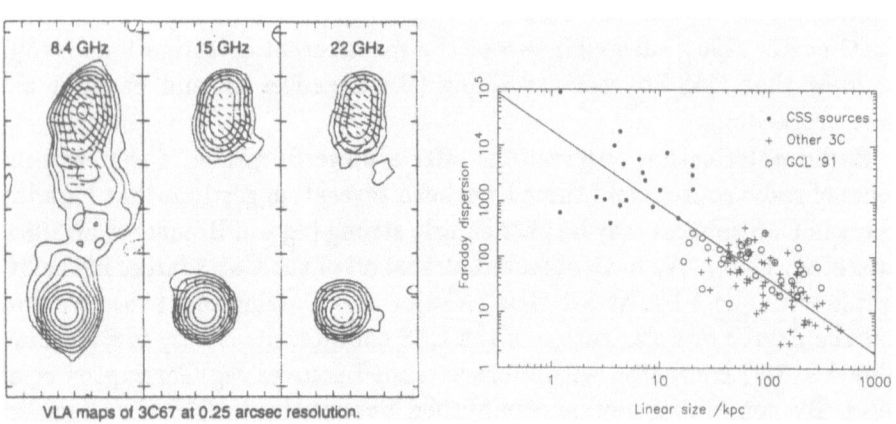

Figure 1. Left: the radio galaxy 3C67: Vectors show the fractional polarization (1 arcsec = 100%)

Figure 2. Right: plot of average Faraday dispersion in cm$^{-3}\mu$Gpc against source linear size for CSS and extended sources.

References

Akujor, C.E. & Garrington, S.T. 1995, *A&AS*, **112**, 235
Garrington, S.T. & Conway, R.G. 1991, *MNRAS*, **250**, 198
Garrington, S.T. et al. 1991, *MNRAS*, **250**, 171
Kato, T. et al. 1987, *Nature*, **329**, 223
van Breugel, W.M.J., Miley, G. & Heckman, T. 1984, *AJ*, 5
van Breugel, W.M.J. et al. 1992, *A&A*, **256**, 56

LARGE ROTATION MEASURES IN CSS SOURCES

W. JUNOR[1], F. MANTOVANI[2], A. PECK[3], D. SAIKIA[4], C. SALTER[5]

[1] *IfA, UNM, Albuquerque, USA.* [2] *CNR, Bologna, Italy.*
[3] *NMIMT, Socorro, USA.* [4] *NCRA, Pune, India.*
[5] *NAIC, Arecibo, Puerto Rico.*

1. Introduction

A majority of compact steep-spectrum (CSS) and GHz-peaked spectrum (GPS) sources show low integrated polarizations ($\lesssim 1\%$) at or below 5 GHz. To investigate whether this is mostly due to large rotation measures (RMs) arising from their being cocooned in dense gaseous envelopes we have observed 26 CSSs and GPSs at 8 and 15 GHz with the VLA A-array. The sample includes equal numbers whose hosts are galaxies and quasars.

2. Results

Preliminary data analysis reveals a number of CSS sources with high RM to add to those contained in Mantovani *et al.* (1994) and Taylor *et al.* (1992). Here we present results for 5 interesting sources.

0538+498 (3C147). For this quasar, the images (convolved to identical resolutions of $\approx 0\overset{\prime\prime}{.}25$) demonstrate considerable position angle (PA) rotation between X and U bands. Strong depolarization also occurs between the two bands, the ratio of the percentage polarizations at 8.1 and 14.9 GHz (DP) being ≈ 0.37. The derived RMs are -1317 ± 8 rad m^{-2} on the main component and $+264 \pm 18$ rad m^{-2} on the extension to the NNE ($RM_{rf} = -3144$ and $+630$ rad m^{-2} in the source's rest frame.) These huge RM's confirm the findings of Kato *et al.* (1987), although the large difference and opposite signs between the RM's on the different components is unexpected. Although the magnitude of the RM for the main component greatly exceeds that of the NNE component, the depolarization of both is

R. Ekers et al. (eds.), Extragalactic Radio Sources, 79–80.

similar, suggesting that much of the extra rotation occurs outside the emission regions. The magnetic-field orientation is remarkably uniform over the source, but bears no obvious relation to its structure.

0127+233 (3C43). The polarized intensity of this misaligned quasar is sufficient to give accurate RM's only for the central component, where values of $< -300 \, \mathrm{rad \, m^{-2}}$ $(\mathrm{RM}_{rf} < -1800 \, \mathrm{rad \, m^{-2}})$ are derived near its peak, with $\mathrm{DP} \approx 1.0$. Away for the peak, the RM's drop to smaller values. It is unclear whether this large RM towards the peak is real; there is a rapid change of the intrinsic PA around this position and spectral index or depolarization gradients along the structure could mimic a high RM at this resolution. The magnetic field in the central component follows the curvature of the source faithfully. For the low RMs likely in the lobes, the fields would lie along the E lobe, and circumferential to the N lobe.

0552+398, 1151–348 and 1634+623 (3C343). These compact sources are cases where polarization VLBI is needed to illuminate the situation further. The CSS quasar, 1151–348, although only marginally resolved, shows PA variations over both the X- and U-band images, suggesting a more-complex structure than revealed by the present resolution. The mean RM is low, but there seems to be a significant RM gradient across the image. The source shows considerable depolarization, with $\mathrm{DP} = 0.68$. The CSS galaxy, 1634+628, displays an RM gradient of $\sim 500 \, \mathrm{rad \, m^{-2}}$ from E to W, despite marginal resolution. Its mean RM is $+650 \, \mathrm{rad \, m^{-2}}$ $(\mathrm{RM}_{rf} = +2570 \, \mathrm{rad \, m^{-2}})$ and the magnetic field is well aligned with VLBI imaging.

The GPS quasar, 0552+398, yields $\mathrm{RM} = +415 \, \mathrm{rad \, m^{-2}}$, $(\mathrm{RM}_{rf} = +4700 \, \mathrm{rad \, m^{-2}})$. In the mid 1980's, O'Dea *et al.* (1990) derived $\mathrm{RM} = -658 \, \mathrm{rad \, m^{-2}}$ between 1.4 and 4.9 GHz. During our 1991 measurements, the source was stronger than at the earlier epoch, and our measured polarization of 1.25% $(\mathrm{DP} = 1.0)$ is higher than obtained from 1.4 to 22 GHz by O'Dea *et al.* (0.48 – 0.8%). It is unclear whether there has been a huge change in the RM of this source over 7 years, or whether there is another explanation for the inconsistency. Considering the radio spectrum of O'Dea *et al.*, their RM was derived in the optically-thick regime, and ours in the optically-thin.

References

Kato, T., Tabara, H., Inoue, M. and Aizu, K. (1987) *Nature,* **329**, pp. 223

Mantovani F., Junor, W. and Bondi, M. (1994) *Procs. of "Compact Extragalactic Radio Sources",* NRAO Workshop No. 23, Eds. Zensus J.A. and Kellermann K.I., pp. 29

O'Dea, C.P., Baum, S.A., Stanghellini, C., Morris, G.B., Patnaik, A.R. and Gopal-Krishna (1990) *Astron. Astrophys. Suppl. Ser.,* **84**, pp. 549

Taylor, G.B., Inoue, M. and Tabara, H. (1992) *Astron. Astrophys.,* **264**, pp. 421

NLR KINEMATICS IN CSS RADIO SOURCES

RICHARD GELDERMAN
National Research Council, NASA/GSFC, Code 681
Laboratory for Astronomy and Solar Physics, Greenbelt, MD

Compact Steep-Spectrum (CSS) radio sources are defined by their powerful yet subgalactic radio structure. High resolution radio interferometers reveal subarcsecond double, triple, and core-jet radio morphologies resembling miniature FR II sources. There are three promising scenarios for how such powerful radio sources could be so intrinsically small: 1) they are young sources which have not had time to grow large, 2) they are confined sources with exceptionally dense gas in the host which either smothers an existing source or frustrates its initial growth, or 3) they are enhanced sources with an intrinsic power much lower than one might expect based on their radio luminosity. To investigate these scenarios, we are continuing a comprehensive optical study of a sample of Compact Steep Spectrum objects.

From the radio selected CSS sources of Spencer et al. (1989) and Fanti et al. (1990) we have observed objects with $m_v < 21$ and $z < 0.9$. Data obtained at the Steward Observatory 2.3^m and Kitt Peak National Observatory 4^m and 2.1^m telescopes include B, V, & R broad band images; narrow band emission line images, low dispersion longslit spectrophotometry covering 3700 Å from [O II]λ3727 to [O III]λ5007; and high dispersion longslit spectrophotometry around Hβ & [O III]. In these proceedings we present the spectroscopic results (Gelderman and Whittle 1994).

The spectra of CSS sources are characterized by strong, high equivalent width emission lines, indicating the presence of a significant amount of ionized gas. The line emission regions for most CSS sources are too compact to be resolved by ground based optical study. Assuming the NLR emission is only associated with the compact radio structure then we derive covering factors which are up to three orders of magnitude higher than values found in typical AGN. However, from our data we cannot confidently distinguish between the youth, frustration, or enhancement scenarios. There is also no conclusive evidence to suggest that the radio jet is directly responsible for the ionization. Emission line flux ratios are typical of gas photoionized by the active nucleus (Veilleux and Osterbrock 1987) and CSS sources fall along the correlation recognized all other powerful radio galaxies and radio

R. Ekers et al. (eds.), Extragalactic Radio Sources, 81–82.

loud quasars when narrow line emission luminosity is plotted against radio luminosity at 1.4 GHz rest frequency.

Even if the ionization process is not directly related to the jets, kinematic evidence suggests the NLR clouds are accelerated through jet/gas interactions. Our high dispersion spectra of CSS sources reveal [O III] emission line profiles which tend to be very broad and show considerable velocity substructure compared to other classes of radio-loud AGN. In low radio power Seyfert and nearby radio galaxies, these kinematic characteristics have been supported by other evidence for jet/gas interactions (see Whittle 1989 for review). Whittle (1992) compares [O III] linewidths for a large sample of Seyfert galaxies to just those Seyferts which contain relatively luminous ($L_{1.4} > 10^{22}$W Hz^{-1}) linear radio sources. For most Seyferts, the [O III] linewidth is correlated with measures of the gravitational potential of the host galaxy. However, Seyferts with luminous linear radio sources seem to have broader lines, indicating an additional acceleration mechanism caused by outflowing bipolar radio ejecta. The profiles also tend to show more velocity structure than the profiles of other Seyferts, with velocity splitting which can sometimes be directly related to the radio lobes. A comparable situation is observed for objects from our CSS sample as compared to other radio loud galaxies and quasars. We suggest that the CSS radio jets are accelerating the ionized gas, either as frustrated sources or as youthful sources working steadily outward to become extended classical doubles.

Finally, while our data do not allow us to determine why CSS sources are so compact, we do conclude that CSS sources represent a real physical class rather than simply being a collection of objects with various explanations for the powerful yet subgalactic radio structure. The sources in our sample have spectra which share similar qualities and are unlike the spectra of typical active galaxies or quasars. In addition, we find no significant optical distinctions based on either the extent or morphology of the radio structure. Thus we are unable to confirm results from studies of compact symmetric objects.

References

Fanti, C., Fanti, R., Schilizzi, R.T., Spencer, R.E., Nan Rendong, Parma, P., van Breugel W.J.M., and Venturi, T. 1990, *AA*, **231**, 333.

Gelderman, R. and Whittle, M. 1994, *Ap. J. Supp.*, **91**, 491.

Spencer, R.E., McDowell, J., Charlesworth, M., Fanti, R., Parma, P., and Peacock, J.A. 1989, *M.N.R.A.S.*, **240**, 657.

Veilleux, S. and Osterbrock, D.E. 1987, *Ap. J. Supp.*, **63**, 295.

Whittle, M. 1992, *Ap. J.*, **387**, 109.

Whittle, M. 1989, in ESO Proceedings: Extranuclear Activity in Galaxies, eds. Meurs & Fosbury.

A WORLD ARRAY VLBI IMAGE OF 3C216 AT 1.6GHZ

C.E. AKUJOR AND R.W. PORCAS

Max-Planck-Intitut fuer Radioastronomie, Bonn, Germany

AND

I. FEJES

FOMI Satellite Geodetic Observatory, Budapest, Hungary

The quasar 3C216 has the distinction of being classified as a blazar as well as a compact steep-spectrum radio source (Akujor et al, 1993). It has a grossly mis-aligned large-scale radio structure. Previous VLBI images include those of the core region (Barthel et al 1988; Venturi et al 1993) with 1 mas resolution, and the sub-kpc scale core-jet with resolution ca. 25 mas (Fejes et al, 1992; Akujor et al, 1993). In order to investigate the connection between these structures seen on different scales, we have made a 1.6 GHz observation of the core-jet region with a "World Array" VLBI network. This consisted of 20 antennas, at Effelsberg (Germany), Jodrell "Lovell" (UK), Westerbork (Netherlands), Medicina and Noto (Italy), Onsala-26m (Sweden), Torun-15m (Poland), Simeiz (Ukraine), Bear Lakes (Russia), Seshan (China), Hartebeesthoek (S.Africa), Green Bank-43m, VLA and the VLBA antennas at HN, NL, PT, KP, LA, BR, OV (USA). The observations were made on 18 November, 1992, using the Mk2 recording system, and these were amongst the last to be correlated using the JPL-Caltech Block2 processor. The NRAO AIPS package was used for the data analysis.

Images of 3C216 made with both uniform and natural weighting of the uv data are shown in Fig 1a,b. They show a core, and jet emission following a gently curving path, with underlying wiggles, ending in a sharp bend ca. 1 kpc from the core. Beyond this point the jet gradually fades in intensity through a string of weak components. The fairly bright knot at the bend appears to be edge-brightened; this, and the flatter spectrum of the knot (Akujor et al, 1993) may support the idea that the jet is being deflected by a gas cloud or clump. The radio emission from the jet fades about half way

R. Ekers et al. (eds.), Extragalactic Radio Sources, 83–84.

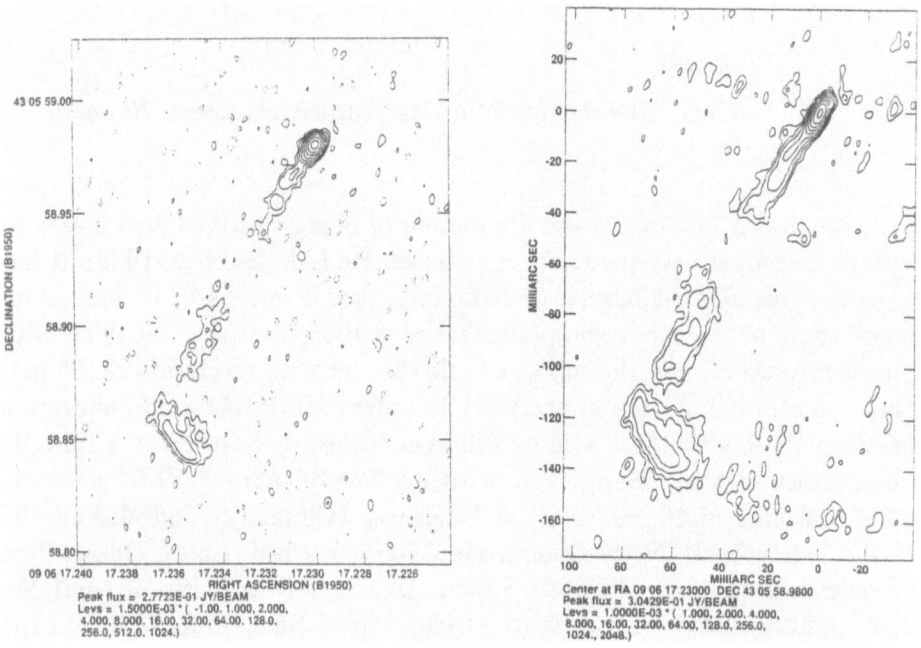

Figure 1. World array images of 3C216 at 1.6 GHz: (a, left) Uniform weighting, beam 4.5 x 4.5 mas; (b, right) Natural weighting, beam 6.7 x 3.5 mas, pa -24.

between the core and this knot, but reappears well before the bend. This feature is prominent in the 0.61 GHz VLBI map of Fejes et al (1992).

References

Akujor, C.E. et al, 1993, in Subarcsecond Radio Astronomie, Ed Davies, R.J. and Booth, R.S., CUP, p265
Barthel, P.D. et al, 1988, *ApJ*, **329**, L51
Fejes, I., Porcas, R.W. and Akujor, C.E., 1992, *A&A*, **257**, 459
Venturi, T. et al, 1993, *A&A*, **271**, 65

5 GHZ EVN POLARIZATION OF 3C286

D. DALLACASA, R.T. SCHILIZZI, H.S. SANGHERA
Joint Institute for VLBI in Europe, Dwingeloo, NL

D.R. JIANG
Shanghai Observatory, Chinese Academy of Sciences, China

E. LÜDKE
Universidade Federal de Santa Maria, Santa Maria, Brazil

AND

W.D. COTTON
National Radio Astronomy Observatory, Charlottesville, USA

3C286 (1328+307) is a powerful radio source identified with a quasar at z=0.849. There is a foreground galaxy responsible for an H I absorption line system at z=0.6922 (Brown & Roberts 1973), centered approximately 2.″5 to the southeast of 3C286. The radio source has a steep spectrum ($\alpha = -0.61$, $S_\nu \propto \nu^\alpha$ between 1.4 and 15 GHz) which turns over at about 100 MHz. Subarcsecond resolution radio images show a misaligned triple structure, dominated by the central component (Spencer et al. 1989) which accounts for at least 95% of the total flux density at all frequencies. 3C286 is one of the strongest extragalactic sources in polarized emission (0.84 Jy at 5 GHz and 1.41 Jy at 1.4 GHz) and with a rotation measure close to 0 rad m^{-2} (Rudnick and Jones 1983). Hence the observed orientation of the electric field vector is essentially independent of frequency.

Polarization data (3 hours) at 5 GHz were recorded on 17 June 1993, in MKIII mode A at six EVN antennas. Effelsberg, Medicina and WSRT recorded in both hands of polarization. Jodrell Bank, Noto and Onsala recorded LCP only. Stokes U and Q images were obtained using the complex clean in order to use also the data with a single crosshand.

The EVN polarization image reveals strong linear polarization, with the magnetic field dominated by the component perpendicular to the source axis (Fig. 1). The peak in polarized emission is located ∼6 mas southwest of the peak in total intensity.

R. Ekers et al. (eds.), Extragalactic Radio Sources, 85–87.

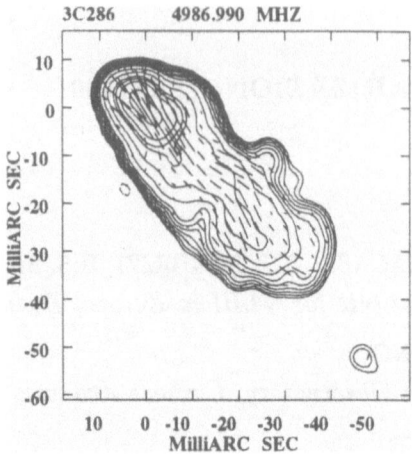

Figure 1.   EVN polarization image of 3C286. Contour levels are -3.0,3.0, 4.2,6,8.4,12,16.8,24,45,90,180,240,360 and 750 mJy/beam and the peak is 1.0 Jy/beam. The vectors represent the projected orientation of the electric vector; their length is proportional to the polarized flux density.

The former interpretation of the morphology of 3C286 was that *the brightest region of 3C286 harbours the core*, and the overall arcsecond morphology is that of a misaligned, asymmetric and core-dominated triple. However, on mas scales the cores of quasars are rarely strongly polarized (at most 2%) and the magnetic fields in the jets of quasars are often quite closely parallel to the jet axes (Cawthorne et al. 1993, Dallacasa et al. 1995). Moreover, the lack of reported significant variability in the total flux density of 3C286 and the failed detection at 43 GHz on the Kashima-Nobeyama baseline (Kameno et al. in preparation) imply that the source orientation is near the plane of the sky and that the core has a steep high frequency spectrum. Finally, the counterjet to the east of the brightest region (seen by Phillips and Shaffer, 1983, and by Spencer et al. 1991) has not been detected at 327 MHz above a surface brightness level of 0.02 mJy/mas^2 (Dallacasa et al. in preparation), nor in our 5 GHz combined EVN+MERLIN image nor in Zhang et al. (1994) at 1.67 GHz.

Here we discuss the possibility that *the brightest region of 3C286 is a lobe with hot-spots in an asymmetric, compact FR II radio source*. In fact, the bright knots of emission are immersed in a large cocoon, as can be seen in the high resolution images at 1.7 and 5 GHz (Zhang et al. 1994), and resemble the morphology of the hot-spots at the jet termination where the radio emitting plasma interacts with the ambient medium, as observed with the VLA for many powerful quasars and radio galaxies on larger scales. Moreover, a steep radio spectrum, high fractional polarization and the perpendicular field orientation, in ensemble, are in agreement with the observed properties of the hot-spot regions of FR II sources (Laing 1988), where the high polarization and transverse field are originated by a compression shock (Meisenheimer et al. 1989). The strong asymmetry between the sizes and flux densities of the two lobes may be due to different local environments, and the main lobe would appear brighter and smaller due to a stronger confinement of the radio emitting plasma (Fernini et al. 1993).

However there is a very small contribution of the lobe in terms of both flux density and polarization (contrary to FR II sources). The core has not been detected at a level of 10-20 mJy in any currently available image at any observing frequency. The upper limit to the fractional contribution of the core to the total flux density would be of 0.3%, eventually similar to those found in radio galaxies, rather than in quasars. Finally, the arcsecond component at 0.″8 to the east should be an unrelated object.

There is also a problem for both interpretations: we would expect ionization of the ambient medium where the interaction with the radio jet takes place. This in turn would trigger significant Faraday rotation also in presence of weak ambient magnetic field. Beyond the local environmental effects in the region where the radio emission is produced, the presence of a gas-rich intervening galaxy (HI column density$\sim2.6\times10^{20}$cm^{-2} for $T_{spin} = 100°$K, as from Brown and Roberts 1973) should produce significant effects. Again, an ad hoc geometry, i.e. a face-on spiral galaxy, would minimize such effects and be also consistent with the narrowness of the observed absorption line, corresponding to a velocity dispersion of only 8 km s^{-1}.

It is therefore clear that a straightforward interpretation of the observational evidence is not possible. We have considered the most relevant arguments for and against the two scenarios, and our conclusion to date is that the "hot-spot" interpretation is more likely acceptable. Further higher resolution polarization images, and a more thorough search for the source core will resolve this dilemma.

References

Brown, R.L., Roberts, M.S., 1973, *ApJ*, **184**, 7.
Cawthorne, T.V., et al., 1993, *ApJ*, **416**, 519.
Dallacasa, D., et al. 1995, *A&A*, **299**, 671.
Fernini, I., et al., 1993, *AJ*, **105**, 1690.
Laing, R., 1988,in Hot Spots in Extragalactic Radio Sources, K. Meisenheimer and H.J. Roser eds., p.27.
Meisenheimer, K., et al. 1989, *A&A*, **219**, 63.
Phillips, R.B., Shaffer, D.B., *ApJ*, **271**, 32.
Rudnick, L., Jones, T.W., 1983, *AJ*, **88**, 518.
Spencer, R.E., et al. 1989, *MNRAS*, **240**, 657.
Spencer, R.E., et al. 1991, *MNRAS*, **250**, 225.
Zhang, F.J., et al., 1994, *A&A*, **281**, 649.

THE EVOLUTION OF EXTRAGALACTIC RADIO SOURCES

A. C. S. READHEAD[1], T. J. PEARSON[1], G. B. TAYLOR[2] AND
P. N. WILKINSON[3]

[1] *California Institute of Technology*
[2] *National Radio Astronomy Observatory*
[3] *University of Manchester*

A series of VLBI surveys of complete samples of radio sources selected at 5 GHz (Pearson & Readhead 1988, hereafter PR; Xu et al. 1995 and references therein) has revealed that $\sim 10\%$ of the objects are "Compact Symmetric Objects" (CSOs), in which high-luminosity radio emission regions are seen on both sides of the center of activity on scales less than one kiloparsec (Phillips & Mutel 1982; Readhead et al. 1984; Conway et al. 1992; Wilkinson et al. 1994). In order to be sure that an object is a CSO, either the center of activity must be pinpointed (Taylor et al. 1996) or compelling morphological evidence of symmetric structure must be found.

CSOs are unusual amongst high radio luminosity galaxies and quasars because the "working surfaces" of the jets range from a few parsecs to a few hundred parsecs from the center of activity, and thus provide a unique probe of the interstellar medium in the inner kiloparsec. The hot spot pressures of CSOs are $\sim 3 \times 10^{-5}$ dyne cm^{-2} and the typical distance of PR CSO hot spots from the nucleus are ~ 50 pc. We assume that the hot spots are confined by ram pressure with the interstellar medium: the pressure in the hot spot is $\rho_{\text{ext}} v_a^2$, where v_a is the advance speed of the hot spot and ρ_{ext} is the density of the external medium. The hot spot pressure implies an external density of $10(v_a/0.02c)^{-2}$ cm^{-3}, an age for the PR CSOs of $\sim 10^4(v_a/0.02c)^{-1}$ yr, and a mass within a 200 pc radius of $\sim 10^7(v_a/0.02c)^{-2} M_\odot$.

Our observations of 2352+495 (Readhead et al. 1996) enable us to place upper limits on the H II and H I density in the external medium of 10^3 cm^{-3}, which implies an age of less than 10^5 yr. Ages significantly greater than this are ruled out by the implied mass within a 200 pc radius. For example, an age of 10^6 yr implies a total mass within a 200 pc radius of over $10^{12} M_\odot$ in

R. Ekers et al. (eds.), Extragalactic Radio Sources, 88–89.
© 1996 *IAU.*

the case of 0108+388, and over $10^{11} M_{\odot}$ in 0710+439 and 2352+495. Thus, although we cannot rule out a dense external medium composed mainly of H_2 by direct observations, we conclude that, regardless of the state of the external medium (HI, HII or H_2) the density must be $< 10^3 \, cm^{-3}$, and the age $< 10^5$ yr. In the case of 2352+495 the supply time of the energy to the lobes is about 10^4 yr and there is no evidence of an extended halo into which energetic particles from the lobes are escaping. Thus the lobe energy supply time is likely the age of the lobes. All of the evidence in the PR CSOs is consistent with an age of 10^4 yr at a size of 50 pc. We will, therefore, assume that this is the typical age of the PR CSOs at this size. It is clear that CSOs are much younger than typical Fanaroff-Riley type II objects, and this raises the interesting question of whether CSOs might evolve into CSS doubles (MSOs), and then into large scale FR II objects (Hodges & Mutel 1987; Fanti et al. 1995). If this unifying evolutionary hypothesis is correct, then it appears that these objects evolve in overall size from ~ 1 pc to ~ 100 kpc with little or no change in expansion speed, in which case the statistics of CSOs, MSOs and large-scale FR II objects show that the luminosity L must decrease approximately according to $L \propto R^{0.3}$, where R is the distance of the jet working surface from the center of activity. This also implies that $\rho_{ext} \propto R^{-1.3}$ over this range of sizes.

If the above evolutionary scenario is correct, then objects such as Cyg A would have been about a factor ten more luminous when they were 100 pc in overall size, and the CSO's which comprise 10% of the high frequency samples we have observed are the precursors of lower luminosity FR II objects. This type of luminosity evolution would have the fortunate consequence that, since objects are more luminous in their early phases, at a given flux density cutoff one is digging deeper into the luminosity function with the smaller sources. This compensates to a considerable degree for the relatively short time that objects spend in this phase, since otherwise there would be very few objects indeed which could be studied in these early phases with high sensitivity.

References

Conway, J. E., et al. 1992, ApJ, 396, 62

Fanti, C., et al. 1995, A&A, 302, 317

Hodges, M. W., & Mutel, R. L. 1987, in *Superluminal Radio Sources*, Cambridge UP, 168

Pearson, T. J., & Readhead, A. C. S. 1988, ApJ, 328, 114

Phillips, R. B., & Mutel, R. L. 1982, A&A, 106, 21

Readhead, A. C. S., Pearson, T. J., & Unwin, S. C. 1984, IAU Symp., 110, 131

Readhead, A. C. S., et al. 1996, ApJ, submitted

Taylor, G. B., Readhead, A. C. S., & Pearson, T. J. 1996, ApJ, submitted

Wilkinson, P. N., et al. 1994, ApJ, 432, L87

Xu, W., et al. 1995, ApJS, 99, 297

PKS 1413+135: A VERY YOUNG RADIO GALAXY

ERIC S. PERLMAN[1], CHRIS L. CARILLI[2], JOHN T. STOCKE[3]
AND JOHN CONWAY[4]

[1] USRA/LHEA, GSFC, Mail Code 660.2, Greenbelt, MD 20771
[2] CfA, 60 Garden St., Cambridge, MA 02138
[3] CASA, Box 389, Univ. of Colorado, Boulder, CO 80309
[4] Onsala Space Obs., S-43992, Onsala, Sweden

PKS 1413+135 is an enigma: while classified as a BL Lac due to its polarized near-IR continuum (Stocke et al. 1992) and optical spectrum (Beichmann et al. 1981), it appears to lie within a spiral host (McHardy et al. 1991, Stocke et al. 1992). In addition, the AGN is highly obscured (Beichmann et al. 1981, Carilli et al. 1992, Stocke et al. 1992, Wiklind & Combes 1994, 1995). Yet there is no evidence that the absorbing gas is being heated and re-emitting the AGN radiation in the form of thermal IR or emission lines (as in, e.g., Sey 2s). This led Stocke et al. (1992) to suggest that the AGN might be background to the optical galaxy.

We present VLBA maps of PKS 1413+135 at 3.6, 6, 13, and 18 cm from observations made on July 10-11, 1994 (Figure 1). Its structure appears similar to classical wide-angle-tail (WAT) radio galaxies, but on a much smaller scale. PKS 1413+135 is likely a CSO (see Readhead, this volume).

Our maps firmly establish the location of the core: Component N has an extremely inverted spectrum, well-fit by a power law of index $\alpha = +1.7$ ($S_\nu \propto \nu^\alpha$). This implies that N is likely self-absorbed (e.g. Miley 1980). While the majority of the extended structure is steep-spectrum ($\alpha \leq -1$), several of the knots and bends are somewhat flatter-spectrum, indicative of reacceleration and/or recollimation. The most far-flung components are the steepest, with power-law indices approaching $\alpha = -2$, perhaps due to synchrotron losses (e.g. Jaffe & Perola 1973).

Comparison with previous data (Perlman et al. 1994) reveals no evidence for superluminal motion. However, beaming is likely present within the core, since radio variability data from the UMRAO public archive yield

R. Ekers et al. (eds.), Extragalactic Radio Sources, 90–91.
© 1996 IAU.

$T_B > 3.6 \times 10^{13}$ K. An alternate explanation, gravitational lensing, is unlikely because we find no evidence for multiple images down to 2 mas.

The outer components of PKS 1413+135 are probably quasi-stationary 'minilobes', where the jets terminate. By assuming minimum-energy conditions and ram-pressure balance, we calculate $t \lesssim 7000n^{0.5}$ years. Alternatively, it is unlikely that a source so small has been 'frustrated' by a dense ISM (e.g. Readhead et al. 1994).

It is likely that CSO, GPS (O'Dea et al. 1991) and CSS (Fanti et al. 1990) sources represent early stages in the lifetimes of powerful radio galaxies. Due to their larger size, the CSS sources may represent a later stage. This is corroborated by their lower turnover frequencies, implying that their jets propagate through a less dense medium. Jet ram pressure, combined with radiation pressure from the AGN, would likely thin the ISM with time.

References

Beichmann et al. 1981, *Nature*, **293**, 711
Carilli et al. 1992, *ApJL*, **400**, L13
Fanti et al. 1990, *A & A*, **231**, 333
Jaffe & Perola 1973, *A & A*, **26**, 423
McHardy et al. 1991, *MNRAS*, **249**, 742
Miley 1980, *ARAA*, **18**, 165
O'Dea et al. 1991, *ApJ*, **380**, 66
Perlman et al. 1994, *ApJL*, **424**, L69
Readhead et al. 1994, *Compact Extragalactic Radio Sources*, pp. 17-22.
Stocke et al. 1992, *ApJL*, **400**, L17
Wilkind & Combes 1994, *A& A*, **286**, L9
Wiklind & Combes 1995, *Absorption Lines in QSOs* (ESO), in press

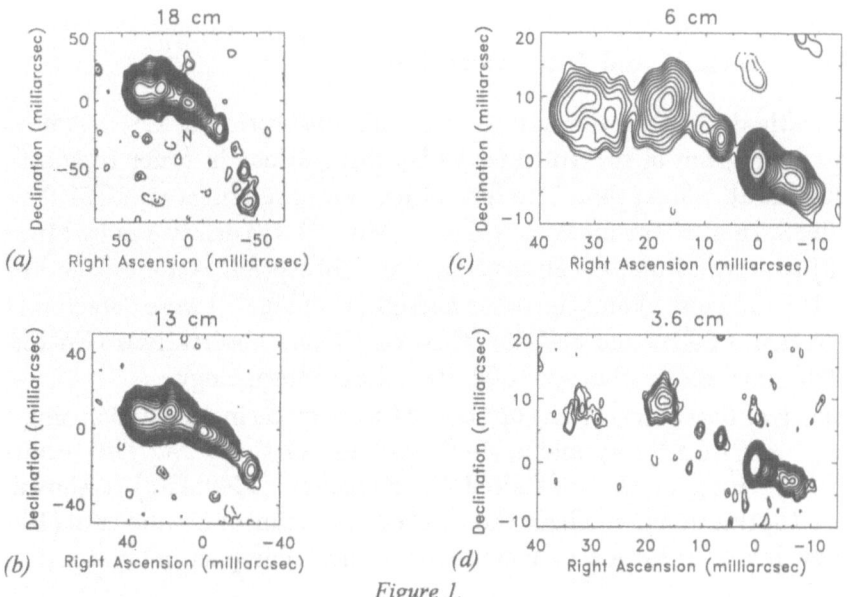

Figure 1.

A CIRCUMNUCLEAR HI DISK IN THE COMPACT SYMMETRIC OBJECT 4C31.04?

J.E. CONWAY
Onsala Space Observatory and J.I.V.E
Onsala, S-43992, Sweden

1. Introduction

Compact Symmetric Objects (CSO's) are strong, compact (<1kpc) objects with radio structure and luminosity similar to classical double radio sources (i.e. two lobes and a weak core), but thousands of times smaller (Wilkinson et al 1994). It has been proposed that CSO's are either young radio sources which will later evolve into classical sources, or sources whose growth is 'frustrated' by dense surrounding gas. In a recent survey of $z<0.1$ CSO's searching for HI absorption, approximately 50% were detected (see Conway et al 1996), a much higher percentage than found in surveys of general radio loud ellipticals with similar sensitivity (see van Gorkom et al 1989).

2. Observations and Interpretation

To investigate further the nature of the HI absorption in CSO's we have begun a program of spectral line VLBI observations in order to spatially resolve the HI absorption. The first object we have observed, 4C31.04, (at $z=0.0598$, 1mas = 1pc for $H_o = 75$kms^{-1} Mpc^{-1}) was discovered by Mirabel (1990) to have strong HI absorption. Two absorption systems, one broad (FWHM 133 kms^{-1}) and the other narrow (< 20kms^{-1}) were detected (Fig 1a). Recent 1.6GHz and 8.4GHz VLBI continuum observations (Cotton et al 1995) have shown that 4C31.04 has a CSO morphology.

Spectral line observations of 4C31.04 were made in July 1995 using the 10 station VLBA array and a single antenna of the VLA. The resulting continuum map at the redshifted HI frequency (1340MHz) is shown in Figure 1b, this image confirms the CSO classification of Cotton et al (1995), with evidence for both a core component and a compact hotspot within a

92

R. Ekers et al. (eds.), Extragalactic Radio Sources, 92–94.

diffuse Western lobe. Figure 1c shows the HI opacity integrated over the deepest part of the broad absorption line, it is large ($\tau \approx 0.07$) and fairly uniform over the E lobe but in the W lobe there is a sharp 'edge' to the opacity, on the core side of the edge the HI opacity is approximately 0.02 while on the other side it is consistent with zero. On the zero side of the HI edge we see strong depressions in the continuum emission (see Figure 1d), possibly caused by free-free absorption by unresolved clouds. Also seen in this region are individual narrow ($< 14\,\mathrm{kms}^{-1}$) HI clouds (not shown), both in the broad and in the narrow, high velocity gas. The visibility of these continuum and spectral features in this region might be partially due to the bright background continuum. High velocity HI and possible free-free absorption features are also seen in front of bright parts of the Eastern lobe.

The fact that the HI edge is roughly perpendicular to the vector from the core to W hotspot strongly suggests that the obscuring gas 'knows' about the radio axis and is associated with the AGN rather than being random foreground gas in the host galaxy. The observations can be explained assuming a disk geometry whose axis is that of the radio jet (see Figure 2), and in which the jet axis is close to the sky plane (as is likely for an isotropically emitting CSO selected on total flux density). The dimensions of the disk required are similar to the disks seen in HST images (e.g. Jaffe et al 1993). In this model the disk completely covers the more distant, Eastern lobe, generating a large opacity, while the HI edge in the Western lobe is caused by the finite thickness of the disk or torus. Matter evaporated from the inner edge of the disk and closer to the AGN could then provide the high velocity HI and ionized clouds which are observed beyond (see Fig 1d) the HI edge. A possible gradient in the centroid of the absorption velocity in directions perpendicular to the radio axis is consistent with this disk model; for a jet inclined at 15° to the sky plane an enclosed mass of 10^8 solar masses within 150pc of the central engine is implied. Satisfying simultaneously the observations of free-free absorption, HI absorption and minimum pressure in the radio emitting plasma gives strong constraints on gas physical parameters and will be discussed elsewhere.

References

Conway, J.E., Readhead, A.C.S., Taylor, G.B., (1996), *ApJ in preparation*.

Cotton, W.D., Feretti, G., Giovannini,G., Venturi, T., Lara, L., Marcaide, J., Wehrle, A.E., (1995), *ApJ*, **452**, 605.

Jaffe, W., Ford, H.C., Ferrarese, L., van den Bosch, F., O'Connell, R.W., (1993), *Nature*, **364**, 213.

Mirabel, I.F., (1990), *ApJ*, **352**, L37.

van Gorkom, J.H., Knapp, G.R., Ekers, R.D., Ekers, D.D., Laing, R.A., and Polk, K.S., (1989), *AJ*, **97**, 708.

Wilkinson P.N., Polatidis, A.G., Readhead, A.C.S., Xu, W., Pearson, T.J., (1994), *ApJ*, **432**, L87.

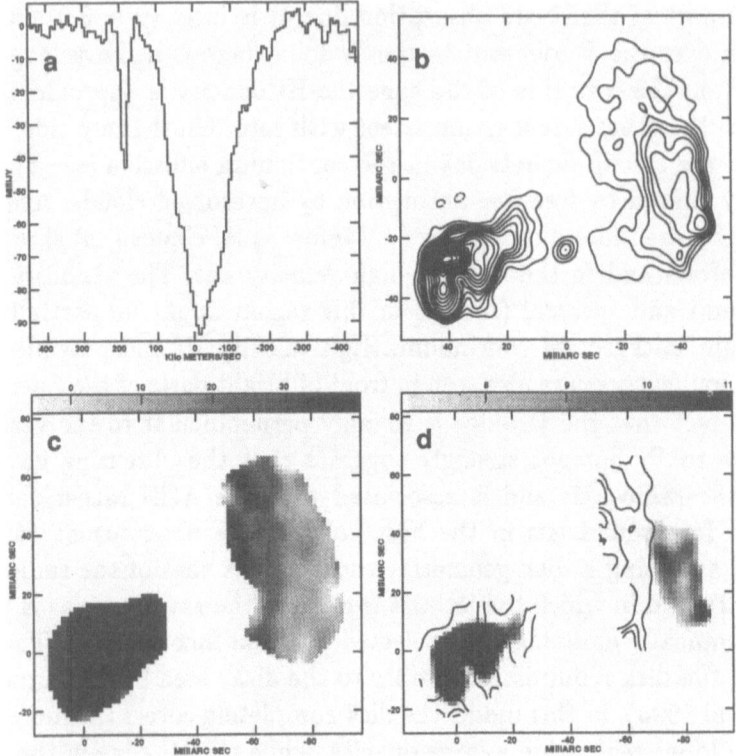

Figure 1. (a) HI Absorption spectrum integrated over the whole source. (b) Continuum (1340MHz) MEM image, 3mas FWHM resolution, first 6 contours are at 1.5mJy/beam intervals, rest at 3.0mJy/beam intervals. (c) HI Opacity averaged over central part of broad absorption line. Greyscale image from -0.005 to 0.040, resolution 12 by 9mas, PA=5 degrees. (d) Greyscale continuum 3mas resolution, plotted from 7 to 11mJy/beam, showing local depressions in the continuum emission. Contours show HI opacities of 0.02, 0.03, 0.04, 0.06, 0.08.

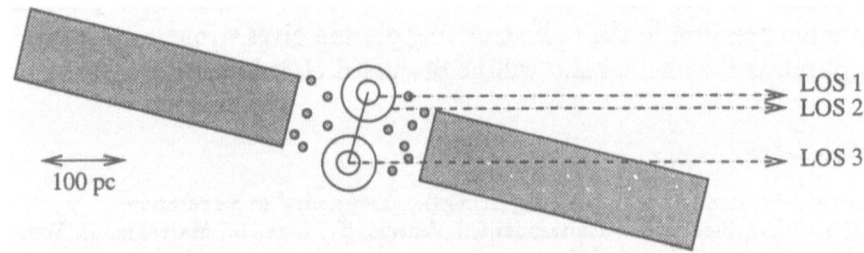

Figure 2. Possible model, grey indicates a gas disk whose axis is parallel to the radio axis, small clouds are shown evaporating off its inner edge. The contours of radio emission from the lobes are shown schematically as two sets of concentric circles, the West lobe in Fig 1 at the top and the East lobe at the bottom. Three lines of sight are shown, to the West hotspot (LOS1), at the edge of the HI absorption (LOS2) and to the East lobe (LOS3).

MULTIFREQUENCY STUDY OF VLBI–COMPACT SOURCES:
1–22 GHZ SPECTRA OBSERVATIONS AND MODEL FITS

YU.A. KOVALEV[1], A.B. BERLIN[2], N.A. NIZELSKIJ[2],
Y.Y. KOVALEV[1,3], AND S.V. BABAK[3]

[1] *Astro Space Center of the Lebedev Physical Institute —*
Profsoyuznaya 84/32, 117810 Moscow, Russia
[2] *Special Astrophysical Observatory —*
N.Arkhyz, Karachaj-Cherkesia 357140, Russia
[3] *Moscow State University — Vorobjevi gori, 119899 Moscow*

Results of instantaneous (with accuracy of a minute) spectra observations for a sample of 113 compact objects, selected from the VLBI survey [1], are presented. Measurements have been made in 1989.9 at the RATAN–600 at 7 wavelengths of 31, 13, 8.2, 7.6, 3.9, 2.7 and 1.4 cm as a set of our spectra monitoring program [2].

Spectra of all sources can be considered as a simple combination of two 'elementary' spectral components: an optically thin at lower frequencies ('the LF–spectrum') and a full spectral shape with a different width from a source to source ('the HF–spectrum') — see Figure 1. A model analysis, like in [3], confirms the hypothesis that such a combination can be explained by a halo–jet structure with the jet at different orientation from a source to source — at least for 20% of sources in our sample (model fitting is in progress).

A statistical analysis does not show any difference between quasars and BL Lac's (see the paper by Y.Y.Kovalev in these Proceedings).

The work of YAK, YYK and SVB was supported in part by ESO grant A–04–049.

References

[1] Preston, R.A., Morabito, D.D., Williams, J.G., Faulkner, J., Jauncey, D.L. and Nicolson, G.D. (1985) *Astron. J.*, Vol. no. **90**, p. 1599.
[2] Kovalev, Yu.A. (1991) *Soobsheniya SAO (in russian)*, Vol. no. **68**, p. 60.
[3] Nesterov, N.S., Kovalev, Y.Y., Babak, S.V. and Larionov, G.M. (1994) *Astron. Rep.*, Vol. no. **71**, p. 850.

R. Ekers et al. (eds.), Extragalactic Radio Sources, 95–96.
© 1996 IAU.

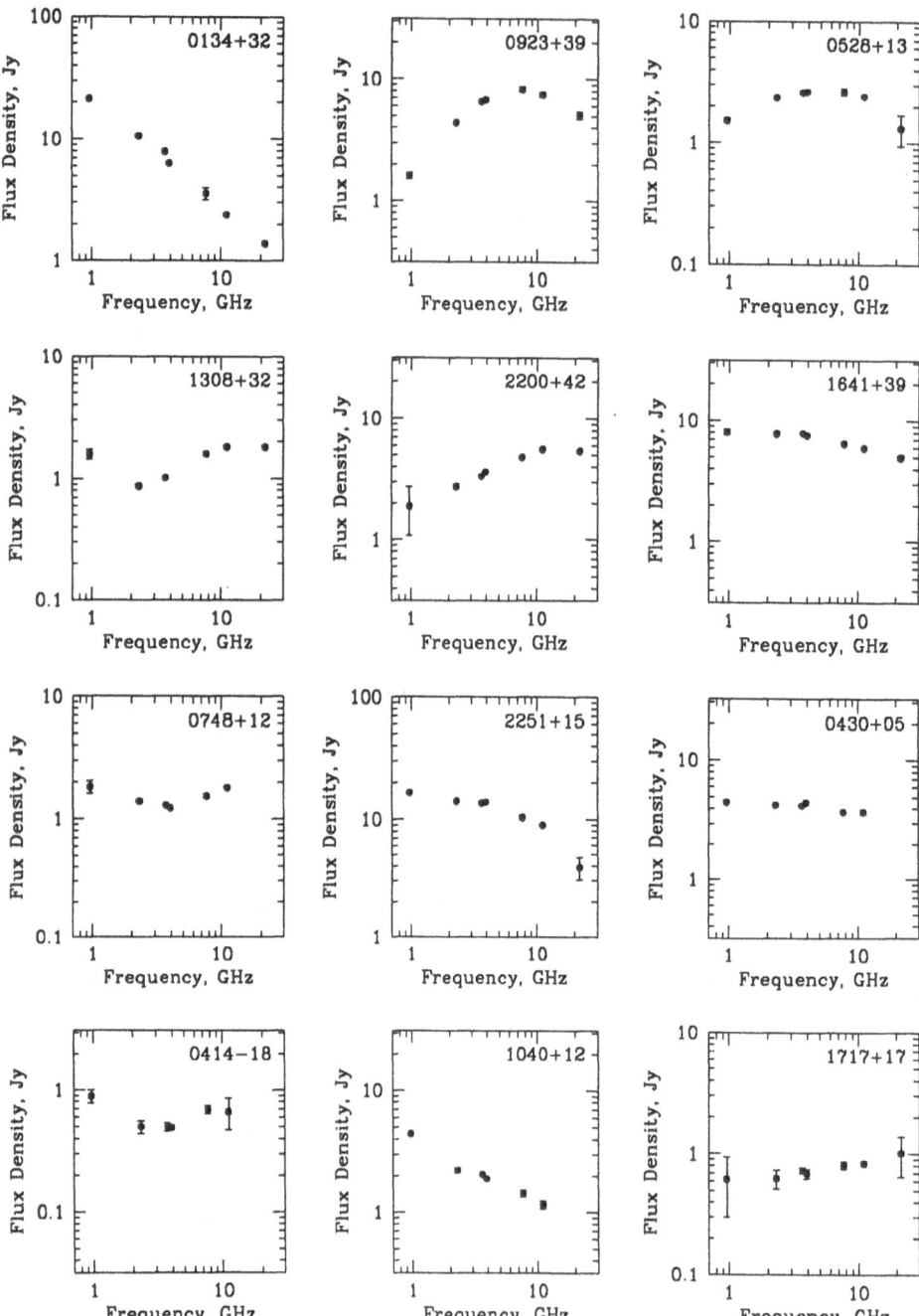

Figure 1. Examples of typical observed instantaneous spectra. All types of spectra in
the columns can be considered as a combination of two 'elementary' spectra in the first
line (the LF–spectrum of 0134+32 with the narrow or wide HF–spectrum of 0923+39 or
0528+13) by shifts on the frequency and flux and changes of the HF–spectra width.

STATISTICAL ANALYSIS OF OBSERVATIONS OF A COMPLETE
SAMPLE OF COMPACT EXTRAGALACTIC RADIO SOURCES

Y.Y. KOVALEV

Moscow State University — Vorobjevi gori, 119899 Moscow;
Astro Space Center of the Lebedev Physical Institute —
Profsoyuznaya 84/32, 117810 Moscow, Russia

Here are analyzed the results of instantaneous multifrequency observations for a complete sample of 113 extragalactic radio sources (see the paper by Yu.A. Kovalev *et al.* in these Proceedings) with declinations $-30° \div +43°$ from the VLBI–survey [1]. It is tested the hypothesis that quasars and BL Lacs have different nature. Statistical distributions for the turnover flux, turnover frequency, spectral indexes, and other parameters are analyzed.

All spectra before calculations had been transformed to the rest frame of the sources. Red shifts and optical identification are used from [2,3] and some other papers. The following curve has been fitted to the spectra as in [4]: $\log F_\nu = C + (\log \nu - B)^2 / 2A$. The interpretation of the parameters A, B and C is in the caption to Figure 1. We didn't use data with big $\sigma_{A,B,C}$ in our analysis. To derive the centimeter spectral index α_{cm} we use data at 2.7, 3.9, 7.6, 8.2 cm.

Strong differences in the distribution of parameters A, B and C are absent for 3 studied subgroups of objects (see Figure 1a,b,c). Quasars, BL Lacs and other sources have one peak in the distribution, which is approximately the same for the different types of objects. Obtained distribution testify in the favour to common physical nature of quasars and BL Lacs.

Comparison of Figures 1a,b,c with Figure 2 from paper [4] gives that the peak values of parameters B and C are similar in these works. It can be concluded that limiting, in the present work, the high frequency at 22 GHz did not influence a correct derivation of the turnover frequency (the highest frequency, used in [4], is much higher than 22 GHz).

Unfortunately (or fortunately) we did not see *any* correlation between logarithms of maximum flux and turnover frequency (see Figure 1 and Table 1). Correlation between (C, A) and (B, A) parameters has been derived (see Table 1). What's it? Is it a selection effect or something more than

97

R. Ekers et al. (eds.), Extragalactic Radio Sources, 97–98.

selection? This result may show a decreasing number of independent parameters, by which the spectra are characterized [4,5].

This work was supported in part by ESO grant A–04–049.

References

[1] Preston, R.A., et al. (1985) *Astron. J.*, **Vol. no. 90**, p. 1599.
[2] Morabito, D.D., et al. (1986) *Astron. J.*, **Vol. no. 91**, p. 1038.
[3] Veron–Cetty, M.-P. and Veron, P. (1993) *ESO Sci. Rep.*, No. 13.
[4] Landau, R., et al. (1986) *Astrophys. J.*, **Vol. no. 308**, p. 78.
[5] Valtaoja, E., et al. (1988) *Astron. Astrophys.*, **Vol. no. 203**, p. 1.

TABLE 1. Correlation Coefficients for the spectral parameters

Optical ID	The number of objects	Correlation Coefficients (Probabilities)		
		(B, A)	(C, A)	(C, B)
Quasars	57	0.331 (98.7%)	0.164 (77.6%)	0.051 (29.2%)
BL Lacs	13	0.161 (40.1%)	0.169 (42.1%)	0.093 (23.9%)
Others	20	0.365 (88.5%)	0.380 (90.3%)	0.079 (25.8%)
All Sample	90	0.333 (99.8%)	0.230 (96.8%)	0.021 (15.5%)

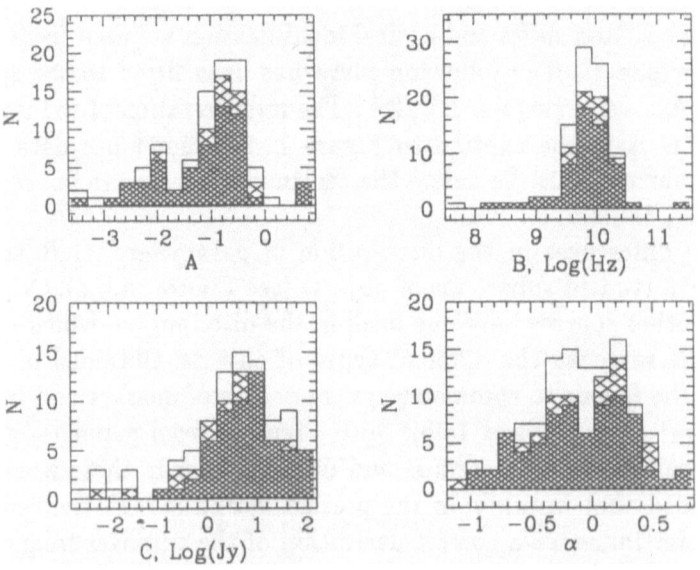

Figure 1. Histograms of the spectral parameters A (a), B (b), C (c) and α_{cm} (d) for sources in the sample. A is an effective width of a spectra, B is a logarithm of the turnover frequency, C is a logarithm of the maximal flux density. It is blacked for quasars, shaded — for BL Lacs, open — for other sources.

HIGH RADIO FREQUENCY STUDIES OF
EQUATORIAL AND SOUTHERN AGN

M. TORNIKOSKI AND E. VALTAOJA
Metsähovi Radio Research Station
Metsähovintie 114, FIN–02540 Kylmälä, Finland

The Swedish-ESO Submillimetre Telescope (SEST) has been used for the high radio frequency observations of our group's AGN monitoring projects since the end of 1987.

Our SEST results from October 1987 until June 1994 will be published in *A&AS* (in press); the data will be available electronically. The data set consists of 155 sources with the signal-to-noise -ratio of at least one observation (at 90 or 230 GHz) ≥ 4.

The sources for which we have data mainly fall into one of the following classes:

- All bright, flat-spectrum quasars between the declination range $0°$ to $-25°$ (results from our analysis of this sample were published in Tornikoski et al. (1993)).
- All southern blazars (BL Lac objects and high polarization quasars).
- A set of bright, variable, nearly-equatorial sources, which are sources of our long-term monitoring campaign plus several multifrequency campaigns.
- A set of sources from the CGRO-multifrequency collaboration.

These data are being used for studies of various radio properties (spectrum, variability, timescales of variability) of the sources, source classification, Doppler boosting effects, etc.

For the most frequently observed sources we have flux curves which enable us to study the growth and decay of radio outbursts in these sources, the possible differences between the outbursts in one source, and the time delays between high radio frequencies (SEST) and lower radio frequencies (Metsähovi) during the radio outbursts. Examples of two sources with relatively long and well-sampled time series of SEST data, 3C 273 and 1510-089, are shown in Figure 1.

R. Ekers et al. (eds.), Extragalactic Radio Sources, 99–100.
© 1996 *IAU.*

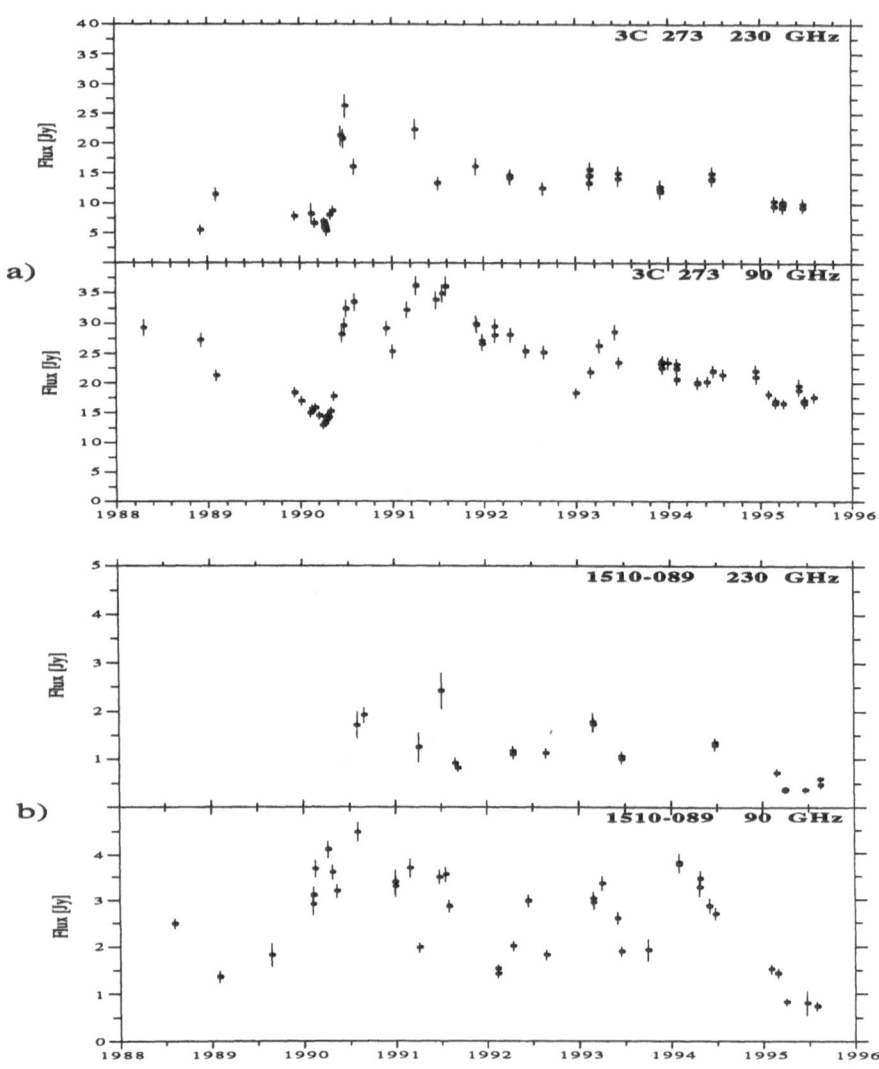

Figure 1. SEST flux density curves of a) 3C 273 and b) 1510-089.

Together with data gathered with other instruments, the SEST data are valuable for understanding the multifrequency behaviour of AGN. The data are used for studies of correlating events at various radiation regimes (cf. Tornikoski et al. (1994), Teräsranta et al. in these proceedings, and several other multifrequency studies).

References

Tornikoski M., Valtaoja E., Teräsranta H., et al., 1993, *AJ*, **105**, 1680
Tornikoski M., Valtaoja E., Teräsranta H., et al., 1994, *A&A*, **289**, 673

A SIMPLE METHOD TO ANALYZE RADIO FREQUENCY OUTBURSTS OF QUASARS

H. TERÄSRANTA AND E. VALTAOJA
Metsähovi Radio Research Station
FIN-02540 Kylmälä, Finland

AND

M. LAINELA
Tuorla Observatory
FIN-21500 Piikkiö, Finland

Quasars have been monitored with the Metsähovi Radio Telescope since 1980 at 22 GHz and 37 GHz. During these years we have recorded numerous outbursts in our sample. Some of the outbursts are simple in nature, while others also seem to have fine structure. Can the outbursts be modelled in a simple way?

We have taken the simplest possible assumption, starting with simple outbursts. The flux is divided into two components, the core flux, which is assumed to be constant, and the outburst flux, which is the varying component. When we subtract the core flux from the total flux, the remainder is the outburst flux. For isolated outbursts (which are rather rare) we find that both the rise and the decay of the flare are well described by the form $lnS = a + bt$, where S is the flux, t the time, and a and b constants. In other words, the outburst flux both rises and decays exponentially.

The flux variations of most sources look somewhat chaotic as several closely spaced flares often overlap. However, we have found out that a combination of self-similar exponential outbursts with variable amplitudes can reproduce the main features of the variations.

The model components can be used in to estimate the zero epochs and the flux density evolution (including predicted future fluxes) of VLBI components. We can also assign the right radio flare phase for correlated events in other wavebands. One example is gamma/radio correlations, where we can determine whether the events are simultaneous or time-delayed relative

101

R. Ekers et al. (eds.), Extragalactic Radio Sources, 101–102.

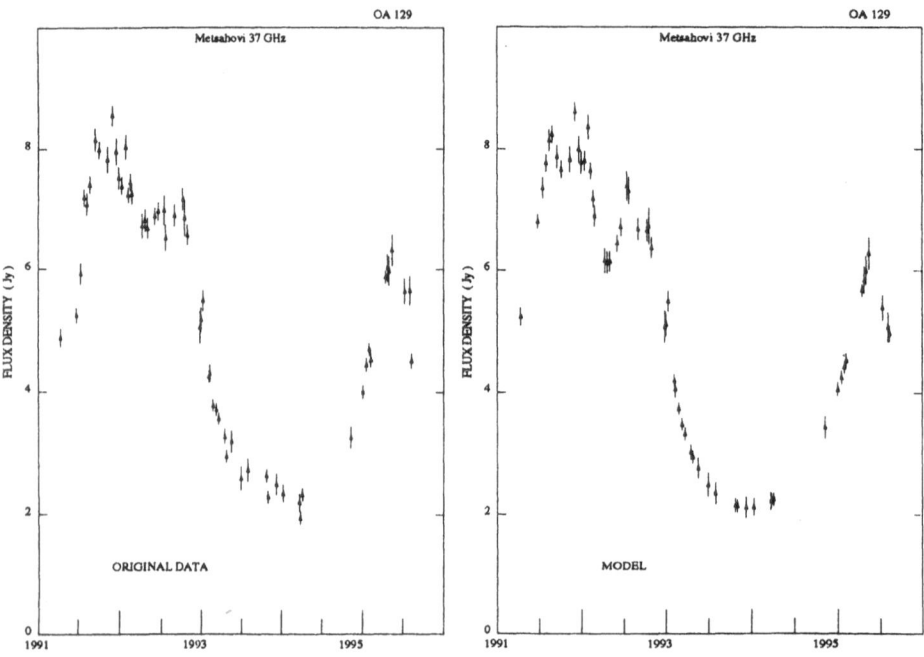

Figure 1. The time series of our 37 GHz observations of OA 129 since 1991 and the associated model.

to each other, and if the gamma-ray emission preferentially occurs at some definite phase of the radio flare evolution.

It is remarkable that although the radio flux density variations of AGN often appear very chaotic, they can to a large extent be modelled with a small number of self-similar flares. This demonstrates that the flux variations are due to series of similar shocks in the relativistic jet, differing mainly in their strength.

LONG-TERM VARIABILITY OF
EXTRAGALACTIC SOURCES AT 843 MHZ

R. W. HUNSTEAD AND B. M. GAENSLER
School of Physics, University of Sydney
NSW 2006, Australia

Time variability is commonly observed in the most compact extra-galactic radio sources. Low-frequency variability (LFV)—at frequencies <1 GHz—is thought to arise through two different mechanisms, *intrinsic* and *extrinsic*. The former is just an extension of the often rapid high-frequency variations, delayed and reduced in amplitude. The latter is usually attributed to refractive interstellar scintillation (RISS; Rickett *et al.* 1984), whereby the variations in intensity are the result of wavefront distortions caused by transverse gradients in electron density. If RISS arises predominantly along the signal path through our Galaxy, we might expect to find evidence for a dependence on Galactic coordinates.

Intrinsic variability is most prominent for $\nu > 1$ GHz, while scintillation tends to dominate below 1 GHz; variability at $\nu \sim 1$ GHz is largely unexplored. The Molonglo Observatory Synthesis Telescope (MOST), operating at 843 MHz, is well suited to studying this intermediate regime. Before and after every 12-hour synthesis, MOST records the flux densities of 5–10 compact extragalactic sources, chosen from a list of 55 calibrators (Campbell-Wilson & Hunstead, 1994). Some 43 000 such measurements have been made over the ten year period 1983–1993, forming the database for the present study (Gaensler & Hunstead, in prep). Some of the MOST 'calibration' sources show quite spectacular variability—see Fig. 1(a).

How do we determine whether a source is varying? The conventional approach is to determine the modulation index $m = \sigma_S/S$, where σ_S is the standard deviation in a source of mean flux density S. We first remove unreliable measurements from each dataset to produce a relatively noise-free light curve for each source. To reduce further the random noise contribution to σ_S we smooth the data by binning into 30-day intervals and then apply a well established test (Fanti *et al.*, 1981) to calculate the χ^2 probability that the flux density has remained constant.

R. Ekers et al. (eds.), Extragalactic Radio Sources, 103–104.

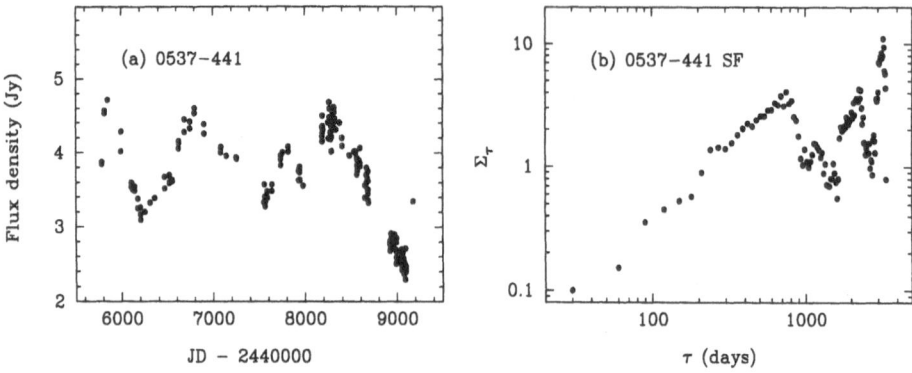

Figure 1. (a) MOST 843 MHz light curve for 0537−441, a peaked-spectrum source identified with a $z = 0.894$ BL Lac/HPQ/OVV and detected as an X-ray and γ-ray source; it is core-dominated in the radio, with a diameter of 1.1 mas at 2.3 GHz. (b) Structure function for a 30-day smoothed version of the dataset shown in (a).

How do we determine a timescale for the variability? The smoothed dataset is mean-subtracted and normalized by the standard deviation to yield a new time series F_t. From this we calculate the structure function $\Sigma_\tau = \langle [F_{t+\tau} - F_t]^2 \rangle$, where τ is a parameter known as the lag (Spangler *et al.*, 1989). At small lags Σ_τ is approximately constant—the noise regime—and as τ increases Σ_τ (when shown on a log-log plot) behaves linearly—the structure regime. Finally there is the saturation regime, where Σ_τ flattens out or turns over. We define the timescale τ_V to be twice the lag at which the structure function saturates.

The structure function for 0537−441 in Fig. 1(b) shows a clear linear regime up to $\tau \sim 700$, where it turns over. At greater lags it begins to oscillate about $\Sigma_\tau \approx 2$; obviously the longest timescales are poorly sampled. If we interpret the first turnover as saturation this implies $\tau_V \sim 1400$ days, which is consistent with the timescales shown in Fig. 1(a).

In all, we find that 17 of the 55 MOST calibrators vary significantly, on timescales ranging from months to years, with 18 non-variables and 20 uncertain. Whereas the non-varying sources are spread uniformly in Galactic latitude, 14/17 variables have $|b| < 30°$. This clear difference between the two distributions suggests that refractive interstellar scintillation is the dominant cause of LFV at 843 MHz.

References

Campbell-Wilson, D. & Hunstead, R.W. (1994), *PASA*, **11**, 33.
Fanti, C., Fanti, R., Ficarra, A., Gregorini, L., Mantovani, F., Padrielli, L. & Weiler, K.W. (1981), *A&AS*, **45**, 61.
Rickett, B.J., Coles, W.A. & Bourgois, G. (1984), *A&A*, **134**, 390.
Spangler, S., Fanti, R., Gregorini, L. & Padrielli, L. (1989), *A&A*, **209**, 315.

GRAVITATIONAL LENSES

J. N. HEWITT
Department of Physics and Research Laboratory of Electronics
Massachusetts Institute of Technology
Cambridge, Massachusetts 01239 USA

Abstract. In approximately half the systems currently recognized as strongly gravitationally lensed, the background object is an extragalactic radio source. Radio observations have played an important role in the identification of lensed systems, and the properties of radio sources allow some of the astrophysical applications of lensing to be realized. High redshift galaxies can be studied through lens modeling and by observing more than one ray path through the lens. The morphological, spectral, and polarization information of high resolution radio images provides strong constraints on the mass distribution in the lensing galaxies. On cosmological scales, radio variability has been applied to the time delay measurement of angular diameter distance.

1. Introduction

Since the discovery of the first gravitational lens in 1979, many new cases of multiple imaging by a gravitational lens ("strong lensing") have been discovered. Other manifestations of gravitational lensing have also been observed, such as the distortion of background galaxies ("weak lensing"), microlensing, a possible statistical excess of quasars near foreground galaxies, and the time delay exhibited in the light curves of images of the same object. In this paper, I restrict my attention to just the strong lenses. Table 1 presents a summary of the strong lenses currently known. A recent summary of the data on these objects is given by Keaton and Kochanek (1995). A large number of the strong lenses are radio sources, and in many cases radio data provided the first evidence for gravitational lensing. The radio data have been important in identifying lenses and providing infor-

R. Ekers et al. (eds.), Extragalactic Radio Sources, 105–110.

TABLE 1. Strong Gravitational Lenses

Object	Number of Images	Lens[1]	Source[1]	z_l	z_s	Radio Source	Ref[2]
0023+171[3,4]	2	G?	G	?	0.95	Yes	H87
0142-100	2	G	Q	0.49	2.72	Faint?	S87
0218+357[4]	2,ring	G	?	0.68	?	Yes	P93b
QJ0240-343[3]	2	?	Q	?	1.4	?	T95
MG0414+0534[4]	4	G	Q	?	2.64	Yes	H92
MG0751+2751[4]	4	G	?	0.35	?	Yes	L93a
BRI0952-0115	2	?	Q	?	4.5	?	M92c
0957+561	2	G, C	Q	0.36	1.41	Yes	W79
LBQS1009-0252[3]	2	?	Q	0.87?	2.74	No	H94
HE1104-1805[3]	2	?	Q	?	2.30	?	W93
1115+080	4	G	Q	0.3?	1.72	No	W80
1120+019[3]	2 3? 4?	?	Q	0.6?	1.46?	No	M89
MG1131+0456[4]	2,ring	G	?	?	?	Yes	H88
1208+101[3]	2	?	Q	2.92?	3.80	No	M92a,b
HST12531-2914	4	G	Q?	?	?	?	R95
1413+117	4	?	Q	?	2.55	Faint	M88
B1422+231[4]	4	G	Q	0.64?	3.62	Yes	P92a
HST14176+5226	4	G	Q?	?	?	?	R95
1429-008[3]	2	?	Q	?	2.08	No	H89
MG1549+3047[4]	ring	G	Q lobe	0.11	?	Yes	L93b
CLASS1600+434[4]	2	?	Q	?	1.61	Yes	J95
CLASS1608+656[4]	4	G	Q	.63	1.39	Yes	M95
1635+267[3]	2	?	Q	?	1.96	No	D84
MG1654+1346[4]	ring	G	Q lobe	0.25	1.74	Yes	L89
PKS1830-211[4]	2,ring	?	?	?	?	Yes	J92
B1938+666[4]	4	?	?	?	?	Yes	P92
2016+112[4]	3	G,G	Q	1.01	3.27	Yes	L84
2237+0305	4	G	Q	0.04	1.70	Faint	H85
2345+007[3]	2	G?	Q	?	2.15	No	W82

[1] "G" denotes galaxy; "C" denotes cluster; "Q" denotes quasi-stellar object.
[2] The reference is to the paper reporting the discovery of the lensed nature of the object.
[3] The only evidence for gravitational lensing in this case is the similarity of the optical spectra of the putative images; therefore, the lensed nature of this source might be viewed as not secure.
[4] Radio observations provided the first evidence for gravitational lensing.

mation on the structure and variability of the images.

As more gravitational lens systems are discovered, we find there are

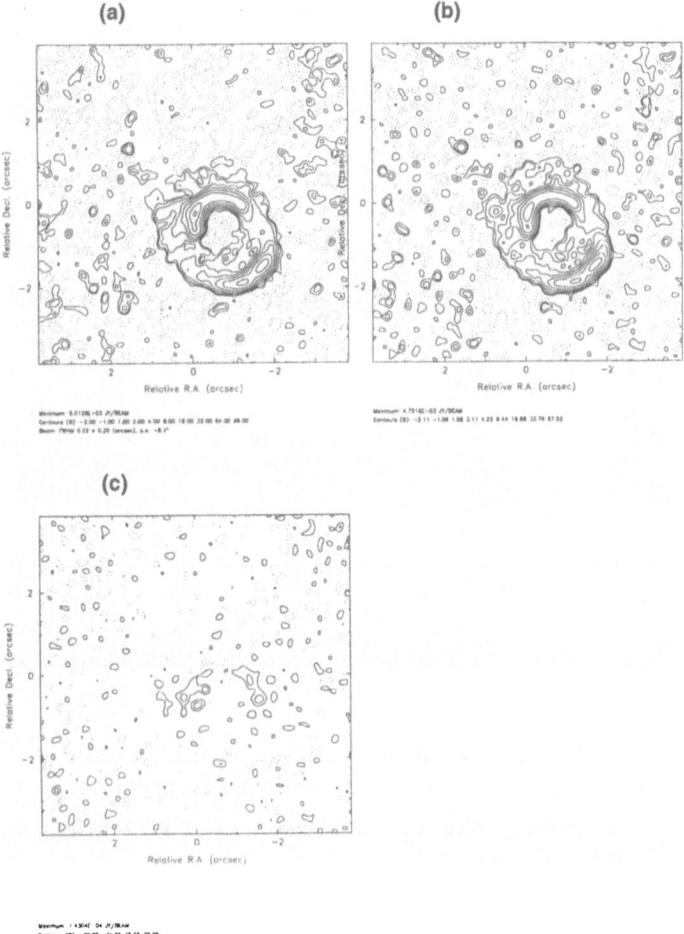

Figure 1. (a) Radio image of MG1654+1346. (b) Model of MG1654+1346. (c) Image computed from the residual visibilities, giving a measure of the goodness of fit of the model. Reproduced from Ellithorpe, Kochanek and Hewitt 1996.

systems that are well suited to particular astrophysical applications. In this paper I give examples of two applications: modeling the mass distribution in the lensing galaxy and measuring the angular diameter distance through observing the time delay between a pair of images.

2. Modeling the Lensing Galaxy

The four-image and ring systems, in which there is *extended* structure associated with the background source, provide the most stringent constraints on the model for the mass distribution in the lensing galaxy. Figure 1 shows three images that illustrate the model fitting procedure in MG1654+1346, using the LensClean algorithm initially developed by Kochanek and Narayan

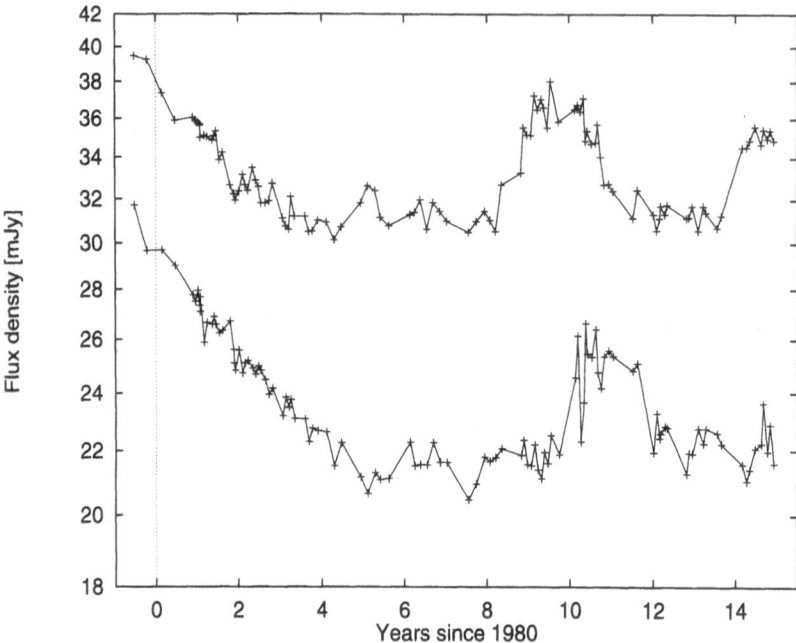

Figure 2. Radio light curves of the two images of 0957+561 (Haarsma *et al.* 1995).

(1992), and modified to operate directly on visibility data by Ellithorpe, Kochanek, and Hewitt (1996). There is evidence for dark matter in this system since the mass profile associated with the best fitting model does not trace the light (Kochanek 1995).

3. Measuring Angular Diameter Distance

The measurement of the time delay associated with the propagation of radiation from two images has long been recognized to give a measurement of angular diameter distance (Refsdal 1964, Narayan 1992). This cosmography provides the opportunity to discriminate among cosmological models and to fit cosmological parameters. To date time delays only for 0957+561 and PKS1830-211 have appeared in the refereed literature. Figure 2 shows the most recent results on radio flux density monitoring of 0957+561 (Haarsma *et al.* 1995). A preliminary analysis of these data gives a time delay of 455 ± 40 days. According to the model of Grogin and Narayan (1996), this implies a Hubble constant of 64^{+23}_{-22} km/(sec-Mpc), where the uncertainty is dominated by the uncertainty in the division of mass between the primary lensing galaxy and its surrounding cluster. Converting this into a measurement of angular diameter distance gives $D = 980 \pm 320$ Mpc. Figure 3 shows angular diameter distance as a function of redshift for different cosmological models. Superimposed is the time delay measurement for 0957+561, as well

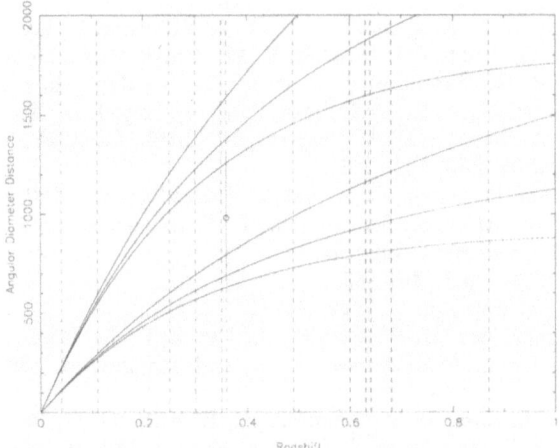

Figure 3. Angular diameter distance as a function of redshift for (in descending order): $H_o = 50$km/sec-Mpc, $\Omega = 0$, $\lambda = 1$; $H_o = 50$km/sec-Mpc, $\Omega = 0$, $\lambda = 0$; $H_o = 50$km/sec-Mpc, $\Omega = 1$, $\lambda = 0$; $H_o = 100$km/sec-Mpc, $\Omega = 0$, $\lambda = 1$; $H_o = 100$km/sec-Mpc, $\Omega = 0$, $\lambda = 0$; $H_o = 100$km/sec-Mpc, $\Omega = 1$, $\lambda = 0$. The derived value for 0957+561 is plotted. The vertical dashed lines represent the lens redshifts tabulated in Table 1.

as vertical dotted lines showing the distribution of the known lenses on this diagram. Perhaps in the future such a plot will display many gravitational lens measurements of angular diameter distance.

References

Djorgovski, S., and Spinrad, H., 1984, *Ap.J.*, **282**, L1 (D84).

Ellithorpe, J. D., Kochanek, C. S., and Hewitt, J. N., 1996, *Ap.J.*, in press.

Grogin, N., and Narayan R., 1996, preprint.

Haarsma, D. B., Hewitt, J. N., Lehár, J., and Burke, B. F. 1995, in *Proceedings of I.A.U. Symposium #173: Application of Gravitational Lensing* (eds. C. S. Kochanek and J. N. Hewitt), Dordrecht: Kluwer Academic Publishers.

Hewett, P. C., Irwin, M. J., Foltz, B. B., Harding, M. E., Corrigan, R. T., Webster, R. L., and Dinshaw, N., 1994, *A.J.*, **108**, 153 (H94).

Hewett, P. C., Webster, R. L., Harding, M. E., Jedrzejewski, R. I., Foltz, C. B., Chaffee, F. H., Irwin, M. J., and Le Fèvre, O., 1989, *Ap.J.*, **346**, L61 (H89).

Hewitt, J. N., Turner, E. L., Lawrence, C. R., Schneider, D. P., and Brody, J. P., 1992, *Ap.J.*, **104**, 968 (H92).

Hewitt, J. N., Turner, E. L., Lawrence, C. R., Schneider, D. P., Gunn, J. E., Bennett, C. L., Burke, B. F., Mahoney, J. H., Langston, G. I., Schmidt, M., Oke, J. B., and Hoessel, J. G., 1987, *Ap.J.*, **321**, 706 (H87).

Hewitt, J. N., Turner, E. L., Schneider, D. P., Burke, B. F., Langston, G. I., and Lawrence, C. R., 1988, *Nature*, **333**, 537 (H88).

Huchra, J., Gorenstein, M., Kent, S., Shapiro, I., Smith, G., Horine, E., and Perley, R., 1985, *A.J.*, **90**, 691 (H85).

Jackson, N., de Bruyn, A. G., Myers, S., Bremer, M. N., Miley, G. K., Schilizzi, R. T., Brown, I. W. A., Nair, S., Wilkinson, P. N., Blandford, R. D., Pearson, T. J., and Readhead, A. C. S. 1995, *MNRAS*, **274**, L25 (J95).

Jauncy, D. L., Reynolds, J. E., Tzioumis, A. K., Muxlow, T. W. B., Perley, R. A., Murphy,

D. W., Preston, R. A., King, E. A., Patnaik, A. R., Jones, D. L., Meier, D. L., Bird, D. J., Blair, D. G., Bunton, J. D., Clay, R. W., Costa, M. E., Duncan, R. A., Ferris, R. H., Gough, R. G., Hamilton, P. A., Hoard, D. W., Kemball, A., Kesteven, M. J., Lobdell, E. T., Luiten, A. N., McCulloch, P. M., Murray, J. D., Nicolson, G. D., Rao, A. P., Savage, A., Sinclair, M. W., Skjerve, L., Taaffe, L., Wark, R. M., and White, G. L., 1991, *Nature*, **352**, 132 (J92).

Keaton, C. R., II, and Kochanek, C. S. 1995, in *Proceedings of I.A.U. Symposium #173: Application of Gravitational Lensing* (eds. C. S. Kochanek and J. N. Hewitt), Dordrecht: Kluwer Academic Publishers.

Kochanek, C. S., 1995, *Ap.J.*, **445**, 559.

Kochanek, C. S., and Narayan, R. 1992, *Ap.J.*, **401**, 461.

Langston, G. I., Schneider, D. P., Conner, S., Carilli, C. L., Lehár, J., Burke, B. F., Turner, E. L., Gunn, J. E., Hewitt, J. N., and Schmidt, M. 1989, *A.J.*, **97**, 1283 (L89).

Lawrence, C. R., Schneider, D. P., Schmidt, M., Bennett, C. L., Hewitt, J. N., Burke, B. F., Turner, E. L., and Gunn, J. E., 1984, *Science*, **223**, 46 (L84).

Lehár, J., 1993, in *Gravitational Lenses in the Universe* (eds. J. Surdej, D. Fraipont-Caro, E. Gosset, and M. Remy), Liège: Université de Liège (L93a).

Lehár, J., Langston, G. I., Silber, A., Lawrence, C. R., and Burke, B. F., 1993, *A.J.*, **105**, 847 (L93b).

Magain, P., Surdej, J., Swings, J.-P., Borgeest, U., Kayser, R., Kühr, H., Refsdal, S., and Remy, M., 1988, *Nature*, **334**, 325 (M88).

Magain, P., Surdej, J., Vanderriest, C., Pirenne, B., and Hutsemékers, D., 1992, *Astron. Astroph.*, **253**, L13 (M92a).

Maoz, D., Bahcall, J. N., Schneider, D. P., Doxsey, R., Bahcall, N. A., Filippenko, A. V., Goss, W. M., Lahav, O., and Yanny, B., 1992, *Ap.J.*, **386**, L1 (M92b).

McMahon, R., Irwin, M., and Hazard, C., 1992, *Gemini*, 36, 1 (M92c).

Meylan, G., and Djorgovski, S., 1989, *Ap.J.*, **338**, L1 (M89).

Myers, S. T., Fassnacht, C. D., Djorgovski, S. G., Blandford, R. D., Matthews, K., Neugebauer, G., Pearson, T. J., Readhead, A. C. S., Smith, J. D., Thompson, D. J., Womble, D. S., Browne, I. W. A., Wilkinson, P. N., Nair, S., Jackson, N., Snellen, I. A. G., Miley, G. K., de Bruyn, A. G., and Schilizzi, R. T., 1995, *Ap.J.*, **447**, 5 (M95).

Narayan, R. 1992, *Ap.J.*, **378**, L5.

Patnaik, A., 1993, in *Gravitational Lenses in the Universe* (eds. J. Surdej, D. Fraipont-Caro, E. Gosset, and M. Remy), Liège: Université de Liège (P93a).

Patnaik, A. R., Browne, I. W. A., Walsh, D., Chaffee, F. H., and Foltz, C. B., 1992, *M.N.R.A.S.*, **259**, 1p (P92).

Patnaik, A. R., Browne, I. W. A., King, L. J., Muxlow, T. W. B., Walsh, D., and Wilkinson, P. N., 1993, *M.N.R.A.S.*, **261**, 435 (P93b).

Ratnatunga, K. U., Ostrander, E. J., Griffiths, R. E., and Im, M., 1995, *Ap.J.*, **453**, 5 (R95).

Refsdal, S. 1964, *M.N.R.A.S.*, **128**, 307.

Surdej, J., Magain, P., Swings, J.-P., Borgeest, U., Courvoisier, T. J.-L., Kayser, R., Kellermann, K. I., Kühr, H., and Refsdal, S., 1987, *Nature*, **329**, 695 (S87).

Tinney, C. G., 1995, in *Gravitational Lenses in the Universe* (eds. J. Surdej, D. Fraipont-Caro, E. Gosset, and M. Remy), Liège: Université de Liège (T95).

Walsh, D., Carswell, R. F., and Weymann, R. J. 1979, *Nature*, **279**, 381 (W79).

Weedman, D. W., Weymann, R. J., Green, R. F., and Heckmann, T. M., 1982, *Ap.J.*, **255**, L5 (W82).

Weymann, R. J., Latham, D., Angel, J. R. P., Green, R. F., Liebert, J. W., Turnshek, D. A., Turnshek, D. E., and Tyson, J. A., 1980, *Nature*, **285**, 641 (W80).

Wisotzki, L., Köhler, T., Kayser, R., and Reimers, D., 1993, *Astron. Astroph.*, **278**, L15 (W93).

SCINTILLATING LENSES

M. A. WALKER
Research Centre for Theoretical Astrophysics, A28
School of Physics, University of Sydney, NSW 2006, Australia

1. Introduction

There are now several examples of distant radio sources being gravitationally lensed by foreground galaxies (e.g. PKS1830–211, Jauncey et. al.
1991). In such cases we know that the line of sight must be passing through
a substantial amount of ionized material, so it is pertinent to ask whether
these free electrons influence the observed properties of the source in any
significant way.

2. Smooth Lenses

Firstly the ionized gas in the lensing galaxy/cluster behaves on average as
a smooth lens which alters the appearance of any background radio source.
Hence one expects that the plasma lens ought to be accounted for when
modeling the radio image structure of any gravitationally lensed source. In
practice though, such effects are very weak indeed and the smooth plasma
lens can be ignored in most circumstances. (Just because the contribution
of gravity to the interstellar refractive index is typically much larger than
the free-electron contribution, at observable radio wavelengths.)

This seems to be at odds with the well known influence of the ISM on
radio wave propagation in our Galaxy contrasted with the apparent lack
of any observational consequences of the gravitational field. This situation
has arisen because of time, frequency or polarization dependence in the
plasma's influence, whereas it is well known that gravitational lensing is
achromatic and polarization-independent. However, it is also the case that
time dependence of the lensing geometry is offset by the long-range nature
of the gravitational field. In contrast, the influence of a plasma is completely
localized to those rays which pass through it; changes in the viewing geom-

111

R. Ekers et al. (eds.), Extragalactic Radio Sources, 111–112.
© 1996 *IAU*.

etry then lead to the phenomenon of interstellar scintillation as a result of inhomogeneities in the plasma.

3. Rough Lenses

Quite generally one can think of the inhomogeneities in the plasma as introducing a distorting, transparent screen – a phase screen – between source and observer. We can immediately identify a number of interesting phenomena arising from the presence of such a phase screen: we expect a complex, evolving pattern of magnifications across the screen; magnifications will be different for small vs. large sources – even if they are coincident – for the same source at different wavelengths, and different again for identical sources in different locations; images will be seen at (wavelength dependent) locations slightly different from those which would be recorded in the absence of the phase screen.

Applying scintillation theory to the case of gravitationally lensed extragalactic sources raises the following points concerning the effect of the ISM in our own Galaxy. Extragalactic sources are too great in angular size to exhibit diffractive scintillation, but often show slow refractive scintillations; even then, only the cores of these sources are sufficiently small to show scintillations — jets and lobes should have no significant magnification introduced by the ISM. For multiple images separated by $\sim$ arcseconds, the scintillations should be essentially independent of each other (as this is much larger than the angular scale of the ISM irregularities which are causing the brightness variations). This is a source of noise in establishing the relative time delay between images. If refractive scintillations are significant then there should also be an associated image (scatter) broadening; temporal smearing associated with the image broadening is negligible.

We can extrapolate this theory into the extragalactic realm if we simply assume that the ISM in the lensing galaxy is similar to that in our own Galaxy. (This is probably not realistic as many lenses seem to be elliptical galaxies, whose ISM is much more homogeneous than spirals.) Using this assumption one finds that the refractive scintillations introduced by a lensing galaxy are typically expected to be small (less than 10% for frequencies above 1.4 GHz). Further, the scintillation time-scale is very long – in excess of a thousand years – so the plasma lens can be regarded as static. Angular broadening of compact sources might, however, be measurable.

References

Jauncey D. L. et al. 1991, *Nature*, **352**, 132

A VLA STUDY OF OOTY SMALL DIAMETER SOURCES

T.K.MENON
Department of Geophysics and Astronomy
University of British Columbia
Vancouver, B.C. V6T 1Z4, CANADA

1. Introduction

In the Ooty lunar occultation survey of about 711 sources at 327 MHz over 200 sources were classified as unresolved with upper limits of about 4″ (Kapahi(1986)). In order to study the nature of these sources I have observed a sub-sample of 71 sources with the VLA in the A-Configuration at 6 cm. The angular resolution ranged from 0.4 to 0.8 arc second. My observations show that of the above 71 sources 56% are well resolved doubles and triples resembling FRII sources with angular separations of 0″.4 to 8″. The flux density ratios of well separated doubles and triples range from 1 to about 8. However the LAS of sources with ratios of less than 2 is found to increase with decreasing flux density . Since most of the low flux density sources are in blank fields in the PSS prints no redshift information is available at present. Being FRII sources they are likely to be very distant sources at comologically interesting ranges of redshift. In order to test whether the above observed increase in the angular size is of cosmological significance we need optical identification and redshift information. These are being planned.

The study has provided detailed maps of a large number of extended sources with sizes less than about 4″. This has led to the discovery of 4 gravitational lens candidates among the above sources. One of the most interesting is a ring structure shown in Fig. 1. The basic source is a typical triple source having a total flux density of about 50.9 mJy at 6cm and with about 12% of the total flux density in a central component. As is seen from Fig. 1 there is a spectacular arc structure extending westward from the main source. In addition it is seen that the contours are also distorted with a noticeable hole on the eastern side suggesting that the arc may indeed be a more complete structure but for the presence of the main source.

R. Ekers et al. (eds.), Extragalactic Radio Sources, 113–114.

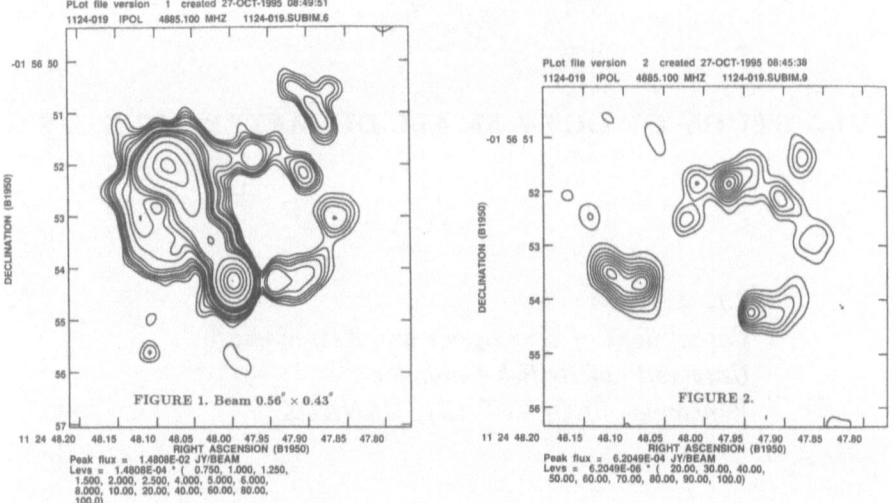

FIGURE 1. Beam 0.56″ × 0.43″

FIGURE 2.

The image of the residual structure, after an approximate subtraction of the main source, shown in Fig. 2 resembles a ring of emission of diameter 3″ with a number of hot spots on the periphery and a bright source at it's eastern end. In the presently known cases of radio ring systems such as MG1131+0456 or 1549+307 (Schneider et al(1992)) the components of an extended background source are suggested to have been imaged by a foreground galaxy into rings and multiple images. However the geometrical situation in the present case appears to be more complex. If the triple source is a background source it is seen as almost undistorted and the center of the ring itself is displaced to the west of the central component of the triple source. Hence it is not clear at present whether the whole image is a superposition of two independent systems one of which is a gravitationally imaged system or the two systems are related by gravitational lensing process. The suggested optical identification for the source is an object which is about 1.5 arc second displaced from the central component towards the southwest. There is at present no information on any other optical object in the vicinity of the radio source. Deeper optical data are being obtained to clarify the situation.

These investigations were supported by a grant from the Natural Sciences and Engineering Council of Canada. National Radio Astronomy Observatory is operated by Associated Universities, Inc., under cooperative agreement with the National Science Foundation.

References

Kapahi, V. 1986, Proc. I.A.U. Symp. 124, A.Hewitt, G. Burbidge, & L.Z. Fang (Eds.), Dordrecht: Reidel, p251

Schneider, P; Ehlers, J; Falco, E.E, 1992, " Gravitational Lensing ", (Springer-Verlag. Berlin)

MULTI-FREQUENCY VLBI OBSERVATIONS OF THE GRAVITATIONAL LENS B0218+357

R. W. PORCAS AND A. R. PATNAIK

Max-Planck-Intitut fuer Radioastronomie, Bonn, Germany

The gravitational lens system B0218+357 comprises 2 image components (A and B) and a radio 'Einstein Ring' (Patnaik et al, 1993). The redshift of the lens galaxy is 0.6847 (Browne et al, 1994) and that of the imaged source 0.96 (preliminary result; Lawrence et al, 1995). The separation of A and B, which are both flat-spectrum radio sources, is only 0.335 arcsec, leading to the hope that the lens is a single galaxy with a relatively simple mass distribution. Refsdal pointed out (1964) that a model of such a distribution, and a measurement of the time difference along the two image paths, leads to an estimate of the Hubble constant, independent of the usual steps in the distance ladder. B0218+357 is one of only a few lensed systems well suited for such measurements. A preliminary value of 12 days has been measured for the A-B time delay, derived from a comparison of the percentage polarisation variations of the images at 15GHz, using the VLA (Corbett et al, 1995).

We have made a detailed study of the mas structures of the A and B images, using a Global VLBI array at 1.6 and 5GHz, and the VLBA at 8.4, 15, 22 and 43GHz (Patnaik et al, 1995, Porcas and Patnaik 1996). At 15GHz both the A and B images are resolved into two components, with a flux ratio between the components which is the same in both. The western component in each image is the more compact, and is identified as the "core". The components in image A (which is ca. 3.5 stronger than B at these frequencies) also exhibit extensions in PA $-40°$; this corresponds to the 'tangential stretching' expected from the geometry of this lens system. We have been able to deduce a relative magnification matrix between these images, and have also shown that the lens potential is non-spherical (Patnaik et al,1995).

At lower frequencies the compact cores become much weaker; both image structures are dominated by extended emission, which increases in size

R. Ekers et al. (eds.), Extragalactic Radio Sources, 115–117.
© 1996 *IAU.*

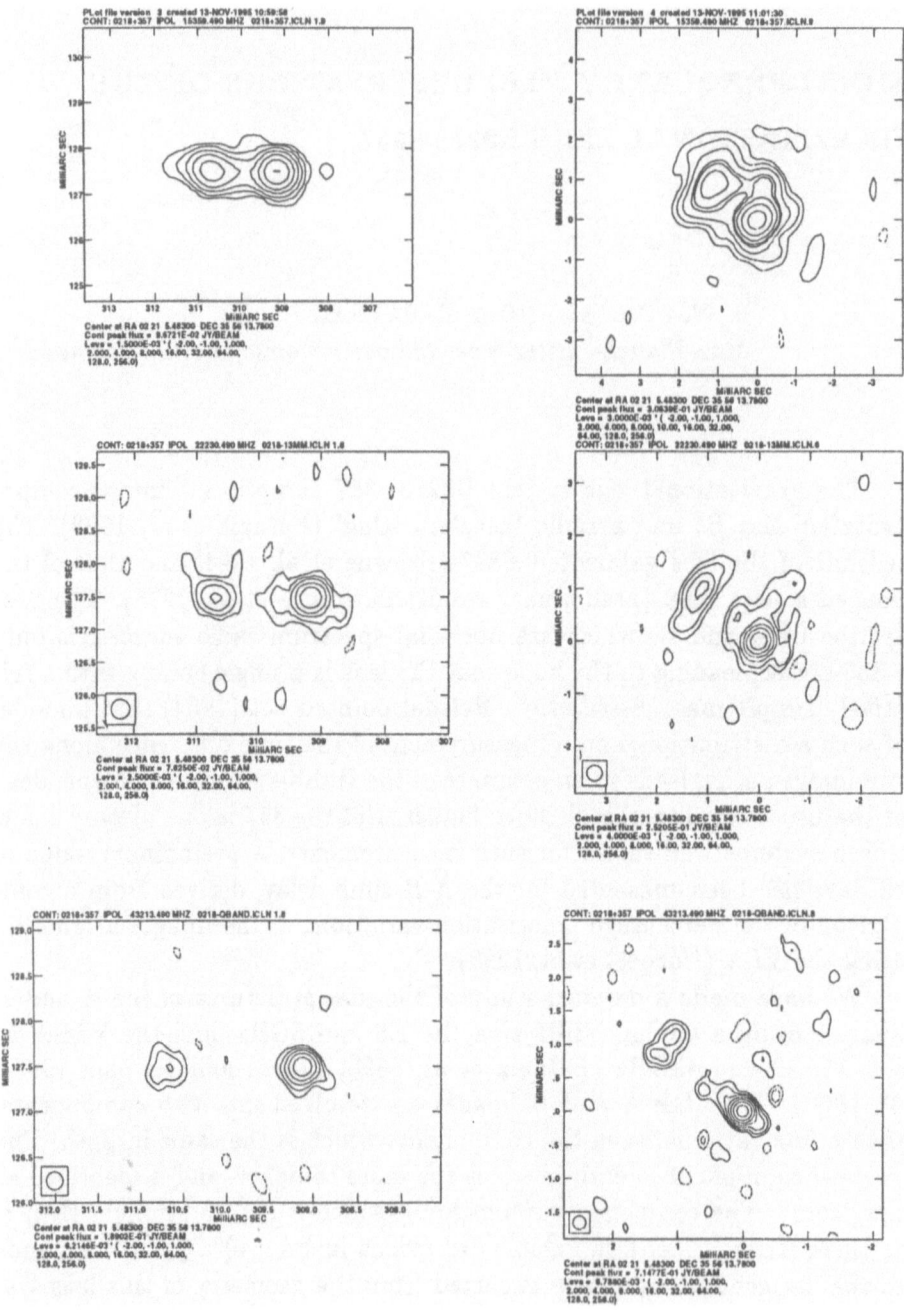

Figure 1.   15 GHz (top), 22 GHz (middle) and 43 GHz (bottom) VLBA maps of B0218+357 made with resolutions 0.5, 0.3 and 0.2 mas, respectively, from observations on 17 July, 1995. The left panel shows image B and the right image A.

with wavelength, but maintains the same directions of extension visible in the higher frequency images. The flux ratio A/B also decreases with wavelength. This can result when the image sizes are not small compared with the scale-size of the lens potential, resulting in gradients of magnification over the extended structure. Nair (1996) has made models of the B0218+357 system which take account of the fact that different parts of the background source are magnified quite differently.

Preliminary results from our latest VLBA observations at 15, 22 and 43 GHz are shown in Figure 1. At 15 GHz the maps of A and B show structures very similar to our previous observations, with two prominent components in each image, and a northwards-pointing extension to the core in A, which has become more pronounced at this later epoch. At 22 GHz and 43 GHz both components are clearly seen in the A and B images, although the core becomes stronger at higher frequencies due to its inverted spectrum, and the northward extension in A (presumably steeper spectrum) becomes weaker. The other component maintains the characteristic elongation in A of PA −40° seen at 15 GHz; it is presumably this component which dominates in the maps of the structures at frequencies of 5 GHz and less in both A and B.

References

Browne, I.W.A. et al, 1993, MNRAS,263, L32

Corbett, E. et al, 1996, Proc IAU Symp. 173,Astrophysical Applications of Gravitational Lensing, ed, Kochanek, C.S. and Hewitt, J.N., Kluwer, (in press)

Lawrence, C. et al,1996, Proc IAU Symp. 173,Astrophysical Applications of Gravitational Lensing, ed, Kochanek, C.S. and Hewitt, J.N., Kluwer, (in press)

Nair, S.,1996, Proc IAU Symp. 173,Astrophysical Applications of Gravitational Lensing, ed, Kochanek, C.S. and Hewitt, J.N., Kluwer, (in press)

Patnaik, A.R. et al, 1993, *MNRAS*, **261**, 435

Patnaik, A.R., Porcas, R.W. and Browne, I.W.A., 1995, *MNRAS*, **274**, L5

Porcas, R.W. and Patnaik, A.R., 1996, Proc IAU Symp. 173,Astrophysical Applications of Gravitational Lensing, ed, Kochanek, C.S. and Hewitt, J.N., Kluwer, (in press)

Refsdal, S., 1964, *MNRAS*, **123**, 307

PKS 1117+146: GRAVITATIONAL LENS OR MICRO LOBES

M. BONDI AND M. GARRETT

NRAL, Jodrell Bank, UK

AND

L. GURVITS

JIVE and NFRA, Dwingeloo, NL
Astro Space Center, Moscow, Russia

PKS 1117+146 is a high power radio source (L_{327MHz}=5.4×10^{26} W/Hz) identified with a galaxy of 20.1 red magnitude at z=0.362 (de Vries et al. 1995). At this redshift 1 mas $\simeq$ 2.9 pc (H_0 = 100 km/s^{-1}Mpc^{-1}). Based on the properties of the radio spectra, PKS 1117+146 is classified as a GigaHertz Peaked Spectrum source (GPS) (Stanghellini et al. 1990). The GPS are powerful but physically small (sub-galactic sizes) radio sources with turnovers in their radio spectra at $\nu \sim$ 1 GHz. They are supposed to be isotropically emitting radio sources confined by exceptional dense circumnuclear gas (O'Dea et al. 1991) or still relatively young (Fanti et al. 1990). PKS 1117+146 is also a low frequency variable (LFV) with no sign of variability at $\nu >$ 1 GHz (Padrielli et al. 1987, Mitchell et al. 1994). The low frequency variability is caused by propagation effects in the interstellar medium of our Galaxy (Mantovani et al. 1990, Spangler et al. 1993). PKS 1117+146 was observed with VLBI global arrays at 608 MHz (Padrielli et al. 1991), at 327 MHz (Altschuler et al. 1995), and at 1667 MHz (Bondi et al. 1996). All the maps are in agreement showing a compact double structure with components separated by about 70 mas. Flux densities and separation of the two components derived from VLBI and MERLIN (see below) maps are listed in Table 1. The flux ratios of the two components from the VLBI observations are very similar, and the spectral index is relatively flat ($\alpha \simeq$ 0.3 − 0.4), even if the strong low frequency variability can introduce uncertainties. The similarity of the VLBI morphology and spectral properties of the two components suggested that 1117+146 could be a possible gravitational lens candidate prompting for higher frequency observations. We observed PKS 1117+146 with MERLIN at 22 GHz in

R. Ekers et al. (eds.), Extragalactic Radio Sources, 118–119.

Frequency MHz	Flux Density SE Comp.	Flux Density NW Comp.	d mas	p.a. degress
327	2.08	1.74	70	-62
608	1.66	1.32	69	-68
1667	1.32	1.03	78	-60
23000	∼ 0.11	∼ 0.06	86	-62

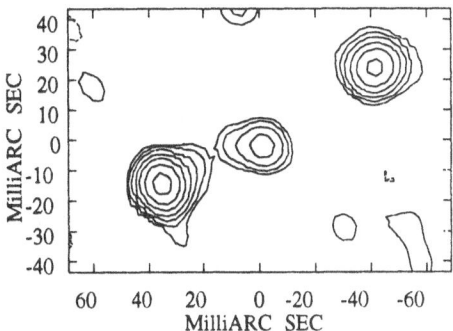

Figure 1.   Merlin 22 GHz map

March 1993. MERLIN observations reveal for the first time a weak central component with a total flux density of about 20 mJy (Fig.1). From Table 1 we can note that the P.A. between the components is constant at all the frequencies while the separation between the peak flux densities significantly increases at higher frequencies. This is the expected behaviour if the 2 components are 2 lobes with hot-spot at the outer edges. The MERLIN map at 22 GHz seems to rule out the possibility that the morphology of PKS 1117+146 is caused by gravitational lensing.

Acknowledgements: MB acknowledges European Union for financial support as an EU Fellow under contract CHBGCT920112.

References

Altschuler, D., et al., 1995, in press
Bondi, M., et al., 1996, in preparation
de Vries, W.H., Barthel, P.D., Hes, R., 1995, *A&A Suppl. Ser.* in press
Fanti, R., et al., 1990, *A&A* **231**,333
O'Dea, C.P., Baum, S., Stanghellini, C., 1991, *ApJ* **380**, 660
Padrielli, L., et al., 1987, *A&A Suppl. Ser.* **67**, 63
Padrielli, L., et al., 1991, *A&A* **249**, 351
Mantovani, F., et al., 1990, *A&A* **233**, 535
Mitchell, K.J., et al., 1994, *ApJ Suppl. Ser.* **93**, 441
Spangler, S.R., et al., 1993, *A&A* **267**, 213
Stanghellini, C., et al., 1990, *A&A* **233**, 379

HIGH PRECISION ASTROMETRY OVER LARGE ANGULAR SCALES WITH CLOSURE CONSTRAINTS: THE TRIPLET 1803+784/1928+738/2007+777

E. ROS[1], J.M. MARCAIDE[1], J.C. GUIRADO[2], T.P. KRICHBAUM[3], R.A. PRESTON[2], M.I. RATNER[4], I.I. SHAPIRO[4] AND A. WITZEL[3]

[1] *Departamento de Astronomía, Universitat de València, Dr. Moliner 50, E-46100 Burjassot (València), Spain*
[2] *Jet Propulsion Laboratory, California Institute of Technology, 4800 Oak Grove Drive, Pasadena, California 91109*
[3] *Max-Planck Institut für Radioastronomie, Auf dem Hügel 69, D-53121 Bonn, Germany*
[4] *Harvard-Smithsonian Center for Astrophysics, 60 Garden St., Cambridge, Massachusetts 02138*

The technique of differential astrometry using the phase-delay VLBI observable promises fractional precisions of $\sim 2 \times 10^{-9}$ in the determination of the separation of sources 5° or 6° apart on the sky (Guirado *et al.* 1995a; Lara *et al.* 1996). In our present research we seek further improvement in this technique through using triplets of radio sources, which provide a closure constraint in the determination of relative angular positions. This constraint not only eases the resolution of the phase-cycle ambiguities (a major problem in the least-squares approach to astrometry with phase delays), but it also strongly constrains the space of allowable parameter values.

The radio sources 1803+784, 1928+738, and 2007+777, hereafter labelled A, B, and C, respectively, were cyclically observed on 20 November 1991 with a 10-antenna VLBI array. These sources are taken from the S5 radio survey (Kühr *et al.* 1981) and are part of an ongoing VLBI monitoring program (e.g. Witzel *et al.* 1988). We have phase-connected the data following a standard procedure (Shapiro *et al.* 1979, Guirado *et al.* 1995a, 1995b) for sets of data corresponding to the pairs of radio sources B-C, C-A, and A-B. For our present analysis, we used only data at λ3.6cm from Bonn,

120

R. Ekers et al. (eds.), Extragalactic Radio Sources, 120–121.
© 1996 *IAU.*

Haystack, VLBA-PT, and VLBA-LA. After the standard phase-connection process we corrected the phase delays for the estimated contribution of the structure of the radio sources, but not for the ionospheric contribution. We have also demonstrated the possibility of estimating simultaneously the three angular separations in a "global" solution, but we have not yet used the full power of demanding zero sky-closure as a constraint in the weighted-least-squares analysis.

We define the sky-closure $C_{\alpha\delta}$ of a triplet of radio sources as a "circular" sum of the estimated relative angular separations for the three pairs of radio sources: $C_{\alpha\delta} = (\sum \Delta\alpha, \sum \Delta\delta)$, where $\Delta\alpha$, $\Delta\delta$ symbolize the differences in right ascensions and declinations for the various pairs of sources. In our case:

$$C_{\alpha\delta} = (\Delta\alpha_{(A-B)} + \Delta\alpha_{(B-C)} + \Delta\alpha_{(C-A)}, \Delta\delta_{(A-B)} + \Delta\delta_{(B-C)} + \Delta\delta_{(C-A)})$$

Preliminary analyses of the data yield the following angular separations (estimated independently for each pair of sources) and statistical standard errors:

$$(\Delta\alpha_{(A-B)}, \Delta\delta_{(A-B)}) = (-1^h27^m02^s.81069 \pm 0^s.00002, \quad 4°30'02''.4481 \pm 0''.0001).$$
$$(\Delta\alpha_{(B-C)}, \Delta\delta_{(B-C)}) = (-0^h37^m42^s.50316 \pm 0^s.00002, -3°54'41''.6773 \pm 0''.0001).$$
$$(\Delta\alpha_{(C-A)}, \Delta\delta_{(C-A)}) = (\quad 2^h04^m45^s.31386 \pm 0^s.00002, -0°35'20''.7707 \pm 0''.0001).$$

$C_{\alpha\delta}$ should be consistent with zero for both coordinates. The above results yield:

$$C_{\alpha\delta} = (0^h0^m0^s.00001, 0°0'0''.0001)$$

The relative separation obtained for the pair 1928+738/2007+777 at 3.6 cm is compatible with the preliminary results of the 3.6/13 cm observations from 1988 (Elósegui 1991) and with the final result of the 6 cm observations from 1985 (Guirado *et al.* 1995a). Minor differences between the solutions are possibly due to opacity effects between 3.6 and 6 cm, and, perhaps also, to small changes in the reference features used in the maps.

References

Elósegui, P., 1991, Ph.D. thesis, Universidad de Granada.
Guirado, J.C. et al., A&A, **293**, 513, (1995a)
Guirado, J.C. et al., AJ, *(in the press)*, (1995b)
Kühr, H. et al., AJ, **86**, 854, (1981)
Lara, L. et al., A&A, *(in the press)*, (1996)
Shapiro, I.I. et al., AJ, **84**, 1459, (1979)
Witzel, A. et al., A&A, **206**, 245, (1988)

EVN PHASE-REFERENCED OBSERVATIONS OF 1308+328

MARÍA JOSÉ RIOJA
JIVE, Dwingeloo, Netherlands

RICHARD W. PORCAS
Max-Planck-Intitut fuer Radioastronomie, Bonn, Germany

AND

JERZY MACHALSKI
Jagellonian University, Cracow, Poland

Machalski and Engels (1994) have drawn attention to the recent outburst in the extragalactic radio source 1308+328. We observed it with VLBI at 2.3 and 8.4 GHz in February 1995, using the EVN, and including antennas at Seshan, Kashima, Hartebeesthoek and Ny-Alesund, all of which provided long baselines. During these observations we switched every few minutes between 1308+328 and the nearby compact source 1308+326, 14 arcmin away. This allowed us to determine a precise relative separation between these sources at both frequencies using phase referencing techniques, and provided a useful comparison between source sizes. We found that the target source, 1308+328, is considerably more compact than the reference source 1308+326, the latter being resolved (correlated flux decreased by 50 %) at 8.4 GHz at 250 million wavelengths. In contrast, 1308+328 appears completely unresolved at this resolution (Fig 1), corresponding to a source size < 0.5 mas.

We have used both AIPS and VLBI3 software to make an astrometric analysis. In AIPS we subtract the interpolated residual phase solution of 1308+326 from that of 1308+328 and estimate the angular separation from the phase-referenced map of 1308+328 (Fig.1) In VLBI3 we use a full theoretical geophysical and relativistic model and a least squares fit to the total visibility phases of both sources. A preliminary estimate of the angular separation from VLBI3, using group delays alone, is given below.

$\Delta\alpha = 30.7389$ s ± 0.0005 s, and $\Delta\delta = 12.84444' \pm 0.00002'$
The values derived from AIPS and VLBI3 are in agreement.

R. Ekers et al. (eds.), Extragalactic Radio Sources, 122–123.
© 1996 *IAU.*

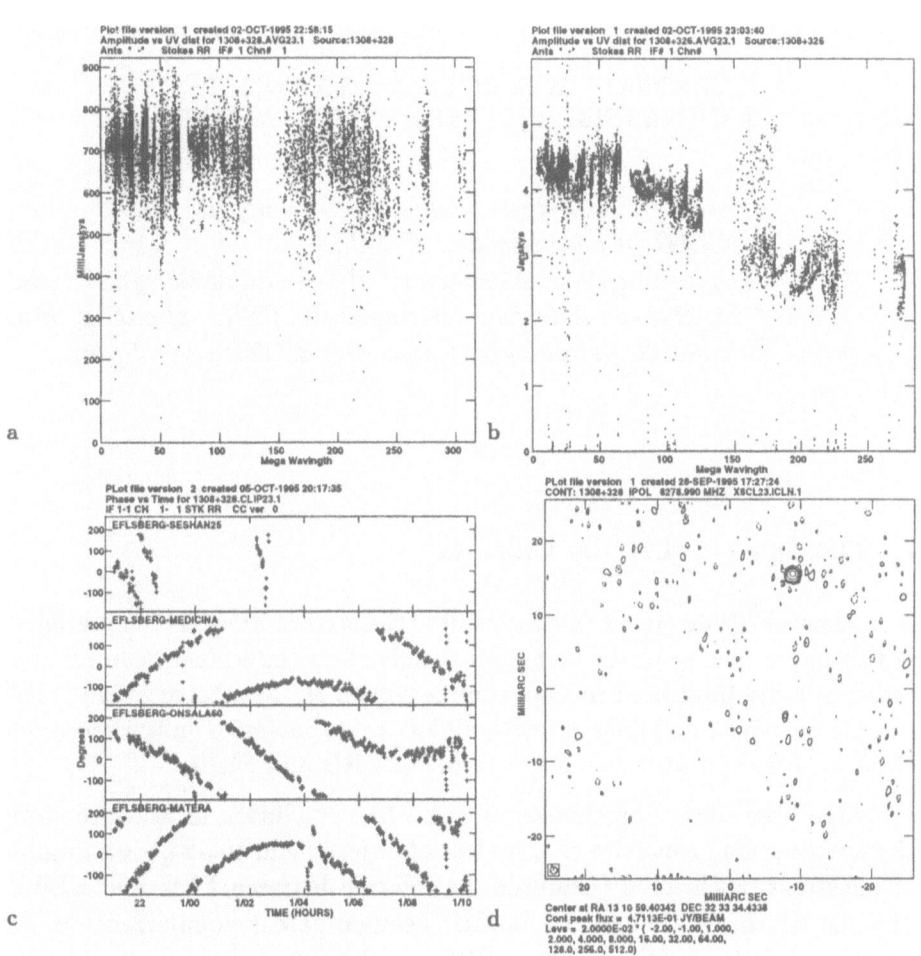

Figure 1. Results from EVN observations at 8.4 GHz: **(a)** Visibility amplitude vs. uv distance for 1308+328; **(b)** the same for 1308+326; **(c)** visibility phase of 1308+328 relative to 1308+326; **(d)** phase-referenced map of 1308+328

References

Machalski, J. and Engels, D., 1994, *MNRAS*, **266**, L69

VLBI OBSERVATIONS OF FRI RADIO GALAXIES

T. VENTURI [1], W.D. COTTON [2], L.FERETTI [1,3],
G. GIOVANNINI [1,3], L. LARA [1,4] AND J.M. MARCAIDE [5]

[1] Istituto di Radioastronomia, CNR- Bologna, Italy
[2] NRAO - Charlottesville, VA, USA
[3] Dipartimento di Astronomia, Università di Bologna - Italy
[4] Instituto de Astrofisica de Andalucia, CSIC - Granada, Spain
[5] D.pto de Astronomia, Universitat de Valencia - Spain

1. The Sample of Radio Galaxies

The Fanaroff-Riley type I radio galaxies (Fanaroff & Riley, 1974) presented in this paper belong to the complete sample of low-intermediate luminosity radio galaxies published in Giovannini, Feretti & Comoretto (1990). This sample includes radio galaxies with different morphologies on the arcsecond scale, such as compact sources, core-halos, FRIs and FRIIs.

We have been observing this sample at VLBI resolution in order to study the parsec-scale properties of this class of objects, and to address a number of questions, such as for example the relation between FRIs and BL-Lac objects, which are thought to be their beamed parent population; the differences and similarities between FRIs and FRIIs on the parsec-scale; the link between the point-like sources in the sample and those in the same power range, but with extended arcsecond scale morphology. The results obtained thus far are summarised in Giovannini et al., 1995.

In this paper we'll concentrate on some of the FRIs in the sample. New 5 GHz VLBI maps will be presented and discussed for the three radio galaxies 3C31, 4C35.03, 3C264. Furthermore we will introduce second epoch observations carried out for four FRIs in the sample, i.e. NGC315, B2 0836+29, 3C338 and 3C465.

R. Ekers et al. (eds.), Extragalactic Radio Sources, 124–126.

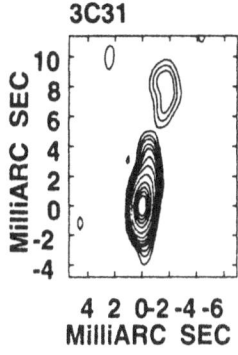

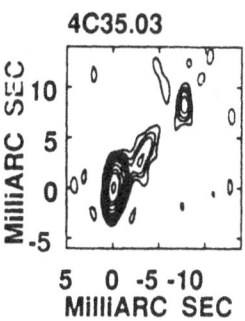

Figure 1. 5 GHz global VLBI contour map of 3C31. Restoring beam is 3.13 × 2.09 mas, in p.a. 22.3°. Peak flux 69.2 mJy/beam, levels -0.4, 0.4, 0.7, 1, 1.5, 2, 3, 5, 10, 20, 30, 40, 50 mJy/beam.

Figure 2. 5 GHz global VLBI contour· map of 4C35.03. Restoring beam is 2.67 × 0.91 mas, in p.a. -4.9°. Peak flux 67.3 mJy/beam, levels -0.4, 0.4, 0.7, 1, 1.5, 2, 3, 5, 10, 20, 30, 60 mJy/beam.

2. Parsec-scale Properties of the New FRI Radio Sources

3C31, identified with the elliptical galaxy NGC383, is a classical FRI. Its arcsecond scale morphology shows an unresolved core and two asymmetric opposite jets. The 5 GHz VLBI map (Fig. 1) shows a strong component, likely to be the core of the radio emission, and a one-sided jet, aligned with the main arcsecond-scale jet. Assuming that the strongest VLBI component is the core, from our map we derive a ratio $R \gtrsim 16$ between the jet to counterjet brightness in the proximity of the core, which leads to a value $\beta\cos\theta \gtrsim 0.504$ ($\beta = v/c$ and θ is the viewing angle).

4C35.03 is identified with the optical galaxy UGC 1651. The source is a classical double on the large scale while the 5 GHz VLBI map (Fig. 2) shows a strong component and a *blobby* jet, pointing towards the main large scale jet. For this source from our map we derive $R \gtrsim 7$ and $\beta\cos\theta \gtrsim 0.371$.

3C264 is optically identified with NGC3862. An optical jet, in position angle $\sim 280°$ was discovered by HST observations (Crane et al., 1993). The large scale radio structure is characterised by a strong core and a distorted northern jet. The 5 GHz parsec-scale morphology (Fig. 3) shows a core and a one-sided jet aligned with the main large-scale jet (and with the optical jet). From our VLBI map we derive $R \gtrsim 28$ and $\beta\cos\theta \gtrsim 0.562$.

The three FRIs presented in this paper show the same properties on the parsec-scale as most of the other FRIs in our sample (see Giovannini et al., 1994; Venturi et al., 1995). Their morphology on this scale is typically one-sided. The constraints derived for the intrinsic Lorentz factor and for

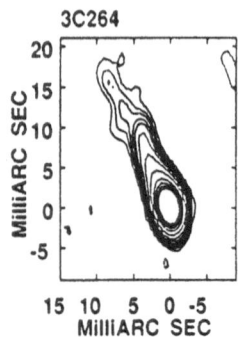

Figure 3. 5 GHz global VLBI contour map of 3C264. Restoring beam is 3.5 × 2.2 mas, in p.a. 10.8°. Peak flux 134.7 mJy/beam, levels -0.4, 0.4, 0.7, 1, 1.5, 2, 3, 5, 10, 20, 30, 40, 50 mJy/beam.

the viewing angle, obtained assuming that Doppler boosting is responsible for the parsec-scale asymmetry and taking into account the constraints imposed by the P_{core}/P_{tot} correlation (Giovannini et al., 1994) are in agreement with the idea that FRI jets start out relativistic, and that FRIs could be the unbeamed population of BLLac objects.

3. Monitoring of FRI Radio Galaxies

The high intrinsic plasma speeds derived for the FRIs in our sample (see also Giovannini et al., 1995) suggest that proper motion of components along the jets should be detectable. In order to test this issue we started a project aiming at the monitoring of the FRIs in our sample. For NGC315, B2 0836+29, 3C338 and 3C465 a 5 GHz second epoch map is already available.

A possible proper motion was found only in the nucleus of 3C338 (see Giovannini et al., this volume). This would confirm the existence of bulk relativistic motion in this source.

References

Crane P., Peletier R., Baxter D., Sparks W.B. et al., *ApJ* **402**, 37

Fanaroff B.L. & Riley J.M., 1974, *MNRAS* **250**, 198

Giovannini G., Feretti L. & Comoretto G., 1990 *ApJ* **358**, 159

Giovannini G., Feretti L., Venturi T., Lara L., Marcaide J.M., Rioja M.J., Spangler S.R., Wehrle A.E., 1994 *ApJ* **435**, 116

Giovannini G., Cotton W.D., Feretti L., Lara L., Venturi T., Marcaide J.M., 1995, in *Quasars and AGN: High Resolution Radio Imaging*, National Academy of Sciences, in press

Venturi T., Castaldini C., Cotton W.D., Feretti L., Giovannini G., Lara L., Marcaide J.M., Wehrle A.E., 1995, *ApJ* in press (Dec. 1st issue)

PARSEC SCALE PROPERTIES OF LOW POWER RADIO GALAXIES

G. GIOVANNINI[1,2], W.D. COTTON[3], L. LARA[1,4],T. VENTURI[1]

[1] *Istituto di Radioastronomia, CNR, Bologna, Italy*
[2] *Dipartimento di Astronomia, Universita' di Bologna, Italy*
[3] *NRAO, Charlottesville, VA, USA*
[4] *Instituto de Astrofisica de Andalucia, CSIC, Granada, Spain*

1. Radio Morphology and Jet Properties

We shortly discuss here the parsec scale information presently available on the extended low power radio galaxy (FR I, Fanaroff and Riley, 1974) NGC6251 (Jones and Wehrle, 1994) and on 11 radio galaxies from the sample presented in Giovannini et al. (1990).

We have used the observational data to derive constraints on the jet velocity and orientation with respect to the line of sight and will use these values to test the predictions of the unified scheme models (see also Venturi et al., this volume).

With the exception of 3C338, (a two-sided symmetric parsec scale jet), and of 1144+35 (a complex structure), we have always an asymmetric parsec scale morphology (core emission and a one-sided jet) where the nuclear emission is always the dominant component. The one-sided parsec scale jets do not show large bends and are always oriented as the main kpc scale jet. This result, coupled with the constraints derived from the jet/counter jet brightness ratio (see also Venturi et al., this volume and Wrobel et al., this volume) and the radio core prominence (Giovannini et al. 1994 and 1995) is consistent with the expectations of unified models. In fact, in the FR I radio galaxies discussed here, observational data are in agreement with the presence of a parsec scale jet with an intrinsic Lorentz factor $\gamma \gtrsim 2$ viewed at angles larger than 30°, even if in a few cases we cannot exclude smaller

127

R. Ekers et al. (eds.), Extragalactic Radio Sources, 127–128.
© 1996 *IAU.*

angles with a lower jet velocity. This is consistent with FR I galaxies being the parent population of BL-Lac objects.

2. Proper Motion

The high jet velocity derived in the previous section, suggests that proper motion should be detectable in parsec scale jet of FR I radio galaxies, if these sources are monitored with a time gap of 1 - 2 years. However a few FR I galaxies have presently at least two observations at different epochs useful to look for the existence of a proper motion while a proper motion is firmly established in strong radio galaxies, quasars and BL-Lac objects (Vermeulen, this volume; Zensus, this volume).

The galaxy 3C274 (M87) shows evidence of stationary knots as well as structures moving at sub-relativistic velocities. Moreover, some sub-structures seem to have a superluminal speed (Biretta et al., 1995). NGC6251 could have both a stationary knot and one moving at v $\sim$ 1.2c (Jones and Wehrle, 1994). The B2 galaxy 1144+35 has a complex parsec scale structure with two unresolved flat spectrum regions which are moving one with respect to the other with an apparent velocity of 1.2c (Giovannini et al., 1995). The symmetric source 3C338 shows a clear change in its radio structure and present data suggest a possible motion corresponding to an apparent velocity of 0.5c (Giovannini et al., 1995).

To better investigate this point, we obtained new maps of 3 FR I radio galaxies: NGC315, 4C29.30 and 3C465. The comparison between the two different epoch maps shows that no proper motion is visible in any of these 3 sources. The derived velocity upper limits are 0.1c, 0.7c and 0.3c for NGC315, 4C29.30 and 3C465, respectively. In this respect the lack of visible motion in FR I parsec scale jets, which are expected to move at a relativistic velocity, could reflect the presence of oblique shocks and/or more complex situations where the apparent knot velocity could be much lower than the jet velocity. This result could suggest that despite the relativistic velocity present in low and high power parsec scale jets, an intrinsic difference in the jet dynamic, could be present.

A Hubble constant $H_0 = 100$ km s^{-1} Mpc^{-1} was used in this work.

References

Biretta, J.A., Junor, W. (1995) *Proc. Natl. Acad. Sci. USA*, **92**, in press
Fanaroff, B.L., Riley, J.M. (1974) *Month. Not. R. Astr. Soc.*, **250**, 198
Giovannini, G., Feretti, L., Comoretto G. (1990) *ApJ* **358**, 159
Giovannini, G., Feretti, L., Venturi, T., Lara L., et al. (1994) *ApJ* **435**, 116
Giovannini, G., Cotton, W.D., Feretti, L. et al. (1995) *Proc. Natl. Acad. Sci.* **92**, in press
Jones, D.L., Wehrle, A.E. (1994) *ApJ* **427**, 221

VLBI OBSERVATIONS OF LOW-REDSHIFT RADIO GALAXIES

S.J. TINGAY[1], D.L. JAUNCEY[2], R.A. PRESTON[3], D.L. MEIER[3],
J.E. REYNOLDS[2], A.K. TZIOUMIS[2], J. LOVELL[4], D.L. JONES[3],
P.M. MCCULLOCH[4], D.W. MURPHY[3] AND G.D. NICOLSON[5]

[1] *Mount Stromlo Observatory, Canberra, ACT 2611, Australia*
[2] *ATNF, Epping, NSW 2121, Australia*
[3] *Jet Propulsion Laboratory, Pasadena, CA 91109, USA*
[4] *University of Tasmania, Hobart, Tasmania 7001, Australia*
[5] *HRAO, Krugersdorp 1740, South Africa*

1. Low-redshift, compact radio sources

Here we will describe briefly some of the VLBI observations we are making
of low-redshift, compact radio sources in the southern hemisphere, using
the Southern Hemisphere VLBI Experiment (SHEVE) array of telescopes
(Jauncey et al., 1994).

2. PKS 0521-365

PKS 0521-365 (z=0.055) is a powerful radio source, with a strong radio core,
extended structure in the form of a radio jet and hot spots (Keel, 1986).
Coincident with the radio jet is an optical jet which has been observed with
HST (Macchetto et al., 1991). PKS 0521-365 is the second lowest redshift
EGRET identified radio source. Our VLBI observations (e.g. Figure 1) show
a strong, unresolved core component and a pc-scale jet which aligns well
with the kpc-scale jet. For more details see Tingay et al., ApJ submitted.

3. PKS 1718-649

PKS 1718-649 (z=0.014) may be the lowest redshift GPS radio source. Recent radio spectra show a turnover between 2.3 and 4.8 GHz. We have
observed low flux density variability and low radio polarisation. Our VLBI

129

R. Ekers et al. (eds.), Extragalactic Radio Sources, 129–130.

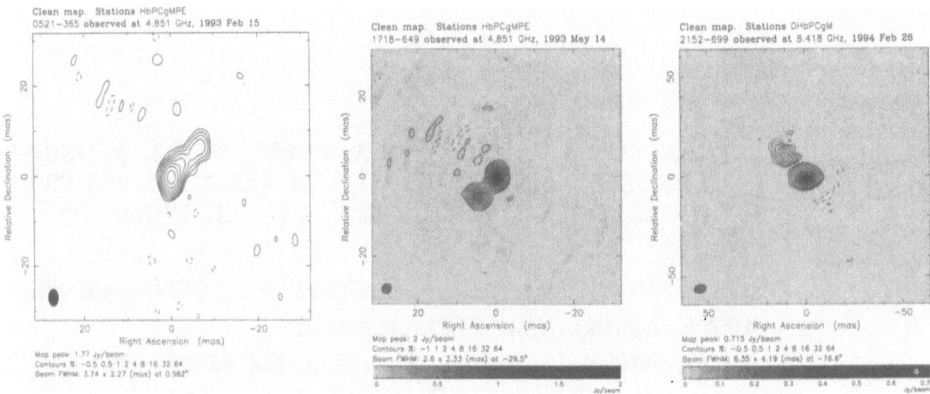

Figure 1. (left) SHEVE VLBI image of PKS 0521-265. (centre) SHEVE VLBI image of PKS 1718-649. (right) SHEVE VLBI image of PKS 2152-699

observations (e.g. Figure 1) show that the total flux density of the source is due to two compact components each approximately 0.5 pc in extent (Tingay et al., in prep). From the HI observations of Véron Cetty et al. (1995) and the optical observations of Filippenko (1985) evidence can be found for confinement of the radio source by a dense, merger-induced environment.

4. PKS 2152-699

PKS 2152-699 (z=0.028) is a powerful FR-II type radio source with an extra-nuclear emission line region which strongly resembles a typical nuclear emission line region (Tadhunter et al., 1988). Our VLBI observations (Figure 1) show that the pc-scale radio jet in this source is exactly aligned with the extra-nuclear emission line region. The jet is likely to be interacting with the galaxy in this region. For a more detailed discussion see Tingay et al., 1996, AJ, in press.

References

Filippenko, A.V., 1985, *ApJ*, **289**, 475
Jauncey, D.L. et al., 1994, in Very High Angular Resolution Imaging (Eds J.G. Robertson and W.J. Tango), p 131
Keel, W.C., 1986, *ApJ*, **302**, 296
Macchetto, F. et al., 1991, *ApJ*, **369**, L55
Tadhunter, C.N. et al., 1988, *MNRAS*, **235**, 403
Véron-Cetty, M.P. et al., 1995, *A&A*, **277**, L79

THE RELATIVISTIC JET IN M84

J.M. WROBEL AND R.C. WALKER

NRAO, Socorro, New Mexico 87801-0387, USA

AND

A.H. BRIDLE

NRAO, Charlottesville, Virginia 22903-2475, USA

1. Introduction

The elliptical galaxy M84 (NGC 4374, UGC 07494) hosts an FR-I radio continuum source (Laing & Bridle 1987, MNRAS, 228, 557) and a dusty, warped optical emission line "disk" (Baum *et al.* 1988, ApJS, 68, 643; Goudfrooij *et al.* 1994, A&ApS, 105, 341). HST imaging shows that the inner dust distribution is not relaxed, but filamentary and complex (Jaffe *et al.* 1994, AJ, 108, 1567). M84 is a member of the Virgo Cluster, at which distance 1 arcsec = 73 pc independent of Hubble's constant (Jacoby *et al.* 1990, ApJ, 356, 332). Our VLA imaging, at 6 cm with a resolution of 500 mas (36 pc) FWHM, shows that the two large-scale jets in M84 are initially asymmetric: the ratio R_I of the intensity of the northern main jet to the southern counterjet exceeds unity. However, these jets symmetrize ($R_I \sim 1$) by a projected nuclear offset $r_p \sim 13$ arcsec (950 pc). Is this VLA symmetrization trend consistent with Doppler boosting in a decelerating, but initially relativistic, jet? To test this, we obtained preliminary VLBA images of M84 at 18 cm and 4 cm, with resolutions of 11 mas (0.80 pc) and 1.9 mas (0.14 pc), respectively, and used them to evaluate R_I on pc scales for comparison with R_I on 100-pc (VLA) scales.

2. VLBA/VLA Symmetrization Trends: $R_I(r_p)$

Fig. 1 shows how R_I in M84 depends on r_p, measured relative to the peaks of the bright VLBA or VLA "cores". On pc (VLBA) scales, R_I is from peak intensities read directly from the digital images every 3 pixels N and S of the nucleus, starting at $r_p = 18$ mas (1.3 pc) at 18 cm and $r_p = 3$ mas (0.22 pc) at 4 cm. On 100-pc (VLA) scales, R_I is the ratio of the areas

131

R. Ekers et al. (eds.), Extragalactic Radio Sources, 131–132.

under the transverse intensity profiles of the main jet and counterjet at matched distances N and S of the nucleus. Profile areas were derived from slices in the E-W direction spaced every 6 pixels (480 mas or 35 pc). These R_I values (i) differ from on-axis ones because the main jet appears wider than the counterjet and (ii) are not sensitive to jet "wings".

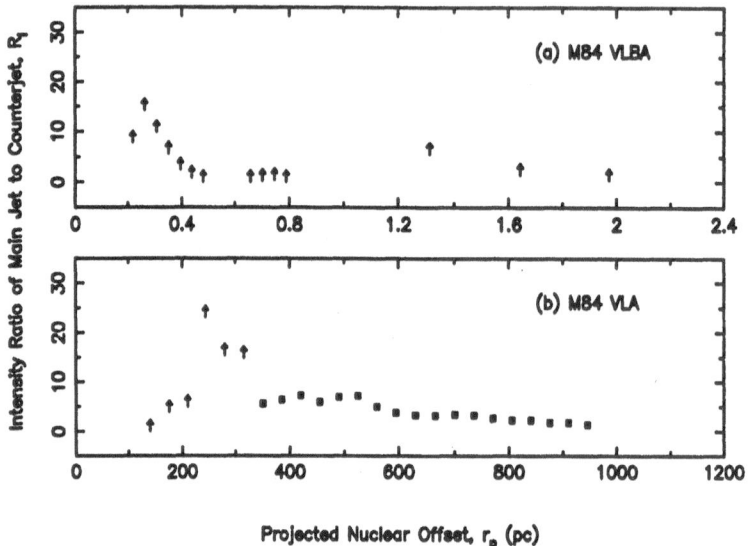

Figure 1. $R_I(r_p)$ in M84.

3. Interpretation: $R_I(\beta_{fluid})$

If R_I differs from unity only through Doppler boosting in identical and opposed relativistic jets, then it will depend on the jet fluid velocity $\beta_{fluid} = v_{fluid}/c$ and on the line-of-sight viewing angle θ. Although θ is unknown for M84 it should be approximately constant with r_p. Within this framework, Fig. 1 implies: (a) $R_I > 15$ at $r_p \sim 0.3$ pc and $r_p \sim 300$ pc, requiring $\beta_{fluid} > 0.49$ and $\theta < 60°$ at these locations. (b) R_I drops gradually between $r_p \sim 350$ pc and $r_p \sim 950$ pc, as expected if twin relativistic jets smoothly decelerate. (c) R_I drops rapidly between $r_p \sim 300$ pc and $r_p \sim 350$ pc, suggesting an abrupt slowing of the jet at this distance. Indeed, the main jet morphology at $r_p \sim 300$ pc resembles an oblique shock. Furthermore, although the main jet is not initially perpendicular to the dust layer seen by HST, a clockwise deflection of the jet at $r_p \sim 300$ pc brings it into closer alignment with the normal to the layer. This hints that the optical line-emitting gas and the dust in M84 trace the medium initially entrained across the jet's boundary layer, both abruptly slowing the jet and deflecting it toward the normal to the gas/dust layer.

THE PARSEC-SCALE NUCLEUS AND JETS OF HYDRA A

GREGORY B. TAYLOR

NRAO

P.O. Box 0, Socorro, NM 87801

1. Introduction

Sensitive, high-resolution VLBA observations of the nuclear region of Hydra A are presented at 1.3, 5 and 15 GHz. Hydra A (3C218) is an outstanding example of a high-luminosity FRI radio galaxy embedded within a cooling flow cluster. VLA observations by Taylor & Perley (1993) have demonstrated extremely high (>5000 radians m^{-2}) Faraday rotation measures (RMs) and a striking RM and depolarization asymmetry between the northern and southern radio lobes. In view of this asymmetry on the kpc-scale Hydra A appears remarkably symmetric on the pc-scale in the radio continuum. Hydra A is also unusual in that the 21 cm atomic Hydrogen line is seen in absorption against the nucleus.

2. Observations

The observations were carried out using the 10 element VLBA of the NRAO[1] on 1995 March 17-18. Both right and left circular polarizations were recorded using 2 bit sampling across a total bandwidth of 8 MHz. The VLBA correlator produced 512 frequency channels in each 4 second integration. The total integration time was ~2.5 hours at each frequency. Amplitude calibration for each antenna was derived from measurements of the antenna gain and system temperature during each run. In addition, the strong calibrator DA193 (0552+398) was observed to refine the amplitude calibration. All editing, imaging, deconvolution, and self-calibration were performed using DIFMAP (Shepherd *et. al*, 1995).

---

[1] The National Radio Astronomy Observatory is operated by Associated Universities, Inc., under cooperative agreement with the National Science Foundation

R. Ekers et al. (eds.), Extragalactic Radio Sources, 133–135.
© 1996 *IAU.*

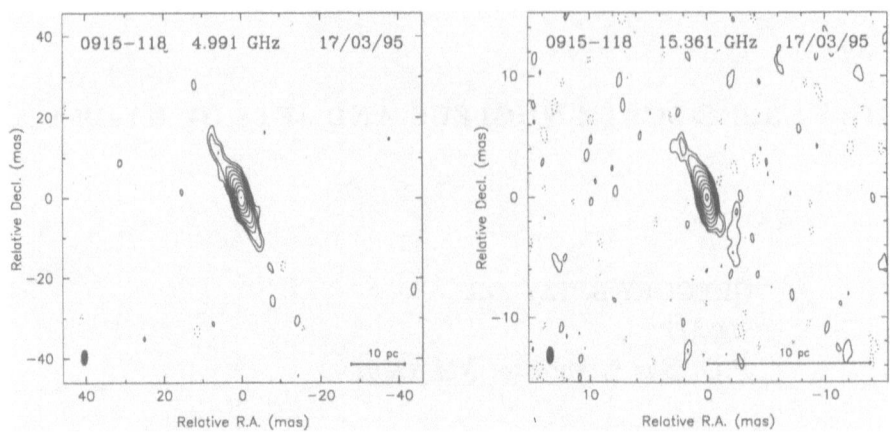

Figure 1.   a. The nucleus of Hydra A at 5 GHz. Contours are drawn at −0.3, 0.3, 0.6, 1.2, 2.4, 4.8, 9.6, 19.2, 38, 77, and 154 mJy/beam where the beamsize is 3.66 × 1.53 mas in p.a. −3°. b. The nucleus of Hydra A at 15 GHz. Contours are at −0.5, 0.5, 1, 2, 4, 8, 16, 32, 64, and 128 mJy/beam where the beamsize is 1.38 × 0.58 mas in p.a. −1°.

3. Results and Discussion

Fig. 1a shows the 5 GHz VLBA image of the nucleus of Hydra A at ∼2 mas resolution. The jet is straight along a position angle of 30° and symmetric about the core with a jet-to-counterjet ratio of 1.12. The northern side, being slightly stronger, is denoted the "jet" side, and the weaker southern jet the "counterjet" side. The jet-to-counterjet ratios are taken from the integrated flux ratios after removal of the core component. Fig. 1b shows the 15 GHz VLBA image at a resolution of ∼1 mas. The inner jet shown in this image is oriented along a position angle of 23°, and gradually curves to match the orientation angle of 30° seen at 5 GHz. The jet-to-counterjet ratio at 15 GHz is 1.15.

Hydra A is only the second FRI source, the first being 3C338 (Feretti *et. al*, 1993), discovered to have symmetric emission on the pc-scale. If all jets start out relativistic and the jet-to-counterjet ratios are ascribed purely to Doppler beaming effects, then such low jet-to-counterjet ratios are expected only for sources very close to the plane of the sky. Observations of the large-scale RM asymmetries in Hydra A predict an inclination angle of 48°(Taylor & Perley, 1993). While this discrepancy can be explained by a large bend between the inner jet and the lobes, this seems unlikely. On the kpc scale, the jet-to-counterjet ratio at 5 GHz is 1.9.

HI was first detected in absorption towards the nucleus of Hydra A by G. Taylor in 1991. Dwarakanath *et. al* (1995) made higher resolution VLA observations confirming this result, and suggested that the HI gas is distributed in a disk within the central few kpc of the galaxy. Here I

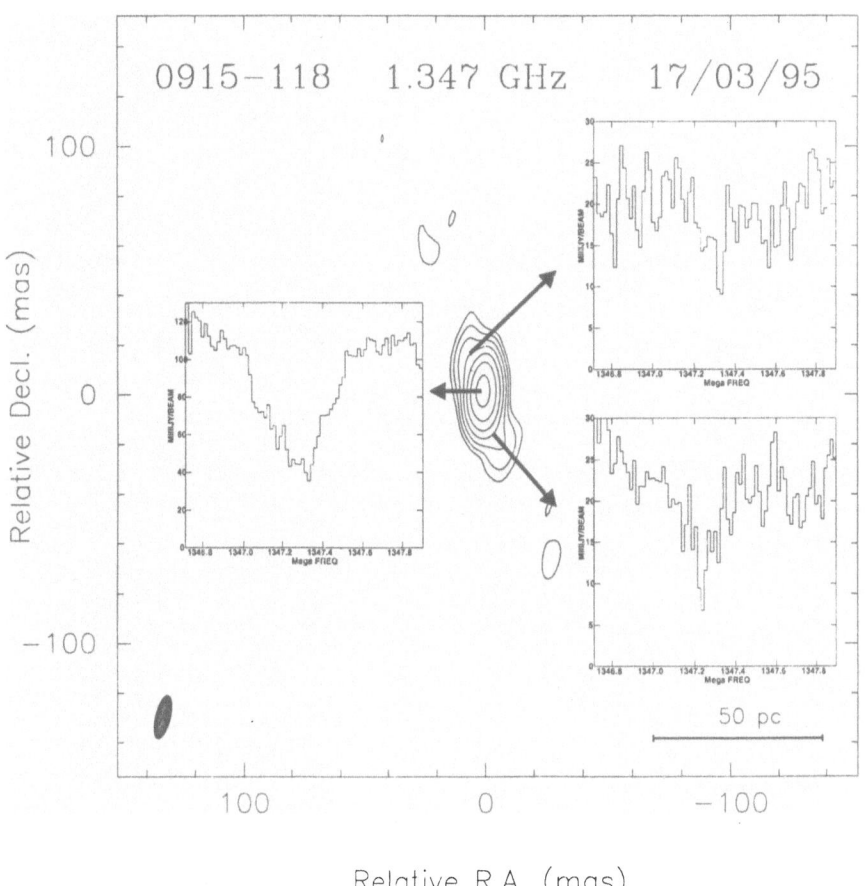

Figure 2. HI absorption against the nucleus of Hydra A. Contours are at -1, 1, 2, 4, 8, 16, 32, and 64 mJy/beam with a beam size of 18×6 mas in p.a. $-14°$. The total velocity spanned by the inset figures is 250 km s^{-1} with a spectral resolution of 3.5 km s^{-1}.

present spatially resolved HI absorption detected against the core and jets of Hydra A (Fig. 2). The two spectra from the northern and southern jet are separated by 23 mas (17 pc), and show absorption peaks shifted by 20 km s^{-1}. The absorption to both the north and south is also considerably more narrow ($\sim$35 km s^{-1}) than towards the core ($\sim$80 km s^{-1}). These observations support the presence of a disk with a thickness of $\sim$20 pc.

References

Dwarakanath, K.S., Owen, F.N., & van Gorkom, J.H. (1995) *ApJL*, **442**, pp. L1-L4
Shepherd, M.C., Pearson, T.J., & Taylor, G.B. (1995) *BAAS*, **27**, pp. 903-907
Taylor, G.B., & Perley, R.A. (1993) *ApJ*, **416**, pp. 554-562
Feretti, L., Comoretto, G., Giovannini, G., Venturi, T., & Wehrle, A.E. (1993) *ApJ*, **408**, pp. 446-451

LOW LUMINOSITY RADIO GALAXIES

P. PARMA
Istituto di Radioastronomia del CNR
via Gobetti, 101, 40129 Bologna

H.R. DE RUITER
Osservatorio Astronomico di Bologna
via Zamboni, 33, 40126 Bologna

AND

R. FANTI
Dipartimento di Fisica, Università di Bologna
via Irnerio, 46, 40126 Bologna

1. Introduction

This review talk will deal almost exclusively with the radio aspects of low luminosity radio galaxies, as the other wavelength bands (optical, X, IR) are covered by others during this conference. By low radio luminosity we mean a radio power at 20 cm in the range 10^{19} to $10^{24.5}$ W Hz^{-1} (Hubble constant of 100, as throughout this talk). At the upper limit occurs the "break" in the radio luminosity (RLF) (Auriemma et al 1977) as well as the transition from Fanaroff -Riley type I to type II (Fanaroff & Riley, 1974). This limitation to weak radio powers does not necessarily imply that such radio sources constitute a homogeneous class of objects; it has been shown by Wrobel & Heeschen (1991) that a number of radio galaxies of very low luminosity have a ratio between far-infrared and radio emission similar to spiral galaxies and this might suggest that these objects are powered by star formation phenomena. Long baseline interferometry is able to pinpoint compact radio nuclei and therefore can distinguish starburst related radio sources from those powered by nuclear activity (see Slee et al 1994).

The aspects of low and high luminosity radio sources are very different, in fact so much so that the classification scheme proposed by Fanaroff and Riley (1974) has often been thought to reflect fundamental physical differ-

137

R. Ekers et al. (eds.), Extragalactic Radio Sources, 137–142.
© 1996 *IAU.*

ences. FR I and FR II are considered to be different in morphology because of different power of the nuclear engine; in addition also the environment is believed to play a role in deciding whether a source will become an FR I or FR II source.

In this talk I will review some of the old and established ideas about FR I sources, but also discuss their properties in the light of recent unification schemes.

2. FR I sources in the eighties

2.1. MORPHOLOGY, JETS

The original Fanaroff-Riley classification turned out to be quite useful; numerous high resolution observations done with the VLA from 1980 onwards have shown that there are a number of characteristic properties common to the large majority of low luminosity sources.

What is an FR I source like? Often 3C 31, with its double jets ending in characteristic plumes, has been put forward as a kind of prototype of low luminosity sources, but in reality this is not quite true. Studies done in recent years with the VLA show that lobe-sources (and not "plumes") are the rule. The lobes are usually "fat", with an average ratio of length to width of 2:1. In the B2 sample of low luminosity radio galaxies 62 % are double sources with lobes, 18 % are tailed sources (WATs and NATs), 11 % compact (radio cores), and only 4 % are "naked" jets (de Ruiter 1990). 3C 31 resembles more a naked jet than any other type. The naked jets occur preferably at the low end of the power range ($\leq 10^{23}$ W Hz^{-1}). Tail sources are typically a factor 10 stronger, up to the break luminosity. At the upper end of the luminosity range ($\sim 10^{25}$ W Hz^{-1}) we find sources with an overall FR I structure but with one-sided jets and hot spots in the middle of the lobes. Often the global structures are distorted. In some sources wiggles or oscillations suggest that non-uniform motion of the parent galaxy or of the power source may be responsible for this. Interaction between the outer parts of the sources and an intergalactic medium is the almost certain cause of the curved tails, which define Wide Angle Tail and Narrow Angle Tail sources, seen in galaxy clusters and groups.

Perhaps the most important characteristic is the presence, in a large majority of FR I sources, of prominent, usually rather symmetric twin jets: in the B2 sample of low luminosity radio galaxies (Parma et al 1987) as many as one half to three quarters have large scale jets that are easily detectable in VLA-type observations.

However, even though the two sides of FR I sources are very symmetric and rarely differ by more than a factor 2 in intensity, the jets tend to be more asymmetric the closer one looks near the nucleus (jet base). This is a very significant fact as we will see later on.

The relaxed morphology of FR I radio sources suggest low flow velocities in the large scale jets. Bicknell (1984) developed a model of turbulent low Mach number jets, where the observed subadiabatic behaviour of the jet brightness is supposed to be caused by entrainment of external material and subsequent deceleration of the jets. He convincingly showed, first for a limited number of sources and later for a larger sample of FR I sources (Bicknell et al 1990), that a low Mach number ($M \sim 2$), turbulent jet does indeed reproduce observed jet properties. The flow velocities far from the base of the jets suggested by the application of his model are indeed low, of the order of a few hundreds of km/s to at most 10000 km/s. Therefore large scale beaming effects should not be important.

2.2. SIZES

Equal power does not imply equal size, even though there exists a well known correlation between the two (de Ruiter et al 1990). The correlation, however, is very loose, and two sources with the same total luminosity may differ greatly in size. Other factors indicate that besides power also the environment must play an important role. Thus we find that if a source is distorted close to the nucleus it will not reach much beyond the size of the parent galaxy.

In order to bring the effects out more clearly we introduce a normalized linear size l_n, which is the ratio between the size of the source and the median size of sources at a given power: thus we take out the dependence of size on radio power. Using the B2 sample of low luminosity radio galaxies (de Ruiter et al 1990) we find that if a source has a large spreading rate (> 0.3) it remains small, $l_n < 1$, but if the spreading rate is small, isolated galaxies tend to become big, with $l_n > 1$. Sources in clusters or groups of galaxies have $l_n \leq 1$, regardless of the spreading rate. These factors listed here therefore point to the properties of the environment (both the nature of the galaxy, its gas content, and the intergalactic medium) as being another decisive factor for the size of a radio source.

2.3. AGES

In principle synchrotron theory provides a direct way to determine the age of the electrons responsible for the synchrotron emission.

For a number of FR I sources in the B2 and 3C samples there are now data of sufficient quality (at 6 and 20 cm) to try and establish an age of the radio emitting material in the lobes. We assume that the spectral index is a direct indication of the time passed since the last acceleration, and use the Jaffe-Perola model (Jaffe & Perola 1973), which assumes an isotropic redistribution of pitch angles, starting from a power-law distribution of electron energies. As usual electron energy losses are considered to be caused by the synchrotron process itself and by inverse Compton process. Taking the steepest spectral index found in the lobes (presumably the oldest region), we used the formulae of Myers & Spangler (1985) to derive the ages (τ) of the sources, and, with the linear size, a velocity of source growth, $v_{growth} \sim D/\tau$.

Practically all ages determined this way fall in the range $10^7 - 10^8$ years, and no correlation with linear size is seen. This immediately implies that the velocity of growth is, statistically speaking, dependent on radio power (through the well known correlation of size and power). A similar correlation was derived by Alexander & Leahy (1987) and Liu et al (1992), who used only 3C sources. The velocities are well below 10^4 km s^{-1}, at least for FR I sources at low redshifts.

3. FR I sources in the nineties

3.1. DEPOLARIZATION

Low luminosity sources appeared to be understood reasonably well, at least qualitatively, when some new developments arose. First, when unification schemes were realized to be quite promising, a number of authors searched for possible non-beamed counterparts of BL Lac objects. The natural candidates, as all agreed, were FR I sources (Urry & Padovani 1995 and reference therein).

Since the large scale low luminosity jets contain slow-moving material, it is not immediately clear how this unification can come about. It was noticed, however, that the depolarization asymmetry found by Garrington et al. (1988), considered to be caused by an orientation effect plus beaming (Laing 1988), was also present in FR I sources, albeit at a much lower level of significance (de Ruiter et al. 1993). These results were in reality based on polarization data at one frequency. At present we have depolarization data and the general trend is fully confirmed: also in FR I sources there is, statistically speaking, an asymmetry in depolarization, but it should be said that this is seen only in a subset, while many other sources are quite symmetric in jet and counterjet depolarization. Therefore the effect is far less dominant than in the sources studied by Garrington et al (1988), which formed the basis of Laing's hypothesis that the depolarization asymmetry reflects beaming. Nevertheless, if the effect in FR I sources is due to the same cause, beaming should occur to some extent also in low luminosity sources.

Subsequently Laing (1993), Komissarov (1994) and Bicknell (1994) proposed a new phenomenological model, in which all jets, including those of FR I sources, are (mildly)-relativistic close to the core (at the inner kpc, or even closer in the lowest luminosity sources), and decelerate quickly to form slow-moving ($< 10^4$ km s^{-1}) large scale jets.

3.2. BRIGHTNESS ASYMMETRIES

A way to substantiate this idea would be to search for brightness asymmetries in the inner parts of FR I jets. This has been done by Parma et al (1994), who found that the brightness asymmetries, as well as the distribution of core powers are consistent with mildly ($\beta \sim 0.7 - 0.8$) relativistic velocities close to the core, while the jet and counterjet gradually become symmetric as the distance from the core increases.

Beyond a few kpc FR I sources are quite symmetric, as has been known for many years now, but the data suggest that asymmetries persist longer for more powerful sources: jet deceleration may be a function of the strength of a source such that close to the transition FR I to FR II Doppler effects can be important. In fact, the residual velocities, after the initial deceleration tend to be higher in the more powerful sources. This follows from energy budget considerations (see Parma et al 1994). The difference FR I/FR II is in this scenario nothing but an effect of jet velocity, and the more powerful radio sources will take up the aspect of FR II's (jets that are asymmetric all the way, hot-spots, etc.).

The question why a source is an FR I or an FR II in the first place, is probably closely linked also with the environment: as Owen & Ledlow (1994) have convincingly shown, the dividing line between FRI and II is a function of the absolute magnitude of the parent galaxy; it is therefore not only the intrinsic strength of the source, i.e. the quantity of energy produced by the central power source, but also the medium and the (gradient of the) gravitational potential the radio material encounters on its way out (see Bicknell 1995).

Summarizing our current ideas about low luminosity sources and unification schemes: FR I jets appear to start (mildly) relativistic, but quickly decelerate, the quicker the lower the power of the source. FR II sources on the other hand remain supersonic until they impact on the external medium and form hot spots.

Considering BL Lacs as the beamed fraction of FR I sources is then a logical consequence, although there remains some doubt as to the Lorentz factors required: in the case of BL Lacs one needs typically $\Gamma \sim 5$ (see e.g. Urry & Padovani 1995), whereas both the jet asymmetries and the distribution of core strengths suggest $\Gamma \leq 2$. Some way will have to be found to remove this discrepancy.

4. Future prospects

It is clear that our ideas about low luminosity radio sources have changed rapidly in the past few years, and this naturally suggests the direction future research will take. Especially the properties of FR I jets close to the nucleus have gained importance, and VLB type observations will have to provide new data on the detailed physics of such jets.

Another new and important development, actually taking place right at the moment, are the new "all-sky" or,—better—, "important-parts-of-the-sky" surveys, carried out both with the VLA and the WSRT. They will provide catalogs of hundreds of thousands of objects to relatively low flux densities at different frequencies (1.4 GHz for the VLA, 327 MHz and, partly 610, MHz for the WSRT). Extraction of huge samples (many thousands) of galaxies from the WENSS survey are already in progress and will lead to samples of low luminosity radio galaxies that contain one to two orders of magnitude more objects than in the existing samples. Statistics will not be a problem anymore!

References

Alexander, P. & Leahy, J.P. 1987, *MNRAS*, **225**, 1

Auriemma, C., Perola, G.C., Ekers, R., Fanti, R., Lari, C., Jaffe, W.J., & Ulrich, M.H. 1977, *A&A*, **57**, 41

Bicknell, G.V. 1984, *ApJ*, **286**, 68

Bicknell, G.V., de Ruiter, H.R., Fanti, R., Morganti, R. Parma, P. 1990, *ApJ*, **354**, 98

Bicknell, G.V. 1994, *ApJ*, **422**, 542

Bicknell, G.V. 1995, ApJ, in press

de Ruiter, H.R., Parma, P., Fanti, C., Fanti, R. 1990, *A&A*, **227**, 351

de Ruiter, H.R., Capetti, S., Morganti, R., Parma, P., Lazzari, L., Fanti, R. 1993, *Jets in extragalactic radio sources*, eds Roser, H.-J., Meisenheimer, K., Springer-Verlag, p.21

Fanaroff, B.L. & Riley, J.M. 1974, *MNRAS*, **167**, 31p

Garrington, S.T., Leahy, J.P., Conway, R.G., Laing, R.A. 1988, *Nature*, **331**, 147

Jaffe, W.J., Perola, G.C. 1973, *A&A*, **26**, 423

Komissarov, S. 1994, *MNRAS*, **269**, 394

Laing, R.A. 1988, *Nature*, **331**, 149

Laing, R.A. 1993, *Astrophysical Jets*, ST Science Institute Symposium Series 6 , eds Burgarella, D., Livio, M. & O'Dea, C.P., Cambridge University Press, p.95

Liu, R., Pooley, G., Riley, J.M. 1992, *MNRAS*, **269**, 928

Myers, S.T. & Spangler S.R. 1985, *ApJ*, **291**, 52

Owen, F.N. & Ledlow, M.J. 1994, *The First Stromlo Symposium: The Physics of Active Galaxies*, ASP Conference Series 54, eds Bicknell, G.V., Dopita, M. & Quinn, P.J., p.319

Parma, P., Fanti, C, Fanti, R., Morganti, R., De Ruiter, H.R. 1987, *A&A*, **181**, 244

Parma, P., de Ruiter, H.R., Fanti, R., Laing, R.A. 1994, *The First Stromlo Symposium: The Physics of Active Galaxies*, ASP Conference Series 54, eds Bicknell, G.V., Dopita, M. & Quinn, P.J., p. 21

Slee, O.B., Sadler, E.M., Reynolds, J.E., Ekers, R.D. 1994 *MNRAS*, **269**, **928**

Urry, C.M., & Padovani, P. 1995, *PASP*, **107**, 803

Wrobel, J.M., & Heeschen, D.S. 1991, *AJ*, **101**, 148

FR I JETS IN SOUTHERN RADIO GALAXIES

PAUL A. JONES AND BEN D. LLOYD
University of Western Sydney, Nepean, PO Box 10, Kingswood,
NSW, 2747, Australia

1. Introduction

Sources flagged as extended or multiple in the Molonglo Reference Catalogue (MRC, Large et al. 1981, 1991), south of $\delta = -30°$, were observed with the Molonglo Observatory Synthesis Telescope (MOST) with a resolution of 44 arcsec at 843 MHz (Jones and McAdam 1992) to give a sample of 193 southern extended sources. Optical identifications were made using the UKST b_J sky survey. We are now using the Australia Telescope Compact Array (ATCA) near Narrabri in Australia to study a subsample of Fanaroff-Riley class I radio galaxies and fit models to the jets.

We use two simultaneous wavelength bands (20/13 cm or 6/3 cm) and measure full polarisation with the ATCA, with its 6 antennas in a 6 km array and for some sources use several 12-hour observations including the 1.5 km configurations for extra shorter uv-spacings. We are also using the MOST data for the largest scales and best surface brightness sensitivity.

2. MOST and ATCA results

The MOST observations of the head-tail radio galaxy B1610−605 (Jones and McAdam 1992, 1994) show that the tail extends 26 arcmin at low surface brightness with width < 1 arcmin. The neighbouring radio galaxy B1610−608 is a wide-angle-tail with the W jet kinked towards the south 1 arcmin from the core. Both galaxies are in the nearby ($z = 0.0143$) Abell cluster A3627 (Abell et al. 1989). We have observed the two galaxies with the ATCA at 20 and 13 cm to study the tail, jets and lobes and how they interact with the intra-cluster medium (Jones and McAdam 1994, 1995). The tail of B1610−605 is narrow (< 10 arcsec within the first arcmin) so that if there are twin FR I jets from the core bent back by ram pressure, this must occur over the scale of a few kpc, within the optical galaxy.

143

R. Ekers et al. (eds.), Extragalactic Radio Sources, 143–144.

The brightness and width of the tail or jets were fitted and deconvolved using the model of a cylinder of Gaussian profile (Killeen et al. 1986). The standard assumptions for synchrotron emitting plasma were used to estimate the magnetic field and minimum pressure. The pressure profiles suggest that the intra-cluster medium is clumpy and that the cluster is not relaxed. This is confirmed by the ROSAT all-sky survey X-ray data (H. Böhringer, private communication).

The radio galaxy B1343−601 (Centaurus B) is one of the brightest extragalactic sources in the sky (240 Jy at 408 MHz), however there has been very little study of this galaxy because of its position (a) close to the Galactic plane (alternative name G309.6+1.7) and (b) in the far southern sky. The galaxy was optically identified by Laustsen et al. (1977) and its redshift measured as 0.01215 by West and Tarenghi (1989). The MOST observations (McAdam 1991) show prominent FR I jets and a diffuse, low surface brightness halo. We have recently (Sept/Oct 1995) observed this galaxy with the ATCA at 6 and 3 cm with two 12-hour observations, to allow detailed analysis and model-fitting of the jets.

Four galaxies which have prominent FR I jets, B1234−723, B1318−434, B1452−517 and B2148−555, selected from Jones and McAdam (1992), were observed with the ATCA at 6 and 3 cm. Fits to the brightness, width, transverse position and polarisation of the jets at the two frequencies are used to determine the equipartition parameters and Bicknell's model for dissipative, turbulent FR I jets (Bicknell 1994) applied. The model takes a spline fit to the jet width and the derived pressure distribution and then predicts the surface brightness, Mach number, velocity, jet density ratio and mass flux. B1318−434 is identified with the elliptical galaxy NGC 5090 which is interacting with the spiral NGC 5091, and is discussed in more detail in Lloyd et al. (1995).

References

Abell G.O., Corwin H.G. and Olowin R. (1989), *Astrophys. J. Suppl.*, **70**, 1

Bicknell G.V. (1994), in Bicknell G.V., Dopita M.A. and Quinn P.J., eds, *The First Stromlo Symposium: The Physics of Active Galaxies*, ASP Conf. Ser. 54, Astronomical Society of the Pacific, p 357

Jones P.A. and McAdam W.B. (1992), *Astrophys. J. Suppl.*, **80**, 137

Jones P.A. and McAdam W.B. (1994), *Proc. Astron. Soc. Australia.*, **11**, 74

Jones P.A. and McAdam W.B. (1995), submitted to *Mon. Not. Royal Astron. Soc.*

Killeen N.E.B, Bicknell G.V. and Ekers R.D. (1986), *Astrophys. J.*, **302**, 306

Large M.I., Mills B.Y., Little A.G., Crawford D.F. and Sutton J.M. (1981), *Mon. Not. Royal Astron. Soc.*, **194**, 693

Large M.I., Cram L.E. and Burgess A.M. (1991), *Observatory*, **111**, 72

Laustsen S., Schuster H.-E. and West R.M. (1977), *Astron. and Astrophys.*, **59**, L3

Lloyd B.D., Jones P.A. and Haynes R.F. (1995), *Mon. Not. Royal Astron. Soc.* in press

McAdam W.B. (1991), *Proc. Astron. Soc. Australia.*, **9**, 255

West R.M. and Tarenghi M. (1989), *Astron. and Astrophys.*, **223**, 61

HIGH-FREQUENCY STUDY OF LOW AND INTERMEDIATE LUMINOSITY RADIO GALAXIES

L. GREGORINI[1,2], P. PARMA[1], U. KLEIN[3], K.-H. MACK[3]

[1] *Istituto di Radioastronomia del CNR - Bologna, Italy*
[2] *Dipartimento di Fisica - Universitá di Bologna, Italy*
[3] *Radioastronomisches Institut der Universität Bonn - Germany*

1. Introduction

All the information on the morphology, spectral index and polarization properties of low and intermediate luminosity radio galaxies were collected in an intermediate frequency range, and some of the basic questions to be elucidated with such measurements are still unanswered. We therefore decided to extend the study of their characteristics towards higher frequencies. From the B2 and 4C catalogue we selected 26 radio galaxies for which VLA or WSRT data are available. These sources were observed at 10.6 GHz using the Effelsberg 100-m telescope (Gregorini & al. 1992; Mack & al. 1994; Klein & al. 1995).

2. The Spectral Index

The integrated radio spectra between 408MHz and 10.6GHz are found to resemble those of strong 3C sources, the average spectral index being $\alpha = -0.69 \pm 0.02 (S_\nu \propto \nu^\alpha)$. Analysis of the individual spectra shows that most of them are straight up to 10.6 GHz.

The spectral index distribution shows a clear difference between low- and high-luminosity radio galaxies: a strong steepening of the spectrum from the outer hotspots towards the inner cores is present on scales of $\sim$ 200 to 300 kpc for a sample of FRII sources (Jägers 1986), while the B2 and 4C galaxies exhibit a steepening from the cores to the outer lobes.

R. Ekers et al. (eds.), Extragalactic Radio Sources, 145–146.

3. The Magnetic Field

The degrees of polarization are higher at 10.6 GHz than what is measured at lower frequencies, suggesting a significant Faraday depolarization. The values are generally low ($\sim 5\% - 15\%$) in sources with complex morphologies, while in those with a simple (core/jet/lobe) structure we find high degrees of polarization, with value of up to 50% locally. The magnetic field orientation is related to the morphology of the source: it is perpendicular in the double jets, longitudinal in the one sided jet and tangential to the edges in the lobes. This is interpreted in terms of their presumed helical structure. The observed integrated rotation measures of the sources are rather low, with values hardly exceeding some 20 rad m^{-2} so that at 10.6 GHz we essentially map the intrinsic (projected) magnetic field orientations.

4. Particle Ageing

We interpreted the results of our spectral analysis of B2 and 4C radio galaxies in terms of particle ages and sources lifetime. Using the Kardashev and Pacholczyk model we derived particle ages from ~ 40 to 70 Myrs for most of the sources. The only exception is 1615+35, which is a head-tail source, the spectrum of which suggests an age of ~ 100 Myrs. Comparison with FRII sources shows that the steepening of the spectra in low-luminosity radio galaxies is not as strong as expected on the basis of studies of 3C radio galaxies. Classical FRII radio galaxies appear to have break frequencies between 1 and 5 GHz, while about half of our sample shows break frequencies higher than this. This could indicate that significant re-acceleration take place in low-luminosity radio galaxies.

References

Gregorini, L., Klein, U., Parma, P., Schlickeiser, R., Wielebinski, R. (1992) High-frequency radio continuum observations of low-luminosity radio galaxies I. A sample of sources with angular sizes > 4', *Astron. Astrophys. Suppl. Ser.*, Vol. no. **94**, pp. 13–35

Jägers, W.J. (1986), The polarization of radio galaxies, *Ph.D. Thesis*, Univ. of Leiden

Mack, K-H., Gregorini, L., Parma, P. Klein, U. (1994) High-frequency radio continuum observations of radio galaxies with low and intermediate luminosity II. Sources with sizes 4' to 5', *Astron. Astrophys. Suppl. Ser.*, Vol. no. **103**, pp. 157–182

Klein, U., Mack, K-H., Gregorini, L., Parma, P. (1995) High-frequency radio continuum observations of radio galaxies with low and intermediate luminosity III. Spectral indices and particle ages *Astron. Astrophys.*, in press

LARGE-SCALE STRUCTURE:
JETS ON KILOPARSEC SCALES

R.A. LAING
Royal Greenwich Observatory
Madingley Road, Cambridge CB3 0EZ, U.K.

1. Overview

This paper examines some of the consequences of the hypothesis that jets in *all* radio galaxies and quasars are relativistic on small scales, in the sense that the flow velocity >0.5c. This idea is suggested by a number of lines of evidence. Firstly, Unified Models (Urry & Padovani, 1995) imply that the relativistic motion required in core-dominated objects must also occur in a larger parent population consisting of most, if not all, extended sources. Secondly, superluminal motion is detected in the nuclei of extended sources and in the kpc-scale jet of M 87 (Hough, 1994; Biretta, Zhou & Owen, 1995). Thirdly, jets are one-sided in the same sense on pc and kpc scales; at all luminosities, the radio emission tends to become more symmetrical on larger scales, as expected if an initially relativistic flow decelerates (Bridle & Perley, 1984; Bridle *et al.*, 1994a; Parma *et al.*, 1994). Finally, depolarization asymmetry occurs in both low (Parma, de Ruiter & Fanti, 1996) and high (Laing, 1988; Garrington *et al.*, 1988) luminosity sources: the implication is that the brighter jet is on the near side of the source. It is likely that the key difference between radio sources in the two morphological classes defined by Fanaroff & Riley (1974) are that relativistic flow persists to the extremities of FRII sources, but that FRI jets decelerate smoothly on intermediate scales (Laing, 1993; Bicknell, 1995). On kiloparsec scales, we can identify structures which we propose should be called *fast jets*. These are well-collimated and generally one-sided (in the sense that the jet/counterjet ratio >4:1). They also have longitudinal apparent magnetic field ($\mathbf{B}_\parallel$). They occur both in FRII sources, and at the bases of FRI jets (Bridle & Perley, 1984). We suggest that they are relativistic flows, and that this fact is crucial to an understanding of their evolution. A framework for the understanding of the variety of ex-

147

R. Ekers et al. (eds.), Extragalactic Radio Sources, 147–152.
© 1996 IAU.

tended structures in extragalactic radio sources in this context is illustrated
in Figure 1, which is an improved version of the diagram presented by Laing
(1993). A fast jet appears to be able to: decelerate and recollimate to form a
slow jet with $\beta \ll 1$ (therefore two-sided unless external effects dominate);
disrupt, as in wide-angle tail sources, or hit the external medium and form
a *hot-spot*. Slow jets are probably formed only when a decelerating fast
jet can be recollimated by the external pressure gradient (Phinney, 1983;
Bowman, Leahy & Komissarov, 1995). This may not be possible for more
powerful sources in flatter pressure gradients and it is likely that wide-angle
tail sources are formed when a fast jet decelerates rapidly but cannot recollimate. Deceleration by entrainment is efficient when the jet is transonic,
and Bicknell (1994) showed that this corresponds to $\beta \approx 0.3 - 0.7$ for a relativistic jet. If the jet does not slow down sufficiently (*e.g.* by mass loading;
Komissarov 1994), then the flow will remain supersonic until it impacts on
the external medium, and an FRII source will result. The radio morphology is therefore determined by a combination of initial jet speed and thrust
and the effects of the environment, via the rate of stellar mass loss and
the pressure gradient. On the largest scales, a *bridge*(backflow) or *tail* (outflow) will be formed. If the jet remains supersonic as far as the end of the
lobe (as in an FRII source), then it is inevitable that a backflow (bridge)
will be generated. As emphasised by Parma, de Ruiter & Fanti (1996), the
majority of FRI sources also show bridges: the residual momentum of the
jets, their density contrast with the external medium and the external pressure gradient are all likely to be important in determining their large-scale
morphologies.

2. FRII Sources: Orientation and Intrinsic Asymmetries

The jet intensity, depolarization, spectral index, arm length and emission-
line sidednesses of FRII sources are all correlated in various ways. In order
to make sense of these relations, we adopt a deductive approach with the
following assumptions:

1. Jets in FRII sources are intrinsically symmetrical and relativistic.
2. Quasars are only seen if their axes are within $45° - 60°$ of the line of
   sight, and a subset of FRII radio galaxies are side-on quasars,
3. Faraday effects are due to magnetic field and/or density irregularities
   in the medium surrounding the radio sources.
4. There is an intrinsic mechanism which associates higher external gas
   densities, smaller lobes and steeper radio spectra (most obviously the
   combination of synchrotron and adiabatic losses).
5. Jets, jet-side hot-spots and some associated material have flatter radio
   spectra than does the lobe emission.

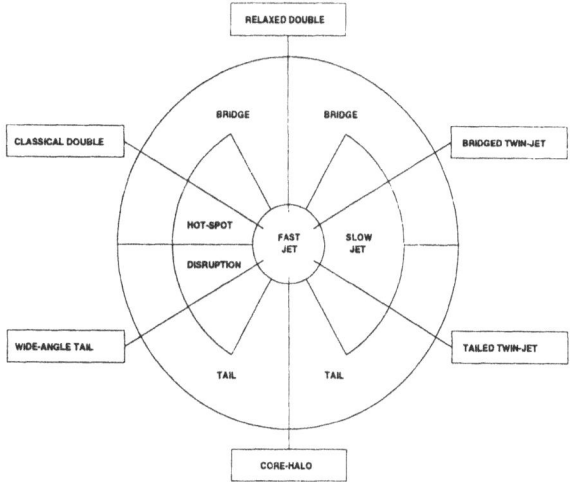

Figure 1. A schematic representation of the main morphological classes of radio source. Each radial line refers to a particular type of source; increasing radius represents the evolution of the jet flow.

6. The average advance speed of the radio source is small ($\leq$ 0.1c), so light-travel effects are small.

From these assumptions, we make the following deductions, which are illustrated in Figure 2:

Sources with strong, one-sided jets are selected to be within about 50° of the line of sight: most quasars satisfy this condition. The differential Faraday depth to the two lobes of such a source is dominated by orientation, so:

the far (counter-jet) side shows stronger depolarization.

The ratio of lobe lengths on opposite sides of the source is determined primarily by intrinsic effects, rather than by light travel, although the latter effect may just be detected in sources close to the line of sight (Scheuer, 1995). Therefore:

depolarization (and jet sidedness) are only weakly correlated with arm length in jetted sources.

By contrast, sources without strong jets are close to the plane of the sky, so the differential Faraday depths to the lobes are determined primarily by variations in external density and

depolarization is stronger on the short side in jetless sources.

In jetless sources, emission on the two sides is (on average) equally boosted. Spectral differences therefore reflect intrinsic effects, so:

The lobe with the steeper spectral index is shorter and more depolarized.

Finally, in jetted sources, the high-brightness emission (jet, jet-side hot-spot, ...) is Doppler-boosted, with a flat spectrum, whilst the spectrum of

the lobe emission is determined by intrinsic effects.
Depolarization is greater in the steeper-spectrum lobe if either effect dominates.

The data in Figure 2 are taken from the references given in Laing (1993), and refer to powerful FRII quasars and radio galaxies. Similar results appear also to hold for CSS sources (Garrington & Akujor, 1996). Bridle *et al.* (1994b) also showed that, whilst the spectral index of high-brightness structure in quasars is correlated with jet sidedness, that of the low-brightness (lobe) emission is correlated with arm length, indicating a mixture of orientation and intrinsic effects in individual sources.

3. FRI Sources as Decelerating Jets: Observations of 3C 31

Parma, de Ruiter & Fanti (1996) describe the evidence that statistical results on asymmetries in FRI jet bases require the flow to be relativistic on small scales. Laing (1993) proposed a simple model in which jets consist of a fast, perpendicular-field *spine* surrounded by a slower longitudinal-field *shear layer*. This model predicts the brightness and polarization structure of the jets as functions of angle and velocity distribution. If jet sidedness is determined entirely by Doppler effects and the jet is axisymmetric, then the three-dimensional velocity field can be deduced from a map of $S(x, y)/S(-x, -y)$ where $S(x, y)$ is the flux density at a point (x, y) on the map. This section describes preliminary results from a detailed study of the low-luminosity radio galaxy 3C31 by Feretti, Giovannini, Parma, Bridle, Perley and the present author. At 8.4 GHz, using a combination of VLA B, C and D-configuration observations (FWHM 0.7 arcsec), the counter-jet is detected all the way into the nucleus, and we can make a detailed sidedness map by dividing the image by a copy of itself rotated through 180° about the nucleus. A profile along the ridge-line of the jet shows that the jet/counter-jet ratio decreases from 15 – 25:1 in the inner region (0 – 5 arcsec from the nucleus) to 1:1 at 30 arcsec, after which the main jet bends significantly. Fine structure in the main and counter-jets causes the sidedness ratio to vary erratically in the innermost region. The ratio is highest along the ridge line, as expected if a slow boundary layer is present, but does not fall to 1 at the jet edges, at least at current sensitivity levels. The inferred velocity for the central spine of the jet falls from $\beta \approx 0.8$ at 2 arcsec from the nucleus to $\beta \approx 0.1$ at 30 arcsec, assuming an angle to the line of sight of 50°. Larger angles are ruled out by the maximum jet/counter-jet ratio, whilst smaller ones are inconsistent with the degree of polarization observed at large distances from the nucleus, assuming the models of Laing (1993). Details of the polarization results (in particular the fact that the flip from $\mathbf{B}_{\parallel}$ to $\mathbf{B}_{\perp}$ occurs closer to the nucleus

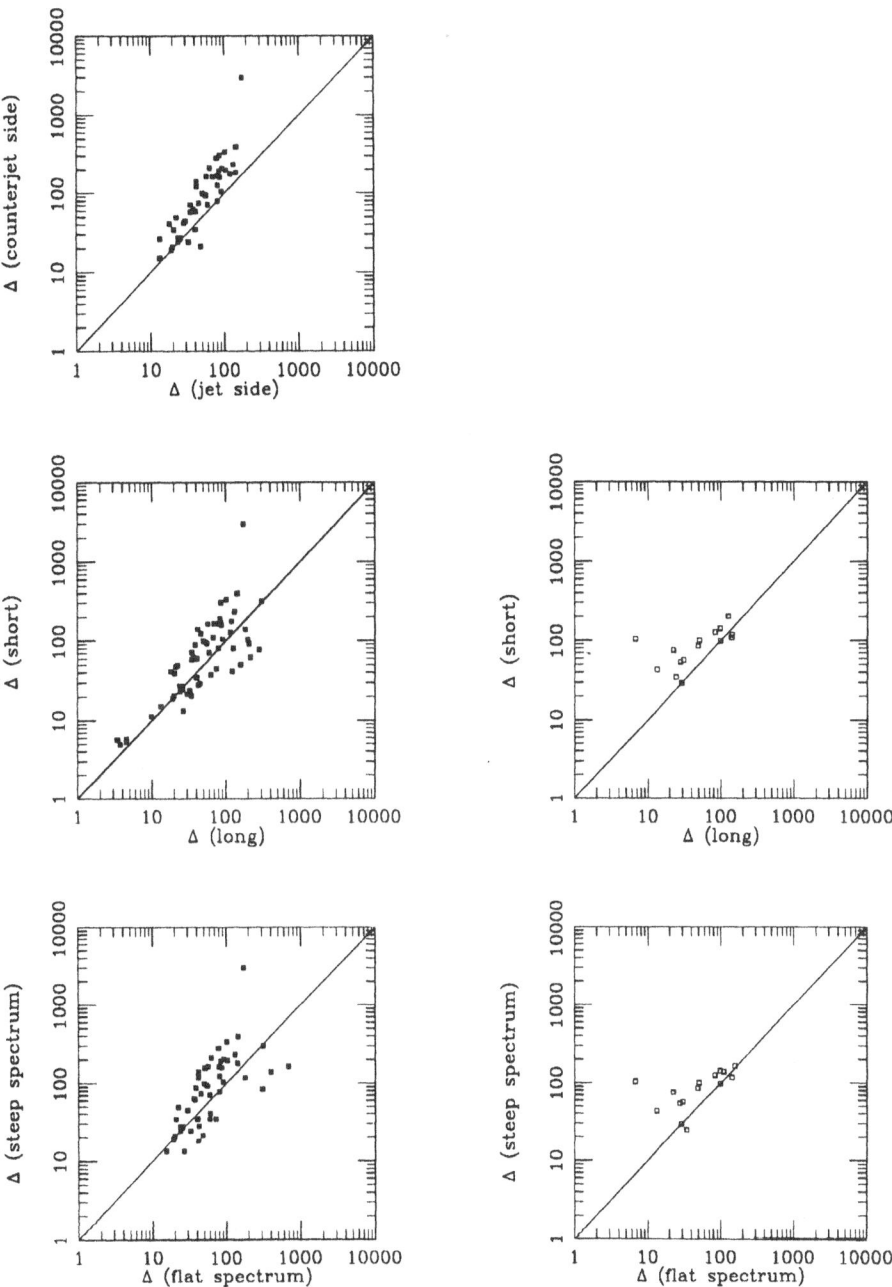

Figure 2. Plots of Faraday depth, Δ (in μGcm^{-3}pc) on opposite sides of FRII radio sources. The left-hand panels represent sources with detected jets; the right-hand panels those without jets. Top: jet side against counter-jet side. Middle: longer lobe against shorter lobe. Bottom: steeper against flatter-spectrum lobe.

in the counter-jet) suggest that the field in the shear layer is not longitudinal, as in the original model, but rather has roughly equal components in the longitudinal and azimuthal directions, with no radial component. This might be produced by averaging over a number of randomly-orientated $B_\parallel$ filaments in the shear layer as, for example, in M87 (Owen, Hardee & Cornwell, 1989). Sufficiently detailed observations might resolve individual filaments, leading to oblique apparent magnetic fields. The initial conclusions of this study are that a decelerating jet model with a $B_\perp$ spine and a shear layer is in good qualitative agreement with the observations, but that further refinement of the shear-layer field structure is needed. Similar conclusions are reached from a study of 3C 66B (Hardcastle et al., 1995). We plan more detailed modelling of the velocity field in 3C 31, together with a comprehensive study of its Faraday rotation properties: it is already known to show a substantial depolarization asymmetry (Burch, 1979; Strom et al., 1983).

References

Bicknell, G.V., 1994. *ApJ*, **422**, 542.
Bicknell, G.V., 1995. *ApJS*, **101**, 29.
Biretta, J.A., Zhou, F. & Owen, F.N., 1995. *ApJ*, **447**, 582.
Bowman, M., Leahy, J.P. & Komissarov, S.S., 1995. *MNRAS*, in press.
Bridle, A.H., Hough, D.H., Lonsdale, C.J., Burns, J.O. & Laing, R.A., 1994. *AJ*, **108**, 766.
Bridle, A.H., Laing, R.A., Scheuer, P.A.G. & Turner, S., 1994, in Bicknell, G.V., Dopita, M. & Quinn, P.J., eds, The First Stromlo Symposium: The Physics of Active Galaxies. ASP, San Francisco, p. 187.
Bridle, A.H. & Perley, R.A., *ARA&A*, **22**, 319.
Burch, S.F., 1979. *MNRAS*, **187**, 187.
Fanaroff, B.L. & Riley, J.M., 1974. *MNRAS*, **167**, 31P.
Garrington, S.T. & Akujor, C.E., 1996. These proceedings.
Garrington, S.T., Leahy, J.P., Conway, R.G. & Laing, R.A., 1988. *Nature*, **331**, 147.
Hardcastle, M.J., Alexander, P., Pooley, G.G. & Riley, J.M., 1995. *MNRAS*, in press.
Hough, D.H., 1994, in Zensus, J.A. & Kellermann, K.I., eds, Compact Extragalactic Radio Sources. NRAO, Socorro, p.169.
Komissarov, S.S., 1994. *MNRAS*, **269**, 394.
Laing, R.A., 1988. *Nature*, **331**, 149.
Laing, R.A., 1993, in Burgarella, D., Livio, M. & O'Dea, C.P., eds, Astrophysical Jets. Cambridge University Press, Cambridge, p. 95.
Owen, F.N., Hardee, P.E. & Cornwell, T.J., 1989. *ApJ*, **340**, 698.
Parma, P., de Ruiter, H.R. & Fanti, R., 1996. These proceedings.
Parma, P., de Ruiter, H.R., Fanti, R. & Laing, R.A., 1994, in Bicknell, G.V., Dopita, M. & Quinn, P.J., eds, The First Stromlo Symposium: The Physics of Active Galaxies. ASP, San Francisco, p. 241.
Phinney, E.S., 1983. PhD thesis, Cambridge University.
Scheuer, P.A.G., 1995. *MNRAS*, **277**, 331.
Strom, R.G., Fanti, R., Parma, P. & Ekers, R.D., 1983. *A&A*, **122**, 305.
Urry, C.M. & Padovani, P., 1985. *PASP*, **107**, 803.
Vermuelen, R.C., 1996. These proceedings.

THE JETS IN FRII RADIO GALAXIES WITH $Z < 0.3$

MARTIN HARDCASTLE
Mullard Radio Astronomy Observatory, Cavendish Laboratory,
Madingley Road, Cambridge CB3 OHE, UK

1. Introduction

If we believe that jets trace the energy transport in radio sources, we might in principle expect to see them in every classical double radio galaxy with compact hot spots. Until recently the detection rate of jets in FRII radio galaxies has been low, although most FRII quasars have bright one-sided jets. However, it seems likely that this is due to lack of sensitivity. Black *et al.* (1992) found jets in up to 70% of a sample of FRII galaxies with $z < 0.15$. We have observed the FRII sources in Laing *et al.* (1983) with $0.15 < z < 0.3$ and discuss results from the combined samples.

2. Observations and results

The sources in the $z < 0.15$ sample were observed at 8 GHz with the VLA as described in Black *et al.*. The sources in the $0.15 < z < 0.3$ were also observed at 8 GHz with the VLA; this choice of frequency gives high sensitivity and resolution. Multi-configuration observations were made.

The jet fraction in the combined sample is high: slightly over half have definite jets and a further 20% have probable jets. As figure 1 (left) shows, there is no significant tendency for the jet flux fraction to fall off as a function of luminosity. This confirms that the result of Black *et al.* was not due to the proximity of his sample to the FRI/FRII luminosity boundary. When jet luminosity is plotted against source luminosity, the correlation is significant at the 99% level. There is clearly a large degree of scatter in the relationship, however. Jets are one-sided; there are only two objects in the sample whose jet and counterjet are of similar luminosity, although several fainter counterjets are detected. It appears to be impossible to generate a strong jet without a strong core, but not all sources with strong cores have strong jets (figure 1, right). There is no significant tendency for broad-line

153

R. Ekers et al. (eds.), Extragalactic Radio Sources, 153–154.

Fraction of flux in jet as a function of luminosity   Core fraction against jet fraction

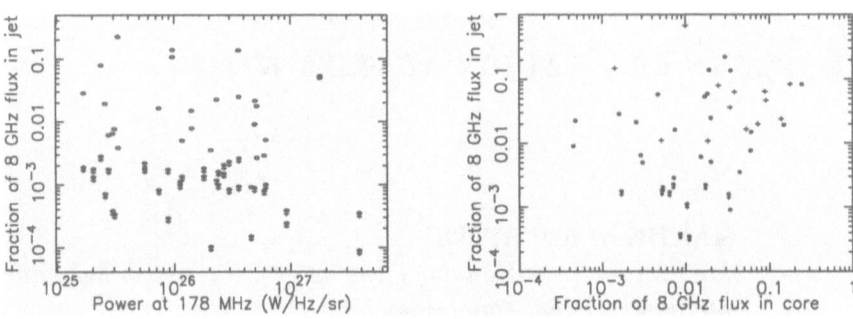

Figure 1. Left: fraction of flux in jet as a function of source luminosity (each lobe plotted individually). Right: fraction of flux in jet as a function of fraction of flux in core (only brightest jet or upper limit plotted). Crosses mark the Bridle *et al.* quasars on the same scale.

radio galaxies to have stronger jets: the very strongest jets in fact come from 'dull' low-emission-line sources (Hine and Longair 1979's class B). If the core is the self-absorbed base of a jet, we would expect core fraction to be an orientation indicator: it is interesting that the thirteen FRII quasars of Bridle *et al.* (1994) lie in the top right of figure 1 (right). Some of the brightly jetted 'dull' FRIIs in the sample may be at small angles to the line of sight in spite of not being identified as quasars.

3. Conclusions

Jets are common in radio galaxies and the luminosity of a jet is broadly correlated with that of the radio source as a whole: thus the jet power is directly related to the power in the beam. The degree of scatter in this relationship must be attributed to some combination of differences in source age, different orientation angles (leading to different effects of relativistic beaming) or large variations in the efficiency of jets. The variation of source age in the sample is probably small. The almost universal one-sidedness of jets suggests that beaming must be significant; however, it is clear that the efficiency of the jet is affected strongly by its environment in particular sources. Work on the statistics of jet sidedness and information on the host galaxies are necessary to allow us to draw quantitative conclusions.

References

Black, A.R.S., Baum, S.A., Leahy J.P., *et al.* 1992, *MNRAS*, **256** 186
Bridle, A.H., Hough, D.H., Lonsdale, C.J., *et al.* 1994, *Astron. J.*, **108** 766
Hine, R.G., Longair, M.S., 1979, *MNRAS*, **193**, 285

LOW-LUMINOSITY RADIO QSOS:

THE B2 SAMPLE

J.I. GONZÁLEZ-SERRANO
Instituto de Física de Cantabria, Santander, Spain

I. PÉREZ-FOURNON
Instituto de Astrofísica de Canarias, La Laguna, Spain

AND

L.GREGORINI, L.PADRIELLI AND H.R. DE RUITER
Istituto di Radioastronomia di Bologna, Bologna, Italy

Abstract. We present preliminary results from imaging and spectroscopy of a complete sample of radio quasars selected from the B2 survey. The optical data, which provide us information on the environment around the quasars, as well as the redshifts, allow us to address several questions related to the origin and evolution of the radio quasars population. Using VLA observations of the sources we investigate the radio morphology and distortions of these low-luminosity radio quasars. We compare these properties with those of powerful radio quasars.

1. Introduction and the sample

Barthel & Miley (1988) found that powerful (3C) radio quasars at high redshift seem to be more distorted than quasars at low redshift, suggesting an epoch-dependence of radio structure and of the properties of the intergalactic medium around QSOs.

We have started an study of a sample of radio quasars which are a factor of 20 less powerful than 3C quasars in order to compare the properties of high and low-luminosity quasars. The sample is the complete sample of B2 quasars which consists of 72 objects identified with blue point-like sources on the POSS plates. The objects are brighter than 21st magnitude and 0.25 Jy at 408 MHz. All have been observed with the WSRT at 6 and 21 cm

155

R. Ekers et al. (eds.), Extragalactic Radio Sources, 155–156.
© 1996 *IAU*.

and with the VLA at 6cm (Rogora et al. 1986, 1987). The sample contains 47 extended sources ($> 10''$) and 25 compact sources ($< 10''$).

2. Observations and results

We have observed the QSOs at the Observatorio del Roque de los Muchachos at La Palma (Canary Islands). We have made CCD imaging at the 2.5m Nordic Optical Telescope (NOT) in the r band in good seeing conditions and reaching a limiting magnitude of ~ 23. Low-dispersion spectroscopy (~ 10Å effective resolution) was carried out at the 4.2m William Herschel Telescope (WHT) using the Faint Object Spectrograph (FOS) in order to complete the redshift information of the sample.

From the VLA maps we have measured the largest angular size (LAS) and the distortion angle (ϕ). This is the angle subtended by the directions hot spot to radio core and radio core to the other hot spot. We have made a similar analysis as was done by Barthel et al. (1988) and Barthel & Miley (1988) with the following results:

• Small sources are more distorted. This was found before by Padrielli et al. (1988) but they had a more limited information on redshifts.

• At high redshift ($z > 2$) only 1 out of 6 sources has $\phi > 20°$, while Barthel & Miley found the relation $\phi \sim (1 + z)^{1.7}$. They concluded that the ambient medium, interacting with the radio jets, is producing the distortions at high redshifts. The difference found by us between low and high luminosity QSOs could be due therefore to a less dense intracluster medium around low-luminosity radio loud QSOs.

References

Barthel et al. 1988, *A&AS*, **73**, 515
Barthel, P.D., & Miley, G.K., 1988, *Nature*, **333**, 319
Padrielli, L., Rogora, A., & de Ruiter, H.R., 1988, *A&A*, **196**, 49
Rogora, A., Padrielli, L., de Ruiter, H.R., 1986, *A&AS*, **64**, 557
Rogora, A., Padrielli, L., de Ruiter, H.R., 1987, *A&AS*, **67**, 267

AN ATLAS OF EXTRAGALACTIC RADIO SOURCES

J.P. LEAHY[1], A.H. BRIDLE[2], R.G. STROM[3]

[1] *University of Manchester - Jodrell Bank, Cheshire, UK*
[2] *N.R.A.O. - Charlottesville, Virginia 22903, USA*
[3] *N.F.R.A. - PO Box 2, 7990 AA Dwingeloo, the Netherlands*

1. Overview

Our *Atlas of Extragalactic Radio Sources* will present high-quality images of the **nearer half** of "3CRR", the sample defined by Laing, Riley & Longair (1983). This is the best-studied complete sample of extragalactic radio sources. All 173 members have secure redshifts and most have been imaged in the radio at high resolution. There is also copious information on their optical line emission, and many have been detected in the sub-mm, FIR, and in X-rays. 3CRR is widely used as a baseline against which fainter, higher-redshift samples can be compared to define the evolution of the population (e.g. Neeser et al. 1995; Law-Green, this conference).

At present, modern radio data on 3CRR is partly unpublished and the rest is spread across many papers. The digital images are not easily accessible, which hinders systematic morphological analysis. The *Atlas* will solve (half) this problem. Many of its images have been contributed by other observers, to whom we are very grateful.

The sample is defined as: extragalactic objects with $S_{178\mathrm{MHz}} \geq 10.9$ Jy, $\delta \geq 10°$ and $b^{II} \geq 10°$ (these define 3CRR), and $z \leq 0.5$; we exclude the starburst galaxy 3C 231 (M 82), to give 85 members, all DRAGNs. The high-redshift part of 3CRR is excluded mainly because the large majority have been imaged at high resolution by Laing & Owen (unpublished).

The sample contains 21 objects with powers below the FR break, and of these only three "fat doubles" lack jets. About a third of the objects above the FR break differ from the "classical" symmetric structure with hotspots at both ends. Twenty-seven of the 59 powerful radio galaxies show jets; another 6 have possible jets. Of these, 12 (possibly 13) have twin jets,

R. Ekers et al. (eds.), Extragalactic Radio Sources, 157–158.

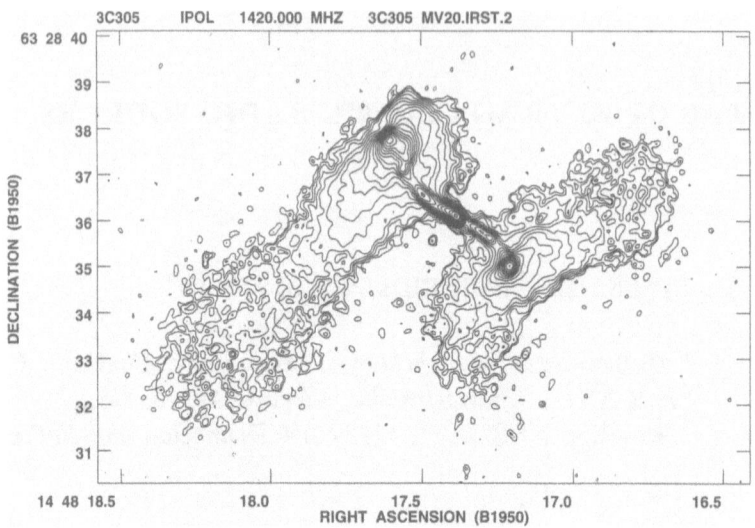

Figure 1. 3C 305 at 20 cm (MERLIN + VLA image). Beam 0.15 arcsec.

suggesting that jets are much less one-sided in powerful radio galaxies than quasars; however, only 4 (5) of these are classical doubles. There are also 5 quasars, all with jets. Cores are now detected in nearly all sample members (the exceptions have usually not been observed to sub-mJy levels at high frequency). Not all these features are visible in the *Atlas* images.

The *Atlas* images have FWHM beam ≤ 5% of the target largest angular size (typically ∼ 2.5%.), and sensitivity and uv-coverage sufficient to show the largest and faintest components in each DRAGN. The images are mostly at 20 cm, and are from the WSRT, VLA, MERLIN or EVN as appropriate for the angular size. The *Atlas* will also contain tables of basic data on each DRAGN, including flux densities from 10 MHz to 1 keV, basic morphological data derived from the images, and possibly optical spectra.

We now have images for nearly all sample members, although in a few cases they do not meet (or barely meet) our target criteria, and improved images are being produced. We hope to make our current collection of images available on the World-Wide Web (look for details on http://www.jb.man.ac.uk/jb.html). Distribution on CD should occur in 1996 as part of the NRAO CD series; publication in print may be some time later. Fig. 1 shows an image recently made for the project.

References

Laing, R.A., Riley, J.M., Longair, M.S., 1983. *MNRAS*, **204**, 151.
Law-Green, J.D.B. et al. 1995. *MNRAS*, **274**, 939.
Neeser, M.J, Eales, S.A., Law-Green, J.D., Leahy, J.P., Rawlings, S. 1995. *ApJ*, **451**, 76.

EXTREME ROTATION MEASURE RADIO GALAXIES

Measuring Cluster Magnetic Fields

C.L. CARILLI[1], R.A. PERLEY[2], G. MILEY[3], H. RÖTTGERING[3], AND R. VAN OJIK[3]

[1] *CfA, Cambridge, MA, USA*
[2] *NRAO, Socorro, NM, USA*
[3] *Sterrewacht Leiden, The Netherlands*

1. Cygnus A: The First Case

Slysh (1966) first pointed out the bizarre properties of the polarized emission from Cygnus A: a strong wavelength dependence on fractional polarization, and a large difference in rotation measures (RMs) for the (integrated) emission from each lobe. These properties were confirmed, but not resolved, in the work of Mitton (1971) and Alexander et al.. (1984).

The resolution of the puzzle came with the multifrequency VLA observations of Cygnus A by Dreher et al.. (1987). They found that Cygnus A is located behind a deep 'Faraday screen' of magnetized, ionized plasma, producing large RMs (up to 5000 rad m^{-2} in magnitude), and gradients in RM (over 1000 rad m^{-2} arcsec^{-1}) across the lobes. The latest measurement of the Faraday screen towards Cygnus A is shown in Perley and Carilli (1996). The RM distribution across Cygnus A is not random, but shows ordered structure on scales up to 30 kpc. The RM structure is typically not correlated with total intensity structure. The tail of the southern lobe shows the largest RM values, and a generally 'noisier' RM distribution than for the rest of the source. This behavior for the southern lobe is consistent with the Laing-Garrington effect and the fact that the southern lobe is the un-jetted lobe in Cygnus A.

Before discussing the location of the RM screen in detail, we first mention two other interesting results from the polarization imaging of Cygnus A. One result is the magnetic field distribution (see Dreher et al.. 1987). The projected fields follow very closely the hard edges of the hot spots and radio lobes, and run parallel to the brighter lobes and filaments in the radio source. This morphology is consistent with simple kinematic dynamo effects for a 'frozen-in' magnetic field. The second result is the fractional

159

R. Ekers et al. (eds.), Extragalactic Radio Sources, 159–162.
© 1996 IAU.

polarization (FPOL) distribution (see Figs. 8 and 9 in Perley and Carilli 1996). FPOLs at 8 GHz range between 10% and 40% in most regions of the source. The one notable exception is the tail of the northern lobe, where FPOLs are 70% over a region almost 20 kpc in size. This is close to the theoretical maximum for an isotropic distribution (in pitch angle) of electrons in a uniform magnetic field, implying almost perfectly ordered fields (in projection) on scales of 20 kpc in the lobes of Cygnus A!

Dreher et al. consider, and reject in a trivial way, a Galactic origin for the large RMs seen towards Cygnus A. They also reject thermal material mixed with the radio lobes as the cause of the large RMs, on the mathematically rigorous basis that the wavelength dependence of observed position angle (PA) remains precisely quadratic for total rotations $>> 2\pi$ radians, without any depolarization with increasing wavelength. This argument has been strengthened with new observations at 8 GHz (Perley and Carilli 1996). They argue that the most likely location of the magnetized screen is the hot, X-ray emitting cluster gas in Cygnus A. The Cygnus A radio source is located at the center of a dense, massive 'cooling flow' cluster atmosphere, with $\approx 10^{14}$ $M_\odot$ of hot gas extending one Mpc from the cluster center (Carilli et al.. 1994). Using the cluster thermal gas density profile and the observed RM distribution, Dreher et al.. derive cluster magnetic field strengths between 5 and 10 μG, for cell sizes between 50 and 10 kpc.

Subsequent analysis of the RM screen towards Cygnus A has revealed that the shocked intracluster medium (ICM) enveloping the supersonically expanding radio source also contributes to the observed rotation measure distribution. The evidence came in the form of detection of an 'arc' of discontinuous change in rotation measure (RM) roughly concentric with the primary hotspot in the northern lobe, with a standoff distance of about 3″ (see Carilli et al.. 1988). This jump in RM signals the point at which the thermal particles and tangential magnetic fields in the ICM are compressed by the shock due to the supersonic advance of the primary hotspot. Detecting this radio quiet bow shock both confirms the double shock structure expected at a jet terminus, and dictates the three dimensional geometry of the source. It also is relevant to the interpretation of the RM screen. The observed RM change at the bow shock implies pre-shock intracluster fields of 8 μG. Hence, Carilli et al.. propose a simple model in which the large-scale RM distribution towards Cygnus A (amplitudes up to 5000 rad m^{-2}, typical scale-sizes $\approx$ 10 to 30 kpc) is caused by the unperturbed cluster atmosphere, while small scale fluctuations (amplitude $\leq$ 1000 rad m^{-2}, scale-sizes $\leq$ 5 kpc) can result from the interaction of the source and the ambient medium.

An alternative model has been proposed by Bicknell et al. (1990), in which the RM screen is located in a thin mixing layer along the contact discontinuity (CD) between the shocked ICM and the radio lobe. In support of this model they point out that the 'striations' in the RM distribution perpendicular to the length of the lobes are suggestive of KH instabilities along the CD. Arguments against this model are the perfect quadratic dependence with wavelength of the observed PAs, and the lack of any 'depolarized skin' along the edge of the radio lobes.

2. Extreme RMs towards Cluster Center Radio Sources

Since the detection of extreme RMs towards Cygnus A, a number of studies have shown that such extreme RMs are characteristic of radio sources at the centers of dense X-ray emitting cluster atmospheres. A summary of these results can be found in Taylor, Barton, and Ge (1994). The most extreme case to date is that of 3C295, which shows RM values up to 20000 rad m^{-2}. A good correlation between cluster X-ray 'cooling flow' rate, and maximum observed RM values, has been demonstrated by Taylor et al... The sources in their study have various radio luminosities and morphological types, ranging from FRI (edge-darkened) sources such as Virgo A and Hydra A, to FRII (edge-brightened) sources such as Cygnus A. The single unifying element for these sources is that they all are located within 100 kpc of the X-ray cluster center. Standard models for radio source hydro dynamic evolution imply that the interaction between the ICM and the radio source is significantly different for FRI vs. FRII sources. This difference can be used to argue that extreme RMs are predominantly a large scale cluster phenomenon, rather than strictly a result of source-ICM interaction.

In parallel with imaging studies of RM screens towards cluster center radio sources, a number of studies have been made of RMs for background radio sources at various impact parameters towards known optical clusters (Kim et al. 1990, Hennessey et al. 1989). Theses studies are different than the imaging studies, in that: (i) the clusters are optically selected, (ii) the impact parameters involved are typically large (out to a few Mpc in some cases), and (3) the RM values are calculated from integrated radio source emission. Still, the results show a clear, and large, increase in the scatter of RM values for sources at impact parameters below a couple hundred kpc. The important implication is that extreme RMs are strictly a cluster center phenomenon. It can be shown (using typical X-ray cluster density profiles) that the bulk of the integrated RM through a cluster occurs in the first 100 kpc or so of pathlength from the cluster center.

In summary, the detection of extreme RMs towards Cygnus A, and subsequently in other cluster-center radio galaxies, shows that the thermal cluster atmospheres must be substantially magnetized, with fields $\geq$ a few μG. In most cases the pressure in the intracluster fields is below the thermal energy density (e.g. the plasma β-factor = ratio of thermal to magnetic pressure is about 30 for the Cygnus A cluster), implying a minor dynamical role for the fields. Even dynamically unimportant fields alter substantially the thermal conductivity of the plasma, and hence are very important when considering the cooling and heating of the ICM (Sarazin 1988). Models for the origin of intracluster fields include: the dynamo action of turbulent wakes of galaxies, injection of fields by previous outbursts of the radio source, ram pressure stripping of fields from galaxies, and amplification of a primordial field (see Sarazin 1988). Another possible mechanism for generating large scale cluster fields is amplification of seed fields during the merger of two clusters. Eilek (1995) presents a detailed 'Zeldovich rope dynamo' model for field generation in a cluster with net-helical turbulence, presumably driven by galaxy motions. She finds that the fields eventually evolve to equipartition strengths. She predicts RM distributions towards cluster center sources which compare well with observations, both in amplitude and structure.

The correlation between extreme RMs and cluster X-ray luminosity raises the possibility of searching for high redshift clusters by using radio polarization measurements. Detection of cluster atmospheres at redshifts ≥ 2 would present a severe challenge to current models for formation of such atmospheres, and for formation of large scale structure in general. We have begun a search for extreme RMs in a large sample of high redshift radio galaxies (Carilli et al.. 1996). In a sample of 40 radio galaxies at $z > 2$ we find a lower limit of 20% to the fraction of sources in the sample which show rest-frame RMs ≥ 1000 rad m^{-2}. The most extreme case thus far is 1138-262 at $z = 2.156$, which shows a maximum RM magnitude of 6250 rad m^{-2} (Röttgering et al. 1995, Athreya et al., this volume). The naive conclusion, based on results for lower redshift radio galaxies, would be that a substantial fraction of high redshift sources are situated at the centers of dense, X-ray emitting cluster atmospheres. This conclusion could be verified through deep X-ray imaging of a few extreme RM sources at high redshift.

References

Alexander, P., Brown, M.T., Scott, P.F. 1984, *MNRAS*, **209**, 851.
Bicknell, G.V., Cameron, R.A., Gingold, R.A. 1990, *ApJ*, **357**, 373.
Carilli, C., Röttgering, H., van Ojik, R., et al. 1996, *Ap.J. (Supp.)*, submitted.
Carilli, C.L., Perley, R.A., Dreher, J.W. 1988, *ApJ*, **334**, L73.
Carilli, C.L., Perley, R.A., and Harris, D.E. 1995, *M.N.R.A.S.*, 270, 173
Dreher, J.W., Carilli, C.L., Perley, R.A. 1987, *ApJ*, **316**, 611.
Eilek, J.A. 1995, *ApJ*, in press.
Hennessey, G., Owen, F., and Eilek, J. 1989, *Ap.J.*, **347**, 144.
Kim, K., Kronberg, P., Dewdney, P., Landecker, T. 1990, *Ap.J.*, 355, 29
Mitton, S. 1971, *MNRAS*, **153**, 133.
Perley, R.A. et al. 1996, in *Cygnus A: Study of a Radio Galaxy*, eds. C.L. Carilli and D.E. Harris (Cambridge: CUP), p. 168.
H. Röttgering, R. van Ojik, G. Miley, and K. Chambers, 1995, *A & A*, submitted.
Sarazin, C.L. 1988, *X-ray emissions from clusters of galaxies*, Cambridge University Press, Cambridge.
Slysh, V.I. 1966, *Sov. Astr.*, **9**, 533.
Taylor, G., Barton, E., and Ge, J.-P. 1994, *AJ*, **107**, 1942.

RADIO POLARISATION STUDIES OF GALAXIES AT Z > 2

R.M. ATHREYA[1], V.K. KAPAHI[1], P.J. MCCARTHY[2]

[1]*NCRA(TIFR) - Pune univ. campus, Pune 411 007. INDIA*
[2]*OCIW - 818, Santa Barbara Street, Pasadena, CA. USA*

AND

W.VAN BREUGEL
IGPP/LLNL - P.O.Box 808, Livermore, CA 94550. USA

We present some interesting results from a radio study of 17 radio galaxies at z >2, selected from a complete sample of Molonglo sources with $S_{408} >$ 0.95 Jy for which optical identifications and spectroscopic redshifts are being obtained in a major observational programme (Kapahi *et al.* 1995).

Observations and results : The 17 galaxies were mapped using the VLA at 1.4, 4.8 and 8.3 GHz with the highest resolution of $\sim 0.3''$ at 8.3 GHz. The integrated spectra are generally quite steep and convex in many sources, steepening to $\alpha \geq 1.2$ ($S \propto \nu^{-\alpha}$) at the highest frequency. Most of the sources have a double structure with sizes ranging up to $22''$. Unresolved radio cores, coincident with the optical galaxies, are detected at the mJy or sub-mJy level in most cases. Several sources also show one-sided jet-like features. Surprisingly, most of the cores appear to have steep spectra between 5 and 8.3 GHz!

Large rotation measures (RM) of a few 100 rad/m^2 are common in these sources; several show RMs of over 1000. The galaxy 1138-252, at z = 2.17, has an RM of 5700. The polarisation at 15-20 GHz (emitted) is typically 5-10%. Most sources show strong depolarisation by 5 GHz (emitted), probably as a result of beam depolarisation. The asymmetry in the radio lobe properties in most sources is, perhaps, a pointer to the irregular and clumpy nature of their environment.

Steep spectrum cores : Eight of the 12 cores detected in the sub-arcsecond images at both 4.7 and 8.3 GHz have a steep spectral index ($\alpha > 0.5$) with $\alpha_{med} = 0.75$. This is in sharp contrast to the flat spectra ($\alpha < 0.5$) of the cores (believed to arise from superposed spectra of syn-

163

R. Ekers et al. (eds.), Extragalactic Radio Sources, 163–165.
© 1996 *IAU.*

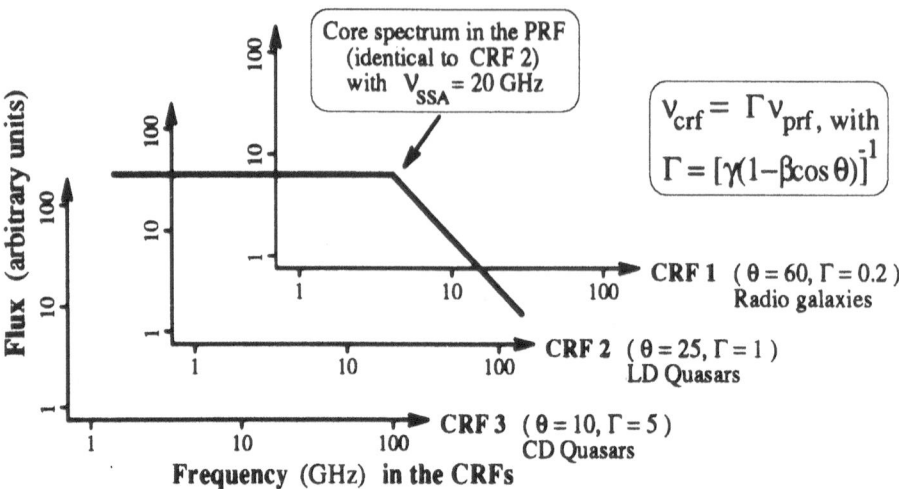

Figure 1. Relationship between the cosmological (**crf**) and the plasma (**prf**) rest frames.

chrotron self-absorbed components) of almost all extended radio sources at lower z and core-dominated quasars (CDQ) at all redshifts.

We identify the steep spectra of cores in high z galaxies with the optically thin synchrotron emission at frequencies above the turn-over of the smallest component in the core. The relation between the frequencies in the cosmological rest frame (**crf**) of the galaxy and the rest frame of the emitting plasma in the relativistic jet (**prf**) is shown in figure 1. Relativistic motion in the cores would lead to a blueshift in objects viewed close to the jet axis and a redshift in objects viewed perpendicular to it. In the context of galaxy-quasar unification (Barthel 1989), the synchrotron turn-over should therefore occur at higher frequencies in the **crf** as one goes from radio galaxies to lobe dominated quasars (LDQ) to CDQs. Indeed, CDQs are known to turn over at ∼50-75 GHz (Gear *et al.* 1994) while some LDQs do so at ∼20 GHz (Antonucci *et al.* 1990) in the **crf**. The corresponding value for our galaxies is estimated to be ∼ 5 GHz which explains why only high z galaxies appear to have steep spectrum cores at the observed frequencies of 5-10 GHz.

We derive physical parameters of the "smallest component" radiating in the cores of these sources using the observed turn-over parameters and the doppler factors. Using $\gamma \sim 5 - 15$ and the typical viewing angles for galaxies and quasars in Barthel's scheme, we obtain a size of ∼1 parsec, magnetic field of ∼1 gauss and electron density of ∼1000 cm^{-3} for the cores of both galaxies and quasars. While these numbers are only order of magnitude estimates, the similar values for quasars and radio galaxies argue for similar physical phenomena and conditions in their cores.

We emphasize that the physical size derived is over a 100 times smaller

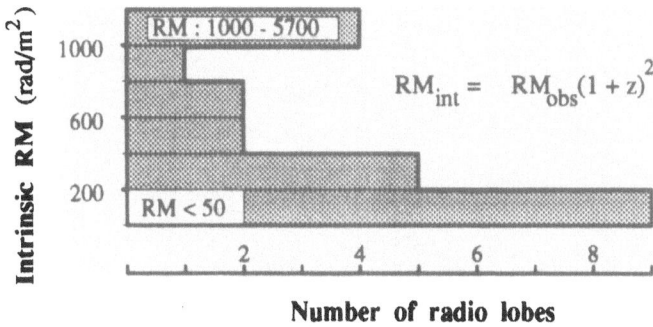

Figure 2.   Histogram of the intrinsic RM of the lobes of high z galaxies.

than the telescope resolution. This phenomenon is therefore very useful for probing radio galaxy cores, most of which are too faint for VLBI studies.

Large rotation measures :   Sources at low z typically have RMs of a few 10 rad/m^2. Only 2 radio lobes in our sample have RM < 50 (another 4 are consistent with RM < 50 within 1σ errors). Five of the 13 galaxies with a measured RM have RM > 800; the corresponding fraction at low z is ≤ 5%. The low z galaxies with large RMs are predominantly in dense x-ray clusters with cooling flows, apart from a few which are compact steep spectrum sources (Taylor *et al.* 1992). The one common feature in high RM sources at all z appears to be a dense environment.

After considering various possibilities for the Faraday material responsible for the RM (from our Galactic ISM to plasma within the radio lobes), we suggest that the most promising option, within the constraints of existing models of galaxy formation, is the dense and magnetized hot gas in the intracluster medium interacting with the bow shock of the radio hotspot. A cluster scale Faraday screen with aligned magnetic field (due to a cooling flow), as has been invoked for lower z objects, is another possibility, though less compelling as it is not clear if there is sufficient time for a cooling flow to start by z ≥ 2.5. However, both these models require magnetic fields of a few μ *gauss* correlated over scales of a few kiloparsec threading the intracluster medium by z = 3. An explanation for the high RMs at high z must necessarily deal with questions regarding the origin of the magnetic field and may have implications for models of galaxy and cluster formation.

References

Antonucci, R., Barvainis, R. and Alloin, D. (1990) *Ap.J*, **Vol.353**, pp.416-418.
Barthel, P.D. (1989) *Ap.J*, **Vol.336**, pp.606-611.
Gear, W.K. *et al.* (1994) *MNRAS*, **Vol.267**, pp.167-186.
Kapahi, V.K. *et al.* (1995) The Molonglo 1 Jy sample of radio galaxies - *This Volume*.
Taylor, G.B., Inoue, M. and Tabara, H. (1992) *Astron.Astrophys.*, **Vol.264**, pp.421-427.

CYGNUS A:

SOME TOPICS FROM THE WORKSHOP

D. E. HARRIS
Smithsonian Astrophysical Observatory
60 Garden St. Cambridge, MA, 02138 USA

1. Introduction

The international workshop on Cygnus A was held in Green Bank WV, USA and was attended by 45 participants who reported on radio, infrared, optical, UV, X-ray, and theoretical results. Rather than attempt to cover all of the papers presented at the workshop, we have selected 8 topics. We apologize to those whose work is not mentioned here. Because of the limitations of length, we have foregone the luxury of figures and we refer to articles in the Cygnus A book (to be published early in 1996 by Cambridge University Press) by first author only. All references to page and figure numbers refer to the Cygnus A book.

2. Central Region

2.1. THE HIDDEN QUASAR HYPOTHESIS

The currently popular model for 'unification' is based on the concept that quasars and radio galaxies are intrinsically the same; only the viewing angle produces the different characteristics by which we assign them to different classes. An integral part of this model is the presence of an obscuring torus which blocks optical and UV emissions from the nucleus and broad line region when the jet emission axis lies more or less in the plane of the sky rather than pointing towards us. Since Cygnus A is a high luminosity radio source, many observers predicted that it would be an ideal test of the unification model and many attempts have been made to look for the observable consequences. The results have been mixed.

   With long slit spectra, Tadhunter and Cabrera-Guerra searched for evidence of an EMISSION CONE OF IONIZING RADIATION, expected to

167

R. Ekers et al. (eds.), Extragalactic Radio Sources, 167–174.
© 1996 *IAU.*

escape along the axis of the torus (i.e. coincident with the radio jet). Tadhunter plots the variation in the [OI]/[OIII] ratio for an off nucleus slit
perpendicular to the radio jet. He finds a 'horseshoe' shaped curve (fig4,
p36), "strong evidence for an anisotropic ionizing radiation field". Cabrera-
Guerra found a similar effect for [OII]/[OIII] (fig5, p30), but Tadhunter
remarks that when calculating the absolute number of ionizing photons,
there is less ionizing radiation from the nucleus than most quasars and
there is no evidence for horseshoe patterns for [NII]/Hβ or [SII]/Hβ.

The FEATURELESS BLUE CONTINUUM has been studied by many
workers to see if it might be evidence for scattered light from the nucleus,
either by dust or by free electrons. Other possibilities for its origin are star
formation regions or free-free emission. Antonucci finds no Balmer edge
(expected from A stars) so he concludes it is not predominantly normal
star light. The polarization is found to be 2.5% rather than the expected
30% for scattered light (Stockton, Tadhunter). Furthermore, the narrow
lines are not polarized. Therefore, scattered light is not the dominant component, and free-free emission is also a minor contributor (Stockton, p3).
The general consensus was that star formation dominated, an explanation
which was bolstered by an HST photo which shows extreme knottiness.

Is there any evidence for a BROAD LINE REGION? So far, Antonucci's
MgII line are the only data which support the notion of reflected light from
a BLR. Stockton failed to find a broad component in some infrared lines
(<800km/s), Ward found no broad Paα, and no scattered component of
emission lines was found (e.g. Hβ or in polarized light; Stockton, Jackson).

There were doubts expressed about the POWER of the hidden quasar.
It was pointed out by several participants that the power from the core
in radio, infrared, and X-rays is characteristic of a rather modest quasar;
Cygnus A can be classified as a high luminosity source only with respect
to its radio lobes. If there is a quasar-like nucleus, it is of medium or low
power.

If there is an obscuring torus, the dust should have associated CO and
other molecular lines in absorption. Barvainis searched for CO in absorption
but failed; $\tau < 0.6$; $N < 10^{16} cm^{-2}$. There is also no ammonia, which is a
more complex molecule than CO, and hence may not be as amenable to
some of the explanations - see below - for the lack of CO; $\tau < 0.1$. Maloney
suggested methods to save the molecular torus via radiative excitation.
Another suggestion for the paucity of CO could be that the gas is mainly
atomic; if NH $< E23.5 cm^{-2}$, Maloney argues that the gas will be atomic.
Blanco detects HI (fig1, p71), but no formaldehyde nor OH. From various
arguments, Blanco deduces that the most likely location of the HI is 4pc
$< r < 1$ kpc, thus it is still an open question if the observed HI can be
associated with a (small) obscuring torus. VLBI observations are planned.

Another necessary consequence of an obscuring torus is ABSORPTION of SOFT X-RAYS. Arnaud reported on GINGA and ASCA data, providing evidence for a hard x-ray component with a power law. These conclusions depend on spectral fitting; both detectors have poor spatial resolution although the characteristic 'X' shaped Point Response Function of ASCA is visible in the 6-9 keV image, showing that the hard source dominates over that of the extended gas at these energies; the inferred diameter is $< 1'$. The visual extinction estimates cover a wide range: Av = 40 from the Lx vs. L(Hα) relation together with Paα/Hα=0.1 (Ward); Av = 120$\pm$25, from a comparison of 3.5 micron and hard X-ray intensities (Ward); and Av > 150 (log N = 3.7E23) from x-ray spectral fitting (Arnaud). A somewhat perplexing complication is that with the ROSAT HRI (resolution 5") we have detected a discrete source coincident with the radio nucleus to 0.5". For the large extinctions found, essentially no photons in the ROSAT band should be visible. One escape from this problem was mentioned: the observed X-rays could be scattered core emission, thereby following a path that evades passage though the torus.

Generally, the evidence for a hidden quasar remains weak. What was once considered as the great hope and test of the unification scheme has failed to deliver conclusive proof. While it is true that various ways have been found to explain the failures, it seems to us that viewing angle is not the sole difference between radio galaxies and quasars. The unification scheme that appears more promising is that which depends on the fundamental source of the energy release around black holes:

Predominantly gravitational/accretion => radio quiet Q/AGN
Predominantly rotational energy => radio galaxy
Both sources are comparable => radio quasar

2.2. DO THE JETS CAUSE OBSERVABLE EFFECTS IN THE ISM?

There was considerable speculation about the cause of the DARK CHANNEL in the NW emission region about 1.3" from the nucleus which coincides with the inner radio jet at a point where the VLA jet intensity reaches its first minimum (Cabrera-Guerra, fig2, p27). Although several of the optical experts opined that its sharp edges suggested dust obscuration, no one offered a plausible explanation for dust enshrouding this segment of the jet and it was later pointed out that there is no evidence for reddening in this area. If the emitting region has been evacuated by the jet, then the region must be 2 dimensional, a possibility that most considered to be unlikely.

Is there evidence in the Extended Emission Line Region for SHOCKS from the RADIO JET? Stockton finds higher excitation (e.g. [OIII]) mainly aligned with the jet; away from the jet, to north and south, the excitation is lower([NII]). From line diagnostics, Clark points out that Cygnus A emission is similar to that found from radio knots close to the nucleus in other powerful radio galaxies rather than that from hotspots and lobes. The former type is characterized by higher ionization and smaller line widths, interpreted to favor photo-ionization over the collisional excitation which would have been evidence for shocks. At IAU175, Bicknell remarked that strong shocks can produce ionizing radiation which thus becomes the intermediary for line excitation which appears to be photo-ionized even though the primary energy is supplied by shocks.

There is also evidence for HIGH VELOCITY GAS. Stockton finds components of [OIII] at 1400 and 1700 km/s blue shifted near the western jet and at 700 km/s redshifted on the other side of the nucleus. He notes however, that the western location is 700 pc from the radio jet. From narrow slit spectra Cabrera-Guerra show that line widths of [NII] increase near the western side of the jet (fig4, p29).

We find these data to be convincing evidence of jet/ISM interaction, although the precise details are far from clear.

3. The Jets

3.1. THEORETICAL ASPECTS

Both Lovelace and Roland develop models which accommodate gamma ray production (as in blazars). Lovelace proposes a Poynting flux jet with eventual pair production and IC radiation from pairs scattering the UV from the nucleus. Roland also uses pairs for gamma ray production, which occurs about 2 light weeks out from nucleus. He suggests a two fluid jet consisting of relativistic e^+/e^- to produce the VLBI structures and gamma rays, and a mildly relativistic p^+/e^- component for the large scale jet and to provide the energy transport to the hotspots and lobes. Hardee's simulations show that a light jet breaks up. However, with a jet density = 4x the external density, the jet is able to propagate for long distances.

For Roland's two fluid jet, the total internal pressure = (thermal+rel protons+rel electrons+B). For the jet to propagate, this must be > external pressure. If the jet is weak, and $\rho v^2 <$ the external pressure, the jet never propagates to large distances, forming instead a bright, short jet like M87. Thus Roland suggests that the difference between FRI and FRII depends on the ram pressure vs. the external pressure; i.e. both the strength of the jet and the ICM (or ISM) pressure are important.

3.2. THE MORPHOLOGY OF THE LARGE SCALE JET

On the kpc scale, the observed morphology of the western jet as it leaves the galaxy and enters the lobe consists of regularly spaced brightenings with a wavelength of 7″, about 10 times the jet radius (Carilli, fig5, p82). These features are inclined about 7°, and could, like M87 be a sort of braiding. Hardee presented the results of simulations of emission and dynamical properties of magnetized jets (a dynamically important magnetic field is included). For a Mach 8 jet with density = 4x the external density, he finds instabilities (pinch, helical, kinks, fluting) which propagate as wave modes down the jet. Both surface waves and body waves are present, and this combination is able to produce oblique filaments (caused by an enhanced B field) from phase differences of helical and elliptical twists (fig1, p115).

Clarke suggests that the physical jet could be much wider than the observed radio jet. This helps the theoretical problem of jet stability. The thin radio jet would then be one edge of the physical jet, and the bifurcated section would just be a piece where both edges were illuminated.

4. The Hotspots

The chief uncertainties for the hotspots are the X-RAY EMISSION PROCESS, the ENERGY DENSITY and MAGNETIC FIELD STRENGTH. The jet composition (i.e. the presence or absence of protons) is part of the answer to these questions. Are the x-rays really synchrotron-self Compton (SSC) photons or might they arise from synchrotron emission (i.e. from the so-called 'proton induced cascade', PIC, process)? What are the energy density and pressure? Does equipartition hold?

The observed spectrum from the radio to optical was reviewed by Roeser (fig2, p125). Hotspot D is empty, A appears to be an M star, and B has another star in the foreground (fig1, p122). Fitting the observed spectra with the usual Fermi acceleration models, Roeser finds, for hotspot A, a break frequency of 5.5 GHz, a cutoff frequency of 9E12 Hz, and a B field of 340 μG. The corresponding values for hotspot D are 10.8 GHz, 8E12 Hz and 400 μG.

Harris reviewed the situation for SSA and PIC as the genesis of the observed X-rays (ROSAT HRI results). For SSA emission, magnetic field strengths are 160 μG (hotspot A) and 245 μG (D). These field strengths are strict lower limits, and they also agree with equipartition fields for the case of little or no energy contribution from relativistic protons. If, however, protons dominate the particle energy density, then either the hotspots are far from equipartition (with a weak field and SSC X-rays) or they would have a field $\geq$ 500 μG and SSC would be a minor contributor to the observed X-rays. In this case, some other process such as the PIC would have to supply the extremely energetic electrons required to produce synchrotron X-rays. Note that the weak field/no proton scenario requires the absence of protons in the hotspot and therefore the jet itself could not contain protons.

An interesting ramification of the PIC process is the generation of ultra high energy protons (to initiate the cascade, protons with Lorentz energy factors > 3E11 are required). Biermann discussed the implications of this for the highest energy cosmic rays observed at the Earth. When the cosmic ray energy is above 3E18eV, they cannot be contained by our galaxy. At this energy, the main constituent changes from medium heavy nuclei to protons. Arrival direction suggests the super galactic plane (fig3, p145).

5. Equipartition in the Radio Lobes

Equipartition between magnetic field strength and energy in relativistic particles has been questioned in three ways. In each case, the suggestion is that the particle energy density dominates over the magnetic field energy density. From numerical simulations with 3D hydro code (plus a trace magnetic field), Clarke is able to generate synchrotron emitting filaments in the lobes (aka 'cocoons'). They appear to be features with enhanced field strength, formed by stretching the B field; therefore Clarke reasons that u(B) < u(gas) + u(particles) and, if filaments are seen (as is the case for Cygnus A), it is probable that the B field is significantly below equipartition (NB: in this case 'equipartition' refers to that between the B field and the relativistic particles + the thermal gas, if the latter is present).

Carilli claims that pressure balance with the external medium requires more than the equipartition field. Furthermore, he calculates that the minimum pressure in the lobes is 7 times less than P(jet). Thus increasing P(lobes) by relaxing equipartition will: confine the jet, balance the external thermal pressure which is known from X-ray data, and allow the jet density > lobe density, providing stability for the jet.

Alexander presents arguments based on ram pressure, advance speed, and aging. Aging arguments give the velocity of separation between the hotspots and lobe material: v = 0.04c to 0.06c; whereas if $\rho v^2 = $ P(hs) - P(ext), and the dentist drill model is assumed, (i.e. the average advance speed is reduced by area hotspot/area lobe), then v(hs)=0.005c. Therefore, the B field in the lobe needs to be below B(equip) to change the aging results. To achieve agreement with the advance speed of the hotspots, P(lobe) > 10 P(min).

While each of these approaches has merit, for the most part, they are model dependent (e.g. if the hotspot internal pressure is much higher than minimum, the advance speed goes up, and the third argument is negated).

6. The Cluster Gas

6.1. THE EFFECTS OF THE LOBES AND THE BOW SHOCK

Clarke has studied various X-ray features evident on the ROSAT HRI image (fig6, p207) with numerical simulations using a 3D hydro code. For a Mach 6 jet with a density = 0.02 x (central density of a King distribution), Clarke was able to reproduce the relative depression in the observed X-ray brightness from the evacuated inner part of the radio lobes. In addition, the higher brightness twin X-ray features most obvious on the eastern side of the source were reproduced in the simulation by thicker parts of the sheath between the bow shock and the radio lobe (fig3, p204). The only X-ray feature not yet understood is the asymmetry between the eastern and western sides. Although the depressions appear on both sides, the twin enhancements of the sheath are almost absent on the western side.

6.2. HOW ROBUST IS THE COOLING FLOW PARADIGM?

The large scale morphology of the X-ray distribution was reviewed by Reynolds from ROSAT PSPC data. Cygnus A lies close to the center of a high brightness, quasi spherical distribution which, in turn, lies at one end of an elliptical region of relatively low surface brightness. The spherical distribution has a diameter of order 150 kpc whereas the long axis of the ellipse is greater than a Mpc. From spectral fits of the x-ray data, the temperature varies from 3 keV (inner part) to 7 keV (outer part). The latter value is in agreement with the GINGA value (fig3, p196). Arnaud reported on the ASCA results: 9 keV for the low brightness gas and 4 keV for the high brightness region. From a deprojection of the ROSAT data, Reynolds finds a central density of $0.03 cm^{-3}$, a cooling radius of 180 kpc (H=50), and a mass deposition rate at the center of 250 $M_\odot$/yr.

There is no question that there is cooler gas in the center and that the cooling time is shorter than the age of the universe. However, in our view, the 'flow' part of the picture remains to be demonstrated. If there is a flow, where does the 250 $M_\odot$/yr go? It would also be useful to see a careful energy budget for the radio source. The main problem with this is that we do not know the actual power of the jet, so we do not know how much energy gets dumped into heating the ICM (by the jet, hotspots, lobes, and bow shock).

7. Conclusions

- HIDDEN QUASAR: Based on the evidence presented at the workshop, there might be a weak quasar-like nucleus but many of the predicted effects are *not* seen.

- EFFECTS of the JETS on the ISM: Probably present, although these effects may be more pronounced in other sources.

- WHAT ARE JETS MADE OF? Not yet determined.

- CAN WE EXPLAIN the JET MORPHOLOGY? Simulations are very promising with various instabilities reproducing regular spacing, etc

- HOTSPOT EQUIPARTITION and ENERGY DENSITIES: If we could have faith in the accuracy of ram pressure estimates, we could get the total pressure inside, and with that, could probably untangle the components.

- ARE the LOBES in EQUIPARTITION? Maybe not. Several lines of argument were presented that the B field is weaker than other components.

- DO WE UNDERSTAND the main X-RAY FEATURES? Yes (via simulations), but some asymmetries remain unexplained.

- Do 'COOLING FLOWS' REALLY FLOW? Is Cygnus A one of the most powerful heat engines in the universe?

We acknowledge support from NASA grant NAG5-2658 and NASA contract NAS5-30934.

The following diagram shows how some of the questions reviewed above are inter-related.

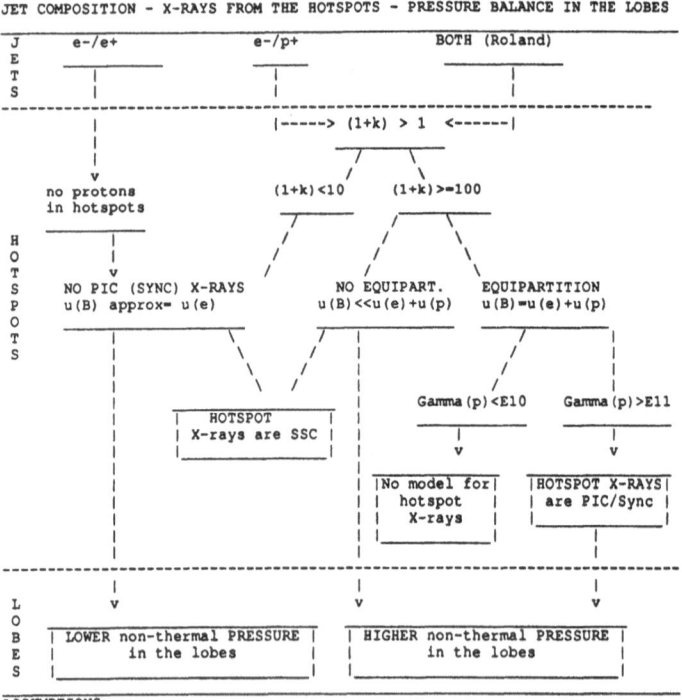

```
JET COMPOSITION - X-RAYS FROM THE HOTSPOTS - PRESSURE BALANCE IN THE LOBES

J     e-/e+                    e-/p+              BOTH (Roland)
E
T      _____                _____            _____
S       |                       |                   |
      --------------------------------------------------------------
        |                |-----> (1+k) > 1 <------|
        |                       /       \
        v                      /         \
     no protons           (1+k)<10    (1+k)>=100
     in hotspots              /       / \
H        |                   /       /   \
O        |                  /       /     \
T        v                 /       /       \
S     NO PIC (SYNC) X-RAYS    NO EQUIPART.    EQUIPARTITION
P     u(B) approx= u(e)       u(B)<<u(e)+u(p)  u(B)=u(e)+u(p)
O      _____              _____          _____
T        |      \          / |                 /     |
S        |       \        /  |                /      |
         |        \      /   |        Gamma(p)<E10  Gamma(p)>E11
         |         \    /    |            |           |
         |      |  HOTSPOT  |  |          |           |
         |      | X-rays are SSC | |      v           v
         |      |_____|  |     |No model for|  |HOTSPOT X-RAYS|
         |                     |     | hotspot    |  | are PIC/Sync |
         |                     |     |  X-rays    |  |_____|
         |                     |     |_____|        |
      ------------------------------------------------------|---------
        |                     |                             |
L       v                     v                             v
O     | LOWER non-thermal PRESSURE |  | HIGHER non-thermal PRESSURE |
B     |      in the lobes          |  |      in the lobes           |
E     |_____|  |_____|
S
```

ASSUMPTIONS:
a) No significant acceleration in the lobes
b) No significant entrainment of ambient gas; i.e. the jets, hotspots, and
   lobes are 3 phases of the same fluid.
NOTES:
a) k is the ratio of energy in relativistic protons to that in electrons.
b) u is the energy density
c) gamma is the Lorentz energy factor; gamma(p) is the maximum energy
   for the power law distribution of relativistic protons.

FIRST MILLIMETER MAPPING
OF THE JET AND NUCLEUS OF M87

V. DESPRINGRE
Observatoire Midi-Pyrénées
14 Avenue Edouard Belin, 31400 Toulouse, France

AND

D. FRAIX-BURNET
Laboratoire d'Astrophysique de Grenoble
BP53X, 38041 Grenoble Cédex, France

An intriguing question about extragalactic jets is why they are so few being seen at optical wavelengths, or equivalently, why the cutoff frequency of the synchrotron radiation is generally not in the optical, but rather in the infrared or even in the sub- millimeter domain. The answer is undoubtedly related to the efficiency of the acceleration of the relativistic electrons responsible for the synchrotron emission. The presence of a break at low frequency somewhere in the synchrotron spectrum is another feature that constrains the model parameters, but its precise location is unknown for most jets, because of the lack of photometry in the millimeter domain. It was thus necessary to fill the gap between radio and optical wavelengths in the synchrotron spectrum of optical jets. The required observation had to be of high sensitivity and high spatial resolution (of the order of $1''$). Another reason for observing at millimeter wavelengths is that molecular lines and thermal emission from cold dust are detectable in this frequency range.

The present work is described in Despringre et al. (1995). The observations were made with the IRAM Plateau de Bure interferometer with three antennas between November 1992 and January 1993 with the BC set of configurations, and with four antennas in the two compact configurations C2 and D in March, 1994. The central frequency was 88.26 GHz. The final map of M87 at 89 GHz after CLEANing is shown in Fig. 1. The beam is $2''9 \times 1''8$ (PA=22°).

R. Ekers et al. (eds.), Extragalactic Radio Sources, 175–176.

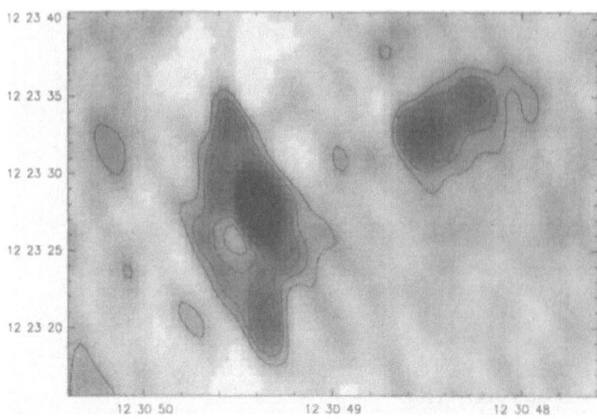

Figure 1.   Grey map of M87 with contour level spacing of 35 mJy/beam and a peak intensity of 1.67 Jy/beam. The beamwidth is 2″9 × 1″8 at PA=22°.

The jet with knots A, B, C and G, is clearly seen. Knots D, E, F, and I, located between the nucleus and knot A, as well as the lobes, are not detected. Within the error bars, our mm intensities for knots A, B, and C fit the linear interpolation between cm and optical wavelengths.

The intensity of the nucleus alone at 89 GHz is 1.88 ± 0.1 Jy. The high resolution spectrum of the nucleus is consistent with a single power law synchrotron spectrum with a break frequency between about 10^{11} and 10^{12} GHz.

It is clear that, in our data, the nucleus is embedded in an extended structure. Since such a component has never been noticed before at any wavelength, we thoroughly looked for any instrumental or deconvolution effect, but found none. Hence, either it is an important artifact caused for a subtle reason not understood at this time, or this feature is real. Indeed, considering other properties of M87, several arguments are in favor of its reality, and its spectrum is rather narrowly peaked at $\simeq 100$ GHz. This extended feature is characterized by an essentially elongated shape ($12″7 \times 6″5$) perpendicular to the jet but not centered on the nucleus, the offset being about $1″5$ to the SE. There are 4 peaks, the maximum brightness being 0.2 Jy/ beam, and a "hole" at the center of the structure. The total intensity (after removal of the nucleus) is 1.4 Jy. Clearly, *thermal radiation from very cold dust at $T \lesssim 10$ K* could explain the properties of this component. Estimating the dust mass from the total intensity for a distance to M87 of 15 Mpc, one obtains: $M_{\rm dust} \simeq 4 \ 10^{12} \ M_{\odot}$.

References

Despringre V., Fraix-Burnet D., Davoust E., 1995, *A&A* in press

UNUSUAL RADIO STRUCTURES IN MARKARIAN 6

M.J. KUKULA[1], A. PEDLAR[2], S.A. BAUM[3], A.J. HOLLOWAY[4]

[1] *Institute for Astronomy - University of Edinburgh,*
Royal Observatory, Edinburgh EH9 3HJ, U.K.
[2] *NRAL - University of Manchester, Jodrell Bank, Macclesfield*
SK11 9DL, U.K.
[3] *STScI - 3700 San Martin Drive, Baltimore, U.S.A.*
[4] *Department of Physics & Astronomy*
University of Manchester, Manchester, U.K.

Markarian 6, a Seyfert Galaxy of type 1.5, is one of several Seyfert nuclei to be observed with MERLIN at 6 and 18 cm as part of a project to investigate the radio structures of these objects on sub–arcsecond scales (corresponding to scales of a few tens of parsecs at typical distances). The angular resolution of MERLIN at 6 cm is equivalent to that of the HST, making the radio images ideal for comparison with HST images of the optical Narrow–Line Region (NLR). In this paper we briefly discuss the results of our MERLIN observations of Markarian 6, along with a 6–cm Westerbork Synthesis Radio Telescope (WSRT) map of the arcsecond (kpc) scale radio emission. The data is discussed in more detail by Kukula *et al.* (1996) and Baum *et al.* (1993).

The 6–cm MERLIN map reveals one of the best examples of a highly-collimated radio jet ever seen in a Seyfert galaxy. This is all the more remarkable considering that Mrk 6 is a Type 1.5 object. Recent studies have shown that whilst approximately 50% of Seyfert 2s feature some kind of extended radio structure, Types 1 and 1.5 almost invariably contain only a single unresolved point source (Kukula *et al.* 1995).

The jet is highly collimated, extending for 645 pc (assuming $H_o = 75$) along PA 180°, and is strikingly similar in size and appearance to the jet in Markarian 3, a nearby Seyfert 2 (Kukula *et al.* 1993). The jet contains 6 distinct radio knots, but none of these are obviously associated with the peak of the optical continuum emission (marked by a cross on the contour

R. Ekers et al. (eds.), Extragalactic Radio Sources, 177–178.

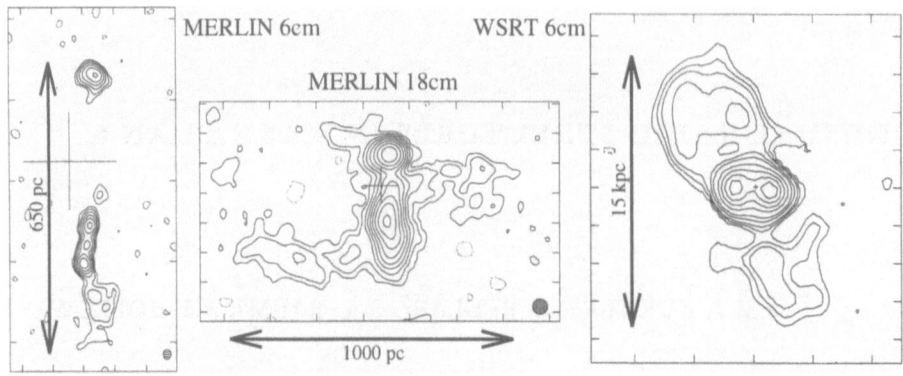

plots). However, the overall shape of the jet corresponds closely to the structure of the optical NLR as shown in the HST image of Capetti *et al.* (1995).

On a slightly larger scale, the 18 cm MERLIN map shows several highly unusual features which emerge from the sides of the jet and extend for several hundred parsecs to the east and west, *ie* orthogonal to the direction of the jet. We know of nothing comparable to these structures in any other Seyfert.

On a still larger scale the 6–cm Westerbork map shows large, edge–brightened lobes of emission to the NE and SW of the bright nuclear component (note that the nuclear component itself has been subtracted from this image), covering a region $\sim 40''$ (15 kpc) in extent. The position angle of the lobes is 30°, similar to that of the minor axis of the host galaxy (40°). We speculate that these structures are in fact shells created by material which is drifting out of the nucleus due to buoyancy effects.

Thus, Mrk6 exhibits complicated radio structures on a variety of scales and position angles. It is interesting to note that if these structures are the relics of a series of outbursts of activity then this suggests that the orientation of the central engine has altered with time.

References

Baum, S.A., O'Dea, C.P, Dallacassa, D., de Bruyn, A.G. and Pedlar, A. (1993) *ApJ*, **419**, 553.

Capetti, A., Axon, D.J., Kukula, M.J., Macchetto, F., Pedlar, A. and Sparks, W.B. (1995), *ApJL, in press.*

Kukula, M.J., Ghosh, T., Pedlar, A., Schilizzi, R.T., Miley, G.K., de Bruyn, A.G., and Saikia, D.J. (1993), *MNRAS*, **107**, 1227.

Kukula, M.J., Pedlar, A., Baum, S.A., and O'Dea, C.P. (1995), *MNRAS*, **276**, 1262.

Kukula, M.J., Holloway, A.J., Pedlar, A., Meaburn, J., Lopez, J.A., Axon, D.J., Schilizzi, R.T. and Baum, S.A. (1996), *MNRAS, in press.*

IRAS 0421+0400: A CURIOUS SPIRAL WITH FLARING JETS

W. STEFFEN AND A.J. HOLLOWAY
Department of Physics and Astronomy, University of Manchester,
Manchester M13 9PL, UK

AND

A. PEDLAR AND D.J. AXON
Nuffield Radio Astronomy Laboratories, University of Manchester,
Jodrell Bank, Macclesfield, Cheshire SK11 9DL, UK

Abstract. We present new VLA observations of IRAS 0421+0400 at 1.4 and 5 GHz, providing dual frequency information and higher resolution than previously available. We find extremely accurate alignment of the central double with the closest (though not brightest) feature in the southern kiloparsec hotspot region. There is a tight relation between the symmetric radio structure and emission line gas in the hotspot regions of this Seyfert 2 type galaxy.

The emission lines in the active galaxy IRAS 0421+0400 show a dramatic ($\approx$ 900 kms^{-1}) increase in the velocity spread at the position of radio hotspots which are located at the beginning of extended radio lobes (Hill et al, 1988; Holloway et al, 1996). In Holloway et al (1996) and Steffen et al (1996) we have investigated a model of a jet crossing a shocked boundary between the ISM/IGM-interface (Loken et al, 1995, and references therein) which could explain the peculiar spectrum and the flaring of the radio structure at approx. 5 arcsec on either side of the galaxy. In this paper we present preliminary results from new VLA observations at 1.4 and 5 GHz, providing dual frequency information and higher resolution than previous observations (Beichman et al, 1985; Hill et al, 1988). The observations were performed on September 1st, 1995, with the VLA in A-array during a 4 hour run.

We find that the hypothetical jets connecting the centre of the galaxy with the lobes stays invisible at the low frequency of 1.4 GHz down to a level of 0.15 mJy/beam. The structure of the southern hotspot region is

179

R. Ekers et al. (eds.), Extragalactic Radio Sources, 179–180.

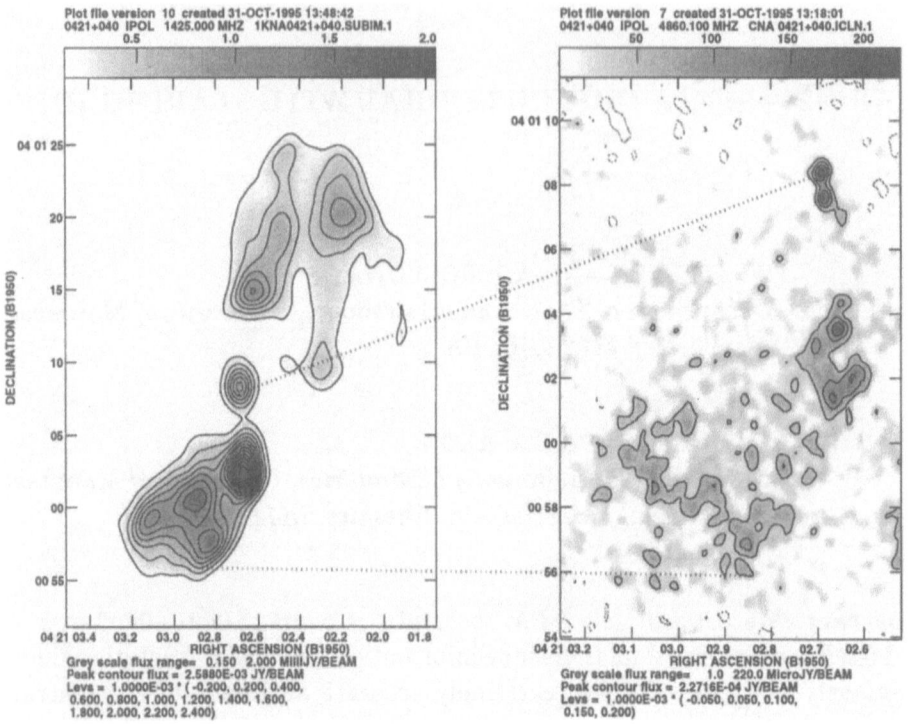

Figure 1. VLA A-array images of IRAS 0421+0400 at 1.4 GHz (left) and 5 GHz. The beamsizes are 1.78 x 1.48 and 0.52 x 0.43 arcsec (PA = −9.5°), respectively. 1 arcsec corresponds to 0.9 kpc (h=1).

well resolved at 5 GHz and is found to be very complex consisting of a bright unresolved hotspot and several filaments. The central double source is extremely well aligned (to within < 1°) with an unresolved low brightness feature located at approximately 1 arcsec inwards from the southern main hotspot. The northern radio emission is consistently of less flux density and no correspondingly well aligned features is found in the northern region.

References

Beichman C., Wynn-Williams C.G., Lonsdale C.J., et al. (1985), *ApJ*, **293**, p.148
Hill G.J., Wynn-Williams C.G., Becklin E.E., MacKenty J.W. (1988), *ApJ*, **335**, p.93
Holloway A.J., Steffen W., Pedlar A., et al. (1996), *MNRAS*, in press
Steffen W., Holloway A.J., Pedlar A., Axon D.J., (1996), Workshop on *Energy Transport in Radio Galaxies & Quasars*, Astronomical Society of the Pacific Conference Series, in press
Loken C., Roettiger K., Burns J.O., Norman M. (1995), *ApJ*, **445**, 80

ANALYSIS OF VLBI SPOT MAP OF H_2O MASERS FOR THE HII REGION COMPLEX IC133 OF SPIRAL GALAXY M33

YU ZHI-YAO
Shanghai Astronomical Observatory, 80 Nandan Road, Shanghai 200030, China

1. Introduction

Interstellar H_2O masers in the Galaxy occur in active star forming regions. The spectrum often shows multiple distinct features. The VLBI maps reveal that each spectral feature corresponds to emission of spatially distinct maser sources (maser spots), whose sizes are $\approx 10^{13}cm$. The maser in M33 is associated with the HII region complex IC133, which has been studied optically by Boulesteix et al. (1974) and Kiwitter and Aller (1981). This maser is the nearest ($\leq 1Mpc$) extragalactic H_2O source visible in the northern sky, although it is not the strongest. It is known to have persisted for over a decade, and its spectra consistently show peak flux densities of $\approx 1.5~Jy$ and at least 10 features spread over $\approx 50~km~s^{-1}$ (Huchtmeier, Eckart and Zensus 1988). Using VLBI Greehill et al. (1990) have obtained the positions of 14 H_2O maser spots.

We have developed a model of multiple distinct rotating and expanding rings, to explain the distribution of those positions.

2. Model and results

The disc geometry and adopted coordinate systems are in Fig. 1. The ring center is assumed as origin of the coordinate system which has the X–Y plane in the plane of the disc and the Z–axis along the ring rotation axis. In general, the disc will be inclined at some angle θ to the line of sight.

For the i-th maser spot in the disc ring we have obtained the following relation (Yu, 1994): $\rho_i^2 = \rho^2 - \frac{\rho^2}{v_e^2} \left(\frac{(v_i - v_0)}{cos\theta} - \omega \rho_i D \right)^2$ where ρ_i is the projected angular distance of the spot from the origin, and v_i its radial velocity, ρ is the radius of the generic disc ring, v_e and ω its expansion and angular velocity respectively, v_0 is the systemic radial velocity of the disc rings and D the distance of M33. The quantities ρ, v_e and ω characterize each disc

R. Ekers et al. (eds.), Extragalactic Radio Sources, 181–182.

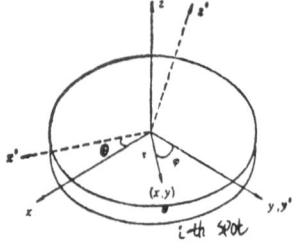

Figure 1. The disc geometry and adopted coordinate system. The primed coordinates represent the observer's system (the X'-axis is the observer's line of sight and the Y'-Z' plane is the plane of the sky.)

ring. Since v_0, D, and θ are known we can derive from the above equation a linear relation between ρ_i^2 and $\Gamma = [\frac{(v_i - v_0)}{\cos\theta} - \omega\rho_i D]^2$

We assume a position for the disc ring center on the map of the H_2O maser spots and measure ρ_i and v_i of every spot. In a plot of ρ_i^2 vs Γ we find that the points are grouped in a few regions and seem to belong to different curves. Thus we fit the data assuming several concentric co-rotating disc rings and adjust the position of disc ring center until we obtain a straight line for every disc ring. A least square fit of the data provides the values of ω, ρ^2 and v_e. The best fit to the data is obtained with two rings with radii differing by 0.4 pc, and expansion velocities differing by 19 km s^{-1}. Adopting $v_0 = + 0.52$ km s^{-1}, D = 270 Kpc, $\theta = 0$ (Greenhill et al. 1990) we find that the position of the disc ring center is 14.5 mas East and 0.0 mas North of the center of the maser spot map. Our results are given in table 1, where the spot numbers are from Greenhill et al. (1990).

Ring No.	Spot No.	$\omega(s^{-1})$	r(pc)	$v_e(kms^{-1})$
1	1,2,3,4,6,7	1.6610^{-14}	10.1	245
2	5,8,9,10,11,12,13,14	1.6610^{-14}	10.5	264

The tangential velocity of the maser spots on rotating and expanding disc ring, $v_t = \omega\ r = 1.66 \times 10^{-14}r$, is $v_{t1} = 5.05\ kms^{-1}$ and $v_{t2} = 5.25$ km s^{-1} for discs 1 and 2 respectively. The 14 maser spots span about 30 mas (3×10^{17} cm) and 12 of them are located in IC133M while spots 11 and 14 belong to IC133SE. Our analysis, instead, indicates that they are distributed in a different way: six of the spots in IC133M on the smaller disc, the remaining plus the two in IC133SE on the other.

References

Boulestei, J., Courtes, G., Lavel, A., Monnet, G., Petit, H., 1974, *A & A*, **37**, 33
Greenhill, L.J., Moran, J.M., Reid, M.J., winn, C.R., Menete, K.M., 1990, *ApJ*, **364**, 513
Huchtmeier, W.K., Eckart, A., Zensus, J.A., 1988, *A & A*, **200**, 26
Kiwitfer, K.B., Aller, L.H., 1981, *MNRAS*, **195**, 939
Yu, Zhi-yao, 1994, *Acta Astrophysica Sinica*, **14**, 66

A SEARCH FOR HI IN STRONG RADIO GALAXIES

S. M. SIMKIN AND J. CALLCUT
Michigan State University
East Lansing, MI 48824-1116, USA

1. Introduction

In the past decade it has become clear that many classical, double radio galaxies have hosts with extensive ionized gas inside the body of the optical galaxy (Hansen *et al.*, 1987) (Baum *et al.*, 1988). We have observed three such sources (PKS349-27, PKS634-20, and Pictor A) in both Hα and HI (with the VLA).

2. Hα and VLA measurements

The Hα images were obtained in 1987/88 with the CTIO 1.5M telescope using narrow band filters for both Hα and the adjacent continuum measurements. The continuum was subtracted from the line filter images and the Hα calibrated in two ways: (a) By comparing the image's nuclear flux with a combined Hα + [NII] emission line flux obtained from spectra which we took with the CTIO 4m during the same observing run (b) By using measurements of flux-standards taken through the same filter-ccd combination. Both methods have multiple sources of possible error (which are difficult to estimate) but agree to within a factor 2.

The VLA observations were taken in 1989 and 1991 with the array in the C/D configuration. The observing band width was set to cover the optical rotation velocity width (see Table 1). The continuum was subtracted from the uv data base and the resulting residual fluxes plotted as spectra. In no case did we detect HI in either emission or absorption.

3. Analysis and Conclusions

Table 1 lists the observed Hα flux and the calculated HII mass derived from these measurements (columns 2 and 3). We have assumed distances

183

R. Ekers et al. (eds.), Extragalactic Radio Sources, 183–184.
© 1996 *IAU.*

of 198, 166, and 106 Mpc for the three objects (PKS 349-27, PKS 634-20, and Pictor A). The range in calculated masses depends on the assumed filling factors for the ionized gas. However, in all reasonable cases (filling factors of 1 to 10%) the ionized HII exceeds 10^9 $M_\odot$.

TABLE 1. HI and HII Parameters

OBJECT	$F_\alpha(10^{40})$ (erg s^{-1})	$M_\alpha/M_\odot$ (10^9)	ΔV (Km s^{-1})	$M_{HI}/M_\odot$ (10^9)	M_α/M_{HI}
P349 (nuc)	3-6	3-6	47	1	3-6
P349 (50x110")	15-30	20-40	800	0.4-0.8	25-100
P634 (nuc)	1.3-2.6	0.4-0.8	46	0.3-0.5	1-3
P634 (70x70")	13-26	15-30	400	0.2-0.4	40-150
PicA (nuc)	6-10	0.1-0.2	11	0.06-0.2	0.5-3
PicA (10x10")	13-26	1-2	190	0.3-1	1-7

The VLA beam size was large relative to the optical extent of the Hα in Pictor A (see column 1); 110 N-S x 50 E-W " for Pictor A but less for PKS 349-27 (36 x 26"), and PKS 634-20 (33 x 24"). We integrated our VLA measurements over the outer regions of these latter two galaxies to try and detect any extended HI. These are listed in the second row for each object. We have estimated three-sigma limits for our VLA HI observations from the rms noise in the residual maps. These estimates, along with the applicable velocity widths (listed in column 4), were used to calculate one of the upper limits to the HI for these objects. Since these noise limits are almost certainly an over-estimate, we also used spectra derived from the continuum-subtracted data to estimate a more stringent upper limit. The limits to the HI mass for each object (both in the nuclear regions and integrated over the disk) are given in column 5. The resulting minimum ratios of HII to HI mass are listed in column 6.

In all three objects there is far less HI than HII. We need much higher resolution 21cm observations to set good upper limits for the HI in the nuclear regions of these objects, but these present observations are sufficient to establish that the extended gas outside the nucleus is almost fully ionized.

References

Hansen, L., N≤rgaard-Nielsen, H.U., and J≤rgensen, H.E. (1987), *A&A Sup.* 71, pp. 465-491

Baum, S.A., Heckman, T., Bridle, A., van Breughl, W. and Miley, G. (1988), *Ap. J. Sup.* **68**, pp. 643-714

A MULTI-RADIO-FREQUENCY STUDY OF A RQQ

K.M. BLUNDELL[1], M. LACY[1]
[1] *Oxford University Astrophysics - Keble Road, Oxford, OX1 3RH, U.K.*

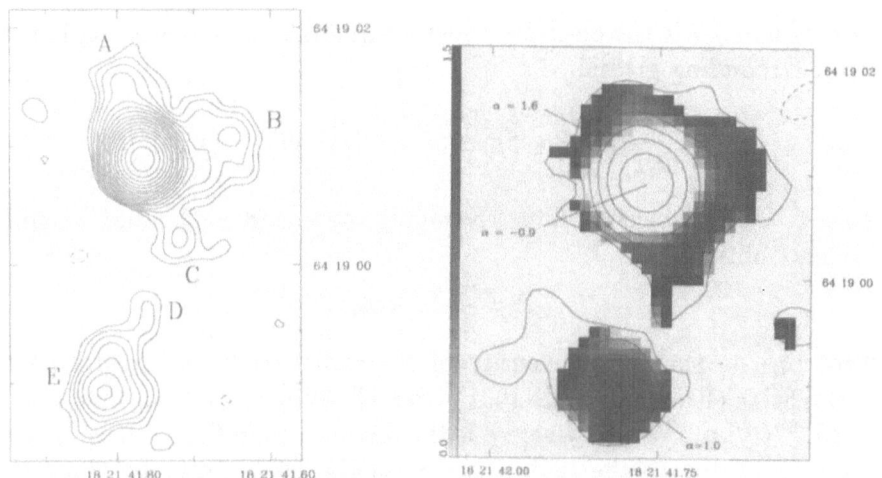

Figure 1. *Left:* E 1821+643 at 8.0 GHz; the circular beam has FWHM of 0.3″. The solid contour levels are logarithmic with ratio $\sqrt{2}$; the bottom contour is at 0.048 mJy/beam and the top contour is at 8.67 mJy/beam. A dashed contour is plotted at −0.048 mJy/beam. *Right:* A map in spectral index between 5 and 8 GHz. Contours in total intensity at 5 GHz are overlaid. The contours are logarithmic with ratio 2; the bottom contour is 0.12 mJy/beam and the top contour is 6.4 mJy/beam. The greyscale shown runs from $\alpha = 0$ (white) to $\alpha = 1.5$ (black). Regions within the two lowest contours which are white have $\alpha < 0$; other white regions have been masked. (We use the convention that $S_\nu \propto \nu^{-\alpha}$.)

1821+643 is exceptionally bright at optical wavelengths with $M_B \approx -27$ (Hutchings & Neff 1991), at a redshift of only 0.297. It is however, a radio-quiet quasar (RQQ) as can be demonstrated by considering *e.g.*, the ratio of the luminosity in [OIII]λ5007 to the radio luminosity (Lacy, Rawlings, & Hill 1992). In contrast with other radio-quiet quasars however, E 1821+643 is located in a very rich cluster (Lacy et al. 1992) and is purported to reside in a gE host galaxy (Hutchings & Neff 1991). The full-resolution 8 GHz map

185

R. Ekers et al. (eds.), Extragalactic Radio Sources, 185–186.

is seen in Fig. 1. A bright unresolved core is seen with extended emission to the north-west (feature B). Features A and C are accurately diametrically opposed on either side of the core and therefore suggestive of jets; this is further supported by the alignment of features C, D and E.

The radiation field of the quasar will contribute significantly to the inverse-Compton losses in the relativistic electron population (*e.g.*, Daly 1992). Following Blandford (1990) we write the rate of energy loss per electron, P as:

$$P = \frac{\gamma^2 \sigma_T}{\mu_0 c} \langle (\vec{E} + c\vec{\beta} \times \vec{B})^2 - (\vec{E}.\vec{\beta})^2 \rangle, \tag{1}$$

where μ_0 is the permeability *in vacuo*, σ_T is the Thomson cross-section, $\vec{E}$ and $\vec{B}$ are the photon electric and magnetic fields respectively, c is the speed of light, $\vec{\beta}$ is the electron velocity in units of c, and γ is the Lorentz factor. Expanding gives:

$$P = \frac{\gamma^2 \sigma_T}{\mu_0 c} \langle \vec{E}.\vec{E} - 2\mu_0 c \vec{\beta}.\vec{J} + c^2 (\vec{\beta} \times \vec{B})^2 - (\vec{E}.\vec{\beta})^2 \rangle, \tag{2}$$

where $\vec{J}$ is the Poynting flux. Averaging over solid angle, and assuming $\gamma \gg 1$ we obtain:

$$P = \frac{4}{3} \gamma^2 \sigma_T c U_{\text{rad}}, \tag{3}$$

where U_{rad} is the energy density of the radiation field. Integrating the quasar SED (Kolman et al. 1991) over all solid angles, we find $U_{\text{rad}} = 4 \times 10^{-9}/(r/\text{kpc})^2 \, \text{J m}^{-3}$ where r is the distance from the nucleus. The energy density due to the radiation field of the quasar may be equated to $B^2/(2\mu_0)$ which gives an "equivalent field" of $\approx 100/(r/\text{kpc}) \, \text{nT}$ compared with the minimum energy field of $\approx 2 \, \text{nT}$ derived from the radio data. Thus for distances $r < 50 \, \text{kpc}$ (9″) from the quasar, the electron ageing will be dominated by the quasar radiation field, provided the regions can see the quasar. Such high equivalent fields could therefore explain the unusually steep radio spectrum of the extended features and partial obscuration of the quasar can explain the non-axisymmetry of the spectrum around the core. Further analysis may be found in Blundell & Lacy (1995).

References

Blundell K.M. & Lacy M., 1995, *MNRAS*, **274**, L9
Blandford R.D., 1990, in Courvoisier T.J.-L., Mayor M., eds, Active Galactic Nuclei. Springer-Verlag, Berlin, p.161
Daly R.A., 1992, *ApJ*, **399**, 426
Hutchings J.B. & Neff S.G, 1991, *AJ*, **101**, 2001
Lacy M., Rawlings S., Hill G.J., 1992, *MNRAS*, **258**, 828
Kolman M., Halpern J., Shrader C.R., Filippenko A.V., 1991, *ApJ*, **373**, 57

PROPERTIES OF RADIO SOURCES (*DISCUSSION*)

Discussion of the paper presented by <u>WALKER</u> (p. 30)

Tsvetanov: Could you remind me of the linear scale on your maps (the +10 to −10 mas bar)?

Walker: For $H_0 = 100 km s^{-1} Mpc^{-1}$, 1mas = 0.26pc

Tsvetanov: Given this linear scale, why would you refer to your model as a disk model?

Walker: Perhaps using the word "disk" is making people think of something more specific than I have in mind. Certainly the scale is at or beyond the maximum for which accretion disk parameters are usually calculated. All I mean to imply is a geometry that has material concentrated toward the center with a distribution such that the near side jet is unobscured all the way to the core while the far side jet is obscured in the inner regions. A disk is merely an obvious example of a structure with this property.

Gurvits: Is this right, that the higher the frequency the larger the angular distance between northern and southern components?

Walker: No, that is not what the data show other than a small effect from the fact that the core is weaker at low frequencies so the centroid of the bright features around the core (resolved at high frequencies) moves down a bit. Certainly the total extent of the bright jet/counterjet system (the features that extend to about 15 mas south of the core) is the same at all of our frequencies.

Krichbaum: When calculating the arm-length ratio and the jet-to-counterjet ratio you need to know where exactly the core is. Did you use the brightest feature as reference or how did you define the location of the core? (see also my poster of Cyg A).

Walker: At the high frequencies, we have much higher resolution images than those shown in this talk. I used the location of the northern-most bright feature in the 1cm image. Actually, the constraints shown in the

R. Ekers et al. (eds.), Extragalactic Radio Sources, 187–194.
© 1996 *IAU*.

talk were based on the 1991 and 1993 data from Vermeulen, Readhead and Backer (1994) and Walker, Romney and Benson (1994).

Condon: NGC 1275 is also a luminous far-infrared source. If the dust is heated by a compact ultraluminous starburst, ionization by that starburst may cause free-free absorption of 3C84 at lower frequencies. The far-infrared spectral index $\alpha(25\mu, 60\mu)$ will be very steep in this case (so much dust implies $T_{25\mu} > 1$. Alternatively, heating of a small amount of dust by the intense AGN radiation field will yield a flat $\alpha(25\mu, 60\mu)$; and the ionization must be less extensive, due to Ly continuum absorption (see Condon et al 1991, ApJ, 378, 65).

Discussion of the paper presented by <u>GABUZDA</u> (p. 49)

Daly: We have studies equipartition Doppler factors and Inverse Compton Doppler factors for a large sample of AGN. We find that the core dominated quasars have high Doppler factors, lobe dominated quasars and radio galaxies have low Doppler factors, but BL Lacertae objects span the full range of Doppler factors, from very low values to very high values; this seems to be consistent with the observations presented.

Gabuzda: Yes.

De Young: Could you elaborate on your reasons why you expect fewer internal shocks (as distinguished from instability) in "stronger" BL Lac jets?

Gabuzda: My suggestion was that BL Lac jets are <u>weaker</u> and <u>slower</u> than quasar jets; this, for example, could explain the lower superluminal speeds observed in BL Lac objects or compared to quasars. The hydrodynamical simulations of Duncan and Hughes (1994) suggest that internal shocks can form more easily in modestly relativistic jets ($\gamma \sim 5$ than in very relativistic jets ($\gamma \sim 10$), so that if BL Lac jets have lower γ's than quasar jets, this might also explain the much more common occurrence of transverse shocks in BL Lacs.

Urry: Do you see any systematic differences in polarization structure or superluminal velocities between low and high-redshift BL Lac objects in the 1 Jy sample?

Gabuzda: No - we see no differences in either polarization properties or superluminal velocities for low and high BL Lacs. This supports the view that low and high B L Lacs really are the same type of source.

Laing: Is it not the case that some quasars (eg 3C279) appear to show $\vec{B}_\perp$ at least in outburst? The distinction between quasars and BL Lac objects in polarization appears to be less absolute in this sense also.

Gabuzda: Yes - integrated polarization monitoring data (for example, of the University of Michigan group) does indicate that transverse magnetic fields dominate in a few quasars at some epochs.

Pohl: Is there evidence for circular polarisation or elliptical in general, in any of these sources?

Gabuzda: It is technically very difficult to extract circular polarization information from the VLBI data; for the few sources (quasars) for which this has been done, no circular polarization was detected. This is certainly something that needs to be investigated further, however.

Discussion of the paper presented by <u>VERMEULEN</u> (p. 57)

Gopal-Krishna: Do you attribute the 'emergence' of broad emission lines in BL Lac to a weakening of the beamed optical continuum?

Vermeulen: Thank you for reminding me of this important point the broad lines we have discovered in B L Lac (Vermeulen et al 1995 Ap J Lett 452, L5) have increased in luminosity by at least a factor of 5, probably 10-20, within the latest 6 years. The lines would have been detectable in earlier published high quality spectra.

Grueff: The correlation between absolute radio power and β could be, at least in part, the result of an observational bias: powerful sources are observed only at high z, and this fact, coupled with a limited range of observed angular speeds, could produce a spurious correlation.

Vermeulen: I believe you are suggesting a possible bias against being able to observe the slower β_{app} at the higher z. However, we do in fact, have plenty of such slower motions at high β, and our limits of detectability are not hampering us here.

Laing: What are the objects with low power and low apparent velocity? Are they galaxies? What do you mean by 'galaxy' in this context?

Vermeulen: They do indeed tend to be galaxies, of course. For the material shown here, I have simply accepted whatever classification was given in the literature for these objects. Once the analysis of the full dataset is done, a thorough, uniform classification scheme, probably based on spectral as well as morphological properties, will clearly be needed.

Urry: The broad emission lines have been seen in BL Lacs before the current result on BL Lac but typically in the higher redshift (higher luminosity) BL Lac objects, so what is unusual is the discovery of broad Balmer lines in a relatively nearby ($z=0.069$) BL Lac.

Vermeulen: I would like to see demonstrated statistically that in the higher z objects call BL Lacs the line luminosity really is lower compared to quasars or galaxies of the same bolometric (or low radio frequency) luminosity. Further, what makes the result in BL Lac particularly interesting is the fact that the line luminosity has increased by a factor ≥ 5, within 6 years (possibly more rapidly; I am not aware of good spectra from the period 1990-1995).

Urry: My question concerns the distribution of superluminal velocities - if you allow for a distribution of Lorentz factors, is $\beta_{patt}/\beta_{bulk} \sim 1$ then allowed (and what range of γ is required)?

Vermeulen: Indeed, $r = \gamma_{patt}/\gamma_{bulk} = 1$ is allowed if there is a distribution of Lorentz factors, ranging from barely relativistic, with a peak perhaps near $\gamma \sim 4$ and a long tail up to $\gamma \sim 20h^{-1}$.

Ekers: 1. Comment: Given the continuity in the distribution of β I am surprised you omit the small β bin in your modelling.
2. Given the evidence just presented for NGC1275 that the "jet" - "counterjet" ratio is caused by an obscuring torus or disk, why do you cling to the beaming aspects of the model instead of testing models which are dominated by pattern rather than bulk motions?

Vermeulen: Regarding your comment: the beaming model explored here assumes relativistic motion, and thus the stationary patterns are not encompassed by the model in the simple form I have used here. However, now answering your question, I believe that the evidence from both large (kpc) and small (pc) scales is overwhelming that beaming has at least some role to play in determining what we see. On pc-scales, the one-sidedness of jets is, I think, much too common, and over too large a range of jet lengths to be dominated by free-free absorption or obscuration in general. However, it is clear that this will have to be one of the ingredients in more sophisticated models, once testing such models is warranted by the data volume and quality.

Woltjer: How did you obtain v/c for "empty" fields?

Vermeulen: Using accurate radio coordinates, we are pursuing spectroscopy to get redshifts [ed. - directly for the "empty" fields], in preference to imaging for POSS "empty fields". Classification of the objects, based on the spectra, and on imaging in progress, remains to be done.

Discussion of the paper presented by <u>GARRINGTON</u> (p. 77)

Burke: L.K. Herold, S.R. Conner and I made a set of observations of CSS's and CSO's with MERLIN – 20 sources from the MG survey, a sample completely independent of the 3CR, and we find the marked asymmetry in polarization that you find in the 3CR sample.

Garrington: That is interesting. Your observations at 5GHz should reveal the depolarisation asymmetry quite clearly in these sources.

Laing: Do the correlations between spectral index and arm length/depolarization found in large FRII's occur in CSS sources?

Garrington: We shall be using our data to investigate these trends.

Discussion of the paper presented by <u>GELDERMAN</u> (p. 81)

Bicknell: This comment is directed towards yours and the previous papers. Bicknell, Dopita and O'Chee (poster paper) have proposed a model in which we propose that the low frequency turnover of CSS and GPS sources is due to free-free absorption in the ionized gas that you are observing. The scatter in the correlation between OIII and radio power may be due to the shock velocity resulting from the expansion of the radio source. The line widths that you observe are typical of the shock velocities required in our model.

Urry: With regard to your histograms comparing properties of CSS radio galaxies and quasars with "normal" radio galaxies and quasars, where do the comparison samples come from, and in particular are they selected in the same way?

Gelderman: The samples include most of the published flux and kinematic data for Seyferts, radio galaxies, and quasars. In particular they include most of the available data for the 3C sample, representing the radio power and redshift ranges from which my CSS sources are drawn. However, complete CSS and comparison samples would be very desirable.

Discussion of the paper presented by <u>*HEWITT*</u> (p. 105)

De Bruyn: In your description of 0957+56 you gave various errors. Do they include the modelling of the cluster of galaxies contributing to delays and potential?

Hewitt: Yes, and in fact it is the dominant source of error at this point.

Discussion of the paper presented by <u>*GIOVANNINI*</u> (p. 127)

Jones D.L. : The 1.2c proper motion you listed for N6251 was derived by assuming the component had moved out from the core; we can not rule out a local brightening of this component, so no specific velocity value is meaningful. All evidence is still consistent with this jet having only subluminal motion.

Discussion of the paper presented by <u>*TAYLOR*</u> (p. 133)

Booth: I think the combination of high resolution line and continuum data is away the most exciting work in current radio astronomy. I am interested, therefore, if you can point out any velocity differences in the line absorption to the N and S of the compact object and any information you might have on the enclosed mass.

Taylor: The velocity difference between the components to the N and S is 20 km/s. I have not yet attempted to use this information to construct a dynamical model.

Ekers: Much higher velocities might be expected - how far does your velocity range extend?

Taylor: What I've shown is just the inner 300 km/s. Our observations covered 8MHz or a total velocity span of 1800 km/s. No higher velocity components were detected.

Laing: What is the jet : counter-jet ratio on kpc scale?

Taylor: This is a very good question. I will put the jet : counter-jet ratio on the kpc scale in my conference proceedings submission (I can't remember it now).

Discussion of the paper presented by <u>PARMA</u> (p. 137)

Laing: Have you mapped the rotation-measure distribution in any of the B2 sources?

Parma: We have mapped the rotation measure only in a few sources. We have data for the whole sample but the work is still in progress.

Woltjer: When you start out with a relativistic jet to make BL Lacs and slow it down to 0.01c, would you not expect to dissipate much energy somewhere? Where does it appear?

Leahy: Bowman, Komissarov and myself (MNRAS, submitted) have discussed this point in detail. Energy released by dissipation appears as heat, but is re-converted continuously into kinetic energy since the jet is propagating down a pressure gradient. The net loss of kinetic energy flux can be quite mild (less than 50%) for deceleration from Lorentz factor of 5 to $\sim$1.2. Deceleration without catastrophic kinetic energy loss is actually much easier to achieve for relativistic jets than for non-relativistic ones. Radiation loss times are much longer than dynamical times for all plasma phases involved.

Discussion of the paper presented by <u>HARDCASTLE</u> (p. 153)

Laing: The correlation between core and jet prominence in Bridle et al (1994) tightened up considerably when straight jet segments were considered. Have you tried this for your sample ?

Hardcastle: No, but we will be trying in the future. (Straight jet segments are not always as well defined in this sample).

Urry: I'd like to raise a point that is relevant to your talk and also to some of the talks this morning, and that is: how do we distinguish between radio galaxies and quasars? We can do this on the basis of optical morphology or perhaps it is more interesting to separate sources according to whether they have broad or only narrow lines. In particular, it appears from recent HST (and also much previous ground based observations), that quasars are in giant elliptical galaxies, so should we call them radio galaxies? Instead, I would call broad -line radio galaxies, for example, the local quasar counterparts of your $z < 0.3$ FRIIs. How would this narrow-line/broad-line division affect your findings?

Hardcastle: I think this is a good point. In my talk I follow the definers of my sample in using "quasar" to mean an object which is stellar on a photographic plate, but a less subjective definition would be useful. As I said, the broad- line objects do not show any significant tendency to have stronger jets. The problem in this sample is the prevalence of so-called "dull" FRIIs which have no line emission and so no orientation indications.

Perley: Elias Fernini at NMSU is studying a similar RG sample but finding jets in only $\sim 10\%$. His sensitivity is about the same as yours, I believe, so why do you find such a higher detection rate? Do you think this is due to using different criteria?

Hardcastle: We believe that at least 50% of our jets meet the Bridle and Perley (1984) criteria. It may be worth noting that Fernini and co-workers are looking at objects matched to quasars, i.e. in unified schemes their sources are all near the plane of the sky. In the same unified schemes our sources would be at all angles to the line of sight and thus might show more beaming and so more jets.

HUBBLE SPACE TELESCOPE OBSERVATIONS OF EXTRAGALACTIC JETS

F. DUCCIO MACCHETTO

Space Telescope Science Institute - 3700 San Martin Drive, Baltimore, MD 21218. On assignment from the Space Science Department of ESA
Based on observations obtained with the NASA/ESA Hubble Space Telescope

1. Introduction

The study of the optical counterparts to the radio jets has been the subject of a number of observing programs with the Hubble Space Telescope (HST). We know that these jets play a fundamental role in transporting energy from the central source to the extended radio lobes. Observations at optical and ultraviolet wavelengths with the HST are essential to obtain spatial resolutions similar to, or better than, those achieved in the radio band and, thus, provide the possibility of directly comparing the sites and mechanisms responsible for the emission at these different wavelengths.

In all cases to date, the emission has been attributed to the synchrotron mechanism, and since the electron lifetime is a strong function of the observed frequency, observations at optical and ultraviolet wavelengths offer the possibility to determine the precise locations where particle acceleration occurs. Comparison of the radio and optical morphologies further allows the study of the confinement mechanisms and diffusion processes within the jet.

A number of important discoveries and observations that place the theoretical models on firmer observational grounds have been published or are about to be published. This review will summarize the results obtained so far.

R. Ekers et al. (eds.), Extragalactic Radio Sources, 195–200.

2. PKS 0521-36

PKS 0521-36 is a relatively isolated radio galaxy at a redshift $z=0.055$ which also harbours a bright $V = 16$ BL Lac nucleus and extended optical line emission. Sparks et al. (1990) reported optical polarization measurements of the jet and nucleus, which confirmed that the emission is due to synchrotron radiation.

The FOC images obtained in 1990 with the FOC (Macchetto et al 1991a) show a bright knot located ≈ 1.8" to the NE and clearly resolved as in the VLA data (Keel 1986). The width of the knot is ≈ 0.8". Beyond this bright knot, the jet has approximately constant surface brightness and a morphology similar to the VLA image with a total length of 6.5". The jet is also resolved in width, 0.6" wide in the fainter regions of the jet, with little or no evidence of structure on a scale of ≤ 0.1". The FOC data appears to show more flux than the VLA data in the region at slightly larger radius from the nucleus but close to the southern tip of the knot. A large bright knot further along the jet is a clear counterpart to the radio knot. The bright knot is an important site where particle acceleration is occurring.

Using standard formulae and values for the relevant parameters, we derive a mean lifetime for the electrons of $t_{1/2} \sim 600$ yrs. This implies that there must be continuous acceleration along the jet of the electrons responsible for the optical emission, since electron diffusion from the bright knot could not account for the observed optical extent.

3. 3C 66B

3C 66B is a relatively nearby bright radio source associated with a 13th magnitude galaxy at a distance of 86 Mpc. At this distance an angular scale of 0.1" corresponds to a projected linear size of 41 pc. Images of 3C 66B were obtained with the FOC and compared with the best VLA map of 3C 66B. Several conclusions can be drawn from these observations (Macchetto et al 1991b).

On the scale of the HST resolution, the jet of 3C 66B is filamentary. Two distinct "strands" can be traced from ≥ 3.7" (1.5 kpc) from the nucleus out to a distance of 7.6" (3 kpc), where they disappear into the noise. The separation between the strands varies between about 0.3" and 0.4", that is about 150pc, and they appear to undergo sharp "kinks" at distances of 2.5" (1.0kpc) and 6.2" (2.5kpc) from the nucleus.

The origin of these kinks is unclear. The fact that they are mimicked in more than one filament suggests that they are not due to an instability mode in an individual filament. They may well trace out irregularities in the ISM of the galaxy and/or be due to time-dependent variations in the power of the nuclear source responsible for producing the jet.

4. 3C 273

3C 273 is one of the nearest and brightest quasars known. Its jet has been extensively studied at radio and optical wavelengths. Using the FOC, Thomson, Mackay and Wright (1993) have carried out high-resolution imaging polarimetric observations of the jet. More recently, Bahcall, et al. (1995) have obtained WFPC2 as well as Merlin observations with matching resolution. The projected jet length is more than 70 kpc ($H_o = 50 kms^{-1}Mpc^{-1}$; $q_o =$ 0.5). The width is only a few tenths of an arcsecond ($\sim 0.5 - 1kpc$). The optical emission is highly confined to the core of the radio jet. It runs along the ridge of the radio emission and is asymmetric compared to the radio. All oblique radio features coincide with optical knots, though there are some optical features without radio counterparts.

Bahcall et al., (1995) suggest that the 'radio jet' consists of two components. The first is the fast-moving jet, shown by the coincident oblique radio and optical features. The second consists of the emission from a surrounding, slow-moving "cocoon". The radio data suggests that the oblique radio features coincident with the optical knots may be in the form of a helix. If the velocity is relativistic, the emission will appear brightest where the velocity vector is closest to the line of sight, the enhancement being independent of wavelength. This would explain the close correspondence between the optical and radio knots. The helical form may arise from a driven Kelvin-Helmholtz instability.

5. 3C 264

NGC 3862 (3C 264) at a distance of 86.2 Mpc is an FR1 source. This galaxy was observed in two bands with the FOC. The images showed a prominent jet-like feature emanating from the nucleus (Crane et al. 1993) morphologically similar to the jets seen in M87 and 3C 66B. Indeed, the bifurcation seen at the end of the jet is reminiscent of the "double-stranded" feature seen in the HST image of 3C 66B (Macchetto et al. 1991b).

Broad-band (F702W) imaging data taken with the WFPC2 shows a very intriguing feature in this galaxy (Sparks et al. 1996). In addition to the very evident optical jet, an almost perfectly circular ring is observed. It is not possible with a single band observation to decide whether this is an emission-line ring, or even if it is a true emission feature or the result of dust absorption closer to the nucleus. The fact that the jet appears to stop suddenly at the position of the ring may be indicative of some physical association between the two features. Observations in other bands will, hopefully, help clarify this mystery.

6. M87

The giant elliptical galaxy M87 contains the closest extragalactic jet, which makes it a prime target for studies of jet structure and kinematics.

Optical and ultraviolet observations of M87 have been carried out with HST and have been extensively reported (Macchetto 1991, Bokensberg et al 1992; Macchetto, Biretta and Sparks 1992).

While the radio and optical images present a remarkable degree of similarity, there are nevertheless significant differences. The optical/UV images show intrinsically higher contrast than the radio, with compact regions of emission localized within the knots. The jet is narrower in the optical/UV, and more concentrated to the jet center in the optical/UV than in the radio. The radio-to-optical spectral index of the inter-knot regions is steeper than that of the knots themselves. There are also differences in the detailed knot structure of the optical emission compared to the radio, and there is a weak overall spectral steepening with distance from the nucleus beyond knot A.

Capetti, et al (1996) have analyzed polarization observation of the M87 jet taken with the FOC in the ultraviolet and with the WFPC1 in the visual. The degree of polarization is typically 30% over most of the jet. At the edges of the jet the polarization is as high as 60%, requiring a highly ordered magnetic field. In the center of the jet the small scale structure of the magnetic field produces significant cancellation reducing the polarization to $\sim 10\%$. The degree of polarization and the polarization pattern are very similar at radio and optical wavelengths. No significant depolarization or Faraday rotation have been detected, in agreement with previous radio determinations. However, the morphology of knot D is considerably different in the VLA observations by Owen et al.(1989) and these HST observations. Knot D1 appears to be relatively brighter and closer to the nucleus in the optical than in the radio images. Capetti, et al (1996) conclude that this component is associated with a shock front. At the location of a shock front acceleration of relativistic electrons occurs, enhancing the synchrotron emission at shorter wavelengths, and the transverse component of the magnetic field is amplified by the compression produced by the shock.

Figure 1 shows FOC images of the nucleus obtained in August 1994 and July 1995 by Biretta et al. (1996). The images, which have been aligned at the bright point source show a feature approximately 150 milliarcseconds (12 pc) from the core, which corresponds roughly to knot M in the 18 cm VLBI image of Reid, et al. 1989. Comparison of the two images clearly shows this feature moves outward by 8.5 mas or 0.66pc, for a velocity of 2.3c+0.3c. From this result it is clear that the jet is already relativistic within the core, and that acceleration is probably unimportant on these scales. Velocity measurements for three features in the region of knot D were also obtained. We find a speed of 2.6c+0.4c for feature DW, and a

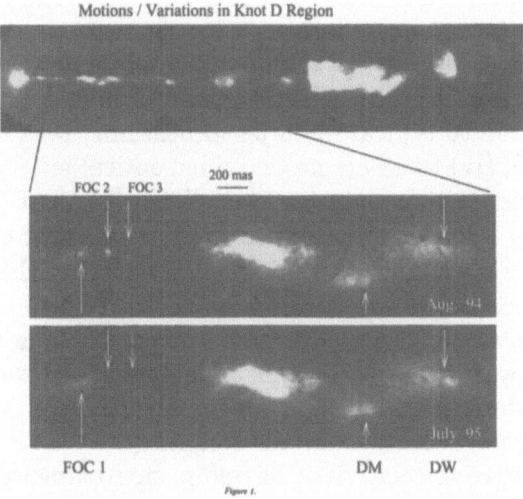

Motions / Variations in Knot D Region

Figure 1.

somewhat larger 3.7c+0.5c for DM. A third feature, FOC1, which was not seen in the radio data, has a speed of 2.1c+0.3c. Two additional features, labeled FOC2 and FOC3, appear to fade from view between 1994 and 1995. This fading rate is much higher than typical synchrotron lifetimes in the jet (100 yrs) suggesting that the adiabatic expansion of the emitting region is likely to play an important role in the evolution.

7. 3C 78

Sparks et al (1995) have very recently reported the discovery of an optical synchrontron jet emanating from the nucleus of NGC 1218, the galaxy hosting the radio source 3C 78. The jet is similar to that of M87, although smaller in projected length. The observations were obtained as part of the HST "snapshot" program to acquire broad-band images with high spatial resolution of 3CR radio sources with the WFPC2 F702W filter.

The visible jet has a total length of 1.37 arcsec which is 0.75 kpc in projection for $H_o = 75$ km.s^{-1} Mpc^{-1}. Within ≈ 0.5 arcsec of the nucleus, the jet is essentially unresolved across its width even at PC resolution, although it is slightly curved and there are distinct knots within that portion of the jet. The gap between the brightest part of the jet and the nucleus appears to be real. At ≈ 0.5 arcsec, the jet fades dramatically, and fans out into a broad plateau of emission. There are also two discrete knots, and a possible third fainter one, within the plateau portion of the jet, seen in the optical data.

8. Conclusions

The HST observations conducted so far have already led to the discovery of two new optical jets (3C 264 and 3C 78) and the identification of at least six other candidates, which will be further studied in the future.

Although few in number, there are common features shared by the optical jet radio sources: (i) all have relatively prominent nuclei in both radio and optical domains; (ii) the nuclei all have flat radio spectra; (iii) the jets are small compared to typical radio jet dimensions, with the arguable exception of 3C 273; (iv) there are no two-sided optical jets - they are all one-sided with a large jet to counterjet lower limit; (v) there is noticeable jet curvature; (vi) in addition, where measured, the optical emission is more localized than the radio and the optical jet is narrower than the radio jet.

An obvious candidate explanation for most of these characteristics is relativistic 'beaming', in which the jet becomes visible in the optical only when pointing towards the observer. In the beaming picture, the jet appears brightened, foreshortened and with a prominent active nucleus. Beaming also blueshifts the synchrotron 'break' frequency.

As an alternative to relativistic beaming, environmental effects may be considered. If the pressure is higher in the vicinity of the optically emitting sources, then the additional confinement may act to enhance their radiative luminosity while suppressing the growth of the source, thereby giving rise to the correlation between size and power. This does not immediately suggest an explanation for the core dominance and one-sidedness; however there may be instabilities which cause jet disruption and optical emission that are sufficiently rapid that only one side is visible at a given time. 'Age' may provide yet a third alternative, with the optical jet sources being young, in the process of forcing their way out through the interstellar medium, and ram pressure playing a similar role in enhancing the visibility.

There are statistical uncertainties at present, however an extensive analysis of many more optical jets should provide results that will be essential in elucidating the nature of extragalactic synchrotron jets.

References

Bahcall, J.N., et al (1995), *ApJ*, submitted
Biretta, J.A., Sparks, W.B., Macchetto, F., Capetti, A. (1996) *ApJ*, submitted
Boksenberg, A., et al (1992), *A&A*, **261**, **393**
Capetti, A., Macchetto, F., Sparks, W.B., Biretta, J.A., (1996), *ApJ*, in press
Crane, P., et al (1993), *ApJL*, **402**, **L37**
Keel, W.C. (1986), *ApJ*, **302**, **296**
Macchetto, F. (1991), *Proc. Physics of AGN*, ed. S.J. Wagner, W.J. Duschl, **325**
Macchetto, F., et al (1991a), *ApJL*, **369**, **L55**
Macchetto, F., et al (1991b), *ApJL*, **373**, **L55**
Macchetto, F., et al. (1992), *Proceedings 182nd AAS Meeting*, **24**, **1183**
Owen, F.N., Hardee, P.E. & Cornwell, T.J. (1989), *ApJ*, **340**, **698**
Reid, M.J., Biretta, J.A., Junor, W., Spencer, R., Muxlow, T. (1989), *ApJ*, **336**, **125**
Sparks, W.B., Miley, G., & Macchetto, F. (1990), *ApJL*, **361**, **L41**
Sparks, W.B., et al (1995), *ApJL*, in press
Sparks, W.B., et al (1996), *ApJ*, submitted
Thomson, R.C., Mackay, C.D. & Wright, A.E. (1993), *Nature*, **365**, **133**

HST OBSERVATIONS OF JETS AND RADIO LOBES

PHILIPPE CRANE
European Southern Observatory
Karl Schwarzschildstraße 2
D85748 Garching
Germany

The Hubble Space Telescope has proved to be remarkably useful for discovering and for studying the optical counterparts of radio lobes and radio jets. Since much of the structure seen in the radio is found on sub-arcsecond scales, it is not surprising that HST, with it's improved resolution relative to ground based observations, would be a major contributor to this field of research. This paper reports briefly on some of these successes, and refers to more detailed descriptions in the literature.

Pictor A Hot Spot

The western hot spot of Pictor A was already known to be highly polarized from the ground based observations(see ref. [1]). Figure 1. shows the polarization map of the western hot spot of Pictor A obtained with HST Faint Object Camera utilizing the F342W filter and the 3 polarizers[2]. The peak count in the F96 images was 13 counts obtained in 900s! Nevertheless, since the FOC is basically a noiseless detector, good quality polarization maps were obtained. Figure 1 shows that the magnetic field vectors are nearly perpendicular to the jet axis. The peak polarization exceeds 60% and indicates that the magnetic field structure in the hot spot has been resolved on a scale of 70 pc.

The Jet in 3C264

This jet was discovered by HST[3]. Here we present an overlay of a recently obtained MERLIN map and the HST image. There appears to be a slight misalignment of the radio jet axis from the optical jet axis. A VLA map at 2 cm and MERLIN maps at 6 cm and at 18 cm have been obtained. The spectral index map between the 2 cm and and 18 cm maps appears to confirm the misalignment above. The details of these observations will be reported in a forthcoming publication[4].

R. Ekers et al. (eds.), Extragalactic Radio Sources, 201–202.

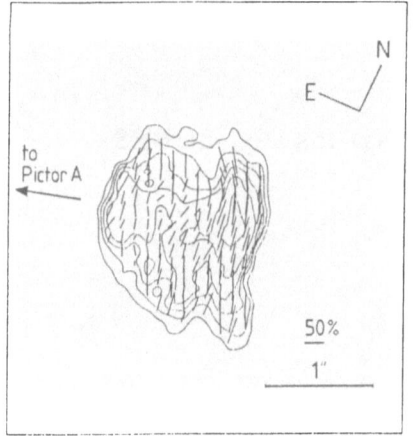

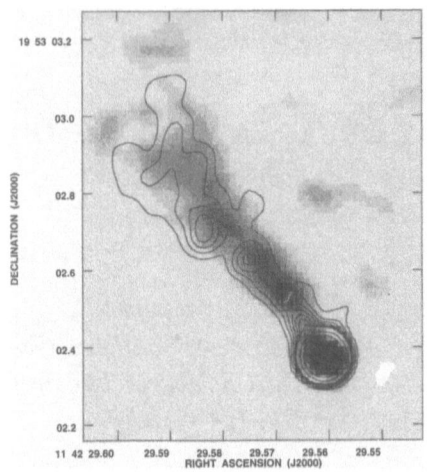

Figure 1. Polarization map of the western hot spot of Pictor A. The resolution is about 70 pc. The lines show the orientation of the *magnetic* field.

Figure 2. Overlay of HST image and MERLIN 6 cm map(contours).

NGC6251–Confirmation of a UV feature

This source which has one of the longest and best studied radio jets[5] has been observed at two different epochs with the preCostar FOC F96 camera and 3 different filters. The second epoch observations confirm the previous report[6] of a UV feature close(≈ 0.2 arcsec) to the nucleus of NGC6251 and at a position angle of roughly $90°$ to the axis of the radio jet. The UV feature does not appear in the V or B band images but only in the UV F342W images. The origin of this feature remains to be determined. It's discovery is a testament to the value of UV high resolution imaging.

PKS 0812+02

This fascinating quasar radio source[7] at a redshift $z = 0.402$ has been observed with WFPC2 with a filter that avoids the main emission line features. The images reveal resolved structure coincident with the end of the radio jet. This may be similar to the radio lobe in 3C33[8]. Further observations are planned which include both continuum and emission lines.

References

[1] Röser H.J., and Meisenheimer, K., (1987) *Ap. J.***314**,70.
[2] Thomson, R.C., Crane,P., and Mackay,C.D., (1995) *Ap.J. Letters*,**446**,L93.
[3] Crane, P., Peletier, R., et al., (1993) *Ap.J. Letters*,**402**,L37.
[4] Crane,P., Pedlar, A., et al.,(1996) in preparation.
[5] Perley,R., Bridle, A.H., and Willis, A.G.,(1984) *Ap. J. Suppl.*,**54**,291.
[6] Crane,P., 1993 in *Jets in Extragalactic Radio Sources*, edited by H.J. Röser and K. Meisenheimer, (Berlin: Spinger)pp 223.
[7] Wyckoff, S., et al.(1983) *Ap. J.***265**,43.
[8] Crane,P., and Stiavelli,M.,(1992) *MNRAS***257**,17P.

MERLIN AND HST OBSERVATIONS
OF THE JET IN 3C273

R.G. CONWAY, S.T. GARRINGTON, T.W.B. MUXLOW,
AND R.J. DAVIS
The University of Manchester, NRAL Jodrell Bank, UK

The jet of 3C273 is the only quasar jet visible at both radio and optical wavelengths. Investigations using ground-based observations suggest that the optical emission, like the radio, is produced by the synchrotron process. We present here a new radio map at 18 cm wavelength made with the improved MERLIN array (Fig. 1b) and compare it with a new optical image obtained with WFPC2 on the Hubble Space Telescope (Fig. 1a) This work has been carried out in collaboration with colleagues from Princeton, Pennsylvania State University and Caltech and is being published by Bahcall et al. (1995).

Both the radio and optical images represent significant advances over those previously available (Thomson et al 1993), due to the recent upgrades to MERLIN and the HST. Furthermore, the images can be accurately aligned because they both include the quasar itself. This allows a detailed comparison of the optical and radio jets at a resolution of approximately 0.15 arcsec.

The HST image shows that the optical jet consists of a series of knots, linked by fainter emission, and tightly confined with a FWHM of 0.3 arcsec. The first feature in the main jet (knot A) is elongated along the main axis. Subsequent knots (B,C,D) lie oblique to the jet axis. The radio jet is broader with width 1.0 arcsec, and the new map reveals considerable structure within the jet. In particular, there are a number of oblique features about 0.5 arcsec long and separated by 1.4 arcsec. Direct comparison of the radio and optical images shows that the oblique radio features all coincide with optical knots.

Because of the difference in radiative lifetimes, the optical emission traces the present location of the active jet, while the radio emission can provide information on its history during the past 1 Myr. We suggest that what has hitherto been called the 'radio jet' is actually two components su-

R. Ekers et al. (eds.), Extragalactic Radio Sources, 203–204.

perposed, firstly the fast-moving jet proper, shown by the coincident oblique radio and optical features, and secondly emission from a surrounding, slow-moving 'cocoon'.

Detailed inspection of the images suggest that the jet may have a helical form. If the velocity is relativistic, the emission will appear brightest where the velocity vector is closest to the line of sight, the enhancement being independent of wavelength. This would explain the close correspondence between the optical and radio knots. We have modelled the projected brightness distribution of a simple relativistic helical jet. Using the velocity and angle to the line of sight implied by the superluminal motion of the VLBI jet (Unwin et al. 1985), the model matches the observed pattern. However, the dynamic range of the present maps does not rule out somewhat slower speeds and larger angles to the line of sight.

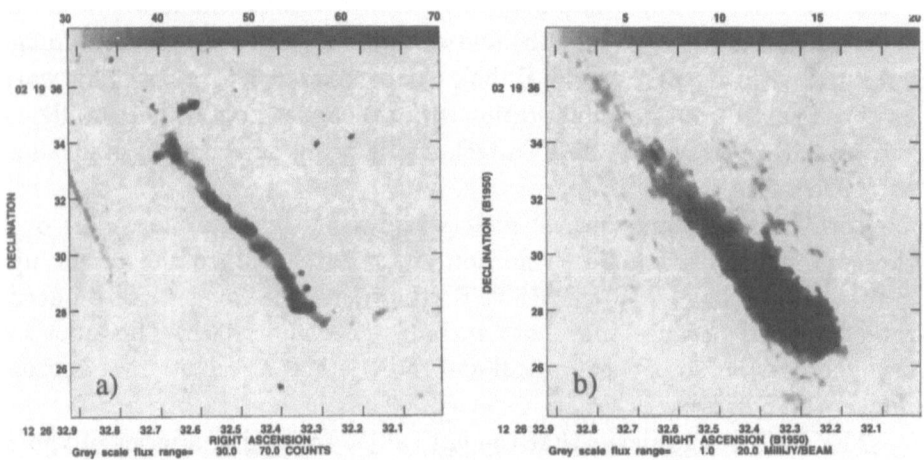

Figure 1. a:) WFPC2 image of 3C273 at $\lambda 5940\text{Å}$. The resolution is limited by the pixel size of 0.1 arcsec. *b):* MERLIN image of 3C273 at λ 18 cm with resolution of 0.14×0.18 arcsec.

References

Bahcall, J.N., Kirhakos, S., Schneider, D., Davis. R.J., Muxlow, T.W.B., Garrington, S.T., Conway, R.G. and Unwin, S.C. 1995 *ApJ Lett.*, **452**, L91.

Thomson R. C., Mackay C. and Wright A., 1993, *Nature*, **365**, 133

Unwin S., Cohen M., Biretta J., Pearson T., Seielstad G.,Walker R., Simon R. and Linfield R. 1985, *ApJ*, **289**, 109

CIRCUMNUCLEAR DISKS IN RADIO-QUIET ACTIVE GALAXIES

A. S. WILSON
Astronomy Department, University of Maryland,
College Park, MD 20742
and
Space Telescope Science Institute, 3700 San Martin Drive,
Baltimore, MD 21218

Abstract. In this paper, I review some of the evidence for gaseous circumnuclear disks in radio-quiet active galaxies (Seyfert galaxies and LINER's). Results from observations of Fe Kα lines, H_2O megamasers, extended structures seen in near infrared observations and ionization cones are discussed.

1. Introduction

The paradigm for active galactic nuclei (AGN) comprises a massive black hole and an accretion disk. In radio-loud objects, the paradigm is strongly supported by the existence of powerful relativistic jets. The evidence for disks in radio-quiet objects is also strong and in this paper I should like to discuss selected recent developments in this area.

2. Evidence for Gaseous Disks

2.1. FE Kα LINES – DISKS AT SEVERAL SCHWARZSCHILD RADII

Tanaka et al. (1995) have recently shown that the Fe Kα emission line in the Seyfert 1 galaxy MCG–6–30–15 is extremely broad, with a full width at zero intensity of 100,000 km s^{-1}. The line is also asymmetric, with a relatively narrow core around 6.4 keV and a broad red wing. Tanaka et al. show that the profile can be modelled in terms of emission from an accretion disk between 3 and 10 Schwarzschild radii from a Schwarzschild or Kerr black hole. This observation is probably the strongest piece of evidence that radio-quiet AGN are powered by massive, relativistic compact objects.

R. Ekers et al. (eds.), Extragalactic Radio Sources, 205–210.

2.2. H$_2$O MEGAMASERS – DISKS AT $\simeq$ 0.1 – 10 PC

2.2.1. *NGC 4258*

Recent VLBI observations (Greenhill et al. 1995a; Miyoshi et al. 1995) of the LINER galaxy NGC 4258 have shown that the water vapor megamasers in this galaxy arise in a thin, edge-on gaseous annulus at a galactocentric radius of 0.13 – 0.26 pc. Maser emission is observed both near systemic velocity, arising from clouds at the near side of the disk and along our line of sight to a central opaque core (hypothesised to be a source of continuum radiation at 22 GHz), and from "satellite lines" with velocities of $\pm$ 900 km s^{-1} w.r.t. systemic (Nakai, Inoue & Miyoshi 1993), arising from gas at the tangent points with rotational velocities directed towards and away from Earth. The satellite lines show an accurately Keplerian rotation curve. The recessional velocities of the near-systemic features are observed to be increasing at a rate of about 9 km s^{-1} yr^{-1} (Haschick, Baan & Peng 1994; Greenhill et al. 1995b). This increase of velocity is believed to result from the centripetal acceleration of clumps of gas in the annulus as they move across our line of sight to the central core (Watson & Wallin 1994; Haschick, Baan & Peng 1994; Greenhill et al. 1995b).

The most important information obtainable from these observations can be summarised in the following simplified way. Suppose that the masers arise in a thin, circular annulus viewed edge-on. Let θ be the observed angular dimension along the projected edge-on disk, V the observed recession velocity of the masers and V$_{gal}$ the systemic velocity. Intrinsic properties of interest include the distance to the galaxy, D, the rotational velocity of the annulus, U, and the radius of the annulus, r. We can relate the information from the various observations to the intrinsic properties as follows.

a) VLBI mapping of the systemic features: $(dV/d\theta)_{syst} = UD/r$.

b) Monitoring of the time dependence of the velocity of individual clumps of gas in the systemic features: $(dV/dt)_{syst} = U^2/r$.

c) VLBI measurement of the angular radius of the satellite lines: $\theta_{sat} = r/D$.

d) Measurement of the velocity of the satellite lines: $V_{sat} - V_{gal} = U$.

e) VLBI mapping of the systemic features at more than one epoch: $(d\theta/dt)_{syst} = U/D$ (this measurement has not yet been made).

Combination of measurements a) through d) gives $D = 6.4$ Mpc, $r = 0.13 - 0.25$ pc, $U = 800 - 1,100$ km s^{-1}, and a central mass $M = rU^2/G = 3.6 \times 10^7$ M$_\odot$ (Greenhill et al. 1995a, b; Miyoshi et al. 1995). What is impressive here is the high degree of internal self-consistency in the model. If instead of obtaining D from the maser observations we assume it is known, measurements a) through d) give four equations for two unknowns. For example, measurement of the angular radius (c) and velocity (d) of the satellite lines allows the angular change in velocity (a) and the acceleration (b) of the systemic components to be correctly predicted. This self-consistency gives confidence in the model and argues strongly against alternative theories in which the masering gas is moving radially.

2.2.2. *NGC 1068*

Using VLA observations, Gallimore et al. (1995) have recently found that the brightest H_2O masers in NGC 1068 trace a 5 pc long structure which is almost at right angles to the local axis of the radio jet. They find that the kinematics may be described by an edge-on disk around a central mass concentration. The inner radius of the disk is 1.3 ± 0.2 pc and the mass within the inner radius $(4.4 \pm 0.8) \times 10^7$ $M_\odot$. Higher resolution observations with the VLBA are desirable to confirm these results and check whether the rotation curve is Keplerian.

2.2.3. *The Population of H_2O Megamasers*

In 1992, five galaxies, all of which are Seyferts or LINER's, were known to contain H_2O megamasers. Surveys of non-active spirals, starbursts and luminous infrared galaxies over the last decade have failed to provide new detections (e.g. Greenhill et al. 1995b). We therefore decided to perform a new survey for H_2O megamasers, targeting exclusively AGN. The results of this survey at present are as follows (Braatz, Wilson & Henkel 1994, 1996):

1) With $\simeq$ 350 galaxies observed, nine new detections have been obtained. For a distance-limited sample (all Seyferts and LINER's listed in the Huchra or Véron-Cetty & Véron catalogs with $cz < 7,000$ km s^{-1}), the detection rate is 5.2%. This fraction is 12% for those sources with $cz < 2,000$ km s^{-1}.

2) All the H_2O detections are either LINER's or Seyfert 2's; none are Seyfert 1's. This result is consistent with the model proposed by Miyoshi et al. (1995) for NGC 4258, in which the molecular disk is viewed edge-on.

3) When measured, the column densities to the nuclei of the detected galaxies (from soft X-ray photoelectric absorption) are high – 10^{22-25} cm^{-2}.

4) The line profiles are mostly narrow ($\approx$ km s^{-1} wide) spikes, with some broad features. NGC 1052, the only elliptical detected, shows only a single, broad (FWHM $\simeq$ 90 km s^{-1}) line.

5) An ongoing VLA project confirms that all masers observed so far are confined to the nucleus of the galaxy.

6) Monitoring of the brightest maser spike in NGC 2639 reveals a redward velocity drift of 6.6 ± 0.4 km s^{-1} yr^{-1} over a period of 1.4 yrs (Wilson, Braatz & Henkel 1995). If this acceleration represents the centripetal acceleration of the near side of an edge-on Keplerian disk, as is the case in NGC 4258, the mass of the central object is 1.5×10^7 $(r/0.1$ pc$)^2$ $M_\odot$ (cf. observation b for NGC 4258). Further observations (i.e. a, c, or d in the NGC 4258 list) are needed to determine r or v, and hence M, uniquely.

Unfortunately, no satellite lines have been detected so far and the maser is too weak for VLBI observations.

7) There is a possible trend for the detections to be highly inclined galaxies. If confirmed, this result would suggest that some of the maser amplification occurs in gas disks coplanar with the galaxy stellar disk.

8) VLBI observations of two of the new detections are planned.

2.3. DUST EMISSION – DISKS AT $\simeq 10^2 - 10^3$ PC

Mkn 348 is a type 2 Seyfert galaxy with broad Hα emission visible in po-larized light (Miller & Goodrich 1990). There is a bi-polar (bi-conical?) nebulosity of high excitation gas in p.a. $\simeq 170°$ (Mulchaey, Wilson & Tsve-tanov 1995), which aligns well with the radio axis in p.a. $\simeq 168°$ (Neff & de Bruyn 1983). These directions are perpendicular to that of the optical polarization in p.a. $\simeq 84°$ (Miller & Goodrich 1990), as is usually the case in Seyfert 2's. Simpson et al. (1995) have recently discovered a red 'bar-like' feature in a J–K color map; this 'bar' aligns in p.a. $\simeq 90°$ and extends about 1 kpc. There is excess emission (after subtracting a model of the galaxy starlight) at K band associated with the 'bar'. This infrared 'bar' is perhaps best interpreted as emission from hot ($\approx 700K$) dust in a torus or disk viewed edge-on. Simpson et al. (1995) speculate that this torus may represent the outer parts of the 'obscuring torus' invoked to hide the broad line region (Miller & Goodrich 1990). These authors also infer that the ob-scuration to the nucleus is $A_V \approx 60$ mag and that the total mass in the torus is M $\approx 3 \times 10^6$ (h/1 pc)(r/500 pc) $M_\odot$, where h is the height of the torus and r its outer radius.

2.4. IONIZATION CONES - SHADOWING BY A DUSTY DISK?

An ionization cone is a region of high excitation, ionized gas within a tri-angular envelope on the sky (presumably conical or wedge-shaped in three dimensions), with one apex at the active nucleus. They are usually inter-preted in terms of illumination of ambient or narrow line gas by a collimated beam of ionizing radiation from the nucleus. Dopita & Sutherland (1995) suggest instead that the emission-line region is excited by local shocks, with the cones resulting from entrainment and pressurisation of cool gas at the boundaries of an outflow driven by a relativistic jet. The first explana-tion is most convincing when the 'cone' has *sharp and straight* edges. Only in such cases can we be reasonably sure that the nebulosity is radiation-, rather than matter-, bounded. Strong evidence that the ionization cone in NGC 5252 is radiation-bounded comes from HI 21 cm mapping by Prieto & Freudling (1993). They find that the neutral hydrogen tends to avoid the ionization cones and there is good evidence for velocity continuity from

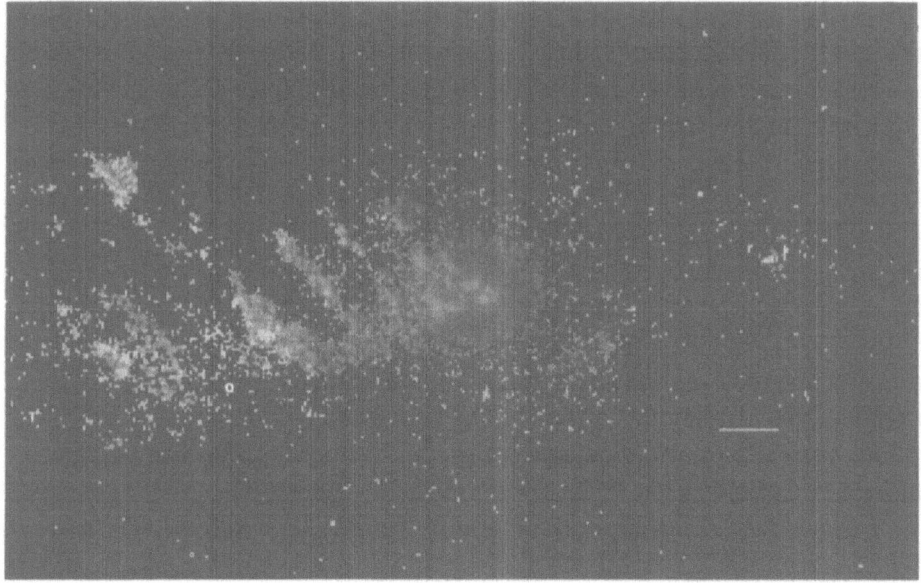

Figure 1. The ratio of the [OIII]λ5007 to the Hα+[NII]λλ6548, 6583 image of the Seyfert 2 galaxy NGC 5643. High excitation gas is white, low excitation dark. The white bar indicates 1 arc sec (≃ 78 pc). The triangular envelope of the high excitation gas is notable. The data were obtained with HST by ASW, C. Simpson, G. A. Bower, T. M. Heckman, J. H. Krolik and G. K. Miley.

the neutral to ionized gases. These results provide strong support to the interpretation (Tadhunter & Tsvetanov 1989) that the ionized arcs within the triangular envelope in this galaxy represent pre-existing ring structures which have been ionized by a collimated, nuclear, radiation field. Still, many nebulosities in Seyfert galaxies are best described as bi-polar, and show strong evidence for interaction with radio jets and lobes (e.g. Wilson 1982; Whittle et al. 1988). In these cases, the morphologies by themselves do not indicate collimated nuclear ionizing radiation. An ionization cone, recently discovered by HST observations, is shown in Fig. 1. The morphological properties of ionization cones are as follows (cf. Wilson & Tsvetanov 1994): 1) Both bi-cones and single cones are seen. Single cones generally project against the far side of the galaxy disk, suggesting the counter-cone is hidden behind the disk's near side. Thus all cases may be intrinsically bi-conical. 2) Observed cone opening angles range between ≃ 40° and 100°. Projection effects can cause the observed opening angle of the ionized gas and the true opening angle of the photon cone to differ (see Mulchaey, Wilson & Tsvetanov 1996 for simulations of this effect). 3) Cone extents range between ≈ 100 pc and 2 kpc. 4) Clearly defined cones have, so far, been seen in only radio-quiet AGN.

The current lack of detections in radio-loud objects (especially Fanaroff-
Riley class II radio galaxies) may reflect observational limitations occa-
sioned by their greater distances than Seyfert galaxies, the lower mean
interstellar gas density, and the possibility that the interstellar gas in these
ellipticals may be mostly hot and any cool gas patchy.

5) Overall there is no relation between the cone and galaxy disk axes.
However, a very strong alignment is found between the cone and radio
axes. Thus the radio plasma and ionizing photons must be collimated by
the same, or coplanar, disks. These disks are not coplanar with, and may
be randomly oriented with respect to, the host galaxy disks (Ulvestad &
Wilson 1984). NGC 4258 is an excellent example of this phenomenon: the
plane of the disk mapped out in H_2O maser emission (Miyoshi et al. 1995)
is almost perpendicular to that of the host galaxy disk. Consequently, the
radio, optical and emission-line jet in this galaxy lies close to the galaxy
disk (e.g. Cecil, Wilson & De Pree 1995).

I thank NASA for support by grants NAGW-3268 and NAGW-4700.

References

Braatz, J. A., Wilson, A. S. & Henkel, C. 1994, ApJ (Letts), 437, L99
Braatz, J. A., Wilson, A. S. & Henkel, C. 1996, ApJ Supplements (submitted) bibitem[]
    Cecil, G., Wilson, A. S. & De Pree, C. 1995, ApJ, 440, 181
Dopita, M. A. & Sutherland, R. S. 1995, ApJ (submitted)
Gallimore, J. F., Baum, S. A., O'Dea, C. P., Brinks, E. & Pedlar, A. 1995, preprint
Greenhill, L. J., Jiang, D. R., Moran, J. M., Reid, M. J., Lo, K. Y. & Claussen, M. J.
    1995a, ApJ, 440, 619
Greenhill, L. J., Henkel, C., Becker, R., Wilson, T. L., & Wouterloot, J. G. A. 1995b,
    A&A, in press
Haschick, A. D., Baan, W. D. & Peng, E. W. 1994, ApJ, 437, L35
Miller, J. S. & Goodrich, R. W. 1990, ApJ, 355, 456
Miyoshi, M., Moran, J. M., Herrnstein, J., Greenhill, L. J., Nakai, N., Diamond, P. &
    Inoue, M. 1995, Nature, 373, 127
Mulchaey, J. S., Wilson, A. S. & Tsvetanov, Z. I. 1995, ApJ Supplements (in press)
Mulchaey, J. S., Wilson, A. S. & Tsvetanov, Z. I. 1995, ApJ (submitted)
Nakai, N., Inoue, M. & Miyoshi, M. 1993, Nature, 361, 45
Neff, S. G. & de Bruyn, A. G. 1983, A&A, 128, 318
Prieto, M. A. & Freudling, W. 1993, ApJ, 418, 668
Simpson, C., Mulchaey, J. S., Wilson, A. S., Ward, M. J. & Alonso-Herrero, A. 1995,
    ApJ (submitted)
Tadhunter, C. N. & Tsvetanov, Z. I. 1989, Nature, 341, 422
Tanaka, Y. et al. 1995, Nature, 375, 659
Ulvestad J. S. & Wilson, A. S. 1984, ApJ, 285, 439
Watson, W. D. & Wallin, B. K. 1994, ApJ, 432, L35
Whittle, M., Pedlar, A., Meurs, E. J. A., Unger, S. W., Axon, D. J. & Ward, M. J. 1988,
    ApJ, 326, 125
Wilson, A. S. 1982, in Extragalactic Radio Sources, IAU Symposium Nr. 97, Eds. D. S.
    Heeschen & C. M. Wade, 179 (Reidel: Dordrecht)
Wilson, A. S. & Tsvetanov, Z. I. 1994, AJ, 107, 1227.
Wilson, A. S., Braatz, J. A. & Henkel, C. 1995, ApJ Letts (in press for Dec 20 issue)

PROBING AGN WITH WATER MASERS

J. M. MORAN, L. J. GREENHILL AND J. R. HERRNSTEIN
Harvard–Smithsonian Center for Astrophyusics
60 Garden St., Cambridge MA 02138

P. J. DIAMOND
National Radio Astronomy Observatory
Socorro NM 86801

AND

M. MIYOSHI, N. NAKAI AND M. INOUE
National Astronomical Observatory
Minamisaku, Minamimaki, Nagano, 384–13, Japan

1. Introduction

Observations of the angular distribution of the water masers in the nucleus of NGC4258 reveal the presence of a thin molecular disk in nearly perfect Keplerian orbit (Miyoshi *et al.*, 1995; Moran *et al.*, 1995). About 300 galaxies have been searched for nuclear water masers to a limiting sensitivity of about 0.1 Jy (e.g., Braatz, 1996); and 16 masers have been detected. The maser imaged by Miyoshi *et al.* (1995) offers the best example of disk structure. VLA data of NGC1068 shows evidence of disk structure (Gallimore *et al.*, 1996; Greenhill & Gwinn, 1996) and NGC2639 has drifting features, which may be due to centripetal acceleration (Wilson *et al.*, 1995).

2. Parameters of the Disk in NGC4258

The morphology and kinematics of the molecular disk in the nucleus of NGC4258 are delineated by water masers, which have unresolved angular extents and narrow linewidths. The accurate definition of the disk properties were made with the Very Long Baseline Array (VLBA), which has both high angular resolution (0.2 mas at 1.35 cm wavelength) and high spectral resolution (0.2 km/s or $\nu/\delta\nu \sim 10^6$). The high resolution image of the maser and its velocity field show that the disk is viewed nearly edge

211

R. Ekers et al. (eds.), Extragalactic Radio Sources, 211–214.

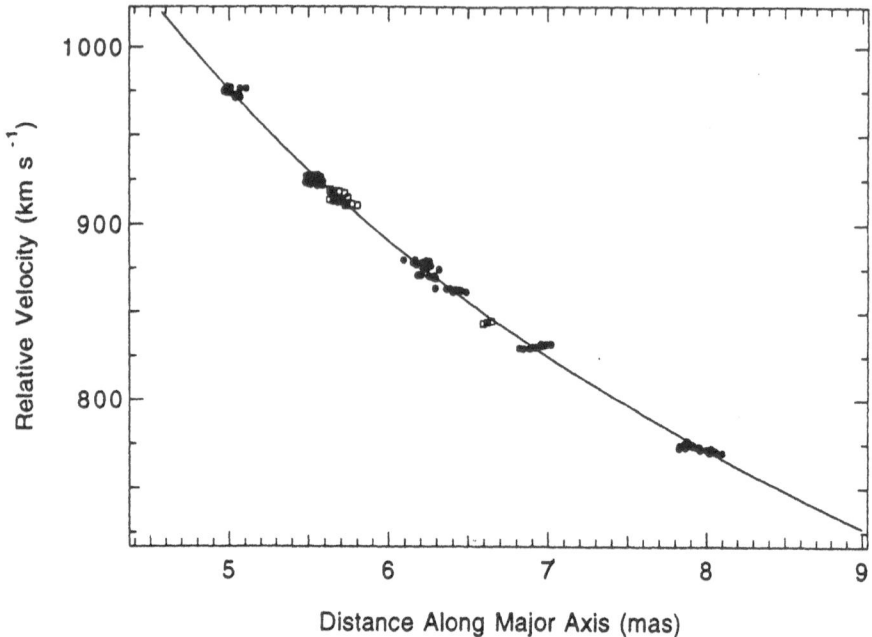

Figure 1. Keplerian rotation curve (magnitude of the line–of–sight velocity with respect to the systemic velocity versus radius from center) of the high velocity masers in the nucleus of NGC4258. Redshifted features, filled circles; blue shifted features, open squares.

on (inclination angle $= 97°$). The high–velocity features have a Keplerian distribution of velocities to an accuracy of better than 1 percent within an annulus of inner radius of 0.13 pc (4 mas and a distance of 6.4 Mpc) and outer radius of 0.26 pc (see figure 1), which requires a binding mass of 3.5×10^7 M$_\odot$. The masers near the systemic velocity of the galaxy show a nearly linear dependence of line–of–sight velocity with impact parameter (apparent solid body rotation). The most reasonable explanation for these features is that they lie near the inner edge of the disk at a radius of about 0.13 pc with a spread of about 0.01 pc.

The disk is slightly warped and its mean position angle is about 83°. The disk axis is parallel, at least in projection, with the synchrotron jet emerging from the nucleus, and inclined by 120° to the rotation axis of the galaxy. Optical emission from the nucleus is linearly polarized in the direction parallel to the disk and may be due to Thomson scattering from electrons located along the spin axis of the disk (Wilkes *et al.*, 1995).

The disk is very thin and we have not yet been able to measure its thickness. The systemic features have a vertical extent of < 0.01 mas, and since their radius is about 4 mas, the ratio of height to radius is <0.0025. The disk is probably in hydrostatic equilibrium with a temperature of <1000K and a toroidal magnetic field of strength <250 mG. The Toomre stability parameter is probably in the range 1–10, implying that the disk is marginally stable (Moran *et al.* 1995; Maoz, 1995). The inward drift velocity is less than 1 m/s and the accretion rate is less that $10^{-3}\alpha M_\odot\, y^{-1}$, where α is the viscosity parameter. Detailed modelling (Neufeld & Maloney, 1995) suggests that the accretion rate might be about $10^{-4}\alpha M_\odot\, y^{-1}$. The emission is extremely sub–Eddington ($L_E = 4.5 \times 10^{45}$ erg s^{-1}, X–Ray luminosity = 4×10^{40} erg s^{-1}) (Lasota *et al.*, 1996).

3. Black hole

The well defined Keplerian curve leaves little doubt that the gravitational binding mass is 3.5×10^7 $M_\odot$, which must be concentrated within a radius of 0.13 pc. The density of a spherical concentration of mass with that radius is 3.8×10^9 $M_\odot$ pc^{-3} or 3×10^{-13} g cm^{-3}. Table 1 gives the minimum inferred central mass density for other massive black hole candidates.

TABLE 1. Galaxies with Possible Massive Black Holes

Galaxy	Distance Mpc	Mass 10^6 $M_\odot$	Diameter pc	Density 10^6 $M_\odot$ pc^{-3}	M/L	Ref
Gal Center	0.0085	2	0.3	90		1
M32	0.7	2	0.7	12		2
M31	0.7	30	0.7	180	100	2
NGC4258	6.4	35	0.26	3800		3
NGC3115	8.4	1000	10	2	40	1
NGC4594	9.2	500	18	0.2	100	1
NGC3377	9.9	80	24	0.01	8	2
M87	15	2400	36	0.1		4
NGC4261	30	1200	14	0.8	5200	5

ref: (1) Genzel & Harris (1994); (2) Kormendy & Richstone (1995); (3) Miyoshi et al. (1995); (4) Harms et al. (1994); (5) Ferrarese et al. (1996)

4. Centripetal Acceleration

The features in the systemic group have long been known to drift in velocity. The interpretation of this phenomenon in terms of a rotating disk is clearly centripetal acceleration. Figure 2 shows spectra at two epochs. The high velocity features drift at less than 1 km s^{-1}y^{-1}, whereas the systemic features drift at about 9 km s^{-1}y^{-1} (see, e.g., Greenhill *et al.* 1995). The amplitude variations and high spectral density of features in the systemic group make a precise estimate of their drifts difficult to estimate.

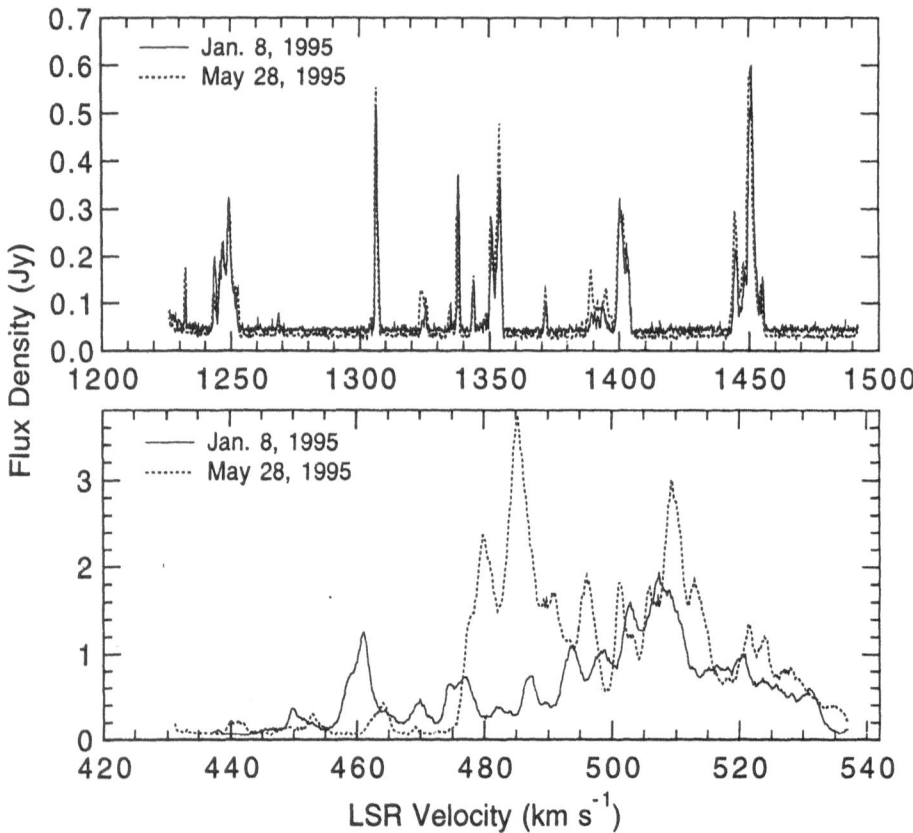

Figure 2. Centripetal acceleration evidenced by drifts in spectral features. (top) high velocity red-shifted features show less than 1 km s^{-1}y^{-1} drift; (bottom) systemic velocity features drift at about 9 km s^{-1}y^{-1}, which is consistent with a distance of 6.4 Mpc

References

Braatz, J. (1996) PhD thesis, U. Maryland, in preparation.

Ferrarese, L., Ford, H., & Jaffe, W. (1996) *Astrophys. J.*, in press

Gallimore, J. F. *et al.* (1995) *Astrophys. J. Lett.*, in press.

Genzel, R. & Harris, A. I. (1994) *The Nuclei of Normal Galaxies*, Kluwer Academic Publishers, Dordrecht.

Greenhill, L. J. *et al.* (1995) *Astron. Astrophys*, **304**, 21–33.

Harms, R. J. *et al.* (1994) *Astrophys. J. Lett.*, **435**, L35–L38.

Kormendy, J., and Richstone, D. (1995) *Ann. Rev. A. & A.*, **33**, 581–624.

Lasota, J. P. *et al.* (1996) *Astrophys. J.*, in press.

Maoz, E. (1995) *Ap. J.*, **455**, L131–L134.

Miyoshi, M., Moran, J., Herrnstein, J., Greenhill, L., Nakai, N., Diamond, P., Inoue, M. (1995) *Nature*, **373**, 127–129.

Moran, J., Greenhill, L., Herrnstein, J., Diamond, P., Miyoshi, M., Nakai, N., and Inoue, M. (1995) *Proc. Nat. Acad. Sci.*, **92**, 5, 11427–11434.

Neufeld, D. A. & Maloney, P. R. (1995) *Astrophys. J.*, **447**, L17–L20.

Wilkes, B. J., *et al.* (1995) *Astrophys. J.*, **455**, L13–L16.

Wilson, A. S., Braatz, J. A., and Henkel, C. (1995) *Astrophys. J. Lett.*, **455**, L127–L129.

WHAT POWERS ULTRA-LUMINOUS IRAS GALAXIES?

R.B.PARTRIDGE, J. MARR AND T. CRAWFORD
Haverford College - Haverford, PA 19041, USA

AND

M. STRAUSS
Princeton University - Princeton, NJ 08540, USA

I report here centimeter–wavelength observations carried out at the Very Large Array (VLA) to help resolve two questions. First, what is the source of the far infrared (FIR) emission in infrared–luminous IRAS galaxies, active nuclei or more widely distributed star formation? And what physics underlies the tight correlation (Helou *et al.*, 1985) between FIR and radio flux? To test potential answers to these questions, we believe it is important to study the most luminous IRAS galaxies. We selected 39 for study from the ultraluminous catalog of Strauss *et al.*(1990 and 1992). All sources had FIR luminosity $\geq 10^{11.4}$ $L_{\odot}$. Radio wavelength observations of these systems provide several advantages. First, in the radio there is no obscuration, so we can "see" the active galactic nuclei, if present. Radio spectral indices can distinguish between synchrotron and thermal emission. And finally, observations at the VLA provide sub–kpc resolution. We observed these sources with the VLA in its C configuration. At 1460 MHz, the effective resolution was $\sim 15''$; and $\sim 4''$ at 4860 MHz. We made follow-up observations on 24 sources in the A configuration with resolution at 4860 MHz of $\sim 0''.5$ (or $300 - 800$ h^{-1} pc for these sources).

All sources were detected in both configurations, with typical $20cm$ fluxes of $5 - 100$ mJy. The $6 - 20cm$ spectral indices had a mean value of -0.65 and a range -0.21 to -0.99, suggesting synchrotron emission as the primary mechanism. The one exception was source 0524+010, a remarkable gigahertz peak spectrum source, with $\alpha = +2.13$ at our frequencies of observation.

Eighteen of the 39 sources are multiple and/or marginally resolved at $4''$ resolution, implying extended radio emission on kpc scales. Sixteen of

R. Ekers et al. (eds.), Extragalactic Radio Sources, 215–216.

the 24 sources observed at higher resolution $(0.5'')$ were multiple and/or resolved, on sub–kpc scales.

The 39 sources we observed have $20cm$ fluxes that correlate well with FIR flux; writing $S_{20} = 10^{-q} S_{FIR}$, we find $q = 2.36$, in excellent agreement with the value of 2.34 found for lower luminosity sources by Condon et al.(1991). We took from the literature radio fluxes for a further 24 ultraluminous sources also found in the Strauss et al. catalog. When these are combined with our measurements, we again obtain a good correlation with $q = 2.40$. These results are reported for the correlation between total radio flux and FIR flux. We separately compared both nuclear radio flux (within the central $0.5''$, or $300 - 800 \, h^{-1} pc$) and extended flux, defined as (total minus nuclear), with the FIR flux. Neither nuclear nor extended flux is as well correlated, so we conclude that the tight correlation we observe is dependent on activity in both the disks and the nuclei of these galaxies.

We thus find that these ultraluminous galaxies obey the same radio–FIR correlation as lower luminosity sources, where the FIR emission is very likely due to star formation (e.g., Helou and Bicay, 1993). Our observed correlation depends on both nuclear and extended radio flux. Further, there is no obvious segregation in the plots of this correlation between sources with optical AGN spectra and sources with HII region or other spectra (Crawford et al., 1996). In addition, few sources at our resolution show convincing evidence of AGN features (jets, double–lobe sources, etc.). Finally, the radio spectral indices suggest synchrotron as the primary emission mechanism at radio wavelengths. The relativistic electrons are presumably generated by supernovae, produced in turn by starburst activity.

All these findings are consistent with the argument that star formation powers these IRAS galaxies, with their large FIR luminosity. Norris et al.(1990) report a similar conclusion from higher resolution studies of luminous starburst galaxies.

This research was supported in part by the National Science Foundation and by the William Keck Foundation.

References

Condon, J. J., Huang, Z.-P., Yin, Q. F., and Thuan, T. X. 1991, Ap. J. **378**, 65.

Crawford, T., Marr, J., Partridge, B., and Strauss, M. 1996, "VLA Observations of Ultraluminous IRAS Galaxies: Active Nuclei or Starbursts?" Ap. J., March 20 issue.

Helou, G., and Bicay, M. D. 1993, Ap. J. **415**, 93.

Helou, G., Soifer, B. T., and Rowan-Robinson, M. 1985, Ap. J. (Letters) **298**, L7.

Norris, R. P., Allen, D. A., Sramek, R. A., Kesteven, M. J., and Troup, E. R. 1990, Ap. J. **359**, 291.

Strauss, M. A., Davis, M., Yahil, A., and Huchra, J. P. 1990, Ap. J. **361**, 49.

Strauss, M. A., Huchra, J. P., Davis, M., Yahil, A., Fisher, K. B., and Tonry, J. 1992, Ap. J. Suppl. **83**, 29.

OPTICAL–RADIO CONNECTIONS

AGN Illumination or Jet/cloud Interactions?

CLIVE N. TADHUNTER
University of Sheffield
Department of Physics, Sheffield S3 7RH

1. Introduction

Many of the most important discoveries in the study of extragalactic radio sources have resulted from investigations of the relationships between optical and radio properties. The optical/radio connections include: **correlations** between optical emission line luminosity and radio power [1,2]; **alignments** between optical/UV and radio structures [1,3]; and **UV excesses** in the spectral energy distributions of radio galaxies compared with normal early-type galaxies [4].

These connections are not only important from a detailed, phenomenological point of view but are also relevant to several of the key issues concerning extragalactic radio sources, including the nature of the central engine, quasar/radio galaxy unification, and the use of radio galaxies as probes of the high redshift universe.

2. Models

Several models have been proposed to explain the connections between radio and optical properties, but the two most promising are anisotropic illumination and jet/cloud interactions.

In the **anisotropic illumination model**[5] the ambient ISM in the host galaxies is illuminated by the radiation cones of quasars hidden in the cores of the galaxies, with the emission lines resulting from photoionization of the ambient ISM by the EUV radiation in the cones, and the extended optical/UV continuum consisting of a combination of the nebular continuum[6] and scattered quasar light[7] (see Figure 1). In this case the relationship between radio and optical properties is indirect: the correlations between optical emission line luminosity and radio power require

R. Ekers et al. (eds.), Extragalactic Radio Sources, 217–222.
© 1996 *IAU.*

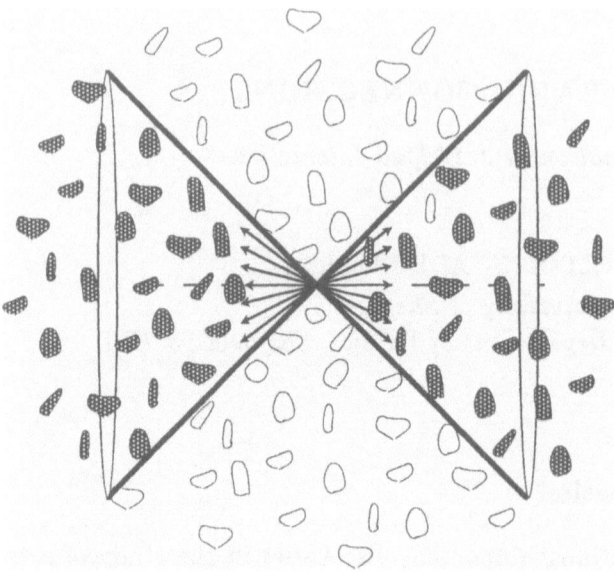

Figure 1. The anisotropic illumination model. A hidden quasar illuminates the ambient ISM in a cone pattern defined by the properties of the central obscuring torus.

that the optical/UV luminosity of the quasar is tied to the radio jet power through the physics of the central engine[8].

In the jet/cloud interaction model the relationship between the radio plasma and the optical/UV emitting components is more direct: as the radio jets plough through the ISM of the host galaxies they shock and sweep up the clouds in their path[9] (see Figure 2). The clouds involved in such jet/cloud interactions will emit both optical emission lines and optical/UV nebular continuum. It has also been proposed that the jets can trigger star formation by compressing the ambient ISM[10,11].

Each of these models is now considered in more detail.

2.1. ANISOTROPIC ILLUMINATION: ADVANTAGES

A major attraction of the anisotropic illumination model is that it is consistent with unified schemes for powerful radio galaxies, which propose that all radio galaxies have quasars hidden in their cores[12]. The putative hidden quasars will have a marked effect on the ISM in the host galaxies: typical ionizing luminosities for 3C quasars fall in the range 10^{44}–10^{46} erg s^{-1} ($H_0 = 50$ km s^{-1} Mpc^{-1}, $q_0 = 0$ assumed throughout), whereas the total emission line luminosities of radio galaxies are 10^{42}–10^{45} erg s^{-1} [8]. Thus, provided that covering factors of a few percent or more are possible, quasar

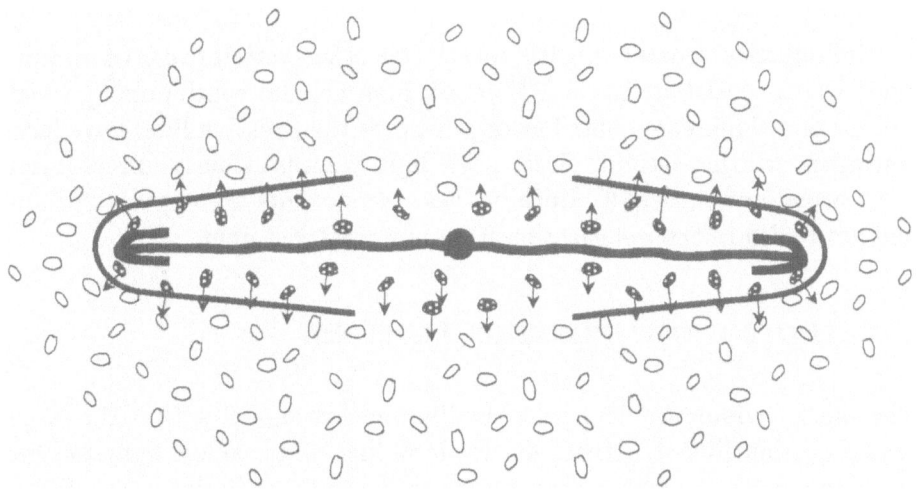

Figure 2. The jet/cloud interaction model. Clouds entering the shocks driven through the ISM by the jets will be compressed, ionized and accelerated.

illumination can supply a substantial fraction of the ionizing energy.

Another point in favour of quasar illumination is that the ratios of most strong emission lines measured in the spectra of low redshift extended emission line regions (EELR) are consistent with photoionization of optically thick, solar abundance clouds by a hard, power-law or hot black-body ionizing continuum [13]. Photoionization has a major advantage over other ionization mechanisms in the sense that the number of free parameters is relatively few. It is possible, for example, to explain the sequence of observed points on the emission line diagnostic diagrams by varying just one free parameter (the ionization parameter, U).

If the quasar is illuminating an undisturbed ISM/ICM, then the kinematics of the emission line gas should reflect the gravitational motions of the clouds in the haloes of the galaxies. In general, this is found to be the case in most low redshift radio galaxies ($z < 0.2$) which have velocity half-amplitudes consistent with the expected mass distributions of early-type galaxies ($\Delta V < 400$ km s^{-1}) [14]. Where larger velocity amplitudes are observed they are invariably associated with EELR along the radio axes, and jet/cloud interactions are suspected.

Perhaps the most compelling evidence for illumination is provided by polarimetry measurements which show that the UV continuum in several powerful radio galaxies is polarized at the 5 – 20% level, with the polarization E-vector aligned perpendicular to the radio/UV axis[7,15]. Recent spectropolarimetry observations also show evidence for scattered broad quasar features in a number of objects[16,17,18]. It is difficult to explain these observations in any other way than quasar illumination.

High-quality optical spectra reveal two other activity-related components which contribute to the UV excess: first, nebular continuum[6], which will make a significant contribution whenever the emission lines have large equivalent widths; second, direct AGN light from previously-unrecognised weak broad-line nuclei[19]. Both of these components are unpolarized and will act to dilute the polarization from any scattered light.

2.2. ANISOTROPIC ILLUMINATION: PROBLEMS

The most obvious problem with the illumination model is that the large-scale emission line structures do not look like cones. While emission line imaging observations have shown clear evidence for radiation cones in several Seyfert galaxies[20], there is no really convincing case of an emission line cone in a powerful radio galaxy. Part of the reason for this is that the gas distribution is sparse and inhomogeneous in the early-type host galaxies and does not provide a uniform flourescent screen against which to see the quasar polar diagram. Although individual low redshift radio galaxies ($z < 0.2$) show broad distributions of emission line gas, the emission line structures are nonetheless still consistent with the broad radiation cones predicted by the unified schemes[5]. The main problem lies with high redshift radio galaxies ($z > 0.5$) in which the linear, almost jet-like optical/UV structures are better collimated than would be expected on the basis of the broad quasar radiation cones[21].

Another problem is that, despite the good general agreement between the standard photoionization model predictions and the observed emission line spectra, some line ratios appear discrepant. A notable example is the key temperature diagnostic ratio [OIII](5007+4959)/4363, which is measured to be significantly smaller than predicted by the standard AGN photoionization models (indicating higher than predicted electron temperatures)[22].

Finally, in many of the high redshift radio galaxies the emission line kinematics are extreme, with larger velocity amplitudes ($500 < \Delta V < 1500$ km s^{-1}) and line widths ($300 < FWHM < 1500$ km s^{-1}) than can be accounted for by gravity alone[23,24]. Again, this is difficult to reconcile with the standard illumination model in which the quasars illuminate the *undisturbed* ISM of the host galaxies.

2.3. JET/CLOUD INTERACTIONS: ADVANTAGES

The jet/cloud interaction model holds the promise of being able to account for many of the features not explained by quasar illumination. Given that the bulk radio jet powers for FRII radio sources are estimated to fall in the range 10^{44}–10^{47} erg s^{-1}[8], we would require the jets to convert $1 - 10\%$ of their power into emission lines in order to explain the total emission line luminosities by jet/cloud interactions alone.

Although the interaction between the warm clouds and the radio jets is likely to be complex, in general we expect the clouds to be compressed, ionized and accelerated as they enter the shocks driven through the ISM/ICM by the jets. Strong evidence for all of these effects has recently been found in detailed studies of low redshift jet/cloud interaction candidates[25,26]. The evidence includes: pressures in the emission line clouds which are orders of magnitude greater than those expected in the hot confining ISM at similar radii; complex emission line profiles with split narrow components ($\Delta V \sim \pm 500$ km s^{-1}), and underlying broad components ($FWHM \sim 1000$ km s^{-1}); [OIII](5007+4959)/4363 and HeII(4686)/Hβ diagnostic line ratios which are more consistent with shocks than with quasar photoionzation; and anticorrelations between line width and ionization state in the EELR.

We do not yet have such compelling evidence for shocks at high redshifts ($z > 0.5$), but it is clear that the high-z radio galaxies are at least qualitatively similar to the low-z jet/cloud interaction candidates, and jet-induced shocks are a promising way of explaining both their highly collimated emission line structures and extreme emission line kinematics.

2.4. JET/CLOUD INTERACTIONS: PROBLEMS

A major problem with the jet/cloud interaction model is that, at low redshifts at least, many of the EELR lie well away from the radio axes, and so it is unlikely that these EELR are directly energized by the jets.

But the most controversial issue concerns the continuum: can jet/cloud interactions explain the UV excess and continuum alignment effect? The possibilities for optical/UV continuum emission in jet/cloud interactions include jet-induced star formation[10,11] and the nebular continuum emitted by the warm, shocked clouds. The nebular continuum will certainly be significant[6], but more careful work is required to determine whether this component *dominates* the continuum in the extended structures aligned along the radio axes. The main evidence for the alternative, jet-induced star formation mechanism comprises the detection of starbursts along the radio axes of a few low redshift radio sources[27,28], but there is currently a lack of direct observational evidence for this mechanism in high redshift radio galaxies.

3. Conclusions

It is clear that no single model can explain all of the properties of radio galaxies across the whole range of radio power and redshift.

At low redshifts ($z < 0.2$) the distribution, ionization and kinematics of the EELR in most radio galaxies are consistent with illumination of the ambient ISM by the anisotropic radiation field of a hidden quasar. It is only in a minority of EELR — invariably those along the radio axis — that we see evidence for something different going on.

The situation is different at high redshifts ($z > 0.5$) where we require *both* quasar illumination and jet-induced shocks: the quasar illumination to explain the polarization properties, and the jet-induced shocks to account for the highly collimated EELR and extreme emission line kinematics observed in many of the high-z objects.

References

[1] Baum, S., Heckman, T., 1989, *ApJ*, **336**, 702.
[2] Rawlings, S., Saunders, S., Eales, S.A., Mackay, C.D., 1989, *MNRAS*, **240**, 701.
[3] McCarthy, P.J., van Breugel, W., Spinrad, H., Djorgovski, S., 1987, *ApJ*, **321**, L29.
[4] Lilly, S.J., Longair, M.S., 1984, *MNRAS*, **211**, 833.
[5] Fosbury, R.A.E., 1989: In: *ESO Workshop on Extranuclear Activity in Galaxies*, Meurs & Fosbury (eds), p169.
[6] Dickson, R.D. *et al.*, 1995, *MNRAS*, **273**, L29.
[7] Tadhunter, C., Scarrott, S., Draper, P., Rolph, C., 1992, *MNRAS*, **256**, 53.
[8] Rawlings, S., Saunders, S., 1991, *Nature*, **349**, 138.
[9] van Breugel et al., 1985a, *ApJ*, **290**, 496.
[10] Rees, M.J., 1989, *MNRAS*, **239**, 1p.
[11] de Young, D.S., 1989, *ApJ*, **342**, L59.
[12] Barthel, P.D., 1989, *ApJ*, **336**, 606.
[13] Robinson, A. et al., 1987, *MNRAS*, **227**, 97.
[14] Tadhunter, C.N., Fosbury, R.A.E., Quinn, P., 1989, *MNRAS*, **240**, 255.
[15] Cimatti, A., di Serego Alighieri, S, Fosbury, R., 1993, *MNRAS*, **264**, 421.
[16] di Serego Alighieri, S., Cimatti, A., 1994, *ApJ*, **431**, 123.
[17] Antonucci, R.R.J., 1984, *ApJ*, **297**, 621.
[18] Antonucci, R.R.J., Hurt, T., Kinney, A., 1994, *Nature*, **371**, 313.
[19] Shaw, M.A. *et al.*, 1995, *MNRAS*, **275**, 703.
[20] Tsvetanov, Z., 1995, these proceedings.
[21] Longair, M.S., Best, P.N., Rottgering, H.J.A., 1995, *MNRAS*, **275**, L47.
[22] Tadhunter, C.N., 1989. In: *ESO Workshop on Extranuclear Activity in Galaxies*, Meurs & Fosbury (eds), p379.
[23] Tadhunter, C.N., 1991, *MNRAS*, **251**, 46p.
[24] Chambers, K.C., Miley, G.K., van Breugel, W.J.M., 1990, *ApJ*, **363**, 21.
[25] Morganti, R., 1995, these proceedings.
[26] Clark, N., Tadhunter, C.N., 1995. In: *NRAO Workshop on Cygnus A*, Carilli & Harris (eds), CUP, in press.
[27] van Breugel, W.J.M. *et al.*, 1985b, *ApJ*, **293**, 83.
[28] van Breugel, W.J.M., Dey, A., 1993, *ApJ*, **414**, 563.

OPTICAL POLARIZATION OF 3C 265

M.H. COHEN, H.D. TRAN AND P.M. OGLE

California Institute of Technology
Pasadena, CA 91125

AND

R.W. GOODRICH

Space Telescope Science Institute
Baltimore, MD 21218

1. Spectropolarimetry, and a Quasar in 3C 265

3C 265 is a high-redshift (z=0.811) radio galaxy showing extended emission line regions (EELR) to 50 kpc from the nucleus (McCarthy et al 1995). However, it does not show the *alignment effect* (McCarthy 1993) that is common in distant galaxies: the EELR is *not* extended along the radio axis.

In December 1994 and January 1995 we observed 3C 265 with the spectropolarimeter at the 10-m Keck Telescope. Figure 1 shows the total flux (F_λ) and the polarized flux (PF_λ) of the nucleus. The Mg II $\lambda2800$ doublet shows broad wings, as noted by Dey and Spinrad (1996). In polarized flux the equivalent width is 58 ± 3 Å (rest frame), which is within the normal range of radio–loud quasars (Wills et al 1995), and the quasar identification is strengthened by the velocity width, about $10,000$ km sec^{-1} (rms). By analogy to Seyfert galaxies we may say that 3C 265 contains a quasar which is largely hidden by a dusty torus; we see its continuum and broad-line radiation by reflection (scattering). The equivalent width in F_λ is less than in PF_λ: $EW(F_\lambda) = 12 \pm 1$ Å, which shows that the scattered quasar light is strongly diluted by starlight plus (probably) an unpolarized continuum (FC2) as in 3C 234 (Tran et al 1995). The narrow lines are all unpolarized, to within noise, so the NLR lies outside the torus and is ionized by photoionization or shock waves.

R. Ekers et al. (eds.), Extragalactic Radio Sources, 223–226.

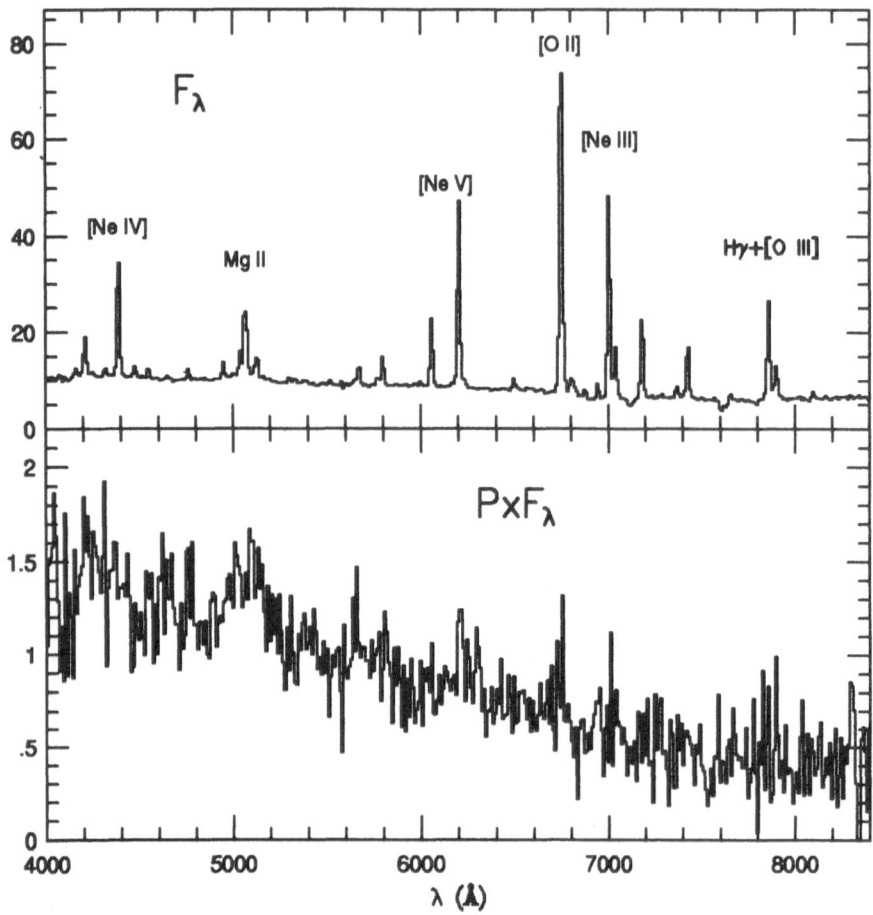

Figure 1. *Top*: 3C 265, Total flux of nucleus. *Bottom*: Polarized Flux. Note broad Mg II in emission in PF_λ, and that the narrow emission lines are unpolarized. The flux units are 10^{-18} erg cm^{-2} sec^{-1} Å^{-1}.

2. Imaging Polarimetry of 3C 265

In May 1995 we made a polarization map of 3C 265 at Keck. The V-band filter passed Mg II $\lambda2800$ and several weak narrow emission lines, but the image is dominated by the continuum. The effective resolution is 1.0″. VLA images (Fernini et al 1993) show a lobe-dominated FR II source. The radio nucleus lies at the peak in optical intensity.

Figure 2 shows 3C 265. The CCD pixels are 0.21″ square and are binned

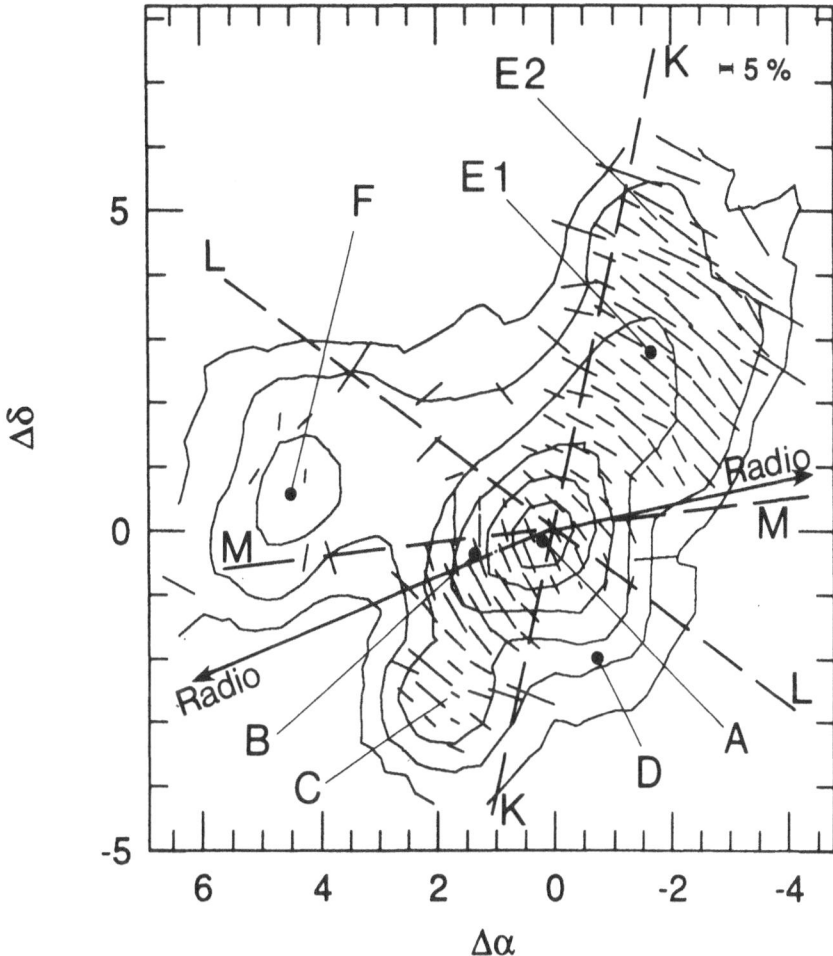

Figure 2. Polarization vectors superposed on intensity image of 3C 265, V Band. Scales are in arc seconds. Contours are logarithmic in factors of 2, starting at 2% of the peak. Radio axes are shown, and possible emission cones; see text.

2×2 in Fig 2. The total exposure is 6480^s, split into 4 settings of the waveplate. The polarization was calculated by averaging in flux and then combining the different waveplate settings. In Fig 2 only those vectors that exceed 2σ in their bin are plotted. Table 1 shows the polarization in a few fields, each 2 bins square (0.85″).

The vectors in Fig 2 (apart from the nucleus) are generally perpendicular to their nuclear radii, and the polarization must be produced by scattering of the nuclear light. 3C 265 is a gigantic reflection nebula. The

TABLE 1. Polarization in 3C 265

Region	P	θ
A	11%	26°
B	18%	31°
C	8%	55°
E1	15%	54°
E2	17%	60°
F	2%	-7°

polarization is 11% at the nucleus (Region A) and increases to the southeast and north (Regions B, E). Regions D and F have much lower polarization, and the value for F is only marginally significant. Region F shows O II λ3727 emission at the same redshift as 3C 265 (McCarthy et al 1995) and might be an interacting galaxy. The foreground galaxy 12″ to the east shows no polarization: $P < 1\%$ (2σ).

Section 1 invokes an obscuring torus, and we expect that there will be a cone of emission from it. Dashed lines KL in Fig 2 show a possible outline of the cone, with half-opening angle 55° and axis along the SE radio axis at 111°. Regions B C and E are in the cone and are strongly polarized. Region F may be weakly polarized by reflection since its mean vector is perpendicular to the radius. Region D is at most weakly polarized and is largely out of the cone. An alternative cone of half angle 35°, KM, catches most of the polarized regions and misses most of both D and F. However, this cone is not symmetric about the radio axis.

Our observations of 3C 265 are in accord with the emerging picture of powerful radio galaxies as quasars seen at high inclination. They also support the obscuring torus paradigm for radio galaxies in addition to Seyfert galaxies.

References

Dey, A. and Spinrad, H. (1996) The Radio Galaxy 3C265 Contains a Hidden Quasar *Astroph. J.*, in press March 1996.

Fernini, I., Burns, J.O., Bridle, A.H. and Perley, R.A. (1993) Very Large Array Imaging of Five FR II 3CR Radio Galaxies, *Astronom. J.*, Vol. no. **105**, pp. 1690–1709.

McCarthy, P.J. (1993) High Redshift Radio Galaxies, *Ann. Rev. Astron. and Astroph.*, Vol. no. **31**, pp. 639–688.

McCarthy, P.J., Spinrad, H. and van Breugel, W. (1995) Emission-Line Imaging of 3CR Radio Galaxies. I. Imaging Data, *Astroph. J. Supp.*, Vol. no. **99**, pp. 27–66.

Tran, H.D., Cohen, M.H. and Goodrich R.W. (1995) Keck Spectropolarimetry of the Radio Galaxy 3C 234, *Astronom. J.*, Vol. no. **110**, in press December 1995.

Wills, B.J. et. al. (1995) The Hubble Space Telescope Sample of Radio-Loud Quasars: Ultraviolet Spectra of the First 31 Quaasrs, *Astroph. J.*, Vol. no. **447**, pp. 139–158.

A POWERFUL JET/CLOUD INTERACTION
IN THE RADIO GALAXY PKS 2250-41

R. MORGANTI[1,2], C.N. TADHUNTER[3], N. CLARK[3] AND
N. KILLEEN [2]

[1] *Istituto di Radioastronomia, Bologna, Italy*
[2] *Australia Telescope National Facility, Epping, Australia*
[3] *Department of Physics, University of Sheffield, UK*

Extended emission line regions aligned with the radio axis are a common feature of powerful radio galaxies and there is much interest in the origin of the extended gas and excitation mechanism. One model that can produce this alignment is photoionization by anisotropic nuclear continuum radiation. However, strong evidence exists, especially in high redshift radio galaxies, for powerful interactions between the relativistic radio jets and the ISM/IGM. Here we present the results of our study of the southern radio galaxy PKS 2250–41 ($z = 0.308$). This object is the most spectacular found in a sample of southern radio sources studied by Tadhunter *et al.* (1993) and it displays particularly clear evidence for such an interaction (Tadhunter *et al.* 1994; Dickson *et al.* 1995).

1. Radio data

In the radio, PKS 2250-41 was observed with Australia Telescope Compact Array (ATCA) at 8 and 5 GHz. At 8 GHz we achieved a resolution of $\sim 1''$ (contours in Fig. 1). At this resolution, the structure of the source is dominated by two bright and slightly resolved components. The two lobes are situated asymmetrically compared to the centre of the optical nucleus. Despite the high frequency of these observations, no radio core was detected ($S_{core} < 0.8$ mJy and $R = S_{core}/S_{ext} < 0.0013$).

Using the 5 and 8 GHz ATCA data, we have studied the polarization, depolarization and Faraday Rotation in the source as follows:
1) very low fractional polarization ($\sim 2.8\%$ at 8 GHz) is observed in the western lobe. In the eastern lobe the fractional polarization is $\sim 10\%$.

R. Ekers et al. (eds.), Extragalactic Radio Sources, 227–229.
© 1996 *IAU.*

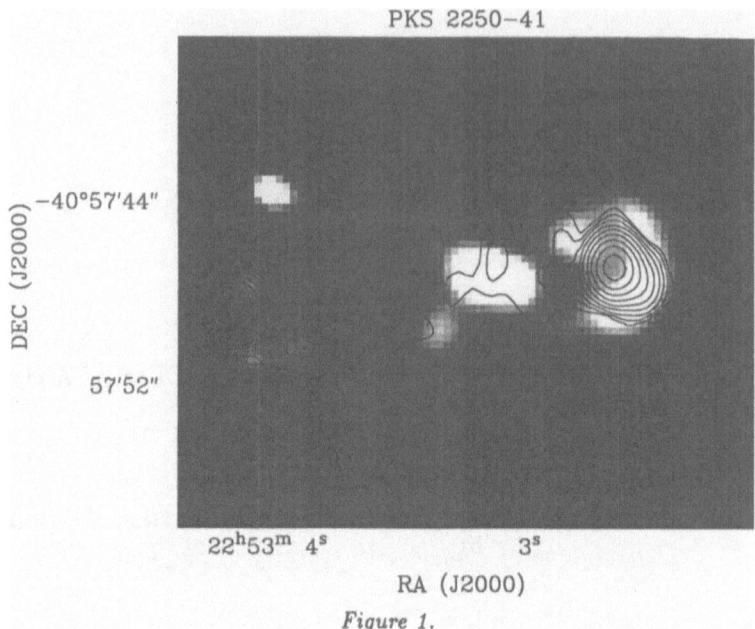

Figure 1.

2) very strong depolarization ($\sim$ 0.4 between 5 and 8 GHz) is observed in the western lobe while none is observed in the eastern lobe.

Depolarization can be attributed to beam smearing or internal Faraday depth effects. Higher resolution radio images will be necessary to disentangle these effects.

2. Optical data

In the optical, PKS 2250–41 was observed with the ESO 3.6 m telescope in July 1993 using EFOSC in broad/narrow–band imaging, spectroscopic and polarimetric modes. Our narrow–band [O III]λ5007 image (grey scale in Fig. 1) reveals a remarkable arc of line emission at radius of 6″ (35 kpc) west of the nucleus. The length of the arc is $\approx$ 8″ (50 kpc), with a projected thickness of 1.″7 (10 kpc). We also see a spatially distinct clump of [O III]λ5007 emission 20 kpc NW of the nucleus. Fainter emission–line arcs are observed to the east of the nucleus, at radii of 8″ (50 kpc) and 10″ (60 kpc).

In Fig. 1, we overlay the [O III]λ5007 image and the 8 GHz total intensity contours. Clearly, the optical and radio emission are closely associated. The western arc of line emission is at the outer edge of the western radio lobe, and the lower surface brightness eastern arc lies at the inner edge of the eastern radio lobe. Optical continuum emission is also seen coincident with the western lobe and at the inner edge of the eastern lobe.

The spectra of PKS 2250-41 show that the extended arc has a much lower ionization state than the nucleus with a typical ratio [OII]/[OIII] between 2 and 3. Moreover, we find that in the western arc the low ionization lines —[NII], [SII] and [OI]— have a velocity dispersion up to $\Delta V \sim 500 \mathrm{kms}^{-1}$ while the high ionization [O III]λ5007 lines have a much lower velocity dispersion. The large velocity dispersions in the extended arc suggest a kinematic disturbance of the emission–line gas by the radio jet.

The spectrum of the western arc can be explained either by photoionization from the active nucleus or by jet-induced shocks (e.g. Sutherland et al. 1993). However, the latter seems to be better at reproducing line ratios (e.g. HeII/Hβ) and the observed difference in linewidth for the different ionic species. Furthermore, the correlation between the linewidth and the ionization state implies that the kinematic disturbance of the emission–line gas and the ionization mechanism are intimately related.

3. Origin of the extended gas

The observed $\geq 10^6 \mathrm{M}_\odot$ mass of warm gas at a distance of 35 kpc from the nucleus of PKS 2250-41 might originate from either (a) jet-induced cooling of a pre-existing hot interstellar medium (Daly 1992; de Young 1989) or (b) pre-existing warm/cool gas clouds interacting with the radio jet. If we assume (a), the phenomenon should be observed in many more radio galaxies. If we assume (b), then where did this gas come from? The approximate mass of emission–line gas is consistent with a merger remnant. Our broad–band images show a companion galaxy to the east of the nucleus which shows signs of gravitational interaction with PKS 2250−41 in the form of tidal tails. Alternatively, we are observing a direct collision between the jet and a companion galaxy. This is supported by the colours of the continuum (following the subtraction of the nebular continuum) that are consistent with what is expected for a late-type spiral galaxy.

References

Daly, 1992 *ApJ*, **399**, 426;

de Young, 1989 *ApJ*, **342**, L59;

Dickson, R.C., Tadhunter, C.N., Shaw, M.A., Clark, N.E. & Morganti, R. 1995 *MNRAS*, **273**, L29;

Sutherland, R.S., Bicknell, G.V. & Dopita, M.A. 1993 *ApJ*, **414**, 510;

Tadhunter, C.N., Morganti, R., di Serego Alighieri, S., Fosbury, R.A.E. & Danziger, I.J. 1993 *MNRAS*, **263**, 999;

Tadhunter, C.N., Shaw, M.A., Clark, N.E. & Morganti, R. 1994 *A&A*, **288**, L21

RADIO OUTFLOWS AND THE ORIGIN OF THE NARROW LINE REGION IN SEYFERT GALAXIES

A. CAPETTI[1], D.J. AXON[1,2], F.D. MACCHETTO[1,2]
AND W.B. SPARKS[1]

[1] *Space Telescope Science Institute - 3700, San Martin Drive, Baltimore, MD 21218, U.S.A.*
[2] *Astrophysics Division, Space Science Department of ESA - ESTEC, NL-2200 AG Noordwijk, The Netherlands*

Seven Seyfert galaxies, Mrk 3, Mrk 78, Mrk 348, Mrk 6, Mrk 573, NGC 3393, IRAS 04210+0400 and IRAS 11058-1131 (all Seyfert galaxies type 2 expect Mrk 6 which is a Seyfert 1.5) were observed with HST. For the first five objects images were taken with the FOC with filters centered on the emission lines [O II] and [O III]. For the remaining three galaxies WFPC2 observations were obtained in the [O III] and Hα lines.

These images allowed us to explore the morphology of their NLRs which extend over less than 200 pc in the case of Mrk 348 to more than 16 kpc in IRAS 04210+0400. In Mrk 3, Mrk 6 and Mrk 348 the line-emission takes the form of a linear structure while in all the other galaxies series of spectacular arcs of line emission are present on both sides of the nucleus (see Fig. 1). In each case the radio and the line emission are very closely associated. In Mrk 3 and Mrk 348 UV ($\sim$ 2100 Å) continuum emission has been detected and it appears to be cospatial with the radio jets.

The observed morphology of the NLR of these Seyfert galaxies can be explained if this results from the interaction of the radio ejecta with the surrounding gas. The line-emitting gas is compressed by the shocks created by the passage of the supersonic jets or by the sweeping-up of gas by the expanding radio lobes. The increase in the density due to the compression causes the line-emission to be highly enhanced in the region where this interaction occurs. The extended UV component may arise from free-free emission of the gas heated by the shocks formed by the radio jets. Ground based long-slit spectroscopy of these galaxies also shows the hallmark of this interaction in their nuclear regions, i.e. very broad and often splitted

R. Ekers et al. (eds.), Extragalactic Radio Sources, 230–231.

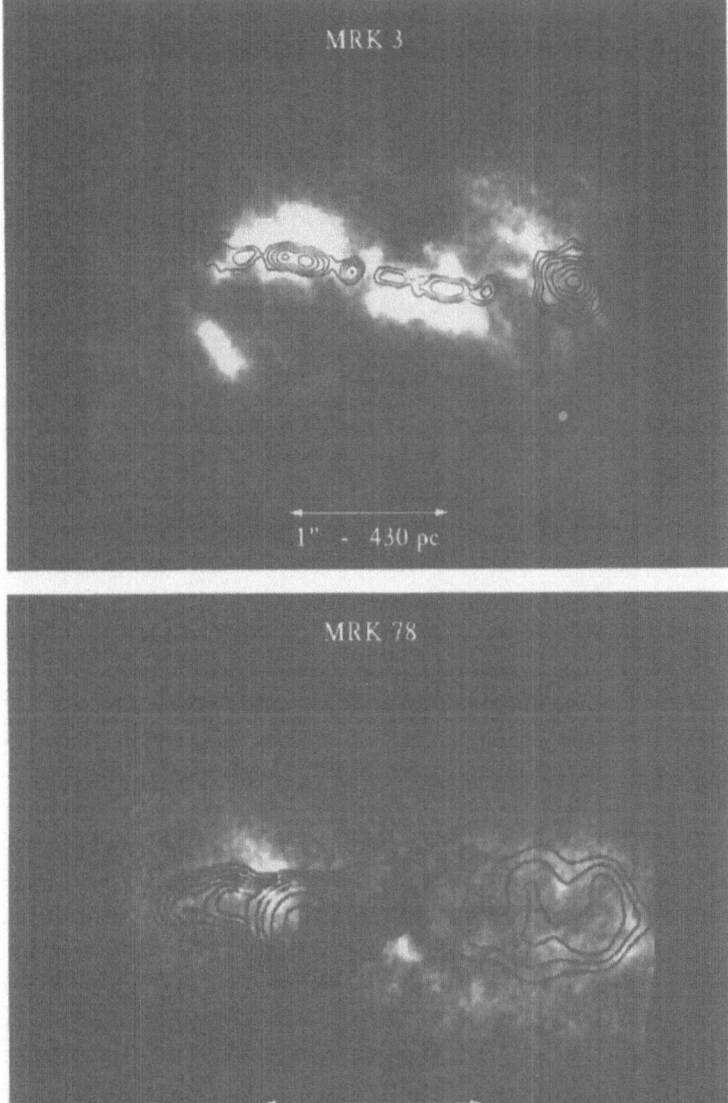

Figure 1. [O III] emission line image of two galaxies observed in this project (Mrk 3 and Mrk 78) with superposed the contour radio maps

lines. It therefore appears that the morphology, the dynamics and possibly the ionization structure of the NLR of these Seyfert galaxies is dominated by the effects of the radio outflows.

MATTER–BOUNDED PHOTOIONIZED CLOUDS

L. BINETTE[1], A.S. WILSON[2], T. STORCHI-BERGMANN[3]

[1] *Observatoire de Lyon - 9 av. Charles André*
F-69561 Saint-Genis-Laval Cedex, France
[2] *Astronomy Department - University of Maryland*
College Park, MD 20742, USA
[3] *Instituto de Fisica - UFRGS, Campus do Vale*
91500 Porto Alegre, RS, Brasil

The *extended* ionized gas in Seyfert and Radio-Galaxies is characterized by large values of the ratio He II/Hβ, which exceeds the value predicted by the standard photoionization model in which the ionizing continuum consists of a power-law. This has lead to the suggestion of considering a matter-bounded (MB) component [3],[5],[2] for explaining such extreme values. We now find[1] that it is also possible to resolve the temperature problem[3] if the thickness and the ionization parameter of the MB is appropriately selected. Adopting a canonical power law ($\alpha=-1.3$) and solar abundances ($Z=1$), we can account for the observed trends in excitation (represented for example by the ratio [O II]/[O III] in Fig. 1)) by varying the relative number of MB clouds (which emit the high excitation lines C IV, [Ne V], He II... and most of [O III]) versus the number of ionization-bounded (IB) clouds (which emit [N II],[S II] [O II], [O I]...). We obtain a one-parameter sequence (solid line) which is function of the weight $A_{M/I}$ of the MB component relative to the IB component. This $A_{M/I}$–sequence successfully reproduces the observed range in He II/Hβ. Note the failure of the traditional U–sequence (long dashed line). Fig. 2 indicates that we can also reproduce the ratio $R_{OIII}=$ [O III]$\lambda4363$/[O III]$\lambda5007$ and therefore resolve the temperature problem. Interestingly, our model indicates a temperature difference of 5 000 K between the IB component ([N II] temperature $\simeq 10\,000$ K) and the MB component ([O III] temperature $\simeq 15\,000$ K) while the traditional U–sequence predicts a difference of only 1 000 K. Such difference of 5 000 K has been reported[4] in the extended gas of Cygnus A.

R. Ekers et al. (eds.), Extragalactic Radio Sources, 232–233.

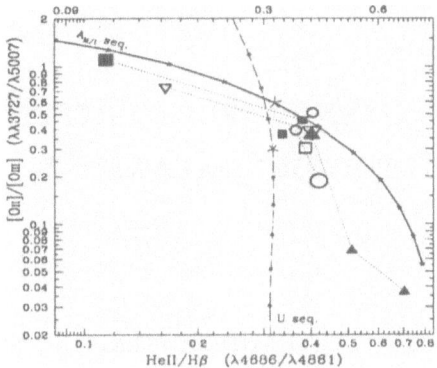

Figure 1. Diagram of the line ratios [O II]/[O III] against He II/Hβ. Filled and open symbols denote Seyfert and radio galaxies, respectively. Larger symbols denote the nuclear values. A dotted line joins measurements at different locations in the same galaxy. The parameter $A_{M/I}$ of our model represents the relative weight of the MB component and *increases* from left to right along the solid line ($0.04 \leq A_{M/I} \leq 16$). The long dashed line represents the traditional U–sequence.

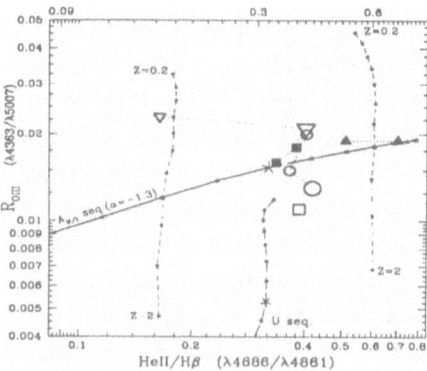

Figure 2. Diagram of the line ratios R_{OIII} (4363Å/5007Å) against He II/Hβ. The symbols have the same meaning as in Fig. 1. $A_{M/I}$ is *increasing* from left to right. The two short-dash lines correspond to metallicity sequences ($0.2 \leq Z \leq 2.0$) at $A_{M/I}$=0.4 and 4.0.

References

[1] Binette, L., Wilson, A. S. and Storchi-Bergmann, T., 1996. *A&A*, submitted.

[2] Morganti, R., Robinson, A., Fosbury, R. A. E., di Serego Alighieri, S., Tadhunter, C. N., and Malin D. F., 1991. *MNRAS*, **249**, 91.

[3] Tadhunter, C.N., Robinson, A. and Morganti, R., 1989. ESO Conf. and Workshop Proc. no.32, Garching, p 293.

[4] Tadhunter, C. N., Metz, S. and Robinson, A., 1994. *MNRAS*, **268**, 989.

[5] Viegas, S. M. and Prieto, A., 1992. *MNRAS*, **258**, 483.

IS THE UV ALIGNMENT EFFECT PRESENT
IN LOW REDSHIFT RADIO GALAXIES ?

ANDREA CIMATTI AND SPERELLO DI SEREGO ALIGHIERI
Osservatorio Astrofisico di Arcetri
Largo E. Fermi 5, I-50125, Firenze, Italy

1. Introduction

When a FRII radio galaxy at $z > 0.7$ is observed in the optical, its continuum appears extended and aligned with the radio axis. This phenomenon occurs actually when the optical bands start to sample the rest-frame UV, and it was called *alignment effect* (McCarthy et al. 1987). The UV continuum of high z radio galaxies shows also strong linear polarization due to scattering of anisotropic radiation escaping from the obscured quasar nucleus (di Serego Alighieri, Cimatti & Fosbury 1994). However, the observations of the UV continuum have been concentrated mostly on high z radio galaxies, leaving open a main question : *is the alignment effect an evolutionary phenomenon present only at high z, or is it simply a selection effect due to the K-correction ?* In order to investigate the origin and evolution of the UV *alignment effect*, we have started a ground–based imaging and polarimetric survey for studying the UV continuum in FR II radio galaxies with $0.1 < z < 0.5$. Depending on the redshift, the filters U and B can be used to sample the UV continuum free from strong emission lines, and in the same rest–frame spectral region observed in radio galaxies at higher redshift. The survey is in progress and here we present only our first results.

2. Results and first implications

We observed 7 galaxies and detected extended and polarized UV continua in 3 of them. Although the sample is small and incomplete, some implications can already be drawn. • Powerful radio galaxies with $0.1 \leq z < 0.5$ *do not* always show the UV *alignment effect*. We recall that the UV continuum

234

R. Ekers et al. (eds.), Extragalactic Radio Sources, 234–235.

of powerful radio galaxies at higher redshift is always aligned with the radio axis. • This results implies that the *alignment effect* is not simply due to a K-correction effect. • The *round* UV continua tend to show lower polarization. In comparison, the *aligned* UV continua are *highly polarized* with the electric vector *perpendicular* to the axis defined by the UV continuum morphology. • As in high redshift radio galaxies, the dominant polarization mechanism is scattering by dust and/or electrons of the anisotropic radiation emitted by an obscured quasar nucleus. In fact, the UV–optical continua can be successfully fitted with scattered quasar radiation + stellar light. • In this respect, powerful radio galaxies which show the *alignment effect* mimic the properties observed at high redshift. The most striking example is provided by 3C 195 (Cimatti & di Serego Alighieri 1995).

3. What triggers the alignment effect ?

Because in low z powerful radio galaxies the *alignment effect* can be absent, we can conclude that it is an *evolutionary* effect, meaning that it is always occurring at $z > 0.7$, but only sometimes at lower z. This evolution can be related to the increase of the radio power at high redshift (Dunlop & Peacock 1990), and/or the evolution of the environment of the radio galaxies (ISM and/or merging companion galaxies). Recent results have suggested that the *alignment effect* at $z \sim 1$ depends on the radio power (Dunlop & Peacock 1993), but more observations at different redshifts are still required. Concerning the degree of polarization, we note that in case of scattering, P depends on the geometrical configuration, on the nature and size of the scattering particles and, in presence of unpolarized radiation, on the luminosity of incident radiation beam. The unified model of powerful radio-loud AGN predicts that anisotropic radiation must be present in all FR II radio galaxies with narrow–lined optical spectra, irrespectively of the redshift. However, we do not observe strong scattered polarized light in all the narrow-lined radio galaxies at low redshift. The origin of this 'evolution' of the UV polarization will be investigated in detail by using our coming sample of low redshift radio galaxies.

References

Cimatti, A., di Serego Alighieri, S., 1995, *MNRAS*, 273, L7
di Serego Alighieri, S., Cimatti, A., Fosbury, R.A.E., 1994, *ApJ*, 431, 123
Dunlop, J.S., Peacock, J., 1990, *MNRAS*, 247, 19
Dunlop, J.S., Peacock, J., 1993, *MNRAS*, 263, 936
McCarthy, P.J., van Breugel, W.J.M., Spinrad, H., Djorgovski, S., 1987, *ApJ*, 321, L29

WEAK RADIO GALAXIES:
NARROW-BAND OPTICAL IMAGING

RENÉ CARRILLO AND IRENE CRUZ-GONZÁLEZ
Instituto de Astronomía, UNAM
Apdo. 70-264, México D.F., 04510, México

1. Introduction

Previous studies show that: a) radio galaxies and radio-loud quasars have emission-line gas (ELG) which is extended on scales of tenths of kiloparsecs; b) there is convincing evidence that the kinematics and excitation of the very extended emission-line gas is governed by its interaction with the outflowing radio plasma; c) the evidence for an interaction is weaker in some radio galaxies. It is argued that the ionization of the ELG may be predominantly produced by the nuclear ultraviolet continuum and the kinematics of the gas due to the gravitational potential of the host galaxy, but it is not yet known whether there is a physical relationship between the ELG and the extended radio jets.

We have obtained optical narrow-band (isolating redshifted Hα+[NII], or [OIII]) images of a radio flux density selected sample of 26 radio galaxies. We also use data available in the literature to study the association between the extended emission-line gas and the radio structure.

The details of our study of extended emitting gas in weak radio galaxies are presented in Carrillo (1995), Carrillo, Cruz-González & Guichard (1995).

2. Observations and Data Reduction

The observations were carried out at the 2.12 m telescope of the Observatorio Astronómico Nacional at San Pedro Mártir, B.C., México. CCD images were taken through narrow-band filter centered on either Hα+[NII]$\lambda\lambda$6548,6583, or [OIII]λ5007. Two Thompson CCD detectors were used: 384×576 and 1024×1024, which with the f/7.5 secondary yield an image

236

R. Ekers et al. (eds.), Extragalactic Radio Sources, 236–237.

scale in the focal plane of 0.3"/pixel and 0.25"/pixel, respectively, so that the field of view is 1.93'×2.89', and 4.25'×4.25' in each case.

Typical observation times at Hα were 30 min and at [O III] 1 hr. A set of standard stars from Oke's list, were acquired each night for photometric calibrations. The reduction of the two-dimensional CCD frames follows the standard procedures. The Image Reduction and Analysis Facility (IRAF) software was used.

3. Results

This work summarizes the results of a study of the optical properties and emitting-line gas nebulae of a representative sample of weak radio galaxies.
Our main results are:

- Spatially extended emission-line gas is quite common in weak radio galaxies. Line emission is detected in all the sources in the representative sample and 81% have resolved emission-line nebulae.
- The size of the emission-line nebulae is ≈ 4.6 kpc, and the emission-line luminosity in Hα+[NII] or [OIII] is ≈ 1.1×10^{41} ergs s^{-1}. The latter is an order of magnitude the luminosity of emission-line nebulae in normal early-type galaxies, and similar to that found in powerful radio galaxies (Baum & Heckman, 1989a, 1989b).
- The emission-line nebulae have a wide range of sizes and morphology. In some sources we observed only small, centrally condensed, kpc scale regions, while in others we detect more extended filaments of line emitting gas at several kpc from the host galaxy nucleus.
- We find very strong correlations of the emission-line luminosity with both the total radio luminosity and the core radio power. WRGs showed to be consistent both in radio luminosity and line luminosity with less luminous sources than powerful radio galaxies or quasars.
- We estimate lower limits to the density of the emission-line gas of ∼ 0.02 to 0.26 cm^{-3}, and upper limits to the total mass in emission-line gas between 1.2×10^8 and 2.6×10^9 M$_\odot$.

References

Baum, S.A., Heckman, T. (1989a) Extended Optical Line Emitting Gas in Powerful Radio Galaxies: Statistical Properties and Physical Conditions, *ApJ*, **336**, pp. 681-701.

Baum, S.A., Heckman, T. (1989b) Extended Optical Line Emitting Gas in Powerful Radio Galaxies: What is the Radio Emission-Line Connection?, *ApJ*, **336**, pp. 702-721.

Carrillo, R. (1995) Ph.D. Thesis, *Facultad de Ciencias*, UNAM.

Carrillo, R., Cruz-González, I., Guichard, J. (1995) in preparation.

OPTICAL PROPERTIES OF FR I RADIO GALAXIES

MICHAEL J. LEDLOW
Dept. of Astronomy, New Mexico State University
Las Cruces, NM 88003-0001, USA

AND

FRAZER N. OWEN
National Radio Astronomy Observatory
Socorro, NM 87801, USA

1. Introduction

From the VLA 20cm survey of $\sim$ 500 Abell clusters reported by (Ledlow and Owen, 1995), we have obtained optical R-Band CCD observations and optical spectra for 265 radio galaxies. The survey is complete for 20cm flux density greater than 10 mJy within 0.3 corrected Abell radii of the cluster center. All Abell clusters with measured $z < 0.09$ were surveyed. This statistically complete sample was supplemented by $\sim$ 200 clusters with $0.09 < z < 0.25$ including sources with flux density > 200 mJy. Only 6% of the sample consists of FR II radio sources, the remainder are twin-jets, tailed, or compact sources associated with the FR I class.

Using optical surface photometry, we have examined the surface brightness profiles, sizes, ellipticities, nuclear luminosities, and isophotal properties of these rich cluster radio galaxies. The frequency of galaxy-galaxy interactions, close companions, and distorted isophotes was also examined. A control sample of 50 radio-quiet ellipticals chosen from the same cluster fields was constructed and analyzed identically. From the surface photometry and optical spectra, we ask the question: *Are radio sources found in a unique population of elliptical galaxies?* If so, what distinguishes radio-loud from radio-quiet galaxies?

From the statistically complete sample for $z < 0.09$, we have constructed the univariate radio and bivariate radio/optical luminosity functions. From these functions we examine how the optical properties of the host galaxy influence radio source lifetimes and evolution. We also compare the lumi-

238

R. Ekers et al. (eds.), Extragalactic Radio Sources, 238–239.

nosity functions for our rich-cluster sample to the non-rich cluster radio galaxy samples from (Auriemma et al., 1977) and (Sadler et al., 1989) to explore the effects of the local galaxy density and environment on radio detection statistics.

2. Conclusions

From the described analysis we make the following conclusions:

• The cluster richness and optical morphology do not affect the radio galaxy detection statistics. The number of radio galaxies simply scales with the number of galaxies surveyed.

• Radio Galaxies trace the normal elliptical galaxy optical luminosity function.

• While there is no strong correlation between optical and radio luminosity, the division between FR I and FR II sources does depend on the optical luminosity ($\propto L_{opt}^2$).

• The surface-brightness profiles, nuclear luminosities, fundamental plane, luminosity/size relationship, and galaxy shape are not significantly different for radio-quiet and radio-loud ellipticals at the same optical luminosity.

• Only 6% of FR I's show evidence for galaxy interactions. 25% show evidence of optical peculiarities such as twisted, non-concentric, or non-elliptical isophotes. However, these statistics are identical for radio-quiet ellipticals selected from the same environment. The frequency of companion galaxies (< 20 kpc projected distance) is the same between the radio galaxy and control sample.

• The break in the differential bivariate luminosity functions corresponds to the FR I/II division and shifts to higher powers as L_{opt}^2.

• At a constant radio power, the fraction of galaxies detected increases with optical luminosity. The optical luminosity, or parameters related to it (such as the total mass, ISM density, pressure, and extent) must influence the initial conditions and subsequent evolution of radio sources.

• From both spectroscopy and optical surface photometry, FR I radio galaxies are indistinguishable from radio-quiet ellipticals. Possibly all elliptical galaxies at some time (or many times) have powerful radio sources. This is consistent with the integrated univariate RLF assuming radio source lifetimes are less than a few times 10^9 years.

References

Auriemma, C.G. et al. 1977, A & A, 57, 41.
Ledlow, M.J. & Owen, F.N. 1995, A.J., 109, 853.
Sadler, E.M., Jenkins, C.R., & Kotanyi, C.G. 1989, M.N.R.A.S, 240, 591.

OPTICAL PROPERTIES OF B2 RADIO GALAXIES

J.I. GONZÁLEZ-SERRANO AND R. CARBALLO
Instituto de Física de Cantabria, Santander, Spain

Abstract. We present a large program of CCD imaging of low-luminosity radio galaxies selected from the B2 survey of radio sources. We aim to study their optical properties: brightness profiles, isophotal distortions, morphological peculiarities. As the sample contains jet and no-jet sources, this investigation will allow us to compare both populations of low-luminosity radio galaxies. In particular, we will study the local environment and test whether the galaxy density around radio sources plays an important role on the presence of jets or not. The analysis of morphological distortions on the host galaxies will provide us with detailed information on past or recent galaxy encounters.

1. The sample

The sample of low-luminosity radio galaxies is described in Parma et al. (1987) (and references therein). It contains ~ 110 sources identified with galaxies in the POSS plates and mapped using the VLA at 1.4 GHz. About 45% of the sources present a radio jet at the kpc scale. The radio power of these sources at 1.4GHz is less than 10^{25} W Hz^{-1} and their redshifts are less than ~ 0.15.

2. Previous results

We continue a study on the optical properties of low-luminosity radio galaxies based on broad-band imaging of this complete sample of B2 radio sources. We have finished the study of those sources containing radio jets at the kpc scale on VLA maps. In González-Serrano et al. (1993) we conclude that low-luminosity radio galaxies with jets inhabit regions of higher galaxy density than powerful radio galaxies. We have found that a high fraction of

R. Ekers et al. (eds.), Extragalactic Radio Sources, 240–241.
© 1996 *IAU.*

this kind of radio sources show indications of past/recent mergers/collisions with another (elliptical) galaxy.

Here we present CCD imaging of this sample of radio sources. The sources are similar to the radio jet sample but they have no jets at the kpc scale on similar VLA maps. In that way we can compare the properties of two populations selected from exactly the same sample.

3. Present results

We are starting now to analyse the non-jet sample in the same way as we did with the jet sample. This will allow to investigate how important is the local galaxy density in the triggering of large scale radio jets, and if merging processes have some role on the presence of jets.

The observations were carried out at the JKT 1m telescope of the Observatorio del Roque de los Muchachos on the island of La Palma. We used a TEK detector and we took 1800 sec integrations in filter V. Pixel scale and field of view are 0.34 arcsec/pixel and 5.8×5.8 arcmin respectively. The observations were done on January and May 1995. Preliminary reduction has been done in the standard way. Isophotal analysis and local galaxy density measurements are in progress.

From visual inspection of the contour maps of the galaxies we notice that there is a fraction of spiral and irregular galaxies which are absent in the jet subsample. We can also see a fraction of distorted galaxies although it seems to be a fraction not as high as it appears to be in the radio jet sample. We also notice that there are few galaxies with companions, while in the jet subsample this is a common feature. In summary, a preliminary inspection of the optical counterparts of the no-jet sources seems to indicate that the environments of both populations are different, being more dense in radio jet sources. Of course, this should be investigated in more detail using the same procedures that were used with the jet sample.

References

González-Serrano, J.I., Carballo, R., & Pérez-Fournon, I. 1993, *AJ*, **105**, 1710
Parma, P., Fanti, C., Fanti, R., Morganti, R., de Ruiter, H.R. 1987, *A&A*, **181**, 244

NEARLY SIMULTANEOUS OPTICAL AND VLBI
POLARIZATION OBSERVATIONS OF BL LAC OBJECTS

D.C. GABUZDA AND P.Y. KOCHANEV
Lebedev Physical Institute, Moscow, RUSSIA

M.L.SITKO
University of Cincinnati, Cincinnati, Ohio, USA

AND

P.S. SMITH
University of Arizona, Tucson, Arizona, USA

1. Introduction

The continua of BL Lacertae objects and other "blazars" are dominated by nonthermal emission that is variable and highly polarized at UV–radio wavelengths (Angel and Stockman 1980; Kollgaard 1994; Allen *et al.* 1993). It is believed that this non-thermal emission is associated with the relativistic jets that are known to exist in these sources, but details of the jet structure and physics are still very uncertain. It is usually expected that the polarization behavior at optical and radio wavelengths should show little or no correlation, even if genuinely simultaneous measurements are compared. It is typically thought that the emission in these two wavebands originates in vastly different parts of the source, where the magnetic field geometries are likely to be quite different. In some inhomogeneous synchrotron source models for blazars, however, depending on the model parameters considered, the radio and UV-optical-IR (UVOIR) emission may be co-spatial (Ghisellini, Maraschi & Treves (1985)). It is thus of interest to search for correlations between the emission of blazars in the UVOIR and radio, to test such models. Our approach to doing this has been to compare simultaneous measurements of the optical and VLBI polarization characteristics of compact AGN. The polarization of the radiation is effectively used as a probe of the magnetic field structures in the regions where the emission at the two wavelengths arises.

R. Ekers et al. (eds.), Extragalactic Radio Sources, 242–243.
© 1996 *IAU.*

2. Results

We will use χ_{opt}, χ_{core}, and χ_{jet} to refer to the optical, VLBI core, and VLBI jet polarization position angles, respectively. Simultaneous measurements of the optical and VLBI polarization for three BL Lacertae objects have recently been analyzed. Comparison of optical and VLBI polarization for five other blazars was presented earlier by Gabuzda & Sitko (1994). In every source, there has been a clear association between χ_{opt} and χ in one of the VLBI components. In six of the eight sources, χ_{opt} is correlated with χ_{core}, in one case (OJ 287) clearly in association with the birth of a new VLBI component; in the remaining two sources, χ_{opt} was correlated with χ_{jet} in a recently emerged new VLBI component very close to the core. Thus, these comparisons point toward a correlation between the optical polarization and the polarization in the VLBI cores and in young, recently emerged VLBI jet components in BL Lacertae objects. This idea is consistent with the suggestion by Gabuzda et al. (1994; see also Gabuzda, Pushkarev, & Cawthorne, these proceedings) that the 6 cm VLBI core polarizations of BL Lacertae objects tend to be dominated by the polarization of new jet components that are not yet resolved from the core.

A more thorough presentation and discussion of these data will be given in a paper by Gabuzda, Sitko, and Smith, to be submitted to the *Astronomical Journal* in November, 1995.

3. Ongoing Work

Simultaneous optical and VLBI polarization data for 10 BL Lacertae objects were obtained in May-August, 1995. These observations will bring the total number of sources for which such data are available to 15. Any optical-VLBI polarization correlations should be much more clearly seen in this larger sample. Work with a larger sample will also enable us to begin comparison of the optical and VLBI degrees of polarization, which we have not yet attempted with our smaller number of sources. We will continue trying to obtain such data, particularly for VLBI experiments at shorter cm wavelengths. Another interesting question that remains to be investigated is whether there is evidence for optical-VLBI polarization correlations for quasars as well as for BL Lacertae objects.

References

Allen *et al.* (1993) *Astrophysical Journal*, **403**, 610.
Angel & Stockman (1980) *Annual Reviews of Astronomy and Astrophysics*. **8**, 321.
Gabuzda & Sitko (1994), *Astronomical Journal*, **107**, 884.
Ghisellini, Maraschi & Treves (1985) *Astronomy and Astrophysics*. **146**, 204.
Kollgaard (1994) *Vistas in Astronomy*. **38**, 29.

CAN BL LAC NUCLEI AND BROAD LINE REGIONS COEXIST?
OR
WHEN IS AN OBJECT A BL LAC?

M.J.M.MARCHÃ
DICE Universidade de Lisboa
Av. Prof. Gama Pinto, 2
1699 Lisboa Codex
Portugal

The problem of BL Lac classification is a long standing one and it is mainly due to the subjectiveness of selection criteria used in the definition of BL Lac samples. For instance, an object will undoubtedly be classified as a BL Lac if it shows flat radio spectrum, high optical and radio polarization, featureless optical continuum with weak or absent emission lines, and variable flux and polarization. However, the problem arises when the object shows some but not all of these properties. In face of this difficulty, different authors (Stickel et al. 1991, Stocke et al. 1991) have tried to make a systematic analysis of the data and it has been common to classify as BL Lacs those objects whose strongest emission lines have equivalent width (EW) $\leq$ 5 Å. Another common criterion is to require the 4000 Å break contrast to be $\leq$ 0.25. Nevertheless, both of these criteria are rather arbitrary and more directly related to practical observational considerations, than they are to any physical distinction between objects. What is proposed here is a slightly different approach; it is proposed that we take a step back from common classification and that instead of imposing strict selection criteria, we create a multi-observational parameter space to investigate any breaks in the distribution of observed properties that will help clarify the distinction between BL Lacs and other flat radio spectrum sources.

With this aim in mind, a new sample containing 55 optically bright flat radio spectrum sources with a flux density limit of 200 mJy has been selected and observed (Marchã et al. 1995: in press) such that the distribution of objects can be discussed according to the percentage of optical and radio polarization, the 4000 Å break contrast, and the EW of emission

R. Ekers et al. (eds.), Extragalactic Radio Sources, 244–245.
© *1996 IAU.*

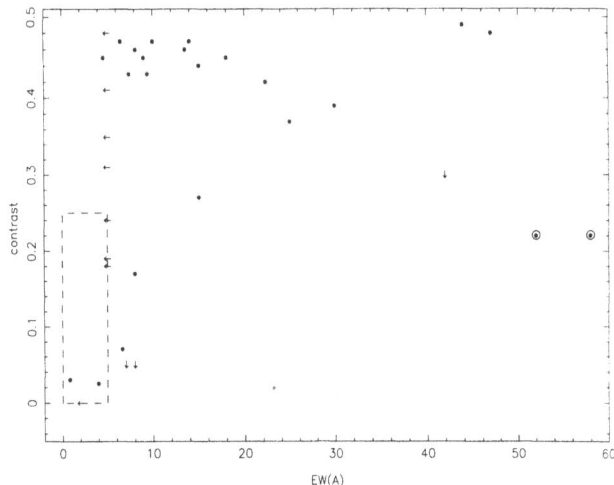

Figure 1. 'Contrast vs. EW' plot for the 200 mJy sample. This plot leaves out 10 sources with Sy-type spectra which have emission lines with EW > 100 Å. The arrows represent upper limits and the centered circles the 2 'hybrid-type' objects mentioned in the text.

lines. Here, however, only the latter two will be discussed. The distribution of break contrast vs. EW of the strongest emission line for EW up to 60 Å is plotted in Fig. 1. This leaves out 10 sources with "Sy-type" spectra whose emission lines have EW > 100 Å. The sample can be divided in two broad categories, one containing the sources with 4000 Å break contrast > 0.4 (galaxies), and the other containing the objects with the break contrast ≤ 0.4. Among this latter category, three types of sources can be identified: the "BL-type" objects (those with EW ≤ 30 Å), the "Sy-type" objects (those with EW ≥ 100 Å), and finally two objects with "hybrid type" properties, i.e., with broad but relatively weak (50 ≤ EW ≤ 60 Å) emission lines. The fact that BL Lacs do not seem to be restricted to the area enclosed by EW ≤ 5Å and break contrast ≤ 0.25, but instead seem to occupy a larger area of the parameter space, makes the point very clear that there is the need for a wider, yet more systematic BL Lac classification. Furthermore, classification should take into consideration a wide range of parameter space involving several observational parameters.

References

Marchã, M.J.M., Browne, I.W.A., Impey, C.D. and Smith, P.S. (1995) Optical Spectroscopy and Polarisation of a New Sample of Optically Bright Flat Radio Spectrum Sources, *M.N.R.A.S.*, in press.

Stickel, M., Padovani, P., Urry, C.M., Fried, J.W., Kühr, H. (1991) The Complete Sample of 1 Jansky BL Lacertae Objects.I. Summary Properties, *ApJ*, **374**, pp.431.

Stocke, J.T., Morris, S.L., Gioia, L., Maccacaro, T., Schild, R.E., Wolter, A., Fleming, T.A., Henry, J.P. (1991) The Einstein Observatory Extended Medium-Sensitivity Survey:II. The Optical Identifications, *ApJS*, **76**, pp.813.

INTRADAY VARIABILITY STATISTICS IN RADIO-SELECTED AND X-RAY SELECTED BL LAC OBJECTS

Implications for Unified schemes

J. HEIDT

Landessternwarte, Königstuhl, 69117 Heidelberg, Germany

1. Introduction

BL Lac objects, characterized by their high variability across the electro-magnetic spectrum, strong and variable polarization in the radio and optical domain and a (nearly) featureless continuum can in general be divided into the radio-selected (RBL) and X-ray selected BL Lac objects (XBL) according to their $\alpha_{ro} - \alpha_{ox}$ spectral indices (Stocke et al., 1985). Attempts to unify both classes within a single model have been suggested e.g. by Ghisellini et al. (1993) or Giommi et al. (1994).

In an attempt to characterize the variability behaviour of the BL Lac class as a whole or to search, whether there are differences between both classes a systematic study of several well defined samples of BL Lac objects have been carried out. The intraday variability characteristics have been examined in the optical, since short-term variability is most likely intrinsic in nature in that frequency regime. The aim of this study was to measure the duty cycle in RBL and XBL, to investigate their variability characteristics and to compare the observations with the prediction of the models.

2. The samples and the observations

The 1 Jy sample of RBL (Stickel et al., 1991) and XBL samples taken from the EXOSAT HGLS (Giommi et al., 1991) and EMSS (Morris et al., 1991) have been observed. The samples consist of 34, 11 and 22 BL Lac objects, respectively. Each object has been observed typically for one week several times per night using telescopes at the Landessternwarte Heidelberg, Calar Alto, Spain, ESO, Chile and Cananea, Mexico. A CCD camera with a R filter was attached to the telescopes in order to perform relative photometry.

246

R. Ekers et al. (eds.), Extragalactic Radio Sources, 246–247.
© 1996 IAU.

3. Results

A χ^2 test has been applied to the lightcurves. This gave the fraction of the variable objects. After subtracting a linear slope, the χ^2 test has been applied to the residuals. This gave the fraction of objects displaying intraday variability. The results are summarized in table 1. The RBLs have a very high duty cycle (at least 0.8), whereas the duty cycle in XBLs is low ($\approx$ 0.4).

TABLE 1. Variability statistics of the samples

The sample	non-variable [%]	variable [%]	intraday variable [%]
1 Jy sample	18	82	82
EXOSAT sample	45	55	36
EMSS sample	73	27	23

By means of structure function and autocorrelation analysis, the typical time-scales, amplitudes and an activity parameter $\dot{I}$ have been investigated. The amplitudes during one week are 28% for the RBL and 10% for the XBL, the time-scales 0.5-3 days in RBL, in a few XBL a time-scales could also be measured lying in the same range. The activity parameter $\dot{I}$ (in %/day) is between 3 and 27 %/day in RBL. Here a trend for higher $\dot{I}$ with increasing absolute magnitude was found. In this sources relativistic beaming may be important. In the XBL $\dot{I}$ was always below 3 %/day.

The different duty cycle is not expected from model, where both classes of BL Lac objects differ mainly by their average inclination angle between jet axis and observer (Ghisellini et al. 1993), but challenges also the suggestions by Giommi et al. (1991), where both classes differ mainly by their high-energy cutoff in their energy distribution in different frequencies. In this model relativistic beaming may be present in both classes.

Acknowledgments This work was supported by the DFG (Sonderfor-schungsbereich 328).

References

Ghisellini, G. et al.: 1993, *ApJ* **407**, 65
Giommi, P. & Padovani, P.: 1994, *MNRAS* **268**, L51
Giommi P., et al.: 1991, *ApJ* **378**, 77
Morris, S.L., et al.: 1991, *ApJ* **380**, 49
Stickel, M., et al.: 1991 *ApJ* **374**, 431
Stocke, J.T. et al.: 1985, *ApJ* **298**, 619

HOST GALAXIES OF RADIO-LOUD & RADIO-QUIET AGN

M.J. KUKULA[1], J.S. DUNLOP[1], G.L. TAYLOR[2], D.H. HUGHES[1]

[1] *Institute for Astronomy - University of Edinburgh,*
Royal Observatory, Edinburgh EH9 3HJ, U.K.
[2] *Astrophysics Group - Liverpool John Moores University,*
Byrom St., Liverpool L3 3AF, U.K.

A clear understanding of both the differences and similarities between the host galaxies of the three main classes of powerful active galaxy – radio-quiet quasars (RQQs), radio-loud quasars (RLQs) and radio galaxies (RGs) – is vital in any attempt to unify or relate the various manifestations of the AGN phenomenon. The unification of RLQs and RGs via orientation effects requires that the hosts of the two types be derived from the same population of galaxies. Meanwhile, the correlation between radio power and host morphology in nearby AGN, with radio-quiet objects (Seyferts) occurring in disc systems and radio-loud sources in ellipticals, is generally assumed to persist at higher redshifts and nuclear luminosities. However, in both cases the evidence remains ambiguous and, moreover, many previous studies have been based on poorly selected samples.

In an attempt to unambiguously determine the impact of the host on the observed properties of an AGN we have assembled three *carefully matched* comparison samples of RQQs, RLQs and RGs covering redshifts $0.1 \leq z \leq$ 0.3. The RLQ and RQQ samples were carefully constructed so that their distributions in the $V - z$ plane are statistically indistinguishable (Dunlop *et al.* 1993) and the sample of RGs was compiled to match as closely as possible the radio-power and redshift distributions of the RLQ sample (Taylor *et al.* 1995). Using ground–based observations we have approached the question of host galaxy properties from two independent directions: infrared (K-band) imaging, to determine the host morphologies, and off-nuclear optical spectroscopy to investigate their star-formation histories.

In any attempt to study the host galaxies of quasars the main problem lies in separating the underlying galaxy light from the PSF of the bright

R. Ekers et al. (eds.), Extragalactic Radio Sources, 248–249.

active nucleus. By observing in K-band (2.2μm), using IRCAM on UKIRT, we were able to maximize the the ratio of host galaxy to quasar light, so that the stellar contribution was more dominant. 2-D modelling of the images has been highly successful (Dunlop et al. 1993, Taylor et al. 1995) and we find that, consistent with the current 'unified scheme', *all* of the RGs and RLQ hosts prefer to be described by an elliptical (de Vaucouleurs) model. Overall the majority of the RQQs do seem to lie in (exponential) disc systems, but for the *most luminous* RQQs in our sample the best fit model is more often an elliptical than a disc. Another highly significant result of this study is that we find no evidence for any systematic differences in the near infrared luminosities of RGs and RQQ & RLQ hosts, all of which have K luminosities $> L_K^*$. Finally, all of the model galaxies have very large half-light scalelengths (typically 20 kpc), consistent with the values for brightest cluster galaxies (Taylor et al. 1995).

We are currently working to complete a programme to obtain deep off-nuclear optical spectra of all the objects in our three samples. Our results demonstrate that it is now feasible to obtain quasar host galaxy spectra from the ground which are of sufficient quality to constrain spectrophoto-metric models of the stellar populations. This was made possible by our deep K–band images, which, because they are uncontaminated by emission lines, allow us to identify where best to place the slit to maximize the level of stellar continuum at least 5 arcsec off–nucleus, thereby avoiding scattered light from the quasar. In all of the objects observed so far the best fit model requires a very old ($\sim$ 13 Gyr) and red underlying stellar population ($B - K$ colours are typically 4 $\rightarrow$ 5). Many of the quasar hosts also require the additional contribution of a blue component (the strength of which varies considerably from object to object) but longward of the 4000Å break the detailed level of the fits is remarkably good, indicating that the spectra are completely dominated by starlight.

Summary: The hosts of all three types of object are large and bright at K-band. The morphologies of the RLQ hosts and the RGs are consistent with unification and, interestingly, we find that for the more luminous RQQs the best-fitting host model tends to be an elliptical. All of the hosts appear to contain an old stellar population and to have red optical - infrared colours. This, coupled with the large scalelengths of the galaxies, probably explains the relative failure of recent attempts to detect quasar hosts with the HST in V-band (*eg* Bahcall et al. 1995).

References

Bahcall, J.N., Kirhakos, S. & Schneider, D.P. (1995), *ApJ*, **450**, 486.
Dunlop, J.S., Taylor, G.L., Hughes, D.H. & Robson, E.I. (1993), *MNRAS*, **264**, 455.
Taylor, G.L., Dunlop, J.S, Hughes, D.H. & Robson, E.I. (1995), *MNRAS, submitted*.

A POSSIBLE FUNDAMENTAL DIFFERENCE BETWEEN RADIO LOUD AND RADIO QUIET AGN

M. CALVANI
Osservatorio Astronomico, Padova, Italia

J.W. SULENTIC AND P. MARZIANI
University of Alabama, Tuscaloosa, USA

D. DULTZIN-HACYAN
Instituto de Astronomía, UNAM, México

AND

M. MOLES
Instituto de Astrofísica de Andalucia, Granada, España

We report on some striking differences between radio loud and quiet emitters that we found in a comparative analysis of the high and low ionization lines for 52 low redshift AGN (31 loud; 21 quiet).

The broad components of $CIV\lambda1549$ and $H\beta$ were chosen as representative of the high and low ionization lines respectively. $CIV\lambda1549$ observations were obtained with the Faint Object Spectrograph on the HST. They were retrieved from the HST data archive and matching optical spectra for the region of $H\beta$ were obtained at several ground based observatories. Details on observations, narrow/broad component deconvolution and profile cleaning from satellite lines (especially FeII) can be found in Marziani et al. (1995). The rest frame for each quasar was determined from the radial velocity of strong narrow lines, typically $[OIII]\lambda\lambda4959,5007$.

Fig. 1 depicts our results: (1) Radio loud AGN show predominantly redshifted, redward asymmetric $H\beta$, while $CIV\lambda1549$ appears broader but predominantly unshifted and symmetric. (2) Radio quiet AGN, on the contrary, show unshifted and symmetric $H\beta$, and blueshifted and blue-ward asymmetric $CIV\lambda1549$.

Interesting additional results are: (1) The peak (or 3/4 intensity level) radial velocity of $H\beta$ almost always exceeds the radial velocity of $CIV\lambda1549$ in both radio quiet and radio loud AGN. (2) There is evidence that *the*

R. Ekers et al. (eds.), Extragalactic Radio Sources, 250–251.
© 1996 *IAU*.

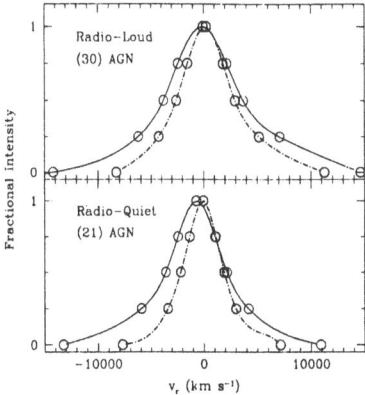

Figure 1. "Median" profiles of Hβ (dot-dashed line), and CIVλ 1549 (solid), for the radio loud quasars in our sample (upper panel) and the radio quiet ones (lower panel). The profiles were built interpolating spline functions to the median values of the radial velocity measured at 0,0.25,0.5,0.75,1.0 peak intensity.

CIVλ1549 and Hβ line profiles are coupled in radio loud AGN. Peak shifts of CIVλ1549 and Hβ appear to be correlated. FWHM(CIVλ1549) and FWHM (Hβ) are also correlated. The ai(1/4) of CIVλ1549 and Hβ, even if not correlated show a marked tendency to occupy one quadrant in the diagram ai(1/4)(CIVλ1549) versus ai(1/4)(Hβ). None of those trends is revealed for radio quiet quasars. (3) The luminosity of Hβ and CIVλ1549 are strongly correlated.

A number of properties suggest a relationship between line profiles and radio loudness: (1) Superluminal sources tend to have low Hβ equivalent width. *If we restrict our attention to sources with* $\beta_{app} \approx 10$, *the Hβ profile is redshifted and redward asymmetric.* (2) The dependence of line profile width on the R (flux of radio core over flux of radio lobe) parameter is the same for both Hβ and CIVλ1549. (3) The shift at the line base appears to be correlated with $L_{\nu,core}$. The largest $L_{\nu,core}$ is associated with the largest zero intensity. (4) The average radio loud quasar is a less efficient FeII radiator than the average radio quiet quasar. There is some indication, however, that core dominated radio sources may have peculiar FeII emission.

Summing up, our results suggest that: (1) *CIVλ1549 and Hβ might well be emitted in the same region in radio loud AGN, at variance with radio quiet AGN, where several lines of evidence suggest that CIVλ1549 is emitted by out-flowing gas, while Hβ is not.* (2) *Radio loudness affects the line profiles.*

References

Marziani, P., Sulentic, J.W., Dultzin-Hacyan, D., Calvani, M. and Moles, M. (1995), *Ap.J. Suppl.*, in press.

THE RADIO AND OPTICAL PROPERTIES OF QUASARS

C.D. IMPEY AND E.J. HOOPER
Steward Observatory
University of Arizona, Tucson, AZ 85721, USA

C.B. FOLTZ
Multiple Mirror Telescope Observatory
University of Arizona, Tucson, AZ 85721, USA

AND

P.C. HEWETT
Institute of Astronomy
Madingley Road, Cambridge CB3 0HA, England

1. Introduction

The relationship between the radio and optical emission of quasars was first studied using radio-selected objects, which generally had high radio luminosities due to the relatively low sensitivity limits of the surveys. More recently, radio follow-up observations have been made of optical surveys surveys in other wavebands. Taken together, the two survey methods have detected quasars with a range of over 6 orders of magnitude in radio luminosity. We report here new observations with the Very Large Array (VLA) of 103 quasars from the Large Bright Quasar Survey (LBQS), with absolute magnitudes in the range $-25 < M_B < -23$ and redshifts in the range $0.2 < z < 1.1$. This can be combined with our previous LBQS observations (Paper I, Paper II) to yield 359 quasars, the largest sample of sensitive radio observations of optically-selected quasars in a single survey.

2. Results

First, the distribution of radio luminosity does not depend on absolute magnitude over most of the range of M_B in the LBQS. The radio-loud fraction remains constant at $\approx 10\%$ for $-28 < M_B < -23$ but rises to $\approx 30\%$ at brighter absolute magnitudes. Second, the radio properties of the LBQS

R. Ekers et al. (eds.), Extragalactic Radio Sources, 252–253.
© *1996 IAU.*

quasars as a function of M_B are consistent with the existence of two radio emission mechanisms, one correlated with optical luminosity, the other independent. The sudden decrease in radio-loud fraction for M_B fainter than -24 observed in the Palomar-Greeen (PG) sample is not present in the expanded LBQS sample. A selection effect to which the LBQS is mostly immune is the likely cause of the decrease in the PG sample.

The radio-loud fraction in optically selected quasars appears to be un-evolving at a value of $\approx 10\%$, aside from a modest increase at $z \sim 1$, from $z = 0.2$ to redshifts approaching 5, based on the LBQS and three high-z optically selected samples. A model based on a radio luminosity function derived from radio-selected quasars (Dunlop & Peacock 1990) matches the data well, including the modest rise around $z = 1$, at all except the highest redshifts. The high radio-loud fraction in the PG sample remains unexplained (Kellerman et al. 1989). Fourth, the radio properties of the X-ray selected EMSS also differ from those of the LBQS (Stocke et al. 1991). The rapid rise in radio-loud fraction observed for $M_B \approx -24$ is primarily an evolutionary effect resulting from a well-established correlation between X-ray and radio luminosity. The rise in radio-loud fraction with redshift is also consistent with the observed correlation between radio and X-ray emission for faint radio sources (Brinkmann et al. 1995).

These results are surprising in terms of the standard paradigm of AGN energy production. The constancy of the radio luminosity distribution over more than two orders of magnitude in optical luminosity is unexpected in the context of the standard black hole model. The radio emission mechanism appears to be only weakly tied to the mass accretion rate which presumably drives the optical and near-UV emission. Also, the radio-loud fraction is essentially constant over 90% of the Hubble time, during which the epoch of quasar formation occurs, and the comoving space density declines by a factor of several hundred to the present value.

This work was supported by NSF grant AST 93-20715 and NASA grant NGT-51152, a NASA Graduate Student Fellowship (EJH).

References

Brinkmann, W., et al. 1995, *Ast.Ap.Supp.*, **109**, 147

Dunlop, J. S., & Peacock, J. A. 1990, *M.N.R.A.S.*, **247**, 19

Hooper, E. J., Impey, C. D., Foltz, C. B., & Hewett, P. C. 1995, *Ap.J.*, **445**, 62 (Paper II)

Kellermann, K. I., Sramek, R., Schmidt, M., Shaffer, D. B., & Green, R. 1989, *A.J.*, **98**, 1195 (PG)

Stocke, J. T., et al. 1991, *Ap.J.Supp.*, **76**, 813

Visnovsky, K. L., Impey, C. D., Foltz, C. B., Hewett, P. C., Weymann, R. J., & Morris, S. L. 1992, *Ap.J.*, **391**, 560 (Paper I)

OPTICAL PROPERTIES OF THE RADIO SOURCE B2 0828+32

M.-H. ULRICH-DEMOULIN AND J. RÖNNBACK

European Southern Observatory, Karl-Schwarzschild-Strasse 2,
D-85748 Garching bei Munchen, FRG

A small number of radio galaxies have two sets of classical double radio lobes with the radio axes aligned in different directions. Furthermore differences in the properties of the radio lobes such as the surface brightness and spectral index indicate that the two sets of double radio structure have different ages. The radio ejection axis has therefore changed direction with time. In the first two known radio galaxies of this type, 3C 315 and B2 0055+26 [1], the host galaxy is a member of a close pair of ellipticals in a common optical envelope suggesting that the complex radio structure is caused by gravitational interaction [1,2].

We recently took deep CCD images in R and V of B2 0828+32 which is another example of radio galaxies with two double radio structures. The R image is shown in Fig. 1. The radio spectral index maps of this radio galaxy allow to set an age to the old pair of radio lobes, 70 Myears, which is thus an upper limit to the age of the young lobes [3]. This is interesting as it constrains the time scale of the phenomenon which changes the direction of the radio axis.

The major results from the study of the optical data can be summarized as follows:

A) The host galaxy of B2 0828+32 has no double core or double nucleus, in contrast with 3C 315 and B2 0055+26, and it has no close companion brighter than $M_v = -16.7$ within 38 kpc (see Fig. 1) assuming $H_0=100$ km s^{-1} Mpc^{-1}. It lies at the edge of a poor cluster.

B) The shape of the luminosity profile is typical for early type galaxies with a de Vaucouleurs r$^{1/4}$ profile, but there is some discrepancy between data and model at radii less than $\sim$ 15 arcsec. This could be a sign of an old merger (transient shell structures or superposition of two mixed components). The relaxation time for a merger of 2 massive galaxies with rotating bulges is of the order 1 Gyr.

R. Ekers et al. (eds.), Extragalactic Radio Sources, 254–255.

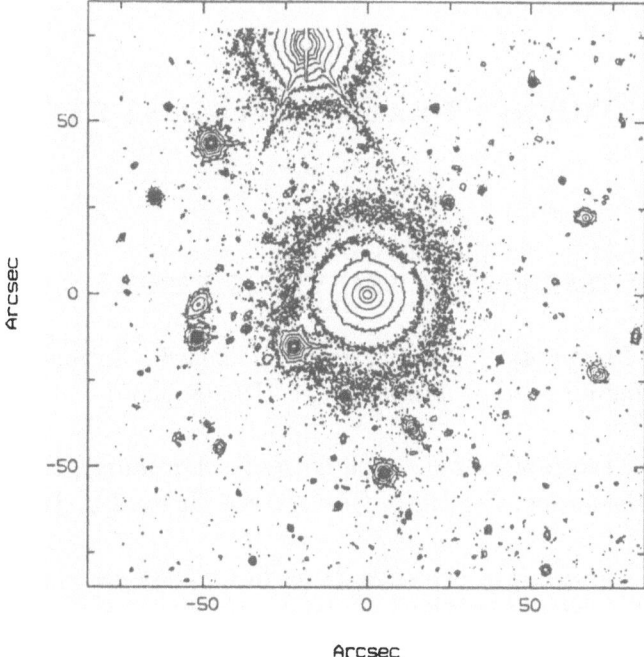

Figure 1.   Contour plot in R of B2 0828+32. North to the top and east left.

The integrated V–R of the host galaxy is blue, V–R = 0.54 ± 0.06, being approximately constant as a function of radius between 5″ and 30″. A typical colour for an early type galaxy is V–R ~ 0.6–0.7.

We conclude that the optical images of B2 0828+32 do not show strong signatures of a recent major merger event. However, the blue colours and weak features in the luminosity profile indicating that the stellar distribution has been disturbed dynamically, may be the last remaining signs of a major merger event some 1 Gyr ago.

This leads us to suggest a model where the change of radio axis direction was caused by the merging of a small galaxy [4] whose gas gathered at the center of the host and was accreted by the black hole (generally believed to power the radio source). We have shown that this could occur in less than 70 Myears if the accretion proceeds at a rate close to the Eddington rate [4]. That event ignited and fueled the younger radio lobes.

References

[1] Ekers, R.D., Fanti, R., Lari, C., Parma, P. 1978, *Nature*, **276**, 588
[2] Parma, P., Ekers, R.D., Fanti, R. 1985, *A&AS*, **59**, 511
[3] Klein, U., Mack, K.-H., Gregorini, L., Parma, P. 1995, *A&A*, in press
[4] Ulrich, M.-H., Rönnback, J. 1995, in preparation

INVERSE COMPTON X-RAYS FROM GIANT RADIO GALAXIE!

D. TSAKIRIS [1], J.P. LEAHY [1], R.G. STROM [2], C.R. BARBER [3]

[1] *University of Manchester, Nuffield Radio Astronomy Laboratories Jodrell Bank, Cheshire SK11 9DL, England*

[2] *Netherlands Foundation for Radio Astronomy, Radiosterrenwacht Dwingeloo, Postbus 2, NL-7990 AA Dwingeloo, The Netherlands*

[3] *University of Leicester, Department of Physics and Astronomy, Leicester, LE1 7RH, England*

1. Introduction

The X-ray radiation from inverse Compton scattering of CMB photons by the relativistic electrons in 'radio' lobes provides a direct measure of their column density at a known energy, unlike synchrotron radiation which also depends on the unknown magnetic field. Thus by combining inverse Compton and radio data we can separately determine the particle energies and field strengths, rather than having to rely on uncertain estimates like minimum energy. The predicted flux is

$$S_{IC} \propto (S_{\nu_{rad}} \nu_{rad}^{\alpha}) B^{-(\alpha+1)} (1+z)^{\alpha+3} \qquad (1)$$

and strong IC signal requires high radio flux and low magnetic field, properties of giant radio galaxies. On the other hand the minimum detectable count rate, I_{min}, increases with the target size due to the larger background contribution. As a result the detectability of IC X-rays for *ROSAT* PSPC B measurements is roughly,

$$\frac{S_{IC}}{I_{min}} \propto (S_{rad} \cdot \theta \cdot z)^{1/2} (1+z)^{15/4} \qquad (2)$$

assuming a spectral index of 0.75. After making detailed prediction of S_{IC} for a number of objects of the 3CR sample, the best candidates were 3C 236, 3C 326, and 4C 73.08.

R. Ekers et al. (eds.), Extragalactic Radio Sources, 256–258.

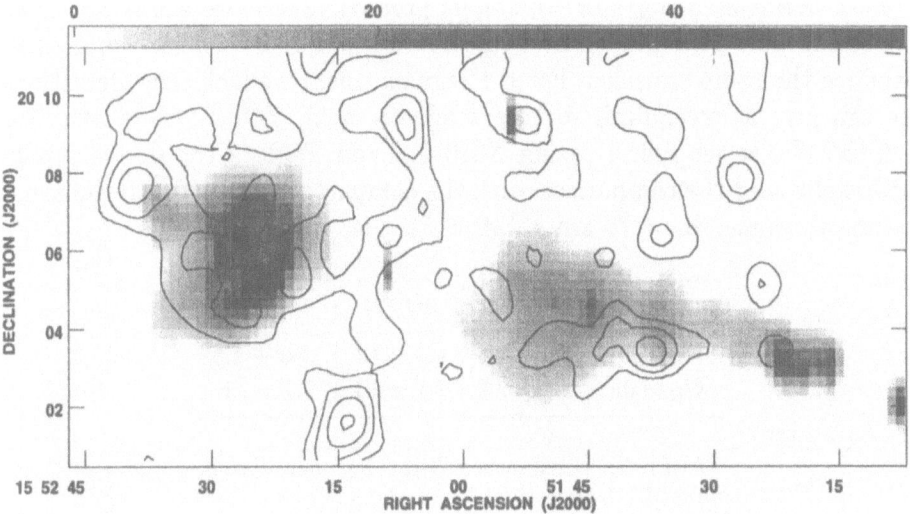

Figure 1. Smoothed X-ray image of 3C 326 in the 0.4–2 keV band overlaid on the radio image. The X-ray image was smoothed with 60×60 arcsec FWHM Gaussian. The contour levels are at 5.04×10^{-4} cts s^{-1} arcmin^{-2} and increasing by factors of $\sqrt{2}$.

2. Observations and Results

We observed the targets with the *ROSAT* PSPC for about 11 ks each. The data were reduced with Dr S. Snowden's programs which model four components of non-cosmic background in the PSPC (scattered solar X-rays, particle background, 'long term enhancements', and 'short term enhancements' (STE's)). Times where our observations were affected by strong solar X-rays and STE's were excluded. To reduce confusion, we masked out all point sources detected at $> 5\sigma$, and restricted analysis to the 0.4–2 keV band. The X-ray images were compared with radio maps from the WSRT. The X-ray flux for each lobe of each object was integrated over a region defined by the outer radio contours. In each case, the average X-ray brightness was slightly higher than the average background in the field. The fluctuations in the background were substantially higher than expected from photon statistics, as expected as much of the background is contributed by individual sources below our detection threshold. To assess the significance of our detection we divided the central region of the image for each map into boxes (excluding the target) with the same number of pixels as the lobe detect cell. The standard deviation of the brightness in these background regions defines our 1σ error.

By observing giants, we avoid the problem of confusion by thermal gas associated with the host galaxy (or even from the group X-ray haloes), as the lobes of the giants are much larger. There are also low upper limits on entrained thermal plasma from the lack of depolarization (Strom & Willis (1980), Willis & Strom (1978), Jägers (1986)). Therefore we assumed we are seeing pure IC emission and calculated the magnetic fields required to produce the radio emission. For the regions where we lack clear detections we can give a lower limit for the magnetic field. Table 1 shows the 0.4 – 2 keV fluxes we found (using XSPEC), and Table 2 the derived magnetic fields and the comparison with the equipartition ones (with the usual assumptions and $H_0 = 75$ km s^{-1}Mpc^{-1}).

Region	3C 236	3C 326	4C 73.08
East lobe	2.9 ± 1.6	9.2 ± 1.4	12.3 ± 2.3
West lobe	9.4 ± 2.2	7.1 ± 2.1	7.3 ± 2.5

TABLE 1. Unabsorbed X-ray flux densities in units 10^{-14} erg cm^{-2} s^{-1}

Target	B_{IC} (nT)		B_{eq} (nT)	B_{eq}/B_{IC}
	mean value	limits		
3C 236 NW lobe	0.08	(0.07 – 0.09)	0.15	1.93
SE lobe	$\geq$0.08		0.12	1.59
3C 326 East lobe	0.14	(0.13 – 0.15)	0.16	1.18
West lobe	0.16	(0.14 – 0.19)	0.16	1
4C 73.08 East lobe	0.07	(0.06 – 0.08)	0.14	2.04
West lobe	$\geq$0.11		0.14	1.29

TABLE 2. Magnetic fields derived from inverse Compton X-rays. The limits for B_{IC} derived using the 1σ errors

3. Conclusions

We have clearly detect X-ray emission from the NW lobe of 3C 236, from both lobes of 3C 326, and from the East lobe of 4C 73.08, indicating magnetic fields at the equipartition values remarkably close given the errors in both estimates. In the other two lobes the measured X-ray flux is above the background, and by 'co-adding' we obtain again a clear detection of $(10.2 \pm 2.8) \times 10^{-14}$ erg cm^{-2} s^{-1}.

References

Strom, R. G. and Willis, A. G., 1980, *Astron.Astrophys.*, **85**, 36

Jägers, W. J., PhD Thesis, The Polarization of Radio Galaxies, University of Leiden, 1986

Willis, A. G. and Strom, R. G., 1978, *Astron.Astrophys.*, **62**, 375

Snowden, S. L., Cookbook for analysis procedures, U.S.ROSAT GO Facility, 1994

ROSAT STUDIES OF EXTRAGALACTIC RADIO SOURCES

W. BRINKMANN, J. SIEBERT, W. YUAN
MPI für Extraterrestrische Physik
Postfach 1603
D - 85740 Garching

1. Introduction

Complementary information over a large energy range appears to be the best way to study the class properties of Active Galactic Nuclei (AGN). While radio data provide superior spatial resolution required for the analysis of source structures and their morphology, optical data are a necessary ingredient for the classification of the objects and for the determination of their redshifts. Finally, X-ray (and more recently γ-ray) observations give vital information about the energetics of the central engines and on the physical conditions of the 'heart of the machine'.

With the recent advent of sensitive X-ray instruments like ROSAT and ASCA it became possible to study AGN in large numbers and at high redshifts. While ASCA has the best energy resolution achieved so far in the X-ray band, ROSAT provides the largest X-ray data base for the near future which consists of two parts:

- The ROSAT All-Sky Survey

  - provides a nearly unbiased (apart from slightly varying exposure and variations in the Galactic absorption) coverage of the whole sky. The limiting sensitivity is a few $\times 10^{-13}$ erg cm$^{-2}s^{-1}$ and the catalogue contains more than 60.000 sources.

- ROSAT Pointed Observations

  - about 75.000 sources in ~ 1.3 sterad of the sky with very deep PSPC exposures ($\geq 2 \times 10^{-15}$ erg cm$^{-2}s^{-1}$). As serendipitous objects from pointed observations these sources have highly non-uniform exposures and selection biases. However, due to the generally long exposures more detailed spectral studies are possible.

259

R. Ekers et al. (eds.), Extragalactic Radio Sources, 259–262.
© 1996 IAU.

2. The X-ray - radio data

The sample of X-ray and radio - loud AGN was obtained by cross correlating large scale radio catalogs with the ROSAT source lists: in the northern sky the Greenbank 5 GHz survey, in the southern sky the Molonglo Reference catalogue, and the PMN radio survey. We thus obtain a list of ≈ 3700 radio - loud X-ray sources of which nearly 2/3 are optically unidentified. The results of broad band studies of the 'known' objects have been published elsewhere (Brinkmann et al. 1994, 1995).

In Fig.1 we show the distribution of the sample in Galactic coordinates. The different types of sources are gray-coded with symbols proportional in size to their X-ray fluxes. Crosses represent the currently unidentified objects.

The more than 1400 unidentified sources on the northern sky have been re-observed with the VLA to obtain arcsec positions and core fluxes. With these positions we are able to find potential optical counterparts from digitized POSS plates and we can study the broad band properties of this population - however, without knowing the nature of the objects and their redshift (Brinkmann et al. 1996).

3. 'Known' objects

The number of 'known' radio - loud ROSAT detected AGN far exceeds all samples known previously. Therefore, statistically more reliable studies of different classes of objects like blazars, radio - loud quasars, radio galaxies - and, amongst them, differentiations with respect to radio properties (steep spectrum, flat spectrum) could be made. In Table 1 we give the currently known (approximate) numbers of objects for which X-ray data are available.

TABLE 1. ROSAT detected radio sources

Type of Object	number	Remarks
quasars	654	Survey: 360
Radio galaxies	245	Survey only
blazars	275	HPQ: 63

Indicated as well is the fraction of sources seen in the Survey only. Clearly, these numbers will increase in the future as more of the ROSAT sources will be identified optically.

With this large database one is able to study the X-ray as well as the broad band properties of various subgroups of objects. For example, assum-

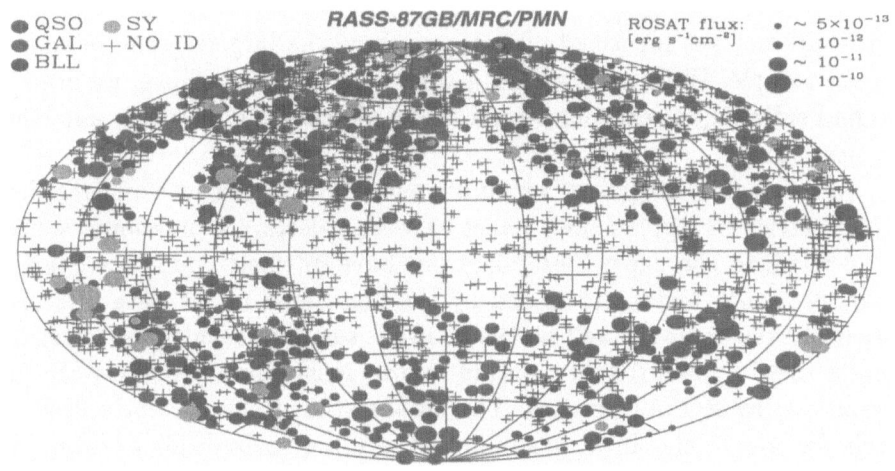

Figure 1. The ROSAT - radio sample of AGN plotted in Galactic coordinates. The symbol sizes are proportional to the X-ray flux of the objects. Crosses represent optically unidentified sources.

ing power law spectra either with fixed Galactic absorption or with free absorption the average spectral properties of the different object classes could be investigated using a maximum likelihood technique (Siebert 1996).

AGN class	$N_{H,galactic}$			$N_{H,free}$		
	N	$\bar{\Gamma} \pm 1\sigma$	Σ^{intr}	N	$\bar{\Gamma} \pm 1\sigma$	Σ^{intr}
Quasars	197	2.14±0.06	0.22±0.06	165	2.20±0.21	<0.34
Galaxies	115	1.92±0.10	0.47±0.08	101	1.96±0.17	<0.24
Seyferts	41	2.02±0.14	0.46±0.11	37	2.04±0.24	0.45±0.23
BL Lacs	91	2.23±0.06	0.25±0.06	80	2.35±0.11	<0.15
Unidentified	938	2.23±0.04	0.40±0.03	815	2.34±0.11	<0.17

Further, the ROSAT data base allows the investigation of well defined complete samples of objects. Even if a certain number of sources in a sample is not detected in X-rays the All - Sky Survey allows the determination of upper X-ray flux limits and the subsequent study of the sample using *Survival Analysis* techniques (p.e., the 2 Jy sample: Siebert et al. 1996).

4. 'Unknown' objects

Optically not yet identified objects represent the largest group of sources in the sample. With accurate VLA positions, as noted above, we are able to find suitable optical counterparts from digitized POSS plates and, thus, optical fluxes.

In flux - flux diagrams, which are successfully used to classify objects and to attribute 'typical' flux values to these objects, the 'unknown' sources tend to fill in the gaps: these diagrams have intrinsic boundaries, not always obviously apparent, caused by the flux limits of the data sets used. As both, the radio as well as the ROSAT surveys reach much lower flux limits these boundaries are shifted or previously empty regions are now populated by ROSAT sources. This implies that some currently used criteria for source classifications might have to be revised: for example, the apparent gap between X-ray and radio selected BLLacs seems to be filled in by the 'unknown' objects. However, most of these sources need to be optically identified as BL Lacs.

On the other hand, unidentified sources placed in phase space volumes in these diagrams typical for certain classes of objects can be optically observed with high rate of success. For example, using the $\alpha_{ro} - \alpha_{ox}$ - diagram for a pre - selection of BL Lac candidates the optical follow up observations reached success rates of more than about 30%.

5. Conclusions

The cross-correlation of the ROSAT All-Sky Survey with large catalogs of radio sources yields a sample of in total more than **3700 radio-loud X-ray sources**. This is the largest and least biased sample of radio-loud X-ray sources available. It allows the statistical analysis of X-ray spectral properties and their dependence on redshift and radio properties for different AGN classes, the relation between soft X-ray luminosity and the emission from other wavebands (radio, optical), the study of the X-ray luminosity function of radio-loud AGN and the connection between different classes of radio-loud AGN in terms of unification schemes. However, optical identifications for the majority of the objects are still needed to get a really unbiased (radio and X-ray) flux limited sample of AGN.

References

Brinkmann W., Siebert J., Boller Th., 1994, *A&A*, **281**, 355
Brinkmann W., Siebert J., Reich W., et al, 1995, *A&AS*, **109**, 147
Brinkmann W., Siebert J., Feigelson E., et al., 1996, *A&A* submitted
Siebert J., 1996, PhD thesis, Techn. University München
Siebert J., Brinkmann W., Morganti R., et al., 1996, *MNRAS* in press

SOFT X-RAY PROPERTIES OF A COMPLETE SAMPLE OF RADIO SOURCES

J. SIEBERT[1], W. BRINKMANN[1], R. MORGANTI[2,3],
C.N. TADHUNTER[4], I.J. DANZIGER[5], R.A.E. FOSBURY[6,7],
AND S.DI SEREGO ALIGHIERI[8]

[1] MPE - Giessenbachstrasse, D-85740 Garching, Germany
[2] IRA - CNR, via Gobetti 101, I-40129 Bologna, Italy
[3] ATNF, CSIRO, P.O. Box 76, Epping, NSW 2121, Australia
[4] Physics Department, University of Sheffield, Sheffield S3, UK
[5] ESO, Karl Schwarzschild Str. 2, D-85748 Garching, Germany
[6] ST-ECF, Karl Schwarzschild Str.2, D-85748 Garching, Germany
[7] Astrophys. Division, Space Science Department, ESA (affiliated)
[8] OA - Arcetri, Largo E. Fermi 5, 50125 Firenze, Italy

Abstract. We investigate the X–ray properties of a complete sample of 88 radio sources derived from the Wall & Peacock 2–Jy sample. We find that L_x correlates well with core radio luminosity for all object classes, whereas the $L_x - L_{total}$ is probably introduced by sample selection effects. Further, evidence for an anisotropic X–ray component in broad line radio galaxies is reported. A full description of the results will be given elsewhere (Siebert et al. 1996).

1. Introduction

It is essential for our understanding of the connection between the X-ray emission of radio-loud AGN and their emission from other wavebands to study complete samples of both quasars and radio galaxies with comprehensive information on the radio and the optical properties of these sources. Here we briefly summarize our findings on the soft X-ray properties of such a sample, consisting of 88 radio sources (68 galaxies, 18 quasars, 2 BL Lacs), selected from the Wall & Peacock 2–Jy sample.

2. Data analysis and results

The X-ray properties have been determined using the ROSAT All-Sky Survey (RASS) and (34) pointed PSPC observations extracted from the public

R. Ekers et al. (eds.), Extragalactic Radio Sources, 263–264.

archive. A source is considered to be detected in X-rays when the significance is greater than 3σ. In case of a non-detection the 2σ upper limit to the X-ray flux is calculated. In total 59 sources are detected (both BL Lacs, all but one (PKS 1151-34) quasar and about 60% of the galaxies). Comparing the RASS and the pointed PSPC observations for 29 sources we find evidence for variability greater than a factor of two in 6 sources (the quasars PKS 0403-13, 3C 279 and PKS 1510-08, the BL Lacs PKS 0521-36 and AP LIB and the radio galaxy 3C 88). The most extreme behaviour is shown by AP LIB, with a decrease in intensity by at least a factor of 20 between the two observations ($\sim$ 3 years). Furthermore, 16 sources show extended X-ray emission, which is either due to an associated cluster of galaxies or the proximity of the source.

2.1. LUMINOSITY DISTRIBUTIONS

From the luminosity distributions for the various radio-morphological subclasses we conclude that FR I radio galaxies are in general less luminous than FR II type sources. None of the compact steep spectrum sources is detected which is most likely due the relatively high redshifts of these sources. Broad line radio galaxies generally exhibit higher X-ray luminosities and a higher detection rate than weak or narrow line radio galaxies. This indicates an anisotropic X-ray emission component and fits nicely into unification schemes for high power radio-loud AGN.

2.2. CORRELATION ANALYSIS

The correlation of L_x with total radio luminosity turns out to be insignificant for all object classes once the effects of the strong L_T - L_c correlation and the redshift bias are properly accounted for via a partial correlation analysis (Akritas & Siebert 1996). On the other hand, the correlation of L_x with core radio luminosity remains significant. The regression slope of the $L_x - L_c$ correlation for quasars is remarkably flat (0.30 ± 0.14), but is consistent with previous findings. Whereas we find an excellent correlation of L_x with optical luminosity for the quasars, the corresponding correlation for galaxies is insignificant. Again the regression slope ($0.84\pm0,16$) is consistent with various previous studies. There is a highly significant correlation of L_x with [OIII]λ5007 line luminosity for quasars with a regression slope close to unity. Although the correlation for the FR IIs alone is not significant, we note a smooth transition from quasars to FR II radio galaxies. FR I radio galaxies exhibit a completely different behaviour with typically $1 - 2$ orders of magnitude lower emission line luminosities, indicating fundamental differences between FR I and FR II type objects.

References

Akritas, M.G., Siebert, J., 1996, *MNRAS*, in press
Siebert, J., Brinkmann, W., Morganti, R., Tadhunter, C.N., Danziger, I.J., Fosbury, R.A.E., di Serego Alighieri, S., 1996, *MNRAS*, in press

X-RAY SPECTRA OF BL LACERTAE OBJECTS FROM THE ROSAT ARCHIVE

G. LAMER, H. BRUNNER AND R. STAUBERT
Institut für Astronomie und Astrophysik, Abt. Astronomie
Waldhäuserstr. 64, D-72076 Tübingen

1. Observations and data analysis

Our sample comprises all BL Lac objects listed in the catalogue of Véron-Cetty & Véron (1993) and which are detected in a ROSAT PSPC observation with at least 50 source counts: 74 objects in total. We reduced the data from the ROSAT archives at MPE and GSFC and fitted single power-law models with photoelectric absorption to the spectra. We calculated the broad band spectral indices α_{rx}, α_{ro}, and α_{ox} from the ROSAT 1 keV fluxes, 5 GHz radio, and optical V band fluxes (Véron-Cetty & Véron 1993).

2. Results

We find that particularly X-ray or radio bright objects (with extreme values of α_{rx}) have considerably harder X-ray spectra than the more intermediate objects (see Fig. 1a). We interpret this finding as a signature of a convex soft (synchrotron) and a hard (Compton) spectral component intersecting each other at different energies below, within, or beyond the soft X-ray band.

We compare the measured spectra with spectral simulations based on a set of simple two component models, including a hard power law and a parabolically steepening soft component with different cutoff energies (Fig. 1b). Figures 1a, 2a, and 2b show that the X-ray spectra as well as the broad band properties are well reproduced by the model. As the new data require a broad range of synchrotron cutoff energies, it is unlikely that the differences of RBLs and XBLs are caused by different viewing angles (e.g. Ghisellini & Maraschi 1989, Celotti et al. 1993); probably intrinsic differences are involved (e.g. Padovani & Giommi 1995).

R. Ekers et al. (eds.), Extragalactic Radio Sources, 265–266.

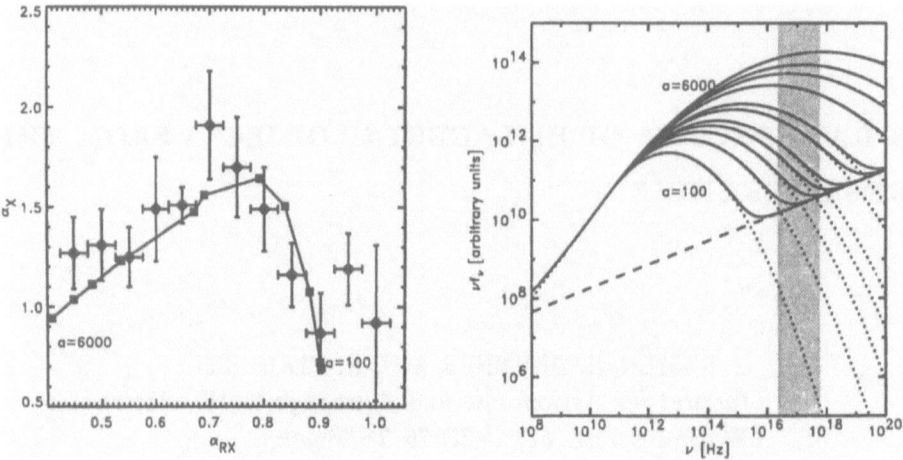

Figure 1.   *a: (left)* Binned X-ray spectral slopes vs. α_{rx} (data points) with simulations (connected points). *b: (right)* Model spectra used for the simulations. Dotted: soft component, dashed: hard powerlaw, solid: total, shaded: PSPC energy range

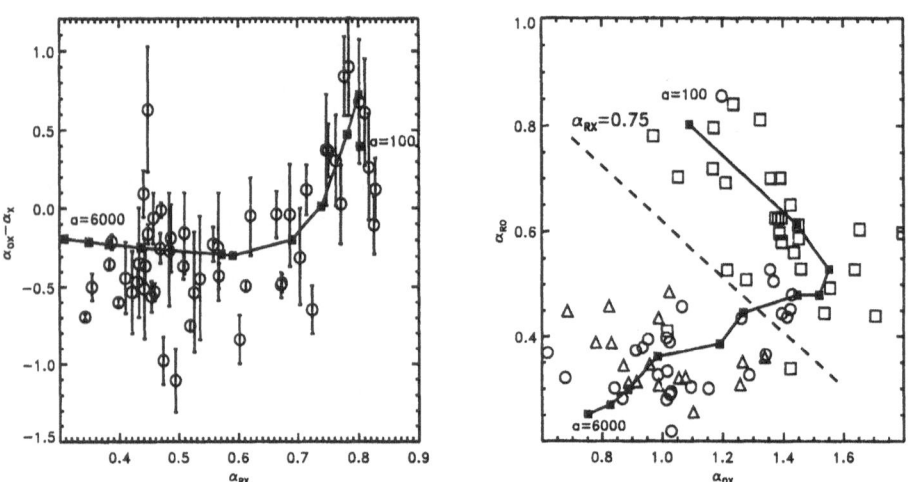

Figure 2.   *a: (left)* Flattening ($(\alpha_{ox}-\alpha_x) > 0$) or steepening of the X-ray spectra relative to the optical– X-ray slope as function of α_{rx} compared with simulations (connected points) *b: (right)* Distribution of the broad band spectra in the $\alpha_{ox}-\alpha_{ro}$ plane. Triangles: EMSS objects, squares: 1 Jy objects, circles: others, connected points: simulated spectra.

References

Celotti, A., Maraschi, L., Ghisellini, G., Caccianiga, A., and Maccacaro, T., 1993, *ApJ*, **416**, 118

Ghisellini, G. and Maraschi, L., 1989, *ApJ*, **340**, 181

Lamer, G., Brunner, H., Staubert, R., 1995, *A&A*, submitted

Morris, S.L., Stocke, J.T., Gioia, I.M., et al. 1991, *ApJ*, **380**, 49

Padovani, P., and Giommi, P., 1995, *ApJ*, **444**, 567

Véron-Cetty, M.P., and Véron, P., 1993, ESO Scientific Report, 13

THE ROSAT X-RAY SPECTRA OF BL LACERTAE OBJECTS

PAOLO PADOVANI
Dipartimento di Fisica, II Università di Roma "Tor Vergata"
Via della Ricerca Scientifica 1, I-00133 Roma, Italy

AND

PAOLO GIOMMI
SAX, Scientific Data Center, ASI
Viale Regina Margherita 202, I-00198 Roma, Italy

1. Introduction

We have analyzed the X-ray spectra of *all* BL Lacs observed (as pointed or serendipitous sources) by *ROSAT*. Spectral indices were obtained from the hardness ratios given in the WGA catalogue [5], a large list of X-ray sources generated from all the *ROSAT* PSPC pointed observations. The selection of the objects was done by cross-correlating the first revision of the WGA catalogue with our recent catalogue of BL Lacs ([3]). This resulted in 163 observations of 85 distinct BL Lacs, which correspond to *about half the confirmed BL Lacs* presently known. This represents the largest number of BL Lacs for which homogeneous X-ray spectral information is available and *the largest BL Lac sample ever studied at X-ray frequencies.*

Following our previous work ([1], [2]), we divide the BL Lacs into HBLs (high-energy cutoff BL Lacs) and LBLs (low-energy cutoff BL Lacs) on the basis of their X-ray-to-radio flux ratios ($f_x/f_r \sim 10^{-11.5}$ being the dividing line), and look for any differences in the X-ray spectra of the two classes. A full presentation of these results will be given elsewhere ([4]).

2. Main Results

BL Lacs have energy power-law spectral indices between 0 and 3 with a mean value $\alpha_x \sim 1.4$ ($f_\nu \propto \nu^{-\alpha}$). Significant differences, however, are present between HBLs (generally selected in the X-ray band: 58 objects), and LBLs (generally found in radio surveys: 27 objects), with $\langle \alpha_{x,HBL} \rangle =$

R. Ekers et al. (eds.), Extragalactic Radio Sources, 267–268.

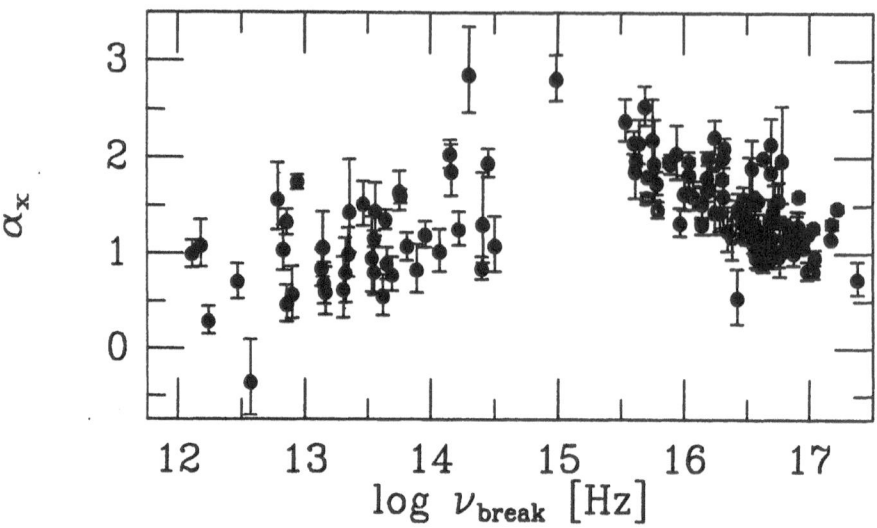

Figure 1. The X-ray spectral index versus the logarithm of the break frequency for BL Lacs. Objects with $\nu_{\text{break}} \gtrsim 10^{15}$ Hz have $f_{\text{x}}/f_{\text{r}} > 10^{-11.5}$ and are, by definition, HBLs.

1.52 ± 0.06 and $\langle \alpha_{\text{x,LBL}} \rangle = 1.06 \pm 0.09$, different at the 99.99% level according to a Student's t-test. Moreover, the X-ray spectral indices of HBLs anti-correlate strongly (99.99% level) with $f_{\text{x}}/f_{\text{r}}$. No correlation is present for LBLs. Similarly, a strong correlation (99.99% level) exists between α_{x} and α_{ox} for HBLs but not for LBLs. The two classes separate extremely well on the $\alpha_{\text{ox}} - \alpha_{\text{x}}$ plane, that is all HBLs apart from one have $\alpha_{\text{x}} \geq \alpha_{\text{ox}}$ (within the errors), while the majority of LBLs have $\alpha_{\text{x}} < \alpha_{\text{ox}}$. The most interesting correlation is, we believe, that between α_{x} and the frequency of the energy cutoff, ν_{break}, which we have estimated from the location of the objects on the $\alpha_{\text{ro}} - \alpha_{\text{ox}}$ plane ([2]). A strong anti-correlation (significant at the 99.99% level) is present between X-ray spectral index and break frequency for HBLs, while LBLs display a weaker, positive, correlation (see Figure 1).

We interpret these results in terms of different mechanisms being responsible for the X-ray emission in the two classes, i.e., synchrotron and inverse Compton for HBLs and LBLs respectively. The observed differences are consistent with the hypothesis that HBLs and LBLs are powered by the same non-thermal engines with different synchrotron cutoff energies.

References

[1] Giommi, P. and Padovani, P., 1994, *MNRAS*, **268**, L51 **310**, **325**
[2] Padovani, P. and Giommi, P., 1995a, *ApJ*, **444**, 567
[3] Padovani, P. and Giommi, P., 1995b, *MNRAS*, in press
[4] Padovani, P. and Giommi, P., 1995c, *MNRAS*, submitted
[5] White, N.E., Giommi, P. and Angelini, L., 1994, IAUC 6100

THE SEARCH FOR A NEW BL LAC SAMPLE

A. WOLTER, A. CACCIANIGA, T. MACCACARO, C. RUSCICA
Osservatorio Astronomico di Brera,
V. Brera, 28; 20121 Milano ITALY

1. Introduction

The number of BL Lacs discovered in the last 20 years is very small (~ 200) if compared to that of quasars. Owing to their featureless optical spectrum, BL Lacs have been discovered mainly in radio (RBL) and X-ray surveys (XBL). The limited statistics available prevents a detailed study of the properties of BL Lacs and, in general, all the conclusions based on the present data sets are affected by large uncertainties. Yet, current results are intriguing, for instance RBL and XBL are found to have a cosmological evolution that differs not only in magnitude but most of all in sign (e.g. Stickel et al., 1991 and Wolter et al., 1994 and reference therein). Therefore, we have initiated a project aimed at discovering a significant number of new BL Lac objects (~ 100), exploiting the fact that they are both radio and X-ray emitters and in particular the fact that they occupy a well defined region in the α_{ox}–α_{ro} plane. The method we use rests on the expertise grown during the construction of the EMSS sample (Gioia et al. 1990, Stocke et al. 1991) and is described in detail in Wolter et al. (1995).

2. The search method and first results

By cross-correlating X-ray and radio source catalogs, we construct a sample of Radio Emitting X-ray sources (REX). The fraction of BL Lacs in this sample is expected to be of the order of 30% of the radio loud ($\alpha_{ro} \geq 0.3$) objects, the remainders being mainly radio emitting quasars and Seyfert galaxies. Optical counterparts are then identified on the Palomar digitized plates, and subsequently classified by optical spectroscopy.

To uniquely pinpoint the optical counterpart, and therefore minimize the required telescope time, we need to have a positional accuracy of the order of a few arcsec for at least one of the two catalogs. This also reduces

R. Ekers et al. (eds.), Extragalactic Radio Sources, 269–270.

the number of spurious X-ray – radio associations. Presently, the available X-ray data over large areas of sky (ROSAT PSPC) have larger error boxes. However, two large surveys of the sky currently under way with the VLA will produce lists of sources with the required positional accuracy. We have thus decided to use all the public and suitable ROSAT PSPC data, of which catalogs have appeared recently (ROSAT NEWS n.32, 1994; and White, Giommi, & Angelini 1994), and the NVSS survey (Condon et al., 1995), that will eventually cover the whole sky north of $\delta = -40°$.

Results of the optical follow up of the first REX observations are very encouraging. Sample spectra obtained at San Pedro Martir (Mexico) in April 1995 are presented in Ruscica et al. (this conference) and will be discussed in detail elsewhere. The success rate of the project is as expected: out of 26 REX observed we found 8 candidate BL Lacs, 14 AGN, 3 galaxies, and 1 object that need further investigation (a probable star). We plot in Fig. 1 the position of the 8 newly found BL Lacs in the α_{ox}–α_{ro} plane.

Among the BL Lac candidates we are finding a mixture of objects of the XBL and RBL type, and this will help in favouring one of two competing theoretical models of the emission in BL Lac objects (cf. Maraschi et al. 1986, Padovani and Giommi, 1995). This sample, unbiased and statistically complete, allowing a direct determination of the absolute properties and relative density of XBL and RBL in the sample, would be instrumental in studying the luminosity function and cosmological evolution of BL Lac objects as a class, and of radio-loud AGNs in general.

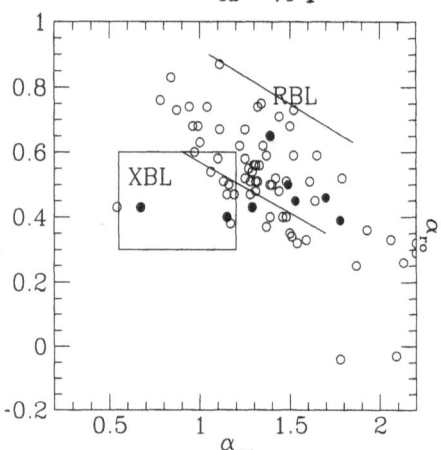

Figure 1. The α_{ox}/α_{ro} plane: empty circles are all the REXs, while black circles are the 8 new BL Lac candidates.

References

Condon, J.J., et al., 1995, in preparation
Gioia, I.M., et al., 1990, *Ap.J. Suppl.*, **72**, 567.
Maraschi, L., et al., 1986, *Ap.J.*, **310**, 325.
Padovani, P., & Giommi, P., 1995, *Ap.J.*, **444**, 567.
Stickel, M., et al., 1991, *Ap.J.*, **374**, 431.
Stocke, J.T., et al., 1991, *Ap.J. Suppl.*, **76**, 813.
White N., Giommi P., Angelini L., 1994, "The Multi-Mission Perspective", Napa Valley.
Wolter, A., et al., 1994., *Ap.J.*, **433**, 29.
Wolter, A., et al., 1995, *MNRAS* submitted

OPTICAL IDENTIFICATIONS FOR A SAMPLE OF REXS: SEARCH FOR BL LAC OBJECTS

C. RUSCICA, A. CACCIANIGA, T. MACCACARO, A. WOLTER

Osservatorio Astronomico di Brera
via Brera 28, I-20121 Milano, ITALY

1. Introduction

Twenty six Radio Emitting X-ray Sources (REXs) have been observed in two observing runs as part of a project aimed at selecting a new and large sample of BL Lac objects (see Wolter et al. this conference). Sources were selected by cross-correlating ROSAT-PSPC public images against VLA radio surveys. Here we present results from the first observations of 14 sources conducted in April 1995 with a B&C spectrograph in longslit mode using a CCD-Tektronix TK-1024 AB detector at the 2.1m San Pedro Martir telescope (Mexico). Table 1 shows the data derived for the 8 confirmed REXs. X-ray fluxes are in units of erg cm^{-2} sec^{-1} in the 0.1-2.4 keV band, radio fluxes are in mJy at 5 GHz, X-ray luminosities are in erg sec^{-1} in the energy range 0.1-2.4 keV and radio luminosities are in erg sec^{-1} at 5 GHz. We assumed $H_o = 50$ km sec^{-1} Mpc^{-1} and $q_o = 0$. A more detailed description of individual sources will be published elsewhere.

2. Results from optical spectroscopy

We have identified 4 BL Lac candidates. For two candidates we were able to measure the redshift from MgI, NaI, G-band and E-band absorption lines (e.g. fig.1) probably due to the host galaxy. We computed also the f_x/f_r ratio as defined in Padovani & Giommi (1995): for the BL Lac candidates we find that the ratios are well below the dividing limit, 2×10^{-11} erg cm^{-2} sec^{-1} Jy^{-1}, indicating that these objects are of the RBL-type (cf. Padovani & Giommi 1995). Finally, we note that the X-ray luminosities of our REXs are in the typical range of their class (Stocke et al. 1991). In the case of the

R. Ekers et al. (eds.), Extragalactic Radio Sources, 271–272.

C. RUSCICA ET AL.

two BL Lac candidates, the values found are at the low end of the range typical for BL Lacs.

TABLE 1. Data for the REXs

Name	f_x L_x	f_r L_r	m_v M_v	z f_x/f_r	α_{ro}	α_{ox}	ID
REX0744.8+2920	$9.62\ 10^{-13}$ $1.38\ 10^{46}$	192 $4.52\ 10^{34}$	15.6: -29.6	1.173 $5.01\ 10^{-12}$	0.41	1.54	quasar
REX0832.0+1953	$5.15\ 10^{-13}$ $5.64\ 10^{45}$	92 $1.59\ 10^{34}$	17.6: -27.3	1.061 $5.60\ 10^{-12}$	0.49	1.34	quasar
REX1136.8+2937	$0.34\ 10^{-13}$ $2.95\ 10^{42}$	32 $3.11\ 10^{31}$	18.2: -21.5	0.136 $1.06\ 10^{-12}$	0.46	1.70	BL cand
REX1137.4+6120	$0.22\ 10^{-13}$ $1.27\ 10^{42}$	326 $2.03\ 10^{32}$	17.9: -21.3	0.111 $6.75\ 10^{-14}$	0.62	1.81	Seyfert 2
REX1309.6+0828	$1.51\ 10^{-13}$ $-$	75 $-$	17.9: $-$	$-$ $2.01\ 10^{-12}$	0.50	1.49	BL cand
REX1416.9+2312	$0.62\ 10^{-13}$ $3.87\ 10^{42}$	43 $2.95\ 10^{31}$	17.0: -22.3	0.116 $1.44\ 10^{-12}$	0.39	1.78	BL cand
REX1503.7+1016	$0.99\ 10^{-13}$ $4.08\ 10^{42}$	113 $5.04\ 10^{31}$	17.6: -21.3	0.095 $8.76\ 10^{-13}$	0.51	1.61	NELG
REX1525.4+4201	$3.30\ 10^{-13}$ $-$	103 $-$	16.8: $-$	$-$ $3.20\ 10^{-12}$	0.45	1.53	BL cand

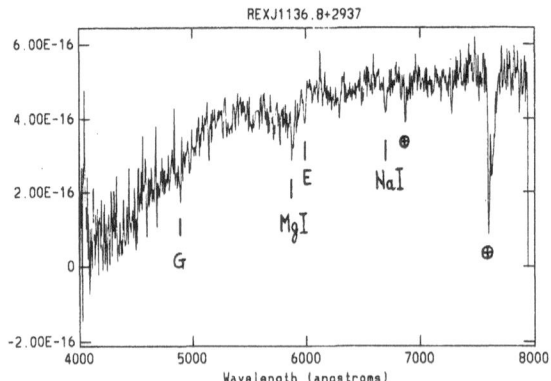

Figure 1. The optical spectrum of one of the BL Lac candidate

References

Padovani P. & Giommi P., 1995, *Ap.J.*, **444**, 567
Stocke, J. T. et. al., 1991, *Ap.J.*, **76**, 813

X-RAY SPECTRAL PROPERTIES OF RADIO GALAXIES

H. INOUE
Institute of Space and Astronautical Science
3-1-1 Yoshinodai, Sagamihara, Kanagawa 229, Japan

X-ray spectral properties of radio galaxies are reviewed in comparison with those of Seyfert galaxies, mainly based on results from Japanese X-ray astronomy satellites, Ginga and ASCA.

1. Spectral Indices in 2 – 10 keV

Figure 1 shows energy-spectral indices in 2 – 10 keV as a function of the cosmological redshift of various kinds of active galactic nuclei (AGN) obtained with Ginga and ASCA. As seen in this figure, all kinds of AGN including radio galaxies have the spectral indices in a fairly narrow range in 0.5 – 1.0 except some BL Lac objects, although radio-quiet quasars seem to have slightly steeper spectra than radio-loud quasars (Williams et al. 1992). Steeper spectra of some BL Lac objects are considered to be due to a large blue shift of a high energy end of a synchrotron spectrum to the X-ray band (e.g. Macomb et al. 1995).

2. Ratio of X-Ray Flux to Far Infrared Flux

2.1. SEYFERT GALAXIES

In the unified model for Seyfert galaxies, a central power house is considered to be partially surrounded by a molecular/dust torus and whether a molecular torus is in or out of the line of sight to a central nucleus determines whether the source is identified to be a Seyfert 2 or 1 galaxy, respectively. Since a part of X-rays should be absorbed and be reemitted in far-infrared-rays by the torus, a far-infrared emission should be observed irrespectively of whether a source is a Seyfert 1 or 2. In fact, far-infrared emissions are generally detected from Seyfert galaxies.

Although no obscuration of optical and X-ray emission from the nucleus is observed from Seyfert 1 galaxies, their far-infrared fluxes are generally

273

R. Ekers et al. (eds.), Extragalactic Radio Sources, 273–276.
© 1996 *IAU.*

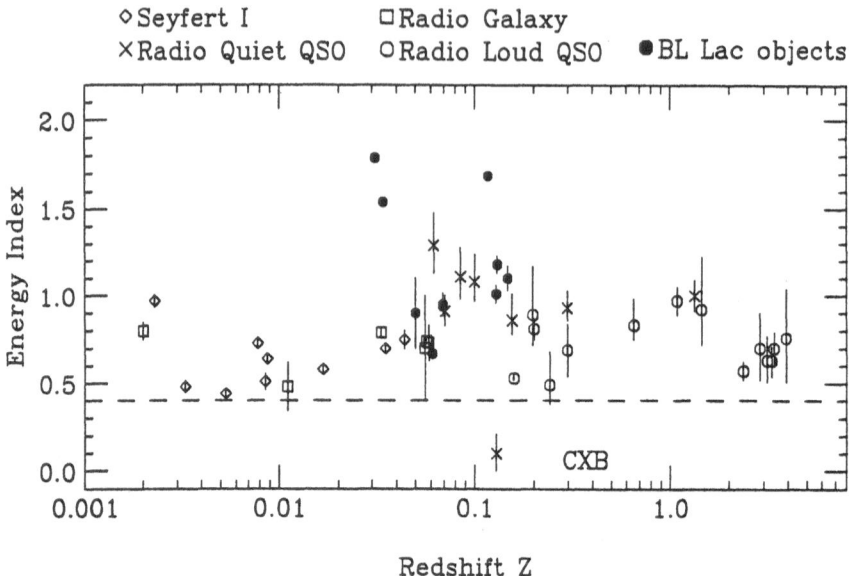

Figure 1. Energy indices in 2 – 10 keV as a function of the cosmological redshift, z, of various kinds of AGNs obtained with Ginga and ASCA.

comparable to the X-ray fluxes. If we calculate the ratio of the X-ray flux (2–10 keV) to the far-infrared flux (60–100 μm flux defined in the IRAS galaxy catalogue, Ver. 2, 1989) for 20 samples observed with Ginga (Nandra and Pounds 1994), we see them distribute in a range from 0.1 to 1 with a mean of 0.5. This is consistent with a conjecture that the far-infrared emissions are from dust tori heated by X-rays from nuclei.

The X-ray to far-IR flux ratio tends to be smaller in Seyfert 2 galaxies than those of Seyfert 1 galaxies. Most of Seyfert 2 galaxies observed with Ginga (Awaki 1992 and references therein) show heavy absorption in the X-ray band and it is consistent with their absence of optical broad lines probably due to obscuration of central broad line regions. The ratios of the observed X-ray flux to the far-IR flux of these sources are in a range $10^{-2} - 10^{-1}$. If we calculate the intrinsic X-ray flux by correcting the flux-decrease by absorption, their ratios of the intrinsic X-ray flux to the far-IR flux become consistent with those of Seyfert 1 galaxies.

Some of the Seyfert 2 galaxies, however, show no large absorption but have ratios of X-ray to far-IR flux as small as $10^{-4} - 10^{-3}$. These exhibit a prominent K-emission line of iron in low ionization states with equivalent width of about 1.5 keV. These iron lines are interpreted to be due to fluorescent process of X-rays by some ambient matter to a central X-ray source and their equivalent width can be calculated on an assumption of a simple geometry (see e.g. Inoue 1985). Then, such a high equivalent width of iron fluorescent line as $\sim$1.5 keV can be obtained only when the central X-ray source is completely obscured by some matter and X-rays scattered by the ambient matter is only observed as continuum. This well explains the very low X-ray to far-IR flux ratios of these sources. In fact, Ginga found presence of heavily absorbed component in the energy range above 10 keV from one of such Seyfert 2 galaxies showing a prominent iron emission line, NGC4945 (Iwasawa et al. 1993). The ratio of the intrinsic X-ray flux after correcting for absorption to the far-IR flux is well in the range of Seyfert 1 galaxies.

These are all consistent with the unified scheme of Seyfert galaxies.

2.2. RADIO GALAXIES

Ratio of X-ray flux to far-IR flux after correcting for absorption was calculated for five radio galaxies observed with Ginga and/or ASCA (3C120, 3C390.3, 3C445, Cen A and Cyg A) and again ranges around 0.1 – 1.0. This is consistent with unification between radio galaxies and Seyfert galaxies.

3. Breadth of iron emission line

An ASCA observation of the Seyfert 1 galaxy, MCG–6–30–15, revealed a broad feature of the iron emission line (Tanaka et al. 1995: see the left panel of Fig.2). Most of the line flux is strongly redshifted from the rest energy and the large full width at zero intensity indicates that the line comes from matter moving with velocity of $\sim$100,000 km s^{-1} (a third of the light velocity). These properties are interpreted to be due to Doppler and general relativistic effect of matter circulating in an accretion disk very close to the central black hole (Tanaka et al. 1995; Fabian et al. 1995). Similar broad line-profiles have been detected from several Seyfert 1 galaxies (Otani 1995; Yaqoob et al. 1995; Iwasawa et al. 1995).

An ASCA observation of the broad line radio galaxy, 3C390.3 (Eracleous et al. 1995), however, didn't show such a broad line feature as in Seyfert 1 galaxy, as seen in Fig.2. The absence or weak appearance of the broad line feature in the radio galaxy might suggest lack of an inner line-reprocessing region in this source and be related to presence of the superluminal jet in this source (Alef et al. 1988).

4. Summary

Similarities and differences in X-ray spectral properties between radio galaxies and Seyfert galaxies have been studied. No large differences have been found in the spectral indices in 2 – 10 keV and in the ratios of the X-ray flux to far-IR flux. The breadth of iron lines may be a difference between

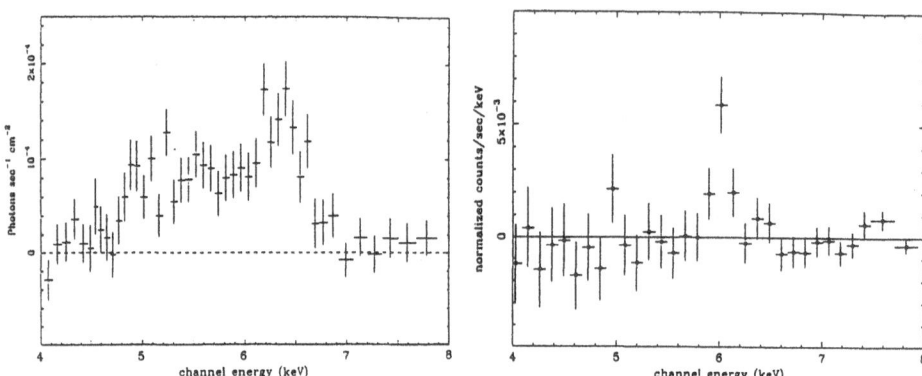

Figure 2. The line profile of iron K_α in the X-ray emission from MCG–6–30–15 ($z=0.008$: left panel) and from 3C390.3 ($z=0.056$: right panel).

them. Several Seyfert galaxies reveal a very broad feature of the iron emission line and the line properties are well interpreted by introducing Doppler and general relativistic effects on the line coming from a region very close to the central black hole. Whereas, the iron emission line from the broad line radio galaxy, 3C390.3 doesn't have such a broad feature. This may relate to presence of jets in radio galaxies, although further X-ray observations of radio galaxies are obviously necessary.

The author thanks H.Kubo, M.Sugizaki, T.Takahashi and A.Yamashita for their help in the preparation of this paper.

References

Alef, W. et al. (1988), *A.Ap.*, **192**, 53.
Awaki, H. (1992), in *"Frontiers of X-ray Astronomy"* ed. Y.Tanaka and K.Koyama, Universal Academy Press, Tokyo, p.537.
Eracleous, M., Halpern, J.P. and Livio, M. (1995), *Ap.J.*, in press.
Fabian, F. et al. (1995), *MNRAS*, in press.
Inoue, H. (1985), *Space Sci. Rev.*, **40**, 317.
Iwasawa, K. et al. (1993), *Ap.J.*, **409**, 155.
Iwasawa, K. et al. (1995), *MNRAS*, in press.
Macomb, D.J. et al. (1995), *Ap.J.*, **449**, L99.
Nandra, K. and Pounds, K.A. (1994), *MNRAS*, **268**, 405.
Otani, C. (1995), *Ph.D.Thesis*, Univ. of Tokyo.
Tanaka, Y. et al. (1995), *Nature*, **375**, 659.
Williams, O.R. et al. (1992), *Ap.J.*, **389**, 157.
Yaqoob, T. et al. (1995), *Ap.J.*, in press.

GAMMA-RAY EMISSION FROM EXTRAGALACTIC RADIO SOURCES: EGRET OBSERVATIONS OF ACTIVE GALACTIC NUCLEI

PETER F. MICHELSON

Dept. of Physics, Stanford University
Stanford, Ca. 94305

Abstract. The Energetic Gamma-Ray Experiment Telescope (EGRET) on the Compton Gamma-Ray Observatory is an imaging high-energy telescope with sensitivity from approximately 20 MeV to 30 GeV. EGRET has observed more than 129 sources during more than 4 years of operation. Among these sources, 51 have been identified with active galaxies. A common characteristic of the AGN sources is that they are all radio-loud, flat radio spectrum sources. Many of them are seen as superluminal radio sources as well. The gamma-ray emission characteristics of these sources are reviewed and some of the proposed emission models are discussed.

1. Introduction.

The Energetic Gamma Ray Experiment Telescope (EGRET) Instrument Team has reported the detection of 129 sources in the Second EGRET Catalog (Thompson et al. 1995). Figure 1 shows the positions of these sources in galactic coordinates. The size of the symbol for each source is proportional to the highest intensity seen for the source by EGRET. These sources include galactic pulsars, a nearby galaxy (the LMC), 51 active galaxies, several high-energy gamma- ray bursts, and 71 other sources not yet identified with known objects. A common characteristic of the AGN sources is that they are all radio-loud, flat radio spectrum sources (von Montigny et al. 1995). Many of them are seen as superluminal radio sources as well. They all appear to be members of the blazar class of AGN (Bl Lac objects, HPQ and OVV quasars), exhibiting one or more of the characteristics of this

R. Ekers et al. (eds.), Extragalactic Radio Sources, 277–280.
© 1996 *IAU.*

source class. One source, Mkn 421, has also been detected by the Whipple telescope in the TeV energy band (Punch et al. 1992). In addition more than two dozen bright, unidentified high-latitude sources have been detected.

Among the most interesting results obtained from analysis of the EGRET AGN observations are the following:

(i) the gamma ray energy flux in many of the sources is dominant over the emission in lower energy bands;

(ii) these sources are distributed over a wide range of redshifts (0.03 to 2.17);

(iii) the photon spectra in the EGRET energy range are generally well-represented by power laws in energy with a wide range of photon spectral indices (-1.6 to -2.6);

(iv) many of the sources exhibit variability, in some cases the gamma-ray flux varies by a factor of 3 in less than a week;

(v) many active galaxies relatively close to earth, including Seyfert galaxies and some notable blazars and superluminal sources, have not been detected in high-energy gamma rays.

2. Gamma-ray emission models.

There are several arguments suggesting that the gamma-ray emission from the EGRET sources is strongly beamed. In particular, the EGRET observations of high fluxes above 1 GeV, short timescale variability, and consideration of absorption via photon-photon pair production appear to rule out emission models in which the gamma radiation is generated in the same volume as the observed X-ray radiation, unless the radiation is beamed towards us (McBreen 1979). This is consistent with the interpretation of blazars as objects in which the non-thermal radiation arises predominantly from a "jet" of relativistic particles directed close to the observer's line of sight (Blandford and Rees 1978). However, the gamma-ray emission in most of the EGRET sources cannot come from radii as large as VLBI sources because of variability constraints. Thus the gamma emission is an important link between radio structure and the putative black hole central engine.

If the gamma-ray emission is from a source moving relativistically with velocity βc and having bulk Lorentz factor $\Gamma = (1 - \beta^2)^{-1/2}$, then a local observer, whose line of sight is at an angle Θ to the direction of the bulk motion, will see emission with apparent luminosity L_{app} and duration Δt_{app} given by

$$L_{app} = L\delta^p, \; \Delta t_{app} = \Delta t\delta^{-1},$$

where $\delta = [\Gamma(1 - \beta cos\Theta)]^{-1}$, L is the comoving luminosity, Δt is the co-moving time duration of the emission, and p is typically between 3 and 4

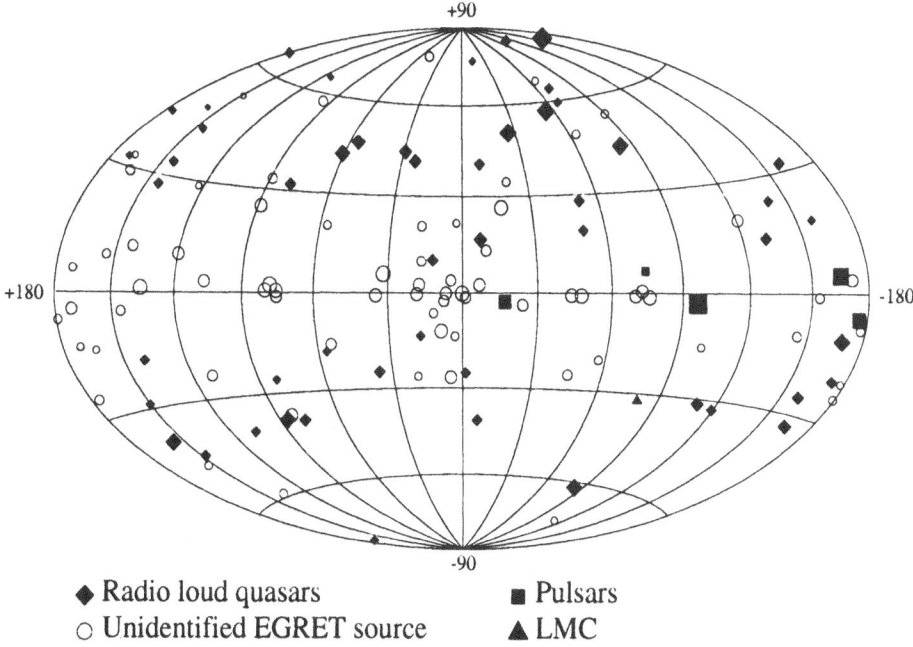

◆ Radio loud quasars   ■ Pulsars
○ Unidentified EGRET source   ▲ LMC

Figure 1.   The second EGRET catalog of high-energy gamma-ray sources.

(Lind and Blandford 1985). Reasonable values of δ result in gamma-gamma opacities near unity. Interestingly, relativistic beaming may also play an important role in gamma-ray bursts. In particular EGRET and COMPTEL observations of high-energy emission from some relatively short duration gamma-ray bursts appears to require that the sources are strongly beamed with δ approaching 1000 if the bursts are cosmological.

Details of models for the AGN gamma-ray emission are as varied as those for the sub-MeV emission, with most involving inverse Compton scattering of low energy photons up to gamma-ray energies by beamed relativistic electrons. The low energy photons come from either a disk (Melia and Königl 1989; Dermer et al. 1992) or are self-generated by the electrons as synchrotron radiation (Jones, O'Dell, and Stein 1974; Marscher 1980; Königl 1981; Marscher and Bloom 1992; Maraschi et al. 1992). It has also been proposed (Blandford and Levinson 1995; Sikora, Begelman, and Rees 1994) that radiation from a disk which has been reprocessed in a halo (eg., broad-line region clouds, disk wind, dust) may be more effectively Compton scattered than radiation coming directly from the disk. This is because in the comoving frame of the gamma-ray producing plasma, the unscattered disk radiation is strongly redshifted while diffuse reprocessed radiation is strongly blueshifted as the jet passes near or through the reprocessing region.

In contrast to the leptonic jet models, the proton-initiated cascade (PIC) model of Mannheim and Biermann (1989, 1992) is based on shock acceleration of protons to energies so high ($10^7 - 10^8$ TeV) that the threshold for photomeson production is exceeded and that the proton cooling timescale becomes equal to the electron cooling timescale. The PIC model has been applied to the spectra of 3C 273, 3C 279, and Mkn 421 (Mannheim 1993). Generally, the model can account for the wide range of spectral indices observed in the EGRET energy range and predicts that below a few MeV, the photon spectrum in the X-ray range should flatten.

In the emission models involving inverse Compton scattering by relativistic electrons, the relative importance of self-Compton scattering and inverse Compton reflection depends on the ratio of the internally generated synchrotron photon density to the ambient photon density, as measured in the rest frame of the jet plasma. Thus, depending on the conditions, either self-Compton scattering or inverse Compton scattering of photons emitted by the accretion disk (and possibly scattered back toward the jet by an ambient medium) might be the dominant gamma-ray emission process. Since all of these photon fields are expected to be present within ~ 1 pc of the central engine, the relative importance of each depends on the value of various parameters such as the location of the Compton scattering region and the ratio of apparent synchrotron luminosity to luminosity of the accretion disk.

References

Blandford, R.D. & Levinson, A. 1995, *ApJ*, **441**, 79

Blandford, R. & Rees, M.J., 1978, in *Pittsburgh Conference on Bl Lac Objects*, ed. A.N. Wolfe (University of Pittsburgh Press), 328

Dermer, C.D., Schlickeiser, R., & Mastichiadis, A. 1992, *A&A*, **256**, L27

Dermer, C.D., & Schlickeiser, R. 1992, *Science*, **257**, 1642

Jones, T.W., O'Dell, S.L., & Stein, W.A. 1974, *ApJ*, **188**, 353

Königl, A., 1981, *ApJ*, **243**, 700

Lind, K.R., and Blandford. R.D., 1985, *ApJ*, **295**, 358

Mannheim, K., & Biermann, P.L., 1992, *A&A*, **53**, L21

Mannheim, K., & Biermann, P.L., 1989, *A&A*, **221**, 211

Mannheim, K., 1993, *Phys. Rev. D*, **48**, 2408

Maraschi, L., Ghisellini, G., & Celotti, A., 1992, *ApJ*, **397**, L5

Marscher, A.P., 1980, *ApJ*, **235**, 386

Marscher, A.P., & Bloom, S.D., 1992, in *Proc. of the CGRO Science Workshop*, NASA Conf. Pub. 3137, 346

McBreen, B., 1979, *A&A*, **71**, L17

Melia, F., & Königl, A., 1989, *ApJ*, **340**, 162

von Montigny, C., et al., 1995, *ApJ*, **440**, 525

Punch, M., et al., 1992, *Nature*, **358**, 477

Sikora, M., Begelman, M.C., & Rees, M.J. 1993, in *Proceedings of the Compton Symposium*, in press

Thompson, D.J., et al., 1995, The Second EGRET Catalog of High-Energy Gamma-Ray Sources, accepted for publication in *ApJ Suppl.*

THE BIG PICTURE FROM RADIO WAVES TO GAMMA RAYS

C.D. IMPEY
Steward Observatory
University of Arizona, Tucson, AZ 85721, USA

1. Introduction

The conventional paradigm of active galactic nuclei (AGN) holds that they are powered by accretion onto a gravitational engine. In addition, the presence of Doopler-boosted radiation from jets and anisotropic obscuration can create large differences between the observed and intrinsic properties of AGN (Antonucci 1994). The "big picture"of extragalactic radio sources must include observations across the electromagnetic spectrum. Relativistic processes produce radiation with a very wide bandwidth. In addition, much of the energy from AGN is reprocessed, and the opacity is a function of wavelength. Finally, it turns out that only a small fraction of the bolometric luminosity of most AGN is emitted in the traditional radio and optical bands.

2. Quasars and Blazars

Samples of strong radio sources selected at high frequencies have the best complementary information at other wavelengths. For example, the roughly 300 quasars in the Kühr (1981) catalog now have virtually complete redshift information. In addition, about 70% of the quasars have optical polarimetry. After controlling for luminosity, strong optical and far infrared emission is correlated with the compactness of the radio source. High optical polarization, a clear indicator of nonthermal emission, is strongly correlated with radio compactness. Sources with these collective properties are called "blazars." It is also possible to distinguish highly polarized quasars, where normal quasar broad lines are superimposed on a polarized continuum, and BL Lac objects, where the lines are weak or absent. Among quasars stronger than 1 Jy at 5 GHz, the proportion of highly polarized objects falls from

R. Ekers et al. (eds.), Extragalactic Radio Sources, 281–282.

60% at $z = 0.3$ to just under 20% at $z = 2.7$. The BL Lac fraction in the 1 Jy catalog falls more steeply, from 30% at $z = 0.3$ to under 10% at $z = 1.5$.

These observations can be explained by a model where the radio and optical emission is beamed with a similar Lorentz factor and geometry, presuming that the optical emission is a sum of power law synchrotron and thermal accretion disk components. The decline in the fraction of highly polarized objects with increasing redshift results from the larger fraction of the emission which is due to the hot thermal component in the optical passband. This simple model works well for a power law spectral index of $\alpha = -1.5$, where $S_\nu \propto \nu^\alpha$. As expected, low polarization quasars are brighter than high polarization quasars, and radio-to-optical spectral index steepens with redshift, due to the dependence of the beamed power on local spectral index. BL Lac objects form a distinct class. No thermal components are seen, even at minimum light, so the rapid decline in redshift must be intrinsic to the population. Compact radio structure is an excellent predictor of nonthermal optical/infrared properties.

3. Strong Gamma Ray Emission

Gamma ray emission has been detected from a significant number of AGN (Montigny et al. 1995); over half of the sources at high galactic latitude in the second EGRET catalog are associated with strong radio sources, typically blazars (Thompson et al. 1995). About 35 of these firm identifications are in the 1 Jy sample. When the spectral energy distributions are normalized to the radio band, it is found that the gamma ray flux is much more closely correlated with the radio flux than with the optical or ultraviolet flux. This may favor models where gamma ray photons are multiply upscattered from much lower energies, rather than from optical/UV seed photons. The correlation between radio flux, S_r, and (non-simultaneous) gamma ray flux, S_γ, is weak but significant. The ratio S_γ/S_r is influenced by the large number of upper limits, but the 95% confidence range is $0.0038 < S_\gamma/S_r < 0.018$. Integrating the source counts, and taking account of variability, the contribution of AGN to the 100 MeV background is 46^{+54}_{-23}%. Given the radio weakness of the more ubiquitous optically selected quasar, they must contribute less than 5% of the gamma ray background.

This work was partially supported by NSF grant AST 93-20715.

References

Antonucci, R.R. 1994, *Ann.Rev.Ast.Ap.*, **31**, 473
Kühr, H. et al. 1981, *Ast.Ap.Supp.*, **45**, 367
Montigny, C. et al. 1995, *Ap.J.*, **440**, 525
Thompson, D.J. et al. 1995, *Ap.J.Supp.*, in press

THE RADIO-GAMMA-RAY CONNECTION: THE RADIO PROPERTIES OF GAMMA-RAY-BRIGHT BLAZARS

M. F. ALLER, H. D. ALLER AND P. A. HUGHES

University of Michigan – Ann Arbor, MI, 48109-1090

1. Introduction

Forty AGN have been detected at high significance level by EGRET in the γ-ray band. Previous studies based on radio observations near or at the times of the EGRET detections suggest that there is a causal connection between individual events in these two wavebands. Here we examine the question of whether the cm-λ and the γ-ray activity are related.

2. Results

The integrated total flux and polarization observations discussed here were obtained with the UMRAO 26-meter telescope operating at 14.5, 8.0 and 4.8 GHz. The time-sampling of these radio data for 32 of the 40 EGRET-detected sources was sufficient for defining the long-term variability, but not all of the objects were observed intensively during the EGRET detection period. The time coverage of the data is up to $\sim$ 30 years. The γ-ray measurements are fluxes, or an upper limit if no detection was measured by EGRET; the time sampling is infrequent; the time coverage is April 1991 through current; and the source material is Thompson *et al.* (1995) for phases 1 and 2 and private communications subsequently (R. C. Hartman 1995).

Even a casual inspection of the radio light curves indicates that these objects are highly active in the cm waveband; when resolved, new events typically occur at 2-year intervals (see also, Hughes, Aller and Aller 1992). As we quantify in Figure 1, **all** the EGRET-detected objects exhibited variability during 1991-1995, ranging from only moderate levels to impulsive large-amplitude variations. Comparison with a flux-limited radio-loud sample clearly shows that the EGRET-detected objects exhibited a higher

283

R. Ekers et al. (eds.), Extragalactic Radio Sources, 283–284.

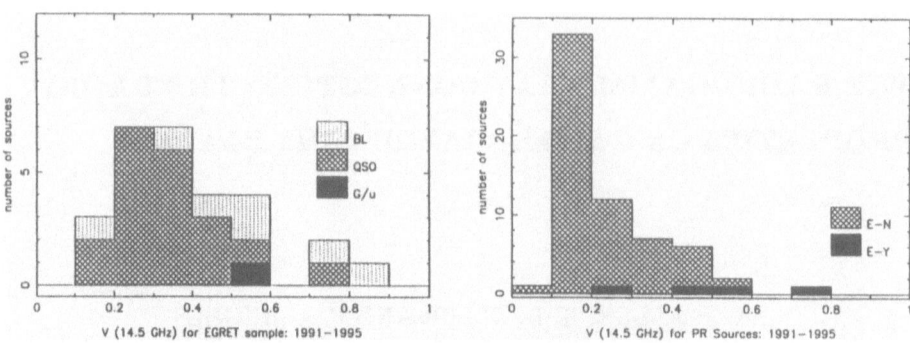

Figure 1. Histograms of a variability index for 1991-1995: a) the EGRET source sample; b) the Pearson-Readhead sample. Black shading denotes EGRET objects.

level of activity. Three of the 4 EGRET-detected sources in the Pearson-Readhead (1988) sample are those with the highest degrees of variability.

Because of the undersampling of the γ-ray data, one cannot carry out a correlation analysis to quantify the existence or absence of associated activity. Inspection of the behavior of the radio-flux light curves at the times of γ-ray detection, shows that: in 20 out of 22 sources detections occurred during burst rises (49 cases), in 2 cases they occurred during plateaus, and in 0 cases they occurred during major declines. The UMRAO data are *integrated* fluxes, and hence not ideally suited for separating contributions from individual evolving source components; thus, it is not always possible to determine unambiguously at what phase of the radio outburst's development the EGRET detection occurred. Nevertheless, the data show that during burst rise they ranged from onset to peak, with no apparent preferred phase. Also, there is no clear correlation between gamma-ray flux and radio flux: 3C 273 which is one of the brightest sources in the radio has been found to be only moderately bright in the γ- ray band, although possibly it has not been viewed at an optimum time; conversely 1156+295 and 1633+382 have been unusually bright in the γ-ray band but relatively weak during the past two decades of observation at Michigan.

We conclude that the data are *suggestive* of causally related activity in the two wavebands, but that better sampling is required to firmly establish this association. We thank R. Hartman for invaluable input, M. C. Aller for help in data preparation, and the NSF for partial support (AST-9421979).

References

Hughes, P. A., Aller, H. D. and Aller, M. F. 1992, *ApJ*, **396**, 469.
Pearson, T. J. and Readhead, A. C. S. 1988, *ApJ*, **328**, 114.
Thompson, D. J. *et al.*, *ApJ Suppl*, in press.

AGN STATISTICS OF SIMULTANEOUS
RADIO AND GAMMA RAY OBSERVATIONS

A. MÜCKE, M. POHL AND G. KANBACH
MPE, Postfach 1603, D-85748 Garching, Germany

AND

P. REICH, W. REICH, R. SCHLICKEISER
MPIfR, Postfach 2024, D-53010 Bonn, Germany

In this paper we present first results of a statistical analysis of correlated variability behaviour of flat-spectrum radio quasars (FSRQ) on the basis of EGRET data and cm-wavelength monitoring data. We use EGRET observations obtained between April 1991 to September 1993 and multifrequency radio observations at 2.8cm, 6cm and 11cm taken with the 100-m-Effelsberg Telescope parallel to the CGRO observations. In the following discussion the observed FSRQ which have not been detected by EGRET yet, are referred to as 'candidates' in contrast to the detected ones, called 'EGRET-sources'. The methods used in this paper are described in more detail in Mücke et al. (1996).

1. The radio state of EGRET sources

Our comparison of the 2.8cm flux density distribution and the 2.8cm-to-6cm spectral index distribution of EGRET sources to candidates is based on 15 highly variable ($|S_{max} - S_{min}|/S_{min} < 100\%$) EGRET-sources and 22 candidates. A total of 18 simultaneous flux and 14 spectral index measurements in the cm-radio and EGRET regime exist. Historical minima and maxima were used to define a linear scale of activity. The analysis is done with likelihood tests and Kolmogorov-Smirnov tests. The probability that the activity distributions of the candidates and EGRET-sources considered on long time scales are statistically identical is 48%. However, at the time of an EGRET-detection the EGRET-sources show a more evenly distribution while the candidates and the monitored EGRET-sources have a higher chance to be detected at low activity. The distribution of the monitored

R. Ekers et al. (eds.), Extragalactic Radio Sources, 285–286.
© 1996 IAU.

EGRET-sources deviate from those simultaneously observed by 2.1σ, while the distribution of the EGRET-sources and candidates differ only by 0.7σ. Therefore, we conclude that γ-ray emission is accompanied by an enhanced radio activity. The spectral index distributions of EGRET-sources and candidates differ with a significance of 2.1σ. Indeed, for EGRET-sources an inverted spectrum seems to be more probable. This might also be true for simultaneously observed EGRET-sources.

2. Correlations between simultaneous measurements in the radio and γ-ray regime

Again we use the high-frequency (2.8cm) radio data. We searched for linear correlations with a χ^2-test using scaled data. No convincing linear correlation is found between the γ-ray flux and the radio spectral index, and between the γ-ray flux and the simultaneous 2.8 cm radio flux density. This is in contrast to the results of other authors (Dondi et al. 1995, Padovani et al. 1993, Salamon & Stecker 1994) who found linear correlations between non-simultaneous observations of radio and γ-ray luminosities which should also be visible in terms of flux. Note, however, that flux-limited samples and the strong dependence of the luminosities on the redshift may simulate a linear correlation. An explanation of the non-correlation of simultaneous radio and γ-ray observations may be a time shift between the radio and γ-ray outburst.

It is found that the γ-ray spectrum tends to harden with increasing γ-ray flux. The χ^2-fit of this correlation has a significance of 73%, and a slope of zero can be ruled out with a probability of 10^{-5}.

3. EGRET detection versus radio state

The characterization of the radio state of each source during an EGRET-observation was performed using 5 classes. The probability of an EGRET-detection compared to an EGRET-non-detection at the different radio states indicates that γ-rays are detected preferably when the radio light curve is increasing. FSRQ have been never seen by EGRET when they are in a decreasing radio state.

References

Dondi, L., Ghisellini, G. 1995, *MNRAS* , **273**, 583.
Mücke, A., Pohl, M., Reich, P., et al. 1996, Proc. 3rd Compton Symposium, A&AS submitted.
Padovani, P., Ghisellini, G., Fabian, A.C., et al. 1993, *MNRAS*, **260**, L21.
Reich, W., Steppe, H., Schlickeiser, R., et al. 1993, *A&A*, **273**, 65.
Salamon, M.H., Stecker, F.W. 1994, ApJ, **430**, L21.

OPTICAL MONITORING

OF GAMMA-RAY LOUD BLAZARS

C.M. RAITERI[1], G. GHISELLINI[1], M. VILLATA[1],
G. DE FRANCESCO[1], S. BOSIO[2], G. LATINI[2], G. CHIUMIENTO[1],
L. LANTERI[1], G. MASSONE[1], F. RACIOPPI[1], AND R.L. SMART[1]

[1] *Osservatorio Astronomico di Torino*
Strada Osservatorio 20, I-10025 Pino Torinese (TO), Italy
[2] *Istituto di Fisica Generale dell'Università*
Via Pietro Giuria 1, I-10125 Torino, Italy

The observations by the *Compton Gamma Ray Observatory* (CGRO) have shown that highly variable and radio-loud quasars emit a significant fraction of their energy in the γ band. According to the Inverse Compton model, the γ-ray emission is due to upscattering of soft (IR-optical-UV) photons by high energy particles. Optical monitoring is thus of great value in providing information on the mechanisms that rule the production of the seed photons for the γ-ray radiation and on the γ-ray emission itself. In particular, detection of variability correlations between optical and γ-ray emissions would be a crucial test for the theoretical predictions.

For the above reasons, we have started an optical monitoring campaign of γ-ray loud blazars in Torino (Italy) since November 1994 (Villata *et al.*, 1995). Our list includes 30 objects. In order to have a better temporal coverage, we entered collaborations with other monitoring groups on some sources [e.g. Massaro *et al.* (1995) on 0422+004; Ghisellini *et al.* (1995) and Latini *et al.* (these proceedings) on 0716+714; Sillanpää *et al.* (1995) and Takalo *et al.* (these proceedings) on 0851+202, as part of the OJ-94 Project, where we are involved also for 0219+428 and 0235+164]. Data have been taken with the Torino 1.05 m R.E.O.S.C. telescope equipped with a 1242 × 1152 pixels CCD camera and standard Johnson *UBV* and Cousins *RI* filters. Observations have been done in the *R* and *B* bands.

Most of the sources show noticeable optical variations on both long and short time scales, and for some of them we could detect also intranight

287

R. Ekers et al. (eds.), Extragalactic Radio Sources, 287–288.
© 1996 *IAU.*

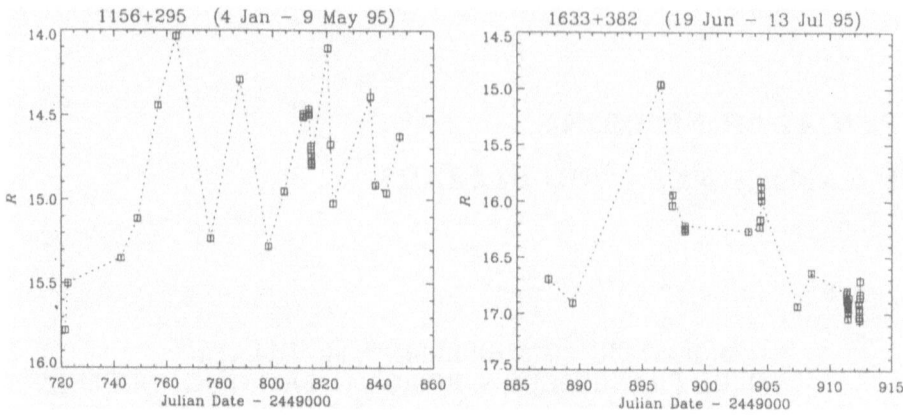

Figure 1. Light curves in the R band for the two quasars 1156+295 and 1633+382

variability.

Figure 1 shows part of the light curves in the R band for 1156+295 and 1633+382. The source magnitude is given as the average value between the magnitudes obtained with respect to two of the three reference stars that we chose in the source field. The optical emission of the highly polarized quasar 1156+295 (4C 29.45) has been rapidly and widely changing during all our monitoring period ($\Delta R_{max} = 1.75, \Delta B_{max} = 1.81$), with the steepest decrease of about 0.9 mag (R) in two days (JD $\sim$ 2449821); a peak was detected at JD = 2449836, during the CGRO observation period (18 Apr – 2 May 95), when a very high γ-ray emission (maybe one order of magnitude greater than the previous detection of $6.3\,10^{-7}\,cm^{-2}\,s^{-1}$ in 1991) was registered. Consistent intranight variability has been found for the quasar 1633+382 (4C 38.41); on JD = 2449904 a variation $\Delta R = 0.34$ mag was detected in half an hour. After a period of quiet emission, this source reached its historical maximum ($R = 14.96 \pm 0.03$) on June 27, 1995 (Bosio *et al.*, 1995); observations in the radio band are strongly encouraged, since a time-delayed correlated peak could be detected.

References

Bosio, S., De Francesco, G., Ghisellini, G. *et al.* (1995) 4C 38.41, *IAU Circ.*, no. **6183**

Ghisellini, G., Villata, M., Raiteri, C.M. *et al.* (1995), *A&A*, (submitted)

Latini, G., Bosio, S., De Francesco, G. *et al.* (these proceedings) Optical Monitoring of the Gamma-ray Loud BL Lac Object S5 0716+714

Massaro, E., Nesci, R., Trevese, D. *et al.* (1995), *A&A*, (submitted)

Sillanpää, A., Takalo, L.O., Lehto, H.J. *et al.* (1995) Confirmation of the 12 Year Optical Outburst Cycle in Blazar OJ 287, *A&A*, (in press)

Takalo, L.O., Sillanpää, A., Pursimo, T. *et al.* (these proceedings) Variability Characteristics of Blazar OJ 287

Villata, M., Raiteri, C.M., Ghisellini, G. *et al.* (1995), *A&AS*, (submitted)

OPTICAL MONITORING OF THE GAMMA–RAY LOUD BL LAC OBJECT S5 0716+714

G. LATINI, S. BOSIO, G. DE FRANCESCO, M. VILLATA
G. GHISELLINI AND C.M. RAITERI
Osservatorio Astronomico di Torino, Pino Torinese, Italy

G. TOSTI AND M. FIORUCCI
Università di Perugia, Via Pascoli, I-06100 Perugia, Italy

AND

M. MAESANO, E. MASSARO, F. MONTAGNI AND R. NESCI
Istituto Astronomico, Università La Sapienza, Roma, Italy

1. Observations

We monitored the γ–ray loud BL Lac object S5 0716+714 between November 1994 and April 1995 (J.D. 2449671–2449838), from the Observatories of Torino, Perugia and Roma. The source was always active in the R and B bands, showing a particularly high optical state during the period of the CGRO pointing (February 14–28, 1995). The lightcurve in the R band is presented in Fig. 1. The complete set of observations is discussed in Ghisellini et al. (1995).

Fast variations are superimposed to longer trends. For the few well sampled peaks in the light curves, the rising and decaying time scales are equal, without 'plateau'. Moreover, the spectral index seems to respond to short time variations of the source, and is rather insensitive to long trends. There is an almost one–to–one correspondence between the changes in relative flux and those of the spectral index α_{BR}: the spectrum is flatter during local maxima of the flux.

2. Synchrotron Self-Compton model: expected γ–ray emission

A simple homogeneous, one–zone SSC model can fit the data, with $B = 0.8$ G, $\delta = 13$ and $R = 3 \times 10^{16}$ cm. The relativistic electron distribution is a

R. Ekers et al. (eds.), Extragalactic Radio Sources, 289–290.
© 1996 IAU.

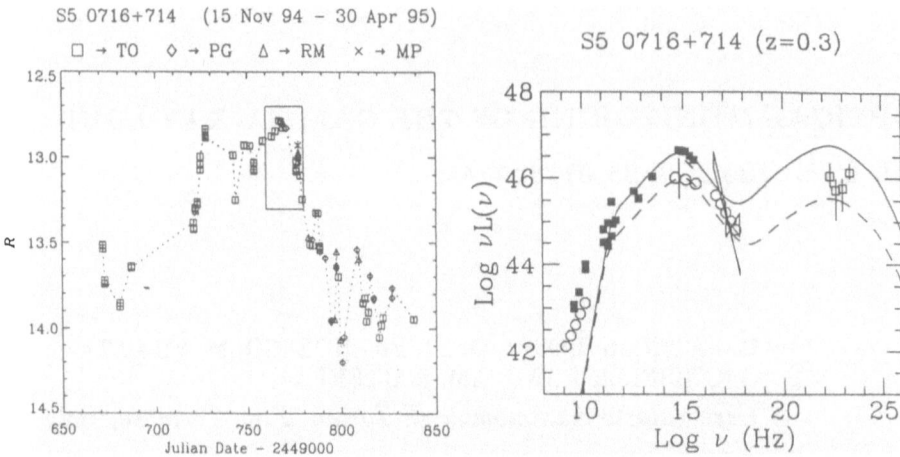

Figure 1. (*left*) Lightcurve in the R band of 0716+714: the box shows the observations simultaneous with the GRO pointing. (*right*) Overall spectrum (assuming a redshift $z = 0.3$, Schalinski et al. 1992) of S5 0716+714 from radio to γ–rays. The solid symbols in the optical (B, V and R) and UV correspond to data simultaneous (Feb. 27 and Mar. 1, 1995) with the period of the GRO pointing, whose data are not public yet. Empty symbols refer to the spectrum published by Wagner et al. (1995). Solid and dashed lines refer to an homogeneous SSC model (see text).

broken power law with the break at $\gamma_b = 5 \times 10^3$, producing a break of the synchrotron spectrum at $\sim 10^{15}$ Hz. The resulting spectrum is shown in Fig. 1 as the solid line. The low optical state has been fitted by changing only the normalization of the electron distribution by a factor 4 (dashed line). This results in a decrease of the synchrotron luminosity of the same factor, while the Compton luminosity decreases quadratically. Note that:

In this one–zone model, the emission is predicted to vary simultaneously at all frequencies between the mm band up to the GeV band, although the amplitude of the synchrotron variations should be smaller than the Compton ones.

We predict that EGRET, in the observation of February 1995, should have observed the source in a very bright state, a factor between 3 and 10 above the flux level of the previous detection.

The plotted γ–ray data are not simultaneous with the low optical state.

References

Ghisellini G. et al., 1995, in preparation
Quirrenbach A. et al., 1991, *ApJ*, **372**, L71
Schalinski C.J. et al., 1992, in Variability in Blasars, eds. E. Valtaoja & M. Valtonen, (CUP), p. 221
von Montigny C. et al., 1995, *ApJ*, **440**, 525
Wagner S. et al., 1995, *AJ*, in press

THE HIGHEST ENERGY COSMIC RAYS

COLIN A. NORMAN

*Dept. of Physics and Astronomy – Johns Hopkins University
and
Space Telescope Science Institute – 3700 San Martin Drive
Baltimore, MD 21218*

Abstract. The current data on the highest energy cosmic rays (UHECRs) is discussed and an understanding of the origin of these particles is reviewed. New and proposed facilities for measurement of UHECRs, neutrinos and γ-rays can interestingly and significantly constrain the physics of the source origin. Cosmic magnetic field strengths are the most uncertain physical parameter.

1. Introduction

Great new data on UHECRs from the Fly's Eye air shower array experiment indicates that at about $10^{18.5}$eV there is both a change in the slope of the energy spectrum and also a change in the composition of the incoming high energy particles from heavy (Fe) to light (p) with increasing energy (Bird et al., 1995, Gaisser 1993). This naively looks like the signature of the extragalactic component accelerated from essentially primordial material and after detailed analysis (Norman, Melrose and Achterberg 1995, hereafter NMA) we argue that this is a solid conclusion. However, recent data from the Akeno group on the muonic component ($\gtrsim$ GeV) for UHECR generated events does not find the expected decrease in the muon to electron ratio as the composition of UHECRs changes from Iron to protons, possibly indicating a smaller composition change than that inferred from the Fly's Eye experiment.

Previous theories of UHECRs have fallen into two classes: (1) Extragalactic Shocks pioneered by Cavallo (1978) and developed in detail by Axford (cf. Axford 1991) and Biermann and Stanev and collaborators (Rachen

291

R. Ekers et al. (eds.), Extragalactic Radio Sources, 291–296.

et al., 1993, Stanev et al., 1995); and (2) Active Galactic Nuclei most recently elegantly discussed by Protheroe and Szabo (1992) and Szabo and Protheroe (1994) and Stecker, Done, Salamon and Sommers (1991). We show here that the radiation field in AGN limits the maximum energy and the propagation length for photopion production to values that exclude AGNs as source for UHECRs. New ideas about the origin of UHECRs include large scale shocks associated with cosmic structure formation (NMA, Kang, Ryu and Jones 1995) and cosmological gamma-ray bursts (Vietri, 1995, Waxman 1995, Milgrom and Usov 1995a,b).

General constraints on the origin of UHECRs were presented in the influential papers of Hillas (1984) and Hill and Schramm (1985) and more recently by Sigl, Schramm and Bhattacharjee (1995) and NMA. Detailed analysis of the current data of the energy spectrum concerning the existing of the GZK cut-off or not and also the implications for a potential new source of UHECRs due to physical phenomena associated with, for example, topological defects associated with cosmic phase transitions have been analyzed by Yoshida and Teshima (1993), Bhattacharjee, Hill and Schramm (1992), Sigl, Lee, Schramm and Bhattacharjee (1995) and Waxman (1995).

One remark about symbolic notation in this paper. Where symbols have thir common or obvious meanings they are not defined due to space constraints. If the reader faces continuing difficulties then they should refer to Norman, Melrose and Achterberg (1995) or Norman and Lacey (1996).

2. Metallicity–Composition Constraints

The protonic component at the highest energies can have two explanations: (1) The source is unenriched or (2) the UHECRs can be accelerated from, say, a plasma of normal, solar metallicity composition and then the heavies can be selectively removed by photo-dissociation into protons on the cosmic background radiation field. Possibility (2) seems to be excluded for the smooth component extending up to $10^{19.5}$eV since at approximately $10^{19.5}$eV the proton energy loss rate is equal to the photodissociation rate for heavies and there is no obvious break in the energy or composition spectrum. However, this metallicity composition argument can be subject to increasingly detailed check as more data on UHECRs becomes available. We conclude that possibility (1) is the most reasonable and consequently rule out an origin of UHECRs in rich clusters, the metal rich environments of galaxies and other interesting models such as the metal-rich environment of colliding galaxies resulting in starbursts (Cesarsky and Ptuskin 1992).

The sources must be local ($z \lesssim 0.3$) since there is a smooth spectrum up to $10^{19.5}$eV and protons cannot travel more than $\lesssim 1$ Gpc due to pair production and pion production on the cosmic radiation background. Extended halos around galaxies have often been invoked but the type of extended halo envisioned by Jokipii and Morfill (1985) has not been observed in general (cf. Wielebinski et al., 1995) although there are interesting special cases such as NGC 4631(Donahue, Aldering and Stocke 1995).

3. UHECRs Cannot Originate in AGN

At first, AGNs seem promising acceleration sites for UHECRs since if one assumes that there is rough equipartition between radiation energy density and magnetic field energy density in AGN (Rees 1987) the maximum energy is $E_{max} \sim 3 \times 10^{19}$ β_{-1} Z L_{46} eV with standard meaning for the symbols (cf. NMA). Generally, AGNs have an intense inner radiation field well parameterized by the compactness parameter $l \sim (\sigma_T L)/(4\pi m_e c^3 R)$ which is generally $l \gtrsim 1$ (Done and Fabian 1989). Then the optical depth for photopion production in IR photons is $\sim 10^3 l$ and the corresponding energy threshold is $\sim 10^{16}$eV. Therefore, UHECRs above 10^{16}eV do not escape AGN. In addition, the isotopic component within the BLR/warped disk/obscuring torus region is of order 10% of the predominately radial photon flow which inhibits use of any models with small angle between photon and particle motion (NMA).

A mention of neutrino astronomy in passing. Although models of neutrino production in AGN are rather uncertain there is a significant prospect that neutrino telescopes will at least place significant constraints on conditions in AGN. The Frejus experiment can already rule out certain classes of models as discussed by Mannhein (1995) for radio quiet AGN. It is not really important what the specifics of these rather uncertain models are, but what is important is that AMANDA, DUMAND, NESTOR and other neutrino facilities in various stages of implementation can easily increase the Frejus limit by orders of magnitude giving potentially quite significant constraints on neutrino producing regions in AGNs (cf. Gaisser, Halzen and Stanev 1995).

4. UHECRs Can Originate in Extragalactic Shocks

The maximum energy achievable in a standard Sedov-Taylor blast wave is $\sim 5\times10^{19}Z$ B_{-6} $E_{61}^{2/5}$ $\rho_{-31}^{-2/5}$ $t_8^{-1/5}$ ln R and in a wind blowing out into the IGM $\sim 10^{20}Z$ B_{-6} $L_{46}^{2/5}$ $\rho_{-31}^{-2/5}$ $t_8^{1/5}$eV. The principal uncertainty here is the magnetic field strength and we have assumed here that microgauss field strengths can be self-generated in such shocks. Primordial field strengths are probably too low to be useful as discussed later.

FRII class radio sources have been frequently discussed as sources of UHECRs (Rachen et al., 1993) and there are places in a typical FRII source where UHECRs can be produced–the hot spot, and the extended cocoon shock. Analysis in NMA gives maximum energies of $\sim 10^{20}\ B_{-4}\ \beta_j$ $R_{j,1\,\mathrm{kpc}}$ eV, and $\sim\ 5\times 10^{19} Z\ B_{-6}\ L_{46}^{1/2}\ t_8^{1/2}\ n_{-4}^{-1/2}\ R_{10\,\mathrm{kpc}}$ eV respectively. These work well, however, FRII sources are generally found at $z \gtrsim 0.3$ and are quite rare within the allowed range of distances for UHECR. More detailed statistical analysis may eventually rule out FRIIs as a plausible source.

5. UHECRs Can Originate from Cosmic Structure Formation

We have analyzed the shocks resulting from cosmic structure formation at the current epoch using a typical CDM model with a typical scale $R_S \sim (2GM/\Omega\delta H^2)^{\frac{1}{3}}$ and shock velocity $V_s \sim (2GM/R_s)^{1/2}$ where $\delta \sim 178$ is the cosmical over density parameter after collapse and virialization of a perturbation that has grown to become non-linear. Having a typical mass scale of $M \sim 10^{15}\,(0.4/b(1+z))^{\frac{6}{n+3}}\ M_\odot$ where n is the index of the intial fluctuation spectrum ($|\delta_k|^2 \propto k^n$) we find a maximum energy of $\sim 3\times 10^{18}\ Z\ B_{-6}\ \Omega^{-1/6}\ h^{-1/3}(1+z)^{-5/2}$eV where again we have assumed self-generated magnetic field strengths of order micro-gauss—consistent with values observed in clusters. Independent of CDM models, typical, collapsing, pancake–like objects give similar estimates of E_{max}. The fluxes expected from these structures are consistent with the observed fluxes of UHECRs. Similar work has been done in this context by Kang, Ryu and Jones (1996).

It is most interesting then that Stanev et al., (1995) have found tentative evidence for a correlation of the arrival direction of the highest energy cosmic rays with very roughly the direction of the super galactic plane. Many selection effects come into play and the exact statistical significance is rather uncertain but it is a most interesting suggestion that follows earlier work of Wdowczyk and Wolfendale (1979) and Giler, Wdowczyk and Wolfendale (1980).

Spectral correlations that have very uncertain significance but again are interesting have been suggested by Stanev et al., (1995) between the arrival directions of the highest energy events and 3C134 (with unknown redshift), Cygnus-A, NGC 315, 3C31, and M87.

6. Gamma Ray Bursts and UHECRs

A few papers have been written recently on the possible association of the source of UHECRs with gamma ray burst sources of cosmological origin (Waxman 1995, Vietri 1995, Milgram and Usov 1995a,b). After brushing aside the initial reaction that this is merely equating the two most unknown sources in high-energy astrophysics, one realizes that the proposal is based on two interesting coincidences of the flux and energy in UHECRs calculated from the standard Mezaros and Rees (1994) model. Relativistic Fireball models for cosmological gamma ray bursts are based on a millisecond rise time, corresponding to a scale of $\sim 10^7$ cm and a cosmological burst luminosity of $\sim 10^{51}$ erg s^{-1} over the burst lifetime of order of seconds. The resulting hyper-relativistic shocks can be used to accelerate particles to energies of order $\sim 10^{20}$ n $\Theta^{-5/3}$eV where n is the density of the ambient ISM and Θ is the beaming semi-opening angle (Vietri 1995). This is standard ping-pong shock acceleration achievable in just a few hits with a spectrum tending toward E^{-2}. For a high efficiency of production of UHECRs such that, not uncommonly, the energy density in UHECRs is a significant fraction of the ram pressure of the shock, the calculated flux in UHECRs is quite consistent with the observed flux of $\sim 3 \times 10^{-8}$ Mpc^{-3} yr^{-1}.

UHECRs are scattered through $\sim 5^0 Z$ E_{20}^{-1} $R_{10\,\mathrm{Mpc}}$ B_{-9} by a coherent magnetic field and $\sim 1^0 Z$ E_{20}^{-1} $R_{10\,\mathrm{Mpc}}^{1/2}$ $l_{coherence,1\,\mathrm{Mpc}}^{1/2}$ B_{-9} for an incoherent structure that they diffuse through. Thus the time delay for a pulse from a GRB at ~ 100 Mpc is greater than $\gtrsim 10^6$ yr. Therefore no close temporal association should be observed.

7. Events Above 100 EeV: Is There New Physics?

The events above 100 EeV are rare and can arise from special local sources. They may even be made of Fe since there is nothing known about their composition. If one takes the point of view that they are a hard component above the GZK cut-off then a question of new physics arises.

A conventional second phase statistical acceleration model calculated by NMA was shown to fail because the acceleration at known sources was insufficient to overcome the losses incurred traveling between acceleration sites. I strongly believe that this type of model could work and only currently fails due to our lack of knowledge of the structure and energy sources in the KM. In fact, the IGM probably has a structure as rich as the ISM. I expect the IGM is pervaded by a rich system of winds, explosions, shocks, structure formation etc. resulting from QSOs, galaxy formation, star burst galaxies, AGNs etc. This structure could produce UHECR spectrum without difficulty. This is probably the most important, albeit speculative, point of this lecture.

An alternative and interesting point of view is that the events above the GZK cut-off herald new physics (Sigl et al., 1995). There is a gap in the data of half a decade in energy between the two highest events and the rest of the data. Future experiments will establish if it is significant. If the gap persists then new physics may be involved. Top-down scenarios where the UHECRs are products of the flat injection spectrum of decaying objects associated with the GUT epoch. Foregoing sentence needs to be corrected for syntax Topological defects have been proposed as a possibility here (Bhattacharjee et al., 1992). Waxman (1995) produces a spectrum in which it does indeed look like the two highest energy events are special.

It is a pleasure to thank my collaborators in the UHECR work, A. Achterberg and D. Melrose for their advice, encouragement, and very stimulating scientific discussions throughout the course of this ongoing work. There were also very useful and interesting discussions on these and related subjects with T. Gaisser, C. Lacey, R. Protheroe, M. Rees, J. Stanev, M. Vietri and A. Waxman.

References

Axford, W. I. 1991, *Ap. J. Suppl.*, **90**, 937

Bhattacharjee, P., Hill, C. T., and Schramm, D. N. 1992, *Phys. Rev. Lett*, **69**, 567

Bird, D. J. et al., 1993, *Phys. Rev. Lett.*, **21**, 3401

Cavallo, G 1978, *A&A*, **65**, 415

Cesarsky, C. and Ptuskin, V. 1992, *Nucl. Phys. B.(Proc. Suppl.)*, **28B**

Donahue, M., Aldering, G. and Stocke, J.T. 1995, *Ap. J.*, **450**, L45

Done, C. and Fabian, A. 1989, **240**, 81

Gaisser, T., Halzer, F., and Stanev, T. 1995, *Phys. Reports*, **258**, 174

Gaisser, T. et al., 1993, *Phys. Rev. P*, **47**, 1919

Giler, M., Wdowczyk, J., and Wolfendale, A. W. 1980, *J. Phys. G. Nucl. Phys.*, **6**, 1561

Hill, C. T. and Schramm, D. N. 1985, *Phys. Rev. D*, **31**, 564

Hillas, A. M. 1984, *Ann. Rev. A. A.*, **22**, 425

Jokipii, J. R. and Morfill, G. E. 1985, *Ap. J. Lett*, **290**, L1

Kang, H., Ryu, D., and Jones, T. W. 1996, *Ap. J. Lett*, in press

Mannhein, K. *Astroparticle Physics*, **3**, 295

Mezaros, P. and Rees, M. J. 1994, *M.N.R.A.S.*, **269**, L41

Milgrom, M. and Usov, V. 1995a, preprint

Milgrom, M. and Usov, V. 1995b, preprint

Norman, C. and Lacey, C. 1995, *Ap. J.*, in preparation

Norman, C. A., Melrose, D. B., and Achterberg, A. 1995, *Ap. J.*, November 20, in press (NMA)

Protheroe, R. J. and Szabo, A. P. 1992, *Phys. Rev. Lett.*, **69**, 2885

Rachen, G. P., Stanev, T., and Biermann, P. L. 1993, *Astro. Ap.*, **273**, 377

Rees, M. J. 1987, *M.N.R.A.S.*, **228**, 47

Sigl, G., Lee, S., Schramm, D. N., and Bhattacharjee, P. 1995, *Science*, in press

Sigl, G., Schramm, D. N., and Bhattacharjee, P. 1995, *Astroparticle Physics*, in press

Stanev, T., Biermann, P. L., Lloyd-Evans, J., Rachen, J. P., and Watson, A. 1995, *Phys. Rev. Lett*, in press

Stecker, F. W., Done, C., Salamon, M. H., and Sommers, P. 1991, *Phys. Rev. Lett.*, **66**, 2697

Szabo, A. P. and Protheroe, R.J. 1994, *Astroparticle Physics*, **2**, 375

Vietri, M. 1995, *Ap. J.*, in press

Waxman, E. 1995, *Ap. J. Lett.*, in press

Waxman, E. 1995, *Phys. Rev. Lett*, **75**, 386

RELATION BETWEEN RADIO AND OTHER WAVELENGHTS
(*DISCUSSION*)

Discussion of the paper presented by <u>CRANE</u> (p. 201)

Perley: The 2 cm VLA radio data of the western hotspot of Pictor A with comparable resolution, show an excellent correspondence between the optical and radio emission.

Crane: I would dearly love to see a comparison with HST data!

Jones D.L.: Have you looked for optical emission further out along the NGC 6251 jet, or just adjacent to the nucleus?

Crane: Unfortunately, the field of view of the FOC is quite limited. So the answer is, no I have not looked further out.

Tsvetanov: Just a brief comment on NGC 6251. Lynds and O'Neil (1994, BAAS) have shown that there is a nuclear disk in this galaxy roughly perpendicular to the radio jet axis, much like in the case of NGC 4261. I have a feeling that what you see is related to that disk, possible some edge effect.

Crane: This may be the case, but the disk you mention is considerably larger than what is seen here.

Discussion of the paper presented by <u>WILSON</u> (p. 205)

Bicknell: Why don't you see the other side of the helix in the NGC 4258 jets?

Wilson: Perhaps because of dust obscuration.

di Serego Alighieri: Comment: The lack of will defined ionization cones in radio-loud objects may also be due to wider opening angles, and stronger radiation pressure, which confused the picture in these objects.

Wilson: I agree that ionizing radiation may escape anisotropically from nuclei if radio galaxies, and that there are several reasons why associ-

R. Ekers et al. (eds.), Extragalactic Radio Sources, 297–304.
© 1996 *IAU*.

ated ionization cones may not be seen. If the cones have similar linear extents to those in Seyferts, they would be hard to resolve spatially at the much longer distances of FRII radio galaxies. Also, it may be that the interstellar gas in ellipticals is mostly hot and not susceptible to photoionization by the nucleus.

di Serego Alighieri: The evidence for non-ionized gas outside the ionization cones would nicely support the fact that these cones are ionization bounded. Is there any such evidence?

Wilson: Prieto and Freudling (1993) have mapped the 21cm HI emission of NGC 5252. They find that the neutral hydrogen tends to avoid the ionization cones and that there is evidence for a continuity of velocity from the ionized gas within the cones to the neutral gas outside them. Their results support the idea that pre-existing neutral gas is ionized by an anisotropic radiation field, i.e. the cones are ionization-bounded.

Macchetto: In the sample of Seyfert 2 galaxies we have observed with HST (see poster by Capetti et al), we find that for the Seyfert 2s with weak jets i.e. where there are lobes well within the galaxy, we appear to need an additional ionizing source to explain the increase in ionization potential with distance from the nucleus and the high-ionization ([OIII]/[OII]) seen in the bright filaments. We believe that in these cases, shocks are the best explanation as they are expected, they reproduce the observed morphology and are the simplest explanation.

Wilson: I think the issue of whether the gas in the NLR and ENLR of Seyfert 2 galaxies is photoionized by a compact nuclear source or by radiation from high velocity shocks in the NLR is still unresolved. The main process may differ in different galaxies - e.g. Seyfert 2's with hidden Seyfert 1's should have strong, compact nuclear ionizing sources, while in Seyfert 2's without Seyfert 1 nuclei (if such objects exist), photoionizing shocks in the NLR may dominate. Of course, the compact nuclear ionizing source may itself result from high velocity shocks.

Discussion of the paper presented by <u>MORAN</u> (p. 211)

Yu Zhi-Yao: Could you find the effect in OH megamaser?

Moran: The effect has not been found in OH megamaser because of the different resolution.

Wilson: As you have shown, it is likely that the maser features near systemic velocity result from amplification of radio continuum photons from a central radio source. What do you think is the source of photons for the high velocity "satellite lines"?

Moran: Of course, it's just a hypothesis that the systemic features amplify a central continuum source, since we have not been able to detect this source directly. We assume that the high velocity features are unsaturated and amplify their own spontaneous emission.

Gelderman: What is the physical reason why the masers are seen only either along the line of sight or at a right angle to the line of sight.

Moran: We are most likely to see masers along directions through the disk where the coherent amplification path length is greatest. This path length is the distance over which the velocity changes by less than the thermal linewidth. For a keplerian velocity field, in a disk viewed edge on, the velocity gradient is zero, and hence the path length maximum, along the direction to the center of the disk and along the disk diameter that is perpendicular to the line of sight.

Discussion of the paper presented by <u>TADHUNTER</u> (p. 217)

Wall: These objects are from flux-limited samples; hence by 'low redshift' objects you mean objects of low radio luminosity, and by high redshift you mean high-luminosity, do you not?

Tadhunter: Yes, the high redshift objects selected from flux-limited samples inevitably have higher radio luminosities and jet powers. I believe that many of the observed differences between high and low redshift objects could be a consequence of this fact, although there remains the possibility that the evolution in the ISM is also important.

Leahy: It seems that one can identify jet-cloud interactions quite easily by their dramatic effect on the radio structure – certainly in 3C171. High redshift objects like 3C294 may be analogous.

Tadhunter: The radio structures observed in 3C171 are certainly unusual in their degree of distortion, but there are also objects like PKS 2250-41 which have relatively normal double structures despite showing compelling evidence for jet/cloud interactions.

Laing: If you attribute most of the aligned continuum in high-z galaxies to jet/shock effects (as in your hybrid model), do you still get the observed polarization, or is there too much dilution of the polarized component?

Tadhunter: In fact, the dilution by the nebular continuum helps to explain why the observed polarization is less than predicted for pure dust scattered light. The uncertainties in the estimated nebular continuum and off-nuclear polarization are such that it is too early to say whether there is too much dilution.

Wilson: Could you tell us the ratios of $[OIII]\lambda4363/[OIII]\lambda5007$ (or alternatively the electron temperatures) that you observe? I think its useful to bear in mind that 4363/5007 ratios at least as high as 0.02 are expected in the matter-bounded plus ionization-bounded photoionization models that Luc Binette described this morning.

Tadhunter: The [OIII] 4363/(5007 + 4959) ratio falls in the range 0.01–0.03 for the EELR, corresponding to electron temperatures in the range 13,000–22,000k (low density limit). There is already good evidence that matter-bounded components are important in some aspects (eg. Tadhunter et al. 1988, MNRAS, 235, 403, PKS 2152-69), but for low ionization EELR there would be a problem with explaining the relatively weak HeII(4686) and the relatively strong [OI](6300) with a model which included a significant matter-bounded component.

Discussion of the paper presented by __MORGANTI__ (p. 227)

Bicknell: The difference in velocity widths of the high excitation over low excitation lines has a natural exploration in terms of shock models. [OIII] for example would arise from the shock precursor whereas the [OI] would come from the more turbulent region behind the shock. The radiative cooling zone would be too small for it to be resolvable.

Morganti: I agree that the kinematic differentiation is difficult to explain in terms of control source photoionization but could be entirely consistent with ionization by fast shocks.

Urry: Your [OIII] image looks very similar to the ionization cones in radio-quiet AGN shown earlier this morning by Wilson and Tsvetanov, for example, and the axis is well aligned with the radio structure as you pointed out – is this the first example of an ionization cone (two-sided, no less!) in a radio-loud object? How does it compare to the ionization cones in radio-quiet AGN?

Morganti: Actually, I have not really looked at it as an ionization cone. I think that the nice coincidence between the [OIII] emission and the radio continuum (not very common in radio galaxies) looks more like the result of an interaction, therefore supporting the important role of jet plasma, more than photoionization, in ionizing the gas.

Discussion of the paper presented by <u>BINETTE</u> (p. 232)

Simkin: How much neutral material will you predict. We find (Simkin & Collaef -poster) that 349-27 has very little HII. Is this consistent with your model?

Binette: I estimate the column density of neutral HI from the ionization bounded component to be less or equal to $3 \times 10^{20} cm^{-2}$. I would need the area of the emitting clouds to convert this into a mass of HI.

Macchetto: In your model the assumption that the "ionization bound" clouds are always behind "matter bound" clouds appears to be contrived. In a natural, real life situation, I would expect a mixing of the two types at all spatial distances. What would be the spectral characteristics in this case?

Binette: We are mixing the two types at all distances. However, what we claim is that the low excitation clouds are such because they see a modified ionizing continuum and not the result of arbitrarily reducing the ionization parameter. This is also the only way to get the low HeII/Hβ observed in some objects. in other words, we propose that whenever a cloud sees a pure power law, this leads to a matter bounded component with high excitation spectrum. This whatever the distance from the source.

Discussion of the paper presented by <u>LEDLOW</u> (p. 238)

Gavazzi: Is there any evidence for a correlation between radio loudness and the presence of some nuclear enhancement in the optical?

Ledlow: No. Comparisons of luminosities measured for small radii and larger apertures show no enhancement in the nuclear luminosities for radio-loud vs. radio-quiet ellipticals.

Laing: What do you think is the source of the discrepancy between your results and those of Disney et al., who claimed that radio galaxies are rounder and redder than "normal" ellipticals, and live in richer environments? Are there real differences (eg in cluster/non-cluster location) or is it that you have been more careful with your control sample?

Ledlow: The analysis from CCD data may be more accurate, but I suspect that there may be some intrinsic differences in the galaxies in and out of rich clusters. This has been suggested by de Carvalho and Djorgovski (1992) and others. However, if true, these differences don't appear to affect the observed radio luminosity function. Upon reflection, I don't think there is a problem with Disney et al.'s control sample.

Menon: Your results are very similar to my results (Menon, 1992 M.N) concerning radio sources in compact groups of galaxies except for one difference. I found a strong correlation between nearest neighbour distance and the occurrence of radio emission in the brightest galaxy in the group. This is understandable since the crossing times in groups are one or two orders of magnitude smaller and velocity dispersions are about 200Km/sec.

Ledlow: I agree that the differences are most likely related to the differences in velocity dispersion and the dynamics of poor group systems.

Discussion of the paper presented by IMPEY (p. 281)

Pohl: Egret sources identified with flat-spectrum radio quasars are a strongly flux-limited sample, both in γ-rays and in the radio regime. The available flux range in both regimes is around one decade, similar to the variability range in the γ-ray range. Comparing simultaneous and simultaneous data there is no direct relation in radio and γ-ray flux; therefore any estimate of the γ-ray background as superposition of AGN is highly uncertain.

Impey: The analysis assumes a factor of 3-4 rms variability in both radio and γ-ray flux, so this uncertainty is incorporated into the analysis. There is formally a correlation for the EGRET 1 Jy sources at the 98% confidence level; the deduced contribution to the γ-ray background of course has a large error bar.

Perlman: To follow up on the previous question: There have now been two detections by EGRET of <1 Jy radio sources MKN421 and 2155-304. There are also a large number (>40) unidentified EGRET sources. These are likely all <1 Jy sources as well.

Impey: That's basically right

Discussion of the paper presented by <u>M.F. ALLER</u> (p. 283)

Ekers: Is the correlation with polarization variability better than with I?

Aller: In a few cases individual events are better resolved in polarized flux than in total flux, but the difficulty in establishing a meaningful correlation is due to the insufficient sampling in the γ-ray region.

Ekers: Is it possible that the polarization variability during "quiescent" times is due to variations in the depolarizing plasma which may be affected by the γ-ray activity?

Aller: No. We have not seen any evidence for time-variable depolarization.

Discussion of the paper presented by <u>NORMAN</u> (p. 291)

Begelman: Granted that the EeV baryons from cosmological distances cannot reach us, wouldn't the distant sources give us a high-energy neutrino background that we might be able to detect?

Norman: Yes, the background limits due to neutrinos from such processes may give interesting limits.

Vietri: The proton mean free path against photo-pion destruction above 10^{20}eV is less than 20kpc combined with the near isotropy of the UHECRs detection of arrival, and their slight concentration toward the Supergalactic plane, indicate that only galaxies can produce UHECRs. Normal galaxies do not produce UHECRs, but they contain GRBs, so I see this as (slight) evidence for GRBs.

Norman: Yes, this is an interesting line of argument in favor of the GRB / UHECR hypothesis.

Gopal-Krishna: Since propagation losses on high energy cosmic rays are so severe, their detection would be biased heavily in favor of the one that travelled straight toward us. Perhaps we need not be so pessimistic about spatial correlation between the detached events and their sources?

Norman: Yes, this is reasonable. The Auger array will provide sufficiently high quality data to confirm this or otherwise.

Rudnick: There is considerable uncertainty in the physical parameters of extragalactic sources. Which are the critical parameters for us to pin down to evaluate their role in high energy cosmic ray acceleration?

Norman: The crucial parameter is the magnetic field strength.

Leahy: Shock acceleration requires a suprathermal component of the initial particle distribution which may be hard to arrange for the unprocessed material you need, except for the case of shocks within radio sources, where the AGN has provided the seed population. After all, these are the only systems you discussed which show synchrotron radiation.

Norman: I assume in our calculation that the injection problem solves itself.

Ekers: You need the mean free path to limit the volume of space for the sources.

Norman: There could be a hard spectrum source beyond the limit as discussed by Sigl et al. 1995.

Woltjer: Even if one believes the composition change at 10^{10}eV, isn't it true that one knows nothing about the composition of the 10^{21}eV events? Moreover, since any cosmic ray acceleration scheme needs injection processes which may be composition dependent can we really be confident that the source region must be metal deficient, even if the CR were?

Norman: First, I agree the metallicity-composition argument is not watertight but it seems interesting and subject to increasingly detailed experimental test with, for example, the Auger array. Second, I also agree the very highest energy events have at present no constraints at all on their composition.

RADIO SOURCES IN CLUSTERS OF GALAXIES

F.N. OWEN
NRAO
Socorro, NM, USA

1. Introduction

Since the Albuquerque IAU meeting, most of the advances in the understanding of radio sources in clusters has come from combining radio observations with x-ray and optical data. The x-ray images from *Einstein* and
ROSAT have been particularly important because they have allowed us to
see the external medium with which the radio sources co-exist and interact.
I will cover three examples of such work in this review.

2. Properties of Radio Sources in and out of Rich Clusters

Much of the work reported in this section confirms the results of the studies of the Bologna group reported by Fanti (1984) but with much larger
samples. Two recent Phd theses by Unewisse (1993) and Ledlow (1994)
have given us large samples, in the southern and northern hemispheres
respectively, with which to study the properties in rich clusters.

One important question is how do cluster properties affect the statistical
properties of radio galaxies ? Both Ledlow's and Unewisse's results show
that radio galaxy densities are very high in the centers of rich clusters, more
peaked than the galaxy distribution. Ledlow's results (Ledlow, 1994; Ledlow
& Owen 1995, 1996) allow us to characterize the properties of cluster radio
galaxies both in the radio and optical. When one considers the bivariate
luminosity function, the observed radial distribution seems to be consistent
with the clustering of the giant ellipticals about the cluster center.

His results also show that the probability of a galaxy being a radio
source does not depend on richness, Bautz-Morgan class, or Rood-Sastry
class. Furthermore, the bi-variate, fractional luminosity function of radio
sources in Abell clusters are consistent in shape and amplitude with the
luminosity function of all radio galaxies. Thus, once we take into account

R. Ekers et al. (eds.), Extragalactic Radio Sources, 305–310.

the optical luminosity of the galaxy, with the current evidence there is no apparent difference between galaxies in Abell clusters (the usual standard) and those in the "field". However, the most massive galaxies tend to live in clusters so some extreme phenomena tend to be found in rich clusters.

How could this be ? Models of extended radio emission rely on an external medium to provide the pressure which keeps the sources together. The radio luminosity depends on the magnetic field strength which in turn should depend on the external pressure. Thus the external medium should be critical to the size, luminosity and evolution of a radio source.

One possibility is that all radio galaxies live in some sort of a cluster and we just have missed them. In particular, Abell did not include them in his catalog. In order to test this idea we have used the ROSAT All Sky Survey (RASS) to look for evidence of extended x-ray emission around a sample of all nearby ($z < 0.05$) 25 radio galaxies, not in Abell clusters, chosen from the bright source list of Wall and Peacock (1985) or from the 3CR. In 22 out of 25 cases we detected an extended x-ray source greater or on-the-order of the size of the radio galaxy. The x-ray luminosity of the median source in this sample was four times weaker than than one finds for the median of Abell richness class 0, also using the RASS. Since we find that the median x-ray luminosity for Abell clusters goes like the square of the number of galaxies Abell counted, then the x-ray luminosities and extents are consistent with cluster richness about half of what is typical for richness class 0, or fifty percent less rich than the lower cutoff of Abell's definition.

Thus these results suggest that all radio galaxies, especially the FR I objects found which dominate our sample, live in some sort of cluster. It is just Abell's definition which has been confusing us. This makes it easier to understand how the luminosity functions and other properties could be so similar. This subject needs more study but these initial results are encouraging.

3. Radio and X-ray Emission from Central Cluster Radio Galaxies

However, the very largest galaxies still are found at the centers of rich clusters. These galaxies, often extending hundreds of kiloparsecs in optical light are often to distinguish from the cluster potential and may not be physically distinct from it.

3.1. CENTRAL GALAXIES IN COOLING FLOWS

One class of such objects are galaxies at the centers of the class of clusters called "cooling flows". In these clusters there is a strong peak in the x-rays on the central galaxy which is caused by a decrease in temperature and a corresponding increase in density to preserve quasi-hydrostatic equilibrium. This structure can be simply interpreted as a slow radial inflow which compensates for the radiative energy losses in the cooling process.

However, at the centers of many strong cooling flows are radio sources, sometimes very powerful like Cygnus A and Hydra A and sometimes less luminous, very peculiar distorted sources. In the cases of Cygnus A (Harris, Carilli & Perley 1994) and Perseus A (Böhringer *et al* 1993) comparison of the x-ray and radio images show evidence for clear interaction of the radio emitting plasma with the hot external medium. In both of these cases it appears that the radio plasma has pushed the hot gas out of its path suggesting the radio jet is energetically important in the center of the cooling flow. To these two we add Abell 2199 (Eilek, Owen, and Zhou 1996). This source has for a long time been one of the most peculiar (see, for example Ge & Owen 1993). The new ROSAT HRI image shows that the filamentary radio structure is wrapped around the base of a "unipolar" x-ray structure suggesting a strong interaction, although it is not clear which medium is dominant.

These examples are starting to make a case that the centers of "cooling flow" clusters are not simple, passive, radial flows but must have a complex velocity field. Furthermore, the nuclear outflow in the form of a radio jet seems to be at least at times energetically important. Since the lifetime of the radio emission is thought to be much shorter than the cooling flow, it seems quite possible that this energy input may be of critical importance in causing the phenomena associated with cooling flows.

3.2. WIDE-ANGLE-TAIL RADIO GALAXIES

Another example of the peculiar radio sources associated with dominate central cluster radio galaxies are the Wide-Angle-Tail Sources (WATS). 3C 465 is the prototype. These sources consist of twin jets (or plumes if you prefer) coming from the center of the galaxies and then bending by an angle typically less than 90 degrees. They resemble the tailed sources attached to smaller cluster galaxies, like NGC 1265, which are understood currently in terms of bending due to the motion of the galaxies in the cluster. However, since WATS are attached to central cluster galaxies which should be at rest, or nearly at rest with respect to the cluster, these bent sources have been harder to understand. See Eilek *et al* (1984) for a discussion of possible models.

However, recently ROSAT observations of the x-ray gas associated with these systems have revealed that the x-ray emission on the scale of the radio sources is almost always asymmetric and aligned with the direction toward which the tails are bending (Gomez et al, 1996). These results have been combined with the simulations of merging clusters first performed by Evrard (1990) and extended by others, e.g Schindler & Müller (1993), Roettiger, Burns & Loken (1993), to produce an explanation. In these simulations, which combine n-body and hydrodynamic codes, the changing gravitational potential produces large scale flows of hot gas in the cluster as well as complex x-ray structures like those often seen in WAT clusters. These flows appear to provide a wind flowing past the central galaxies which bend the WATS, although a detailed model of the interaction still needs to be constructed.

Thus in both the cooling flows and the WAT clusters, evidence now exists for strong interactions between the radio source and the hot cluster gas which are important for our understanding of the the the total system.

4. Radio Galaxies and the Butcher-Oemler Effect

The cluster-cluster mergers simulations discussed in the previous section were motivated by cosmological arguments which suggest that such activity should be an ongoing part of the evolution of large scale structure in the universe. Both the common nature of substructure in nearby clusters and more general cosmological arguments suggest that galaxy clustering increases with time. Another possible outcome of this process is changing galaxy evolution with cosmological epoch. This could be due to changes with epoch in the cluster intergalactic medium and/or the probability of galaxy-galaxy mergers.

One possible piece of evidence for such processes affecting galaxy evolution is the systematic blueing of galaxy populations with epoch reported by Butcher and Oemler (1978), that is the Butcher-Oelmer (BO) effect. The blue population they report is supposedly dominated by galaxies that have undergone strong star formation in the last Gyr or so. On this timescale the galaxies have had time to move completely across the cluster and thus the location of the activity which caused the star formation is hard to pin down. Also once one gets away from the core of the cluster it is hard to pick out galaxies from the field which might be part of the activity. Thus if we want to look for star bursts and their cause actually in progress, we need some other signature of the process.

One good indicator should be the centimeter continuum emission at luminosities below 10^{23} W Hz^{-1} (at 20cm, $H_0 = 75$ km s^{-1} Mpc^{-1}) where star formation dominates the galaxy luminosity function. The VLA can reach about 10^{22} W Hz^{-1} at 20cm at a $z \sim 0.4$ in 24 hours integrations and thus can probe star formation to interesting redshifts. Furthermore the VLA primary beam is large enough, about 30 arcmin FWHM, at this wavelength to cover supercluster scales. The optical ID's with galaxies which would be bright enough to be star burst candidates found in such deep surveys are rare enough in a random field to pick out a good sample for further optical study from the background clutter, on supercluster scales.

To date we have studied two clusters with this technique: Abell 2125 (richness=4, $z = 0.25$, blue fraction=0.19) and Abell 2645 (richness=4, $z = 0.25$, blue fraction=0.03). The first results of this study are reported by Dwarakanath & Owen (1996). In addition to the VLA data we have obtained optical spectra and colors at KPNO for a representative fraction of the objects. For Abell 2125, we have observed about 70% of the IDs brighter than about R=19. We find 20 radio galaxies so far in the cluster. For A2645, we have observed all but one of the IDs to a similar magnitude limit and find only four radio galaxies. The limiting radio luminosity is about 10^{22} W Hz^{-1}. Thus there seems to be a large difference between the number of IDs in these two apparently similar clusters. Also the number of radio galaxies associated with Abell 2125 appears greatly to exceed any cluster known nearby.

However, it is not clear that all this radio activity in Abell 2125 is due to star formation. Only about half the IDs and less of the confirmed IDs have the very blue colors and/or the emission-line spectra to resemble nearby, strongly star forming, galaxies. The rest are red galaxies without obvious star formation signatures. They resemble FR I radio galaxies found nearby, although they make be slightly bluer than an uncontaminated old stellar population and we only have spectra covering wavelengths below about 6000 Å in the galaxy rest frame. Thus on current evidence, the excess radio galaxy population seems to be made up of both star forming objects and objects one would guess are FR I-like.

A second interesting result is that the IDs in Abell 2125 are spread over a supercluster scale of at least 6 Mpc, where the VLA primary beam cuts off. The IDs occupy a band running from NE to SW exclusively. Furthermore, the most of the red galaxies are clustered nearer to the cluster center while there is a patch of blue radio galaxies centered about 2 Mpc SW of the central galaxies. This pattern suggests to us that we could be looking at activity related to a cluster-cluster merger and that the Butcher-Oemler effect could be occurring on a much larger (supercluster) scale than the optical observations have so far indicated.

So far these results can only be looked at as suggestive. We need to see if the pattern repeats in more clusters and as a function of redshift. If it does it may indicate that the evolution of large scale structure not only stimulates star formation but also activity related to the nuclear sources and black holes which most of us believe are responsible for radio jets. Maybe the timescale for the jet activity to be stimulated is just longer than the star formation, which may begin before the galaxy-galaxy merger process is complete. Furthermore, if radio emission in galaxies is stimulated by large scale structure formation, it may be that this process is responsible for the strong evolution in the radio source luminosity function with epoch, not some local process isolated in the galaxy nucleus related to black hole formation.

5. Conclusions

1) Almost all radio galaxies seem to live in some sort of cluster-like environment.

2) The distorted structures of central cluster radio galaxies are due to their complex interaction with the thermal x-ray gas.

3) The evolution of radio source populations may be due to the evolution of the clustering environment with epoch.

References

Böhringer, H., Voges, W., Fabian, A.C., Edge, A.C., & Neumann, D.M. (1993) *MNRAS*, **264**, L25

Butcher, H.R. & Oemler, A. (1978) *ApJ*, **219**, 18

Dwarakanath, K.S. & Owen, F.N. (1996) Butcher-Oemler Effect and Radio Continuum in *Westerbork Synthesis Radio Telescope 25th Anniversary Workshop, Cold Gas at High Redshift* in press

Eilek, J.A., Burns, J.O., O'Dea, C.P., & Owen, F.N. (1984) *ApJ*, **278**, 37

Eilek, J.A., Owen, F.N., & Zhou, F. (1996) in preparation

Evrard, A.E. (1990) *ApJ*, **363**, 349.

Fanti, R. (1984) Non Thermal Radio Sources in Clusters of Galaxies in *Clusters and Groups of Galaxies*, Reidel, Dordrecht p. 185

Ge,J-P & Owen, F.N. (1993) *AJ*, **105**, 778.

Gomez, P.L., Pinkney, J., Burns, J.O., Wang, Q., & Owen, F.N. (1996), *ApJ*, submitted

Harris,D.E., Carilli, C.L., & Perley, R.A. (1994) *Nature*, **367**, 713

Ledlow, M.J. (1994) An Optical/Radio Study of Radio Galaxies In Abell Rich Clusters *Phd thesis* University of New Mexico

Ledlow, M.J. & Owen, F.N. (1995) *AJ*, **109**, 853

Ledlow, M.J. & Owen, F.N. (1996), *AJ*, submitted

Roettiger, K., Burns, J.O. & Loken, C. (1993) *ApJ*, **407**, L53

Schindler, S. & Müller, E. (1993) *A & A*, **272**, 137

Unewisse, A.M. (1993) Radio Emission from Southern Clusters of Galaxies *Phd thesis* University of Sydney

Wall, J.V. & Peacock, J.A. (1985) *MNRAS*, **216**, 173

GENERAL PROPERTIES OF GIANT RADIO GALAXIES

U. KLEIN AND K.-H. MACK
Radioastronomisches Institut der Universität Bonn - Auf dem Hügel 71, 53121 Bonn, Germany

AND

L. SARIPALLI
Indian Institute of Astrophysics - Koramangala, Bangalore - 560034, India

1. Introduction

So far, the number of so-called giant radio galaxies (GRGs) is small. We define them as radio sources with linear sizes larger than 1 Mpc ($H_0 =$ 75 km s^{-1} Mpc^{-1}). On-going low-frequency surveys may come up with many more candidates for this species. Their very existence may ultimately be connected with their environment. For instance, for a source like 3C236, which exhibits a linear and undisturbed structure over some 4 Mpc, the surrounding medium must have a low density. GRGs may thus serve to probe large volumes of the most tenuous intergalactic medium (IGM), e.g. via depolarization studies.

Previous investigations of the large-scale properties of GRGs were restricted to frequencies ≤ 1 GHz. Owing to their large (angular) sizes, aperture synthesis techniques become quickly insensitive to their large-scale components as one goes to higher frequencies. The need for single-dish work is then obvious.

2. Effelsberg projects

Extensive mapping programs have been carried out in the recent past with the Effelsberg 100-m telescope in order to obtain sensitive images in all Stokes parameters of the northern sky GRGs. These high-dynamic range maps complement ideally the low-frequency data obtained with the WSRT (see e.g. Willis and O'Dea, 1990). Maps at 2.7, 4.75 and 10.55 GHz are now available for the GRGs NGC315, NGC6251, DA240, 3C130, 3C236, 3C326,

R. Ekers et al. (eds.), Extragalactic Radio Sources, 311–312.
© 1996 *IAU.*

4C73.08, 8C0821+695, 0503-286, 1331-099, 1245+67, 4C34.47, 4C39.04, 4C74.26, and 1358+305. For most of these WSRT measurements at $\lambda\lambda 92$ and/or 49 cm have also been obtained (partly previously published by other authors). These can be used to derive relevant spectral and polarization parameters. Here we report first results on the depolarization characteristics of the lobes of GRGs as derived from the comparison of low- and high-frequency data. Part of our high-frequency mapping programme of GRGs and first results have been published by Klein et al. (1994), Saripalli et al. (1995) and Parma et al. (1996).

3. First results

A first glance at the radio maps of GRGs show that they are individuals. They are rarely symmetric, some have complex structures (e.g. NGC6251), some are narrow and straight (e.g. 3C236), some are fat doubles (e.g. DA240). The question arises what makes them look so different, and why they often show asymmetries. If it is the influence of the IGM, our multifrequency measurements should allow us to investigate this. We have therefore begun to analyze their depolarization as a function of wavelength, in order to look for any asymmetries which could be linked to any other asymmetry, viz. morphology, jet sidedness, magnetic field structure or spectral index. The fact that linear polarization is generally rather strong in GRGs even at large wavelengths already implies that the matter density of the surrounding IGM must be relatively low, inferring μG magnetic field strengths. Our preliminary analysis indicates, however, that there is not a clear relation of depolarization asymmetries with any other differences in the lobes of GRGs. The deduced number densities of thermal electrons are in the range between 1 and a few 10^{-5} cm^{-3}. An important conclusion at this stage is that the density of the environment around GRGs is not extraordinarily tenuous as one might expect. Detailed analyses of the maps are underway, and final conclusions may not be drawn before higher resolution polarization studies (e.g. with the VLA), which may eventually resolve any existing Faraday screens, will be made. These studies will also be aided by ongoing evaluations of the ROSAT data base, which aim at detecting hot gas that may surround GRG host galaxies.

References

Klein U., Mack K.-H., Strom R., Wielebinski R., and Achatz U. (1994) *A&A*, **283**, 729
Parma P., de Ruiter H. R., Mack K.-H., van Breugel W., Dey A., Fanti R. and Klein U. (1996), *A&A*, in press
Saripalli L., Mack K.-H., Klein U., Strom R. and Singal A. K. (1995) *A&A*, in press
Willis A. G. and O'Dea C. P. (1990) *IAU Symp.*, **140**, 455

MORPHOLOGIES IN MPC-SIZE RADIO GALAXIES

L. SARIPALLI[1,4], R. SUBRAHMANYAN[2,4], R. W. HUNSTEAD[3]

[1] *Indian Institute of Astrophysics, Koramangala, Bangalore 560 034, India*
[2] *Raman Research Institute, Sadashivanagar, Bangalore 560 080, India*
[3] *Astrophysics Department, School of Physics, University of Sydney, NSW 2006, Australia*
[4] *Australia Telescope National Facility, CSIRO, P O Box 76, Epping, NSW 2121, Australia*

1. Introduction

The extended radio structures or lobes found in edge-brightened radio galaxies represent interactions with the ambient medium over the source lifetimes. Their study probes the temporal evolution in the radio sources and the properties of the ambient medium encountered at different locations. Previous studies (Leahy & Williams, 1984; Leahy *et al.*, 1989) involved radio galaxies with sizes $\sim$ 400 kpc, and revealed radio morphologies with a variety of off-axis distortions that could be interpreted as due to different ways in which the lobes interact with the galactic halos. Such studies are lacking for radio galaxies having Megaparsec sizes, which extend to distances well beyond the observed galactic halos and are suspected of evolving in a different regime (Baldwin, 1982). Towards learning the evolution of radio galaxies in this size regime, we have carried out a study of a complete sample of Mpc-size radio galaxies; details are presented in Subrahmanyan, Saripalli & Hunstead (to appear in MNRAS, 1996).

2. Sample Selection

Our sample of eight Mpc-size radio galaxies was selected from the MRC extended source sample of Jones & McAdam, 1992. This sample is complete

313

R. Ekers et al. (eds.), Extragalactic Radio Sources, 313–314.

for sources with $S_{843} > 1$ Jy and angular size > 1.5 arcmin. We have further restricted the sample to sources with angular sizes in the range 3–12 arcmin to ensure that the imaging would reproduce all structural components in the sources. The selected sources are all edge-brightened with linear sizes $> 750h_{50}^{-1}$ kpc and not located in any rich clusters or crowded fields. The sources were observed with the Australia Telescope at 1.4 GHz and imaged with a resolution of 10 arcsec.

3. Results and Discussion

We have compared the radio-lobe morphologies in our sample of eight giants with the lobe morphologies in smaller-sized radio galaxies (the sample of Leahy & Williams 1984) that have the same power and redshift range.

1. Sharp lobe boundaries are frequently seen in the Mpc-size radio galaxies. This indicates that the lobe plasma is confined and not diffusing freely into the ambient medium.

2. The axial ratios of the giants are found to be the same as the axial ratios in the smaller radio galaxies. The Mpc-size structures are simply scaled versions of smaller radio galaxies. This result is indicative of self-similarity in the evolution of radio galaxies.

3 The lobe off-axis distortions, that are frequently seen in small-size radio galaxies, are also present in the giants and are similar; however, the deviations from axial symmetry in the giants are seen to occur at distances as far as 500 kpc. This points to galaxy halos whose properties continuously fall with distance, as against having abrupt cutoff at sizes of ≈ 100 kpc—the seen boundaries of the halo X-ray emission.

4. The lobes in the Mpc-size radio galaxies appear more fragmented and non-uniform except in cases where the neighbouring galaxy density is higher. The minimum pressures in the lobes are a factor seven lower than in the smaller sources, somewhat higher than expected considering the similarity in the axial ratios. This indicates smaller lobe filling factors for the giants.

5. A number of large galaxies show indications for discontinuous activity in the past: recessed hotspots, minima just upstream from the hotspots, and symmetrically-located, recessed enhanced emission regions.

References

Baldwin J.E., 1982, in Andernach H., Wielebinski R., eds, *Proc. IAU Symp. 97, Extragalactic Radio Sources*, Reidel, Dordrecht.
Jones P.A., McAdam W.B., 1992, *ApJS*, **80**, 137.
Leahy J.P., Williams A.G., 1984, *MNRAS*, **210**, 929.
Leahy J.P., Muxlow T.W.B., Stephens P.W., 1989, *MNRAS*, **239**, 401.

TOWARDS A COMPLETE SAMPLE OF GIANT RADIO GALAXIES AT Z > 0.4

A.P. SCHOENMAKERS[1,2], H. RÖTTGERING[2], H. VAN DER LAAN[1] AND A.G. DE BRUYN[3]

[1] *Sterrenkundig Instituut, University of Utrecht, The Netherlands*
[2] *Sterrewacht Leiden, University of Leiden, The Netherlands*
[3] *NFRA, Dwingeloo, The Netherlands*

1. Introduction and source selection

Giant Radio Galaxies are the largest radio sources associated with galaxies. They have linear sizes exceeding 1 Mpc (e.g. Saripalli *et al.*, 1986). They used to be known only at redshifts below 0.2, and it has been argued that this was not only due to a selection effect, but that there is a real cutoff in the space distribution of these objects (e.g. Gopal-Krishna & Wiita, 1987). Recently, a small number of giants have been found at redshifts largely exceeding 0.3, indicating that such a cutoff does not exist (e.g. Cotter *et al.*, 1995).
Studying GRGs is important for a number of reasons, including:

- GRGs are the only sources that probe the Inter Galactic Medium (IGM) on Mpc scales: the physical conditions in the diffuse part of the radio lobes (the so-called bridges) strongly reflect the physical state of the ambient medium (e.g. Subrahmanyan & Saripalli, 1993). If GRGs are so large because they reside in low-density environments, they could be populating voids in the large-scale galaxy distribution. If this were indeed the case, they provide us with unique and otherwise not obtainable information on the physical state of the IGM there.
- The extreme sizes raise questions such as: How do these sources evolve? How do their progenitors look like? What can we learn about their host galaxies? How do high redshift GRGs compare to their low redshift kin?
- Do these sources obey the orientation dependent unification scheme (e.g. Barthel, 1989)? If they do, then a sample selected on basis of a

315

R. Ekers et al. (eds.), Extragalactic Radio Sources, 315–316.
© 1996 *IAU.*

large linear size should contain only a few radio quasars.

With the advent of sensitive surveys such as the WENSS (De Bruyn *et al., these proceedings*) and the FIRST (Becker *et al.*, 1995; Becker, *these proceedings*), we can, for the first time, select large samples of candidate GRGs.

From the WENSS we selected all sources with a flux density between 80 and 300 mJy at 325 MHz, and with an angular size exceeding 1 arcminute. Subsequently, we used the FIRST survey to select only sources with an FRII morphology and with an angular size between 1.5 and 4 arcminutes. Since FRII radio sources always have radio powers larger than $\sim 10^{26.2}$ W Hz^{-1} at 325 MHz, our imposed flux limit should only select sources with $z > 0.4$ and thus with linear sizes above 700 kpc (for $H_0 = 50$ km s^{-1} Mpc^{-1}).

2. First (preliminary) results

- In an area of roughly 1000 square degrees we have found 103 GRG candidates, indicating that there should be ~ 4000 of such sources distributed over the sky. This means that there is a substantial population of large FRII radio sources.
- Based on the maps from the FIRST survey, we find that 75% of the candidate GRGs contain cores, down to the FIRST flux limit of ~ 0.2 mJy at 1.4 GHz.
- There is no relation between flux and angular size in this sample.
- There is no strong indication that the slope of the differential source counts of GRG differs to within the errors to that of the whole population of radio sources at these flux levels. This indicates that evolution of the space density of GRGs follows that of the entire radio source population at these flux levels.

References

Barthel, P.D., 1989, *ApJ*, **336**, 606
Becker, R.H, White, R.L., Helfand, D.J., 1995, *ApJ*, **450**, 559
Cotter, G., Rawlings, S., Saunders, R., 1995, *MNRAS*, submitted
Gopal-Krishna, Wiita, P.J., 1987, *MNRAS*, **226**, 531
Saripalli, L., Gopal-Krishna, Reich, W., Kuhr, H., 1986, *A&A*, **170**, 20 Subrahmanyan, R., Saripalli, L., 1993, *MNRAS*, **260**, 908

SPECTRAL INDICES AND PARTICLE AGEING STUDIES ON THE GIANT RADIO GALAXY NGC6251

K.-H. MACK[1], U. KLEIN[1], L. SARIPALLI[1,2], A. G. WILLIS[3] AND C. P. O'DEA[4]

[1] Radioastronomisches Institut der Universität Bonn - Auf dem Hügel 71, 53121 Bonn, Germany
[2] Indian Institute of Astrophysics - Koramangala, Bangalore 560034, India
[3] Dominion Radio Astrophysical Observatory - P.O. Box 248, Penticton, BC, Canada V2A 6K3
[4] Space Telescope Science Institute - 3700 San Martin Drive, Baltimore, MD 21218, USA

1. Introduction

In the framework of our high-frequency survey of giant radio galaxies with the Effelsberg 100-m telescope (Klein et al., 1994; Saripalli et al., 1995) we have obtained radio continuum maps of NGC6251, a source of 1.5 Mpc size ($H_0 = 75 km\ s^{-1}\ Mpc^{-1}$). Together with low-frequency WSRT observations (Willis & O'Dea, 1990), these measurements form a unique data base which for the first time allows thorough studies of the spectral index over a large frequency range. Theoretical models of particle ageing have been fitted to the spectrum to determine particle ages and other relevant physical parameters. Because of the immense size of NGC6251 these numbers provide information about the physics of the surrounding intergalactic medium.

2. Spectral index distribution and particle ageing

At low frequencies, steep spectra are found in the outer lobe regions with mean values of 1.4 ($S_\nu \sim \nu^{-\alpha}$) in the south-eastern lobe. Significant spectral asymmetries could not be found despite the asymmetric morphology and in contrary to other sources of this species. The core has a typical flat spectrum ($\alpha = 0.3$). The jet spectrum is very characteristic ($\alpha_{mean} = 0.6$). Along the jet a gradual steepening is seen up to $\alpha = 0.8$ where it

317

R. Ekers et al. (eds.), Extragalactic Radio Sources, 317–318.

widens and mixes with older lobe particles. Both hot spot regions possess spectra of $\alpha = 0.55$, close to the theoretical values expected for shock-accelerated particles. The high-frequency maps qualitatively show a similar behaviour, with a slight indication of increasing spectral indices. Similar to the work of Carilli et al. (1991) we have tried to fit a model spectrum to all data points in order to derive the break frequency ν_B across the source. The break frequency is used to evaluate the particle ages at various positions over the source, assuming that the dominant energy losses are due to synchrotron and Inverse Compton processes, neglecting adiabatic expansion and particle reacceleration. We have applied the model by Jaffe & Perola (1973) which allows for permanent pitch angle isotropization. The lowest break frequency which could be derived from three frequencies lies around 3 GHz and is found at the southern edge of the north-western lobe. Since at lower frequencies the emission is much more extended and could not be detected at high frequencies, we have to take this value as an upper limit. Using a minimum energy magnetic field of 0.6 μG for the lobes and a field equivalent to the microwave background of 3.4 μG we obtain a minimum source age of $\tau = 6 \cdot 10^7$ yrs.

3. NGC6251 as probe for the intergalactic medium

Because of the immense size of NGC6251 it can serve as a probe of the surrounding intergalactic medium (IGM). Following the example of Lacy et al. (1993) we have estimated the density of the IGM by balancing the momentum flux of the jet against the ram pressure of the external medium. The most uncertain parameter is the size of the effective area, the piston impinging on the IGM. We identify the effective area with the hot spot area, whose radius is assumed to be 2.5 kpc, which is an upper limit. The source advance speed was determined by dividing the source radius by the source age yielding a mean velocity of 0.045c. Using these parameters we obtain a particle density $n_e \sim 6.2 \cdot 10^{-6} \ cm^{-3}$ for the ambient medium. This value is lower than what is generally inferred for GRGs, but may be in accord with the fact that the jet of NGC6251 is obviously subject to strong influences by the surrounding medium, as evidenced by the various distortions and changes in brightness.

References

Carilli C. L., Perley R. A., Dreher J. H., and Leahy J. P. (1991) *ApJ*, **283**, 554
Jaffe W. J. and Perola G. C. (1973) *A&A*, **26**, 423
Klein U., Mack K.-H., Strom R., Wielebinski R., and Achatz U. (1994) *A&A*, **283**, 729
Lacy M., Rawlings S., Saunders R., and Warner P. J. (1993) *MNRAS*, **264**, 721
Saripalli L., Mack K.-H., Klein U., Strom R., Singal A. K. (1995) *A&A*, in press
Willis A. G. and O'Dea C. P. (1990) *IAU Symp.*, **140**, 455

EVOLUTION OF POWERFUL EXTENDED RADIO SOURCES

Implications for Cosmology and Cosmogony

R. A. DALY
Princeton University – Department of Physics
Princeton, NJ 08544 – USA

Abstract. Powerful extended radio sources are observed out to relatively large redshift. They may be used to study the properties and redshift evolution of the gaseous environment in the vicinity of each source, the active galactic nucleus (AGN), and the source size. This information may then be used to study and constrain cosmological and cosmogonical models. It is interesting to note that the rate of change of quantities with redshift allows constraints to be placed on global cosmological parameters and on models of structure formation and evolution that are completely independent of those inferred using the cosmic microwave background or local dynamical studies, and thus provide an important complement to these studies. The method does require that we understand the physics of the sources well enough to account for intrinsic source evolution. The physics of powerful extended radio sources that propagate supersonically appears to be relatively straight-forward. We have used the radio properties of the sources to deduce the ambient gas density, the beam power of the AGN, the characteristic time a particular AGN is on, and a characteristic source size that allows the sources to be used to probe global cosmological parameters. We plan to use the radio properties of the sources to deduce the ambient gas temperature, which will be combined with the ambient gas density to constrain cosmogonical models, that is, models of structure formation and evolution.

1. Results

In the short space allotted here we can only summarize results that are presented elsewhere; for details please see the papers by Daly (1994, 1995),

R. Ekers et al. (eds.), Extragalactic Radio Sources, 319–320.

Guerra & Daly (1995, 1996), Wan & Daly (1996a,b,c), Wellman & Daly (1995, 1996a,b), and references therein.

The data of Leahy, Muxlow, & Stephens (1989) and Liu, Pooley, and Riley (1992) were reanalyzed, and the lobe width a_L, the lobe magnetic field B_L, and the lobe propagation velocity v_L were estimated for each radio lobe in a consistent way (Wellman & Daly 1995, 1996). The final sample consists of 27 radio lobes from 14 radio galaxies, and 14 radio lobes from 8 radio loud quasars. The sources have projected core-lobe separations between 25 and 200 h^{-1} kpc, and redshifts between zero and two.

The beam powers are similar to that of Cygnus A, and are proportional to the radio power of the source. The ambient gas density in the vicinity of the radio lobe and hotspot range from about 10^{-2} to 10^{-3} cm^{-3}; galaxies and quasars are in similar gaseous environments; the composite density profile resembles that in low-redshift clusters of galaxies (such as the gas around Cygnus A); and the fit is significantly improved if the core density is allowed to evolve with redshift in the sense that the core density decreases with increasing redshift. A statistically significant result is obtained when the sources are used as a cosmological tool: the present data favor a low density universe with or without a cosmological constant over a flat matter-dominated universe. In addition, it appears that the characteristic time the AGN puts out a collimated outflow decreases with increasing redshift, or with increasing beam power, suggesting that higher redshift sources are smaller because they are on for a shorter period of time.

It is a pleasure to thank my students and collaborators Eddie Guerra, Lin Wan, and Greg Wellman for interesting discussions. This work was supported in part by the US National Science Foundation.

References

Daly, R. A. (1994), *ApJ*, **426**, 38

Daly, R. A. (1995), *ApJ*, **454**, in press

Guerra, E. J., & Daly, R. A. (1995), *ApJ*, submitted

Guerra, E. J., & Daly, R. A. (1996), *Cygnus A Workshop*, eds. C. Carilli & D. Harris (Cambridge: Cambridge University Press), in press

Leahy, J. P., Muxlow, T. W. B., & Stephens, P. W. (1989), *MNRAS*, **239**, 401

Liu, R., Pooley, G., & Riley, J. M. (1992), *MNRAS*, **257**, 545

Wan, L., & Daly, R. A. (1996a), *ApJ*, in press

Wan, L., & Daly, R. A. (1996b), *Cygnus A Workshop*, eds. C. Carilli & D. Harris (Cambridge: Cambridge University Press), in press

Wan, L., & Daly, R. A. (1996c), *Energy Transport in Radio Galaxies and Quasars*, eds. P. Hardee, A. Bridle, & A. Zensus (ASP Conference Series), in press

Wellman, G. F., & Daly, R. A. (1995), *ApJ*, submitted

Wellman, G. F., & Daly, R. A. (1996a), *Cygnus A Workshop*, eds. C. Carilli & D. Harris (Cambridge: Cambridge University Press), in press

Wellman, G. F., & Daly, R. A. (1996b), *Cygnus A Workshop*, eds. C. Carilli & D. Harris (Cambridge: Cambridge University Press), in press

RADIO SOURCE ENVIRONMENTS AT REDSHIFTS > 0.5

M. LACY[1], S. RAWLINGS[1], M. WOLD[2], A. BUNKER[1],
K.M. BLUNDELL[1], S.A. EALES[3], P.B. AND LILJE[2]

[1] *Astrophysics, Keble Road, Oxford OX1 3RH*
[2] *Institute of Theoretical Astrophysics, University of Oslo*
[3] *Physics & Astronomy, University of Wales, Cardiff*

1. Introduction

The most powerful radio sources in the local Universe are found in giant elliptical galaxies. Looking back to a redshift of 0.5 ($\approx$ half the age of the Universe for $\Omega = 1$), we see that these host galaxies are increasingly found in moderately rich clusters [1,2]. This fact gives us hope that radio sources can be used as tracers of high density environments at high redshift. By exploiting radio source samples selected over a wide range in luminosity (Blundell et al., these proceedings), we will also be able to test whether the luminosities of radio sources are correlated with their environments.

2. Methods of studying environments at high redshifts

For redshifts $0.5 < z \lesssim 0.8$, methods based on measuring the excess number of galaxies in a CCD frame relative to a comparison frame seem to still be viable, despite contamination by foreground and background galaxies. In an on-going project using the NOT on La Palma, we have imaged the fields of seven radio-loud quasars in this redshift range through either R and I or R and V filters so as to straddle the 4000Å break in the rest-frame of the companions. Three of our seven frames show an excess of galaxies relative to the comparisons significant at $> 3\sigma$.

Radio techniques based on depolarisation/rotation measure studies may prove useful for studying the gaseous environments of objects at redshifts higher than those accessible to current X-ray studies. Correlation with the richness of the optical environments is expected if indeed Faraday rotation arises in an external halo.

R. Ekers et al. (eds.), Extragalactic Radio Sources, 321–322.

Redshifts $\gtrsim 1$ require multicolour or narrow-band imaging to find companions in all but the very richest fields. Companions at $z > 2$ are of great interest as they are "normal" galaxies at high redshift, very few of which are currently known. Although ≈ 100 radio galaxies are now known above $z = 2$, their use in studying galaxy evolution is limited as high-redshift galaxies are known to have their optical light contaminated by contributions from the active nucleus, including reddened quasar light (e.g. [3]), emission lines [4], jet-induced star formation (e.g. [5]) and possibly others. Companions should be relatively unaffected by such processes and thus are of much more value for studying galaxy evolution. We have used narrow-band imaging in the near infrared around either the redshifted Hα or [OIII]4959/5007 lines in four fields at $2.1 < z < 2.7$. We typically find 2-3 objects per ≈ 1 arcmin^2 field with significant excesses in the narrow band, which we now need to follow-up spectroscopically.

A further technique which becomes viable at $z > 3.4$ is that of imaging across the Lyman continuum break, which is redshifted beyond the bandpass of the U filter in the optical. This is potentially a very robust way of finding high-z galaxies as *almost no* flux shortward of the redshifted Lyman limit is able to penetrate the clouds of intervening Lyman limit absorption systems. A pilot study of the field of 4C41.17 ($z = 3.8$) has revealed three potential companions [6].

Our discovery of two radio galaxies at $z > 4.2$, 8C1435+635 ($z = 4.25$; [7]) and 6C0140+326 ($z = 4.41$; Rawlings et al., in preparation) allow us to use B-band to sample below the redshifted Lyman limit, a much more attractive option given the far higher observing efficiency in this band. We also hope that Lyα imaging, unsuccessful at $z \sim 2$, might be more successful at such high redshifts. Lyα photons are destroyed by resonant scattering from HI followed by absorption onto dust grains. By catching galaxies early enough, before they form dust, we might hope to detect Lyα emission. We have one Lyα emission-line candidate in the field of 6C0140+326, which again requires spectroscopic confirmation.

References

[1] Hill G.J., Lilly S.J., 1991, *ApJ*, **367**, 1
[2] Ellingson E., Yee H.K.C., Green R.F., 1991, *ApJ*, **371**, 49
[3] Rawlings S., Lacy M., Eales S., Sivia D.S., 1995, *MNRAS*, **274**, 976
[4] Eales S.A., Rawlings S., 1993, *ApJ*, **411**, 67
[5] Lacy M., Rawlings, S., 1993, *MNRAS*, **270**, 431
[6] Lacy M., Rawlings S., 1995, *MNRAS*, in press
[7] Lacy et al., 1994, *MNRAS*, **271**, 504

THE ENVIRONMENTS OF RADIO GALAXIES

J.E. PESCE[1], R. FALOMO[2], G. FASANO[2], AND R. SCARPA[3]

[1] STScI, 3700 San Martin Dr., Baltimore, MD 21218 USA
[2] OAP, vicolo dell'Osservatorio 5, 35122 Padova, Italy
[3] Univ. di Padova, vicolo dell'Osservatorio 5, Padova, Italy

1. Introduction

In the unified schemes of AGNs BL Lac objects are believed to be Fanaroff-Riley (FR) type I radio galaxies with a relativistic jet aligned to the observer's line of sight (e.g. Urry & Padovani 1995). Kollgaard et al. (1992) and Owen et al. (1995) suggest some FR II sources can also be BL Lac parents. Clearly, isotropic properties such as the galaxy environment of both beamed and unbeamed objects should be identical.

We are currently studying BL Lac environments (e.g. Pesce et al. 1995) and plan to compare them to those of radio galaxies. Previously, based on statistical analysis of sources at $\langle z \rangle = 0.05$, Prestage & Peacock (1988) concluded that FR I galaxies inhabit richer clusters than FR II galaxies.

Higher redshift samples were studied by Yates et al. (1989; $\langle z \rangle \sim 0.49$) and Hill & Lilly (1991; $\langle z \rangle \sim 0.45$). These FR I and II galaxies are found in similar environments. For FR Is there is no change in environment with redshift. Few objects are found in clusters richer than Abell class 1.

These samples are not complete, however, and the analyses were inhomogeneous. We thus carried out a study of environments and host galaxies of a large sample of FR I and II sources. To date we analyzed 29/95 objects with $\langle z \rangle \sim 0.07$ selected from Wall & Peacock (1985) and Ekers et al. (1989). Fasano et al. (1995) discuss the properties of this sample.

2. Analysis

We use the digitized ESO-SERC survey plates (cf. Lasker et al. 1990). The FOCAS software (Valdes 1982) was used to detect and classify all objects around the target. We calculate $N_{0.5}$, the number of galaxies within 0.5 Mpc of the target with $m \leq m_3 + 2$ where m_3 is the 3rd brightest (cluster) galaxy, corrected for background (Bahcall 1981).

323

R. Ekers et al. (eds.), Extragalactic Radio Sources, 323–324.

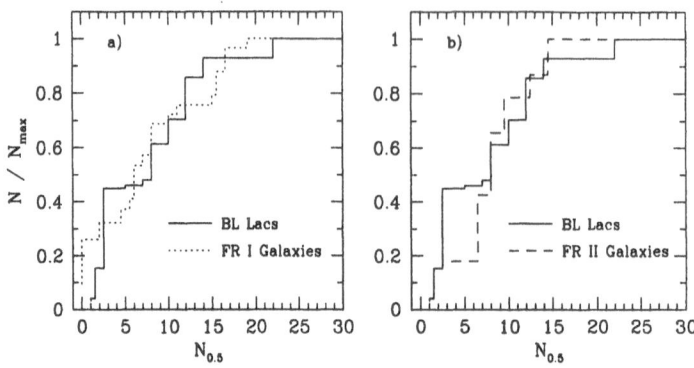

Figure 1.   (a) The cumulative distribution of $N_{0.5}$ for BL Lac objects (solid line; 11 objects) compared to that of FR I galaxies (dotted line; 21 objects). (b) The BL Lac $N_{0.5}$ distribution (solid line) compared to that of FR II galaxies (dashed line; 6 objects). Two sources are hybrid and have been excluded.

3. Results

We compared the radio galaxies to BL Lacs studied by us previously (Pesce et al. 1995). In Figures 1a and 1b we show the cumulative distribution of $N_{0.5}$ values for BL Lacs and FR I sources and BL Lacs and FR II sources, respectively. A KS test gives the probability that the BL Lacs and FR I galaxies are from the same population as 91.4%, slightly significant, while the probability for BL Lacs and FR II galaxies is only 63.3%. We note the numbers are small, however, and need to be confirmed using the whole dataset. We are also analyzing 30 BL Lacs with redshifts < 0.15 to be compared directly to the radio galaxies.

Using CCD images we determined the incidence of close companions in our sample. Within a radius of 100 Kpc the average number of companions in excess of the background is 1.1 ± 1.6. For FR I and II galaxies the numbers of excess companions are 1.2 ± 1.8 and 0.8 ± 1.3, respectively.

References

Bahcall, N.A. 1981, *ApJ*, **247**, 787
Ekers, R.D., et al. 1989, *MNRAS*, **236**, 737
Fasano, G., Falomo, R., & Scarpa, R. 1995, *MNRAS*, in press
Hill, G.J. & Lilly, S.J. 1991, *ApJ*, **367**, 1
Kollgaard, R., et al. 1992, *AJ*, **104**, 1687
Lasker, B.M. et al 1990, *AJ*, **99**, 2019
Owen, F.N., Ledlow, M.J., & Keel, W.C. 1995, *AJ*, **109**, 140
Pesce, J.E., Falomo, R., & Treves, A. 1995, *AJ*, **110**, 1554
Prestage, R.M., & Peacock, J.A. 1988, *MNRAS*, **230**, 131
Urry, C.M., & Padovani, P. 1995, *PASP*, **107**, 803
Valdes, F. 1982, in: *Instrumentation in Astronomy IV*, SPIE **331**
Wall, J.V., & Peacock, J.A. 1985, *MNRAS*, **216**, 173
Yates, M.G., Miller, L., & Peacock, J.A. 1989, *MNRAS*, **240**, 129

CLUSTERS OF GALAXIES AT INTERMEDIATE REDSHIFTS: A SAMPLE SELECTED AT RADIO WAVELENGHTS

A. ZANICHELLI[1,2], R. SCARAMELLA[1], M. VIGOTTI[2],
G. VETTOLANI[2], G. GRUEFF[3,2]

[1] *Osservatorio Astronomico di Roma, Monteporzio Catone,
Italy*
[2] *Istituto di Radioastronomia, C.N.R., Bologna, Italy*
[3] *Dipartimento di Astronomia, Università di Bologna, Bologna,
Italy*

In order to gather a sample of intermediate redshifts ($z \sim 0.1 \div 0.2$) clusters avoiding evolutionary effects or biases induced by limited sensitivity of instruments and optical plates that affect samples selected through inspection of optical plates, color diagrams or X-ray emission properties, we plan to use radio galaxies as suitable tracers of dense environments (e.g. Allington-Smith *et al.*, 1993). This would allow us to effectively test different environments and population properties, and would also give valuable information on the effect of environment on the radioemission phenomena. Moreover, it would not impact on the X–ray or optical properties of clusters, since there is no significant correlation between the radio properties of galaxies within a cluster with its L_X (Feigelson, Maccacaro and Zamorani, 1982), or with richness of the cluster (Zhao, Burns, and Owen, 1989).

1. The Radio Sources Catalogue and Optical Identifications

From the new 1.4 GHz NRAO VLA Sky Survey (NVSS) (Condon *et al.*, 1994) maps we extracted a catalogue of 11922 pointlike radio sources and 3371 double radio sources down to a flux limit 2.5 mJy, on an area of ~ 550 square degrees in the region of the South Galactic Pole. The reliability of source positions in the catalogue was found to be consistent within the errors predicted by Condon *et al.* (1994), which vary from 5" for sources at the flux limit, to < 0.5" for sources brighter than 20 mJy.

R. Ekers et al. (eds.), Extragalactic Radio Sources, 325–326.

Optical identifications were made with galaxies in the Edimburgh-Durham Southern Galaxy Catalogue (EDSGC) (Heydon-Dumbleton, Collins and MacGillivray, 1988), up to a limiting magnitude $b_J = 20.0$, above which the EDSGC is no more complete and the stellar confusion exceeds 10%. Restricting our search to a radio-optical distance $\leq 7.0"$, which corresponds to a contamination from chance coincidence $< 5\%$ and to an identification rate $\sim 5\%$ at each flux, we found a preliminary sample of 609 optically identified pointlike radio sources.

2. The Cluster Sample

The selection of cluster candidates was performed through the search of excesses in galaxy density near the position of the radio galaxies. For each ESO plate in the region covered by our catalogue we calculated a gaussian smoothed matrix of counts of optical data in the EDSGC, with bin size 30" and smoothing length $2'$. We looked for excess densities inside a circular region of radius $3'$ centered at the position of each identified radio galaxy, having defined the *"excess density"* on each plate as $\rho_{crit} \geq 3\sigma + \rho_m$, where ρ_m and σ are the mean value and sigma of galaxy density fluctuations calculated on the whole plate.

Neglecting all those radio galaxies which fall inside the Abell radius of an already known ACO or Edimburgh-Durham Clusters Catalogue cluster, and those identified with galaxies brighter than $b_J = 17.5$, we found a list of 59 cluster candidates. Spectroscopy will be taken for each radio galaxy and $\sim 10 - 15$ nearby companions in order to have positive cluster identifications and to improve the results of Allington-Smith *et al.* (1993), who took only photometry of fields centered on radio galaxies at intermediate redshifts.

Assuming for the candidates a redshift range $0.1 \div 0.2$, the power range covered by our sample would be $2.4 \times 10^{22} < P_{1.4GHz} < 2.5 \times 10^{24}$ ($H_0 = 100$km/sec Mpc, $q_0 = 1$), thus well representing FR I population of extragalactic radio sources.

References

Allington-Smith,J.R., Ellis,R.S., Zirbel,E.L., Oemler,Jr., 1993. *Astroph. J.* **404**, 521.

Condon,J.J., Cotton,W.D., Greisen,E.W., Yin,Q.F., Perley,R.A., Broderick,J.J., 1994. In "Astronomical data analysis software and systems III.", *Astron. Soc. Pac. Conf. Ser.* **vol. 61**, p. 155, eds Crabtree,D.R., Hanisch,R.J., Barnes,J.

Feigelson,E.D., Maccacaro,T., Zamorani,G., 1982. *Astroph. J.* **255**, 392.

Heydon-Dumbleton,N.H., Collins,C.A., MacGillivray,H.T., 1988. In "Large-Scale Structure in the Universe - Observational and Analytical Methods", p. 71, eds Seitter,W.C., Duerbeck,H.W.& Tacke,M., Springer-Verlag, Berlin.

Zhao,J.H., Burns,J.O., Owen,F.N. 1989. *Astron. J.* **98**, 64.

QUASAR RADIO MORPHOLOGY AND ENVIRONMENT AT
Z ∼ 1/2

TRAVIS A. RECTOR, JOHN STOCKE AND ERICA ELLINGSON
Center for Astrophysics and Space Astronomy
University of Colorado - Boulder
Boulder, CO 80309-0389 USA

Abstract.
    The radio morphology of thirty radio-loud quasars at $z \sim 1/2$ is studied as a function of cluster galaxy density using data available in the literature. While correlations between several morphological parameters of the radio emission are consistent with interactions with an intracluster medium (ICM), we do not see the correlation between galaxy density and quasar radio morphology expected from ICM interaction. Therefore, distorted radio morphology is **not** an indicator for the presence of a distant cluster as previously suggested. Because such correlations are found at $z \sim 0$, these results are suggestive that the rich clusters containing quasars at $z \sim 1/2$ have not yet developed a dense ICM. An inelastic collision with neighboring galaxies and intergalactic clouds can account for the morphology of several of the sources in this sample and is a viable alternative explanation.

1. Introduction

Previous studies have confirmed that a dense ICM is responsible for distorted radio sources at $z \sim 0$. Hintzen and Scott (1978, ApJ 224,L47) proposed that distorted quasars may be tracers for a dense ICM at high z. To test this hypothesis we studied a sample of 30 "classical triple" radio-loud quasars at $0.26 < z < 0.62$ to detect any correlation between the quasar-galaxy spatial covariance amplitude B_{gq} and four morphological characteristics: hotspot position in lobes, projected bending angle θ, projected physical size l, and length-asymmetry Q. Correlations between these parameters and core and total radio powers were also tested.

R. Ekers et al. (eds.), Extragalactic Radio Sources, 327–328.

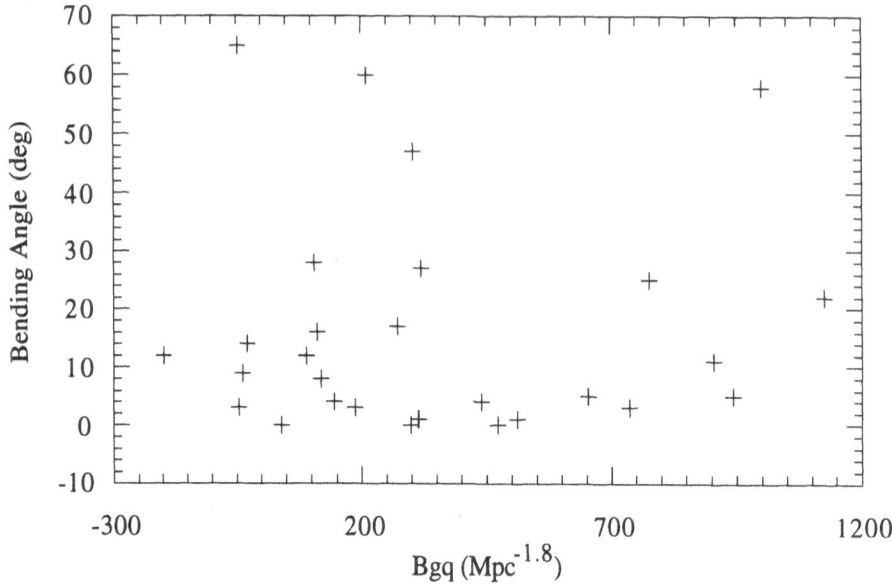

Figure 1. Projected bending angle θ as a function of B_{gq}. No correlation is present; in fact, many highly bent sources are not in dense clusters.

2. Results

A correlation was found between B_{gq} and z. This is believed to be quasar luminosity evolution; i.e. radio sources in clusters at low redshifts have faded and are now called radio galaxies (Ellingson *et al.* 1991 ApJ 371,36). Anti-correlations between l and θ and l and Q are believed to be merely projection effects. No correlation is detected between B_{gq} and θ (Fig. 1) or B_{gq} and l. This suggests that a dense ICM is not solely responsible for the distorted morphology of these quasars. No correlation is seen between radio power and morphological parameters, confirming that variations in the radio power are not responsible for the observed morphologies.

3. Conclusions

Interaction with an ICM is not the sole determinant of the extended radio morphology of a quasar; alternate mechanisms for bending sources at $z \sim 1/2$ must exist. It is incorrect to assume that quasars with distorted structure are always members of rich clusters as previously suggested. An inelastic collision between one of the lobes and a nearby galaxy halo or intergalactic cloud is a likely explanation for the "dogleg" structure seen in some of our sources (e.g. 3C 275.1; Stocke *et al.* 1985, ApJ 299, 799). These results also suggest that rich clusters of galaxies surrounding quasars at $z \sim 1/2$ do not possess a dense, hot ICM as found at $z \sim 0$.

QUASAR RADIO STRUCTURE IN CLUSTER ENVIRONMENTS

J.B.HUTCHINGS
Dominion Astrophysical Observatory
Victoria, B.C., Canada

AND

A.C.GOWER, S.RYNEVELD AND A.DEWEY
Dept of Physics and Astronomy
University of Victoria, B.C., Canada

We have VLA snapshots at 6cm and 20cm of $\sim$50 QSOs of redshift ≤ 0.7 with a range of known optical galaxy environments. The radio sources were characterised by measures of flux, size, and shape. The cluster density of the QSOs is given by the B_{gq} number from galaxy companion data largely published by Yee and Ellingson (e.g. ApJ 411, 43, 1993).

Number of sources. In several of the fields, one or two other sources were detected. We limit our discussion of these sources to within 200 arcsec of the map centre (the QSO) to avoid (50 MHz) bandwidth smearing at larger angular distances.

All the extra sources lie at distances beyond the known galaxy companions. Thus, we do not yet know if they are associated or background sources. As a comparison, we count the sources detected in our maps from a separate program of 6 very rich (X-ray selected) clusters. In all cases the central cD galaxy was one of the strongest sources. We thus counted extra sources around the cD galaxies, with the same limiting flux and area limits as in our QSO clusters.

The numbers aside from the central source are: mean QSO cluster 0.7; typical rich cluster 1.6; exceptional rich cluster (A2390) 5. Thus, the rich clusters appear to contain on average more sources than the QSO clusters. Here too we do not yet know in general if they are cluster members or some background population (enhanced by lensing?). However, in the exceptional cluster A2390, most of the extra sources are not identified with cluster members (Abraham et al Ap.J. 1996 preprint), or any optically visible object .

R. Ekers et al. (eds.), Extragalactic Radio Sources, 329–330.

The mean B_{gq} of the rich clusters is 1400, compared with 300 for the QSO sample, although some of the QSO clusters have B_{gq} near to 1000. There is no correlation between B_{gq} and number of sources in the QSO sample.

Cluster environment. The distribution of QSO B_{gq} values has a roughly Gaussian spread about zero, with $\sigma \sim 350$, and a tail to higher values. Values below 350 are consistent with zero, and the remainder are significantly clustered. In order to use a subset matched in redshift for high and low B_{gq}, we use QSOs with redshift >0.4. The table shows values of properties in this matched sample of 37.

Property	$B_{gq} > 350$ mean (& median)	$B_{gq} < 350$	Difference (high B_{gq} – low B_{gq})
B_{gq}	833 (792)	37 (0)	(defining cut)
z	0.55 (0.58)	0.52 (0.51)	—
m_v	17.6 (17.0)	17.6 (17.2)	—
Lobe dominance (1-4)	3.5 (4)	2.7 (3)	~ 1
Size (kpc)	162 (134)	142 (135) 100	resolved sources: — all sources: larger
Bend angle (^{o})	8 (6)	16 (9)	less bent?
Lobe length ratio	0.72 (0.72)	0.71 (0.65)	—
Source complexity (1-5)	2.8 (3)	3.0 (3)	—
α_{core}	-0.08 (-0.12)	-0.12 (-0.03)	—
log P (W/Hz) mean core	25.4 (25.3)	25.6 (25.5)	-0.2 (-0.2)
mean lobe	25.6 (25.7)	25.2 (25.2)	0.4 (0.5)

The data suggest that the high B_{gq} group has:
1. More lobe-dominated type (*all* compact sources have low B_{gq})
2. Higher lobe luminosities (by a factor 2-3)
3. Lower core luminosities (by a factor <2)   *and possibly*
4. Larger source size (if unresolved sources are included)
5. Less bent sources (mainly because two very bent ones are low B_{gq})

K-S tests indicate that 1 - 5 above are significant at the 2σ to 3σ level. There are also significant continuous correlations with B_{gq} for the powers. The lobe length ratio, source complexity, core spectral index, and number of additional sources show no dependence on B_{gq}.

These results will be published in full elsewhere.

RADIO-OPTICAL ALIGNMENTS IN NEARBY COOLING FLOW CLUSTER CENTRAL GALAXIES

B.R. MCNAMARA

Harvard-Smithsonian Center for Astrophysics, Cambridge, MA, USA

1. Summary

The centers of dominant cluster galaxies in cooling flows are often unusually blue, they have spatially extended nebular line emission and bright, FR I radio sources (Fabian 1994). As a class, they are the most rapidly evolving giant elliptical galaxies known. Among the most interesting of these objects, the Abell 1795 ($z = 0.06$) and Abell 2597 ($z = 0.08$) central cluster galaxies have very blue, lobe-like structures that are located along their FR I radio lobes (McNamara & O'Connell 1993). This discovery was surprising because correlations between the radio source and blue optical continuum were thought to occur exclusively in powerful, FR II radio galaxies at redshifts $z > 0.6$ that show the alignment effect. By analogy with the distant radio galaxies, the blue lobes are thought to be regions of star formation that were triggered by the passage of the radio source (De Young 1995), or scattered light from an obscured, anisotropically radiating active nucleus that is beaming its light obliquely to the line of sight (Sarazin & Wise 1993; Crawford & Fabian 1993; Sarazin et al. 1995). Scattered light is usually polarized. Therefore, polarization measurements of the aligned optical continuum should provide a strong test of the scattering hypothesis.

McNamara et al. (1995) have obtained U-band polarimetry of the blue lobes in the Abell 1795 cluster central galaxy. They found an upper limit to the degree of polarization of the light emitted from the lobes of $< 7\%$. The accuracy of this measurement is limited by the presence of diluting background starlight. This limit is inconsistent with the lobes being scattered light that originated in an obscured, anisotropically radiating nucleus, unless the radiation is beamed and is viewed at an angle $< 22^{o}$ to the line of sight, which is unlikely. The absence of a detailed correspondence between

R. Ekers et al. (eds.), Extragalactic Radio Sources, 331–332.

the radio lobes and optical lobes and the absence of a polarized signal is also inconsistent with synchrotron light.

The blue optical lobes are probably regions of vigorous star formation. If a burst of star formation were triggered by the expanding radio lobes, the age of the burst population should be $\sim 10^7$ yr. The star formation rate in both lobes, assuming the Local IMF, would then be $\sim 20\ M_\odot$ yr^{-1} and the stellar mass of the lobes would be $\sim 10^8\ M_\odot$. The large cooling flow in A1795 may be fueling the star formation and the radio source or the fuel may have originated from one or more gaseous cluster galaxies that recently fell into the cluster's core. This result strongly suggests that the radio sources in central cluster galaxies may be a significant factor driving the evolution of their stellar populations.

References

Crawford, C.S. & Fabian, A.C. 1993, *MNRAS*, **265**, 431

De Young, D.S. 1995, *ApJ*, **446**, 521

Fabian, A.C., 1994, *ARAA*, **32**, 277

McNamara, B.R. & O'Connell, R.W. 1993, *AJ*, **105**, 417

McNamara, B.R., Jannuzi, B.T., Elston, R. Sarazin, C.L., & Wise, M. 1995, *ApJ*, submitted

Sarazin, C.L. & Wise, M.W. 1993, *ApJ*, **411**, 55

Sarazin, C.L., Burns, J.O., Roettiger, K., & McNamara, B.R. 1995, *ApJ*, **447**, 559

DIFFUSE CLUSTER RADIO SOURCES

L. FERETTI AND G. GIOVANNINI

Istituto di Radioastronomia CNR, Bologna, Italy
Dipartimento di Astronomia dell'Università, Bologna, Italy

1. Introduction

Diffuse radio sources in clusters remain a poorly understood phenomenon. They are very extended sources (0.4-0.6 Mpc), of low surface brightness and steep spectrum, which cannot be identified with any active radio galaxy. They are a rare phenomenon, as they have been found so far in few clusters of galaxies. This paper reviews the current findings about this kind of sources, and the suggestions about their formation and evolution.

A Hubble constant H_0=100 km/s Mpc is used throughout.

In the effort of investigating the properties of diffuse cluster sources, we classify them in three classes: cluster-wide halos, relics and mini-halos. Cluster-wide halos include sources located at the cluster centers, while relics are those diffuse extended sources located at the cluster peripheries. Moreover, in some clusters with a central dominant galaxy, the relativistic particles can be traced out quite far, forming what is called a mini-halo. The existence of these features surrounding the central strong radio galaxies is well established in the Perseus and Virgo cluster. Mini-halos have a size much smaller than the cluster-wide radio halos and relics (1%-20%). Since the active galaxy responsible for the emission in these sources is known, problematics are different in mini-halos from halos and relics. Therefore, we will not consider this class here.

2. Cluster-wide halos

The prototype example of this class of sources, residing at the cluster centers, is Coma-C, belonging to the Coma cluster (A1656). Halos show an essentially regular radio morphology and no or little polarization. An up-

R. Ekers et al. (eds.), Extragalactic Radio Sources, 333–338.

per limit to the polarized flux in Coma-C is ~10%. Significant polarization was detected only in the highest brightness region of A2256, at the level of ~20% at 20 cm [1]. The clusters presently known to harbor a cluster-wide radio halo are given in Table 1, where we give largest linear size, monochromatic power at 1.4 GHz, total power between 10 MHz and 10 GHz, spectral index, minimum energy density, minimum total energy and equipartition magnetic field. Equipartition parameters were computed under the standard assumptions ($\Phi = 1$ and $k = 1$), integrating the spectrum between 10 MHz and 10 GHz with the proper spectral index. The last column gives the reference to recent papers.

The existence of a radio halo has been suggested in the literature for few other clusters. From new VLA data [2], we exclude the presence of a halo in A401, A754, A1609 and A2142, where the previous detection is due to discrete sources, while more sensitivity is needed to draw conclusions for A610, A665 and 0016+16.

TABLE 1. Parameters of halos

Clus	Size kpc	$P_{1.4}$ W Hz^{-1}	P_{tot} erg s^{-1}	α	u_{min} erg cm^{-3}	U_{min} erg	H_{eq} μG	Ref
A1656	550	3.2×10^{23}	6.1×10^{40}	1.34	2.4×10^{-14}	5.0×10^{58}	0.5	[3]
A2163	600	-	3×10^{41}	-	5.4×10^{-14}	8.0×10^{58}	0.8	[4]
A2218	250	7.9×10^{22}	9.0×10^{39}	1.1	4.3×10^{-14}	5.3×10^{57}	0.7	[5,2]
A2255	725	2.5×10^{23}	1.6×10^{41}	$\gtrsim 1.5$	3.1×10^{-14}	5.3×10^{58}	0.6	[6,2]
A2256	700	1.2×10^{23}	1.6×10^{41}	1.9	1.1×10^{-14}	1.1×10^{58}	0.5	[7,8]
A2319	660	5.1×10^{23}	9.2×10^{40}	1.3	3.7×10^{-14}	7.1×10^{58}	0.6	[2]

3. Relics

Diffuse cluster sources, similar to the central halos, located in the cluster peripheral regions were suggested to be *relics* of currently inactive radio galaxies. However, no compelling evidence of this has been found so far. The prototype of this class is 1253+275, in the Coma cluster. Moreover, the Coma cluster is permeated by diffuse radio emission of very low brightness, in the region between the central halo Coma-C and the peripheral source 1253+275. This diffuse source is referred to as Coma bridge. The relics known so far are listed in Table 2, with their distance from the cluster center, and the parent cluster. The source parameters are in Table 3, where columns have the same meaning as in Table 1. The properties of relic radio sources are comparable to those of central halos. They have similar largest size, but their morphology is generally more irregular and elongated. Moreover, in 1253+275 and in 0917+75, a significant percentage of polarized flux is detected. The difference in the polarization properties could be naturally explained from the larger number of magnetic field cells along the line of sight at the cluster center, compared to the outer regions.

TABLE 2. Relics

Name	Distance from c.c.	Cluster
0038-096	320 kpc	A85
0917+75	2.5 Mpc	A786, Rood#27
1253+275	1.35 Mpc	A1656
Coma Bridge	1.1 Mpc	A1656
1401-33	90 kpc	S753
2006-56 region	1.04 Mpc	A3667

TABLE 3. Parameters of relics

Name	Size kpc	$P_{1.4}$ W Hz^{-1}	P_{tot} erg s^{-1}	α	u_{min} erg cm^{-3}	U_{min} erg	H_{eq} μG	Ref
0038-096	200	5.1×10^{23}	1.6×10^{41}	>1.5	1.5×10^{-13}	1.8×10^{58}	1.3	[2]
0917+75	780	2.0×10^{24}	1.7×10^{41}	1.0	3.5×10^{-14}	2.8×10^{58}	0.6	[9]
1253+275	580	1.7×10^{23}	2.0×10^{40}	1.1	3.4×10^{-14}	1.1×10^{58}	0.6	[10]
Coma Br.	970	4.9×10^{22}	1.6×10^{40}	1.5	9.8×10^{-15}	2.2×10^{58}	0.3	[11]
1401-33	220	1.0×10^{23}	2.5×10^{40}	1.4	9.7×10^{-14}	6.2×10^{57}	1.0	[12]
2006-56	870	2.6×10^{24}	3.7×10^{41}	1.2	2.3×10^{-14}	1.5×10^{58}	0.5	[13]

4. Confinement

It is evident from Tables 1 and 3 that the minimum energy densities in halos and relics are very similar. If these radio sources are in pressure equilibrium with the ambient medium, we would expect much higher energy densities in halos with respect to relics, reflecting the pressure profile of the cluster intergalactic medium. Typically, the minimum internal pressure in halos is about 3 order of magnitude lower than the external pressure of the ambient gas, obtained by X-ray data. For peripheral relics, the X-ray data are scarce. However, an imbalance of about a factor of 10 was found for 1253+275 in the Coma cluster [10]. Our conclusion is that the energy density in these sources is likely to be higher than the minimum value.

5. The Coma-C spectrum and the Coma cluster magnetic field

The halo Coma-C is the best studied example of a cluster-wide radio halo. The spectral index distribution [3] shows a central *plateau* with an almost constant value $\alpha \sim 0.8$, and an outer spectral steepening, with values of α increasing up to 1.8. This behaviour provides strong evidence that the particle reacceleration mechanism is more efficient at the cluster center. Since the central *plateau* is comparable in size to the region within one cluster core radius, the turbulent acceleration mechanism is likely to be related to wakes produced by the galaxy motions.

An important information on the magnetic field in the Coma cluster can be obtained from the rotation measure studies of the tailed radio galaxy NGC4869, located in the cluster central region and therefore seen through the intracluster medium. The polarization properties of this radio galaxy [14] indicate the presence of local fluctuations in the rotation measure (RM), occurring on typical scales of $\approx 2.5''$, i.e. 0.85 kpc, and arising in a screen external to the source. The dispersion of RM observed across NGC4869, explained by assuming that the magnetic field is tangled on typical scales of the same size as the RM fluctuations, leads to a strength of magnetic field associated with the intergalactic medium between 8.3 and 11.7 μG, depending on the location of the radio galaxy along the line of sight (see also [15]). Even allowing for uncertainties related with this determination, the Coma magnetic field is more than one order of magnitude larger than the equipartition value (Table 1).

The evidence for a magnetic field larger than the equipartition value has been suggested by [16] for the diffuse source 0917+75: from the absence of X-ray synchro-self Compton emission a lower limit of 1 μG was found, which should be compared with the 0.6 μG given in Table 3. These data favour the hypothesis that the energy density is not minimum, as already suggested by the confinement arguments.

6. Properties of clusters with halos

The properties of clusters containing radio halos and relics are presented in Table 4. X-ray data are from [17] and [18]. We remind here that the Coma cluster contains both a central halo and peripheral relics. The clusters with halos are characterized by: i) high richness, ii) no strong galaxy concentration, iii) high X-ray luminosity, iv) no cooling flow, v) high temperature (7-14 Kev). The cooling flow in A2319 is not confirmed by recent ASCA results (Yamashita, private communication). Also, the central dominant galaxy in this cluster is not very dominant [19]. The fact that halos form in clusters with high temperature, i.e. large mass, is not of obvious interpretation. Moreover, high temperature and high X-ray luminosity imply a high central pressure. This fact seems in contrast with the pressure imbalance mentioned in Sect. 4, and would support the suggestion that internal pressures within halos are not minimum.

TABLE 4. Cluster properties

Name	z	RS	BM	Rich.	L_x(2-10 keV) erg s^{-1}	T keV	Cool flow
Coma	0.0232	B	II	2	2.1×10^{44}	8.11	NO
A2163	0.2010	I	-	2	1.8×10^{45}	13.9	-
A2218	0.1710	-	II	4	2.5×10^{44}	6.72	NO
A2255	0.0809	C	II-III	2	1.3×10^{44}	7.3	NO
A2256	0.0601	B	II-III	2	2.9×10^{44}	7.51	NO
A2319	0.0564	cD	II-III	1	4.4×10^{44}	9.9	WEAK
A85	0.0518	cD	I	1	1.9×10^{44}	6.2	YES
A786	0.1241	F	-	0	-	-	-
S753	0.0139	-	I	0	-	-	-
A3667	0.0552	-	I-II	2	2.6×10^{44}	6.5	NO

From X-Ray observations of A2256, it was suggested that the presence of a halo source could be related with the existence of a cluster merger process [20, 21]. According to [22], the interaction between the intergalactic medium of a subgroup and that of the main cluster may produce shocks which can accelerate the electrons radiating in the halos, and amplify the magnetic field. The merger hypothesis is consistent with the presence of halos in clusters without a cooling flow, since a major merger event is expected to disrupt a cooling flow.

We found indeed that all the clusters in Table 1 show the evidence of a recent merger, either from X-ray or optical data. In the Coma cluster, the radio halo could be associated with the ongoing merger at the cluster center, between the two groups centered on the two dominant galaxies NGC4874 and NGC4889, close in projected distance but offset by about 800 km/s [23]. The merger between the NGC4839 group and the main cluster, which was originally suggested to be responsible for the halo energy supply [24], is in our opinion not related with Coma-C, because it occurs in a peripheral region. Also in the case that the group has already passed through the cluster core about 2 Gyrs ago [25], it is difficult to reconcile the merger energy release with the spectrum of Coma-C. However, this merger could be relevant for the maintenance of the Coma bridge, and of the relic 1253+275. Therefore, an ongoing merger process seems to be crucial for the energy supply to the radio halo, in addition to the energy possibly provided by galactic wakes (Sect. 5).

We note, however, that the merger alone cannot account for the origin of halos, as the existence of a merger process seems to be common in clusters of galaxies, while the halo sources are a rare phenomenon. For example, the cluster A754 shows a subcluster collision both from optical and X-ray data [26,27], but no radio halo is detected (Sect. 2). In the case of Coma-C, it was suggested [3] that the tailed radio galaxy NGC4869, which is orbiting around the cluster center is responsible for the relativistic electron supply. The need of tailed radio galaxies residing at the cluster centers, as the origin of relativistic particles, could explain the rarity of halo type radio sources. In conclusion, the formation of a halo can be due to a cluster merger event, which amplifies magnetic fields and reaccelerates the existing particles, deposited by tailed radio galaxies. The origin of relics is more puzzling, but we point out that relics have many similarities with halos.

References

[1] Bridle, A.H., Fomalont, E.B., Miley, G.K., Valentijn, E.A. (1979) A&A, 80, 201
[2] Feretti, L., Giovannini, G., L., Böhringer, H. (1995), in preparation
[3] Giovannini,G.,Feretti,L.,Venturi,T.,Kim,K.T.,Kronberg,P.P.(1993) ApJ, 406, 399
[4] Herbig, T., Birkinshaw, M. (1994) BASS, 26, 1403
[5] Partridge, R.B., Perley, R.A., Mandolesi, N., Delpino. F. (1987) ApJ, 317, 112
[6] Burns, J.O., Roettiger, K., Pinkney, J., et al. (1995) ApJ, 446, 583
[7] Kim, K.T. (1995) ApJ, Submitted
[8] Röttgering, H., Snellen, I., Miley, G., et al. (1994) ApJ, 436, 654
[9] Harris, D.E., Stern, C.P., Willis, A.G., Dewdney, P.E. (1993) AJ, 105, 769
[10] Giovannini, G., Feretti, L., Stanghellini, (1991) A&A, 252, 528
[11] Kim, K.T., Kronberg, P.P., Giovannini, G. Venturi, T. (1989) Nature, 341, 720
[12] Goss,W.M.,McAdam,W.B.,Wellington,K.J.,Ekers,R.D.(1987) MNRAS, 226, 979
[13] Goss, W.M., Ekers, R.D., Skellern, D.J., Smith, R.M. (1982) MNRAS, 198, 259
[14] Feretti, L., Dallacasa, D., Giovannini, G., Tagliani, A. (1995) A&A, 302, 680
[15] Felten J.E. (1996) in Clusters, Lensing and the Future of the Universe, ed. V. Trimble, ASP Conference, in press
[16] Harris, D.E., Willis, A.G., Dewdney, P.E., Batty, J. (1995) MNRAS 273, 785
[17] David, L.P., Slyz, A., Jones, C., et al. (1993) ApJ, 412, 479
[18] Edge, A.C., Stewart, G.C., Fabian, A.C. (1992) MNRAS, 258, 177
[19] Hanisch, R.J. (1982) A&A, 116, 137
[20] Briel, U.G., Henry, J.P., Schwarz, R.A., et al. (1991) A&A, 246, L10
[21] Fabian, A.C., Daines, S.J. (1991) MNRAS, 252, 17P
[22] Tribble, P.C. (1993) MNRAS, 263, 31
[23] Colless, M., Dunn, A.M. (1995) ApJ, in press
[24] Briel, U.G., Henry, J.P., Böhringer, H. (1992) A&A, 259, L31
[25] Burns, J.O., Roettiger, K., Ledlow, M., Klypin, A. (1994) ApJ, 427, L87
[26] Henry P.J. & Briel U.G. (1995) ApJ, 443, L9
[27] Zabludoff A.I. & Zaritsky D. (1995) ApJ, 447, L21

ELECTRON TRANSPORT IN THE COMA CLUSTER

J.G. KIRK[1], P. DUFFY[1], R.O. DENDY[2]
[1] Max-Planck-Institut für Kernphysik, Postfach 10 39 80,
D-69027 Heidelberg, Germany
[2] UKAEA Government Division, Fusion,
UKAEA/Euratom Fusion Association
Culham, Abingdon, Oxon OX14 3DB, UK

1. Introduction

The inner regions of the Coma cluster of galaxies contain a source of diffuse synchrotron emission ('Coma C') which has a linear dimension of at least 500 kpc. There has been interest in the question of where the relativistic electrons responsible for this emission originate, and how they are transported through the intra-cluster medium (Tribble, 1993). It is widely thought that one or more of the radio galaxies in the centre of the cluster provides a likely source for particles which then diffuse out into the halo (Giovannini et al., 1993); a process which depends critically on the structure of this field in the intra-cluster medium. Recent observations of the emission from NGC4869 (Feretti et al., 1995), which occupies a central position in the Coma cluster, indicate the presence of a magnetic field which is both stronger ($B \approx 8\mu$ G) and tangled on much shorter scales $\lesssim 1$ kpc than had been thought previously (Kim et al., 1990). These new results suggest not only a shorter cooling time for energetic electrons, but also a slower rate of diffusive transport. In this paper we show the constraints that the new observations place on transport theories of the relativistic electrons.

2. Electron transport

We consider the magnetic field to be a superposition of a regular field, B, and two kinds of fluctuation; microscopic and macroscopic. Microscopic fluctuations, on the scale of a particle gyroradius (r_g), result in spatial diffusion parallel and perpendicular to the magnetic field $\kappa_\parallel = \kappa_B/\epsilon$ and $\kappa_\perp = \epsilon\kappa_B/(1 + \epsilon)$ where $\epsilon \ll 1$ describes the amplitude of the fluctuations and

R. Ekers et al. (eds.), Extragalactic Radio Sources, 339–340.

$\kappa_B = r_g v/3$ is the Bohm coefficient with v the electron speed. Macroscopic fluctuations, with a scale length much larger than r_g, cause the field lines to wander. In the quasilinear regime $b \equiv \langle \delta B^2 \rangle^{1/2}/B \ll 1$, this can be described as a magnetic diffusion perpendicular to the average field direction with coefficient $D_M = b^2 \lambda_\parallel/4$, where $\lambda_\parallel$ is the correlation length along the field direction.

The effective diffusion coefficient of an electron depends on whether the mean free path for scattering on the microscopic fluctuations is longer or shorter than $\lambda_\parallel$ (Duffy et al., 1995). In the former case, with $c/2$ the average speed along a field line and $b = 1$, the spatial diffusion coefficient is $D_{\text{eff}} = D_M v = 10^{31}(\lambda_\parallel/1\,\text{kpc})\,\text{cm}^2\,\text{s}^{-1}$ independent of the magnetic field strength. Diffusive transport over a distance R requires $R^2/D_{\text{eff}} < t_{\text{synch}}$ for $B > 3\,\mu G$ where t_{synch} is the synchrotron cooling time. Combining the above requires that $\lambda_\parallel > 40\,\text{pc}$, in contradiction with the recent measurements, provided one identifies the correlation length with the scale of field reversals. This limitation can be avoided if the electrons are continually accelerated while they diffuse. If the microscopic fluctuations responsible for diffusion are Alfvén waves, acceleration is to be expected (Schlickeiser et al., 1987). Compensation for the synchrotron cooling requires $\kappa_\parallel = t_{\text{synch}} v_A^2/9$. In the presence of microturbulence sufficient to give the required $\kappa_\parallel$, the mean free path is shorter than the correlation length $\lambda_\parallel$, and the effective diffusion coefficient results from diffusion along field lines which themselves diffuse, a process known as 'compound diffusion' (Duffy et al., 1995): $D_{\text{eff}} = b^4 \kappa_\parallel/2$. The time required to propagate a distance of 500 kpc is then $b^{-4} \times 4 \times 10^{11}$ years, which substantially exceeds the age of the cluster.

3. Conclusions

The constraints placed on quasilinear theories of electron transport in the tangled magnetic field of the Coma cluster are that with or without continuous reacceleration particles cannot be transported from the centre to the edge of the diffuse emission region. Plausible alternatives, although not without their own difficulties, include acceleration models beyond quasilinear theory or local injection and acceleration of electrons.

References

Duffy, P., Kirk, J.G., Gallant, Y.A., Dendy, R.O. 1995 A&A, **302**, L21
Feretti, L., Dallacasa, D., Giovannini, G., Tagliani, A. 1995 A&A, **302**, 680
Giovannini, G., Feretti, L., Venturi, T., Kim, K.-T., Kronberg, P.P. 1993 ApJ, **406**, 399
Kim, K.-T., Kronberg, P.P., Dewdney, P.E., Landecker, T.L. 1990 ApJ, **355**, 29
Schlickeiser, R., Sievers, A., Thiemann, H. 1987 A&A, **182**, 21
Tribble, P.C. 1993 MNRAS, **263**, 31

COMA CLUSTER: CR ACCELERATION AND GAMMA-RAY PRODUCTION

A.V. DOGIEL

P.N.Lebedev Institute, Leninski pr.53, 117924 Moscow, Russia

The Coma cluster is situated near to the North Pole and its galactic coordinates are: $l \sim 50^0$ and $b \sim 87^0$. The distance to the cluster is about 138 Mpc. Observations in soft X-rays (see e.g. Briel, 1992) discovered there the thermal emission from the diffuse hot intracluster gas whose density and temperature are: $n \sim 3 \cdot 10^{-3}$ cm^{-3} and $T \sim 10^8$K. Measurements of hard X-ray flux from Coma showed controversial detection in different experiments (see Bazzano et al. 1990 and Rephaeli et al. 1994).

Radio investigations of Coma show an extended halo (see e.g. Schlickeiser et al. 1987, Kim et al. 1990, Giovannini et al. 1993). Radiosize of the halo is 40'. The radio flux is generated by relativistic electrons through their synchrotron losses. The origin of these electrons in the halo is unclear. They cannot be ejected by one of radio galaxies in the cluster since the luminosity of the likeliest candidate of the electrons is by a factor 10 less than that is needed to supply the Coma halo. Therefore two models of for the electron origin are usually discussed: a) in-situ acceleration and b) production of secondary electrons in the halo (see Schliekeiser et al. 1987).

Parameters of charge particle propagation in the halo are determined by the structure of magnetic fields there.

As it follows from the observations of Faraday depolarization there are two magnetic field components in the Coma halo: the tangled magnetic field has a strength of about 8.5±1.5 μG, while the uniform component is much weaker, about 0.2±0.1 μG. The scale of magnetic fluctuations is less than 1 kpc (Feretti et al. 1995). On the other hand, Kim et al. (1990) obtained that the magnetic field reversal occurs on a scale size of 14 –40 kpc and its strength is 2μG.

Particle propagation in these tangled magnetic fields is described as

R. Ekers et al. (eds.), Extragalactic Radio Sources, 341–342.

diffusion with coefficient (see Dogiel et al. 1987)

$$D = c^2 \int\limits_{-\infty}^{\infty} \int\limits_{-\infty}^{t} \left(\mathbf{h}(\mathbf{r},t)\delta(\mathbf{r}' - \rho(\mathbf{r},\tau,t))\mathbf{h}(\mathbf{r}',t') \right) d^3r\, dt'$$

where ρ is the trajectory of magnetized particles, $\tau = t - t'$ and $\mathbf{h} = \mathbf{H}(\mathbf{r},t)/\mid \mathbf{H}(\mathbf{r},t)\mid$. For the spectra of magnetic field fluctuations in Coma the diffusion coefficient can be approximated by $D_{xx} \geq (\pi c L_{cor})/6 \sim 10^{31} cm^2/s$, where L is the correlation length of magnetic field fluctuations.

Assuming in-situ acceleration we immediately obtain the parameter of Fermi II acceleration if the characteristic velocity of magnetic fluctuations is of the order of the thermal velocity u: $D_{pp} \sim u^2/D_{xx} \leq 10^{-15}s$

The accelerated electrons generate in the halo radio and gamma-ray fluxes. From the observed radio flux we could estimate the Coma gamma-ray emission. Its value strongly depends on the strength of magnetic fields there.

If the field strength is about of 10^{-5}G then the inverse Compton gamma-ray flux near the earth at 100 keV is $\sim 10^{-8} ph/cm^2 s$ and bremsstrahlung radiation: $\sim 10^{-6} ph/cm^2 s$. In both cases this flux cannot be detected by the OSSE telescope.

If the magnetic field strength is $\sim 10^{-6}$G the inverse Compton flux is $10^{-5} ph/cm^2 s$ (still not seen by OSSE). The bremsstrahlung flux, however, is large enough ($\sim 2 \cdot 10^{-4} ph/cm^2 s$) to be detected by this telescope.

In the case of secondary electron production we could estimate the gamma-ray flux from π^0-decay. It equals $10^{-9} \div 10^{-8} ph/cm^2 s$ for both cases that is less than the EGRET telescope sensitivity.

References

Bazzano, A., Fusco-Femiano, R., Ubertini, P., Perotti, F., Quadrini, E., Court, A.J., Dean, A.J., Dipper, N.A., Lewis, R. (1990) Hard X-rays from Coma cluster region,*Astrophys.J.*, **362**, pp. L51–L54.

Briel, U.G., Henry, J.P., Böhringer, H. (1992) Observations of the Coma cluster of galaxies with ROSAT during the All-Sky Survey, *Astron.Astrophys.*, **259**, pp.L31–L34.

Dogiel, V.A., Gurevich, A.V., Istomin, Ya.N., Zybin, K.P. On relativistic particle acceleration in molecular clouds, *Mon.Not.Roy.Astron.Soc.*, **228**, pp.843–868.

Feretti, L., Dallascasa, D., Giovannini, G., Tagliani, A. (1995) The magnetic field in the Coma cluster, *Astron.Astrophys.*, **302**, pp.680–690.

Giovannini, G., Feretti, L., Venturi, T., Kim, K.-T., Kronberg, P.P. (1993) The halo radio source Coma C and the origin oh halo sources, *Astrophys.J*, **406**, pp.399–406.

Kim, K.-T., Kronberg, P.P., Dewdney, P.E., Landecker, T.L. (1990) The halo and magnetic field of the Coma cluster of galaxies, *Astrophys.J*, **355**, pp.29–37.

Rephaelli, Y., Ulmer, M., Gruber, D. (1994) OSSE search for high-energy X-ray emission from the Coma cluster, *Astrophys.J.*, **429**, pp.554–556.

Schlickeiser, R., Sievers, A., Thiemann, H. (1987) The diffuse radio emission from the Coma cluster, *Astron.Astrophys.*, **182**, pp. 21–35

THE RADIO LOBES OF VIRGO A

H. ROTTMANN, K.-H. MACK, U. KLEIN
Radioastronomisches Institut der Universität Bonn - Auf dem Hügel 71, 53121 Bonn, Germany

R. WIELEBINSKI
Max-Planck Institut für Radioastronomie - Auf dem Hügel 69, 53121 Bonn, Germany

N. KASSIM
Naval Research Laboratory, Washington DC 20375-5351, USA

AND

R. PERLEY
NRAO, P.O. Box 0, Socorro, New Mexico 87801, USA

1. Introduction

In the framework of our multi-frequency study of Virgo A we have performed observations of Vir A at 10.55 GHz with the Effelsberg 100-m telescope. Using our improved CLEAN procedure for single dish data we have increased the dynamic range to some 40 dB.

By applying our newly developed polarization CLEANing technique we are able to diminish instrumental polarization effects. Since Faraday rotation is negligible at $\lambda 2.8$ cm the measured linear polarization is a direct trace of the projected magnetic field in Vir A. In combination with low-frequency data obtained with the VLA it is possible to determine parameters like spectral indices, break frequencies, and spectral ages.

2. The Radio Maps

Sensitive maps of Virgo A have been obtained at $\lambda 2.8$ cm using the Effelsberg 100-m telescope, including all Stokes parameters to derive the linear polarization. For spectral investigations maps have been obtained with the VLA at frequencies of 74 MHz, 333 MHz, and 1.46 GHz (Kassim et al.,1993). The lobes surrounding the central core show a roughly 'S-shaped' structure, with two brighter components, one to the east and one to the

343

R. Ekers et al. (eds.), Extragalactic Radio Sources, 343–344.

southwest of the core. The perception that the extended components are radio lobes, rather than forming a diffuse halo surrounding the core (as was believed in the past), is strongly supported by the polarization data. The map of the polarized intensity exhibits the same dual structure, with an eastern and southwestern component. The polarized emission is found to be strongest at the outer edges of these components. This coincides with regions of an increased degree of polarization, with maximum values exceeding 70%, which is close to the theoretical limit. This is a feature frequently seen in double-lobed radio galaxies and can be explained in terms of compression of the magnetic field as the jet material is impinging on the intracluster medium (ICM) and is deflected. Virgo A exhibits a distinct asymmetry between the southern (which is the side of the approaching jet) and the northern lobe (the counterjet side) in three respects :

- The southern region of the source shows an increased average brightness by a factor of 1.2 .
- The mean degree of polarization is higher by a factor of 1.15 on the southern side compared to the northern side of the source.
- The spectrum on the jet side is flatter than on the counterjet side.

The observed asymmetries are prevalent in powerful double radio sources with one-sided jets as described by Laing et al.(1988).

3. Spectral investigations

We have constructed maps, showing the distribution of the spectral index across Vir A. For this purpose three frequencies (333 MHz, 1.4 GHz, and 10.55 GHz) were used. The spectral indices range from $\alpha=0.4$ in the core region to $\alpha=2.2$ in the outer areas of the source ($S_\nu \sim \nu^{-\alpha}$). These values are noticeably lower than the ones previously published (Andernach et al., 1979). The spectral steepening, caused by synchrotron and inverse Compton losses, can be used to calculate break frequencies ν_b, by fitting spectral models to the data, following the method proposed by Carilli et. al. (1991). We have calculated ν_b for the JP (Jaffe & Perola, 1973) model. The spectral age of the relativistic particles is computed using a relation of Alexander et al. (1987) for both radio lobes.

Eastern Lobe : $\nu_b = 7.5$ GHz $t = 36$ Myrs
Southern Lobe : $\nu_b = 11.9$ GHz $t = 29$ Myrs

References

Alexander,P., Leahy,J.P.: 1987, *MNRAS* **225**, 1
Andernach,H.,Baker,J.R.,von Kap-herr,A.,Wielebinski.R.: 1979, *A&A*, **74**, 93
Carilli,C.L.,Perley,R.A.,Dreher,J.W.,Leahy,J.P.: 1991, *Ap.J.*, **383**, 554
Jaffe,W.J.,Perola,G.C.: 1973, *A&A*, **26**, 423
Kassim,N.E.,Perley,R.A.,Erickson,W.C.,Dwarakanath,K.S.: 1993, *AJ*, **106**, 2218
Laing,R.A.: 1988, *Nature*, **331**, 149

THE RELIC RADIO SOURCE B2 0924+30

U. KLEIN AND K.-H. MACK
Radioastronomisches Institut der Universität Bonn
Auf dem Hügel 71, 53121 Bonn, Germany

AND

L. GREGORINI AND P. PARMA
Istituto di Radioastronomia del C.N.R.
Via P. Gobetti, 40129 Bologna, Italy

1. New radio maps

The double radio source B2 0924+30, associated with the luminous E/S0 galaxy IC2476, may be considered a prototypical genuine relic of a 'dead' radio galaxy as it seems to perfectly fulfill the following criteria: It has a rather steep overall radio spectrum (Ekers et al., 1981; Cordey, 1987). Its core luminosity is by far the lowest known so far (Giovannini et al., 1988). No coherent jet structure or other signs of activity are visible. Since only four possibly genuine relics of radio galaxies are known so far (Harris et al., 1993), a study of the archetypical source B2 0924+30 is of eminent importance for the understanding of this rare species of radio galaxies.

We have obtained maps of B2 0924+30 at $\lambda\lambda$49 and 6.3 cm with the WSRT (HPBW 29″×57″) and with the Effelsberg 100-m telescope (HPBW 150″), respectively. A bright source located at the periphery of the western lobe, which is not related to the radio galaxy and identified with a quasar (Ekers et al., 1981; Arp, 1977) was subtracted from the maps prior to further analysis. We have smoothed the λ49 cm map to the 150″ beam size in order to carry out a spectral comparison. A strong decrease of the lobe width at the higher frequency is noticed.

2. Spectral index and particle ages

We have compiled the total flux densities of B2 0924+30 by integrating the two maps in elliptical rings adapted to the overall source shape, with careful inspection of the map zero levels, and the quasar subtracted. A

345

R. Ekers et al. (eds.), Extragalactic Radio Sources, 345–346.

pronounced steepening is seen beyond about 2 GHz. Between 151 and 408 MHz the spectral index is $\alpha = 0.91 \pm 0.13$ ($S_\nu \sim \nu^{-\alpha}$), while between 4.75 and 10.55 GHz it is $\alpha = 2.25 \pm 0.52$, implying a steepening by $\Delta\alpha = 1.34$, which is significant in spite of the relatively large error at the highest frequency.

We have computed the spectral index distribution across B2 0924+30 between $\lambda\lambda 49$ cm and 6.3 cm. No significant asymmetry of the spectrum of the lobes is evident. A pronounced steepening of the spectrum is seen away from the major axis, where the spectrum steepens to $\alpha = 1.6$. The latter behaviour is reminiscent of the source B2 1321+31 (Klein et al., 1995), although with a less pronounced steepening in that case.

The integrated spectrum of B2 0924+30 and its spectral index distribution have been analyzed in order to determine the characteristic break frequency of the synchrotron emission and corresponding particle ages, assuming that synchrotron and inverse Compton losses are at work. Since particle injection is presumably no longer taking place, we have fitted the model of Jaffe & Perola (1973) to the integrated spectrum yielding a break frequency $\nu_B = 6.5$ GHz and an 'injection' spectral index of $\alpha_i = 0.84$. With an equipartition field strength of 1.4 μG this results in an average particle age of 50 Myrs.

Since we have only two frequencies at our disposal we have applied the Myers-Spangler algorithm (Myers & Spangler, 1985) to the maps. With the inferred equipartition magnetic field we obtain a break frequency of around 4 GHz along the source's main axis, implying an age ≈ 70 Myrs, while away from the lobes these values are 3 GHz and ≈ 80 Myrs, respectively.

The particle ages derived above are not very high. Since they reflect the lifetime of the relativistic particles after the cease of energization of the former hot spots by the central engine, this means that once the energy transfer to the hot spots is stopped the whole radio source dies relatively quickly. The relatively short time scale resulting for the switch-off of B2 0924+30 naturally explains the relative paucity of such relic sources.

References

Arp H. (1977) *IAU Coll.* **37**, 377, Editions du centre national de la recherche scientifique, Paris

Cordey R. A. (1987) *MNRAS*, **227**, 695

Ekers R. D., Fanti R., Lari C., and Ulrich M.-H. (1975) *Nature*, **258**, 584

Ekers R. D., Fanti R., Lari C., and Parma P. (1981) *A&A*, **101**, 194

Giovannini G., Feretti L., Gregorini L., and Parma P. (1988) *A&A*, **199**, 73

Harris D. E., Stern C. P., Willis A. G., and Dewdney P. E. (1993) *AJ*, **105**, 3

Jaffe W. J. and Perola G. C. (1973) *A&A*, **26**, 423

Klein U., Mack K.-H., Gregorini L., and Parma P. (1995) *A&A*, in press

Myers S. T. and Spangler S. R. (1985) *ApJ*, **291**, 52

RADIO SPECTRA AND PARTICLE AGES OF THE HEAD-TAIL RADIO GALAXY NGC1265

L. FERETTI AND G. GIOVANNINI

Istituto di Radioastronomia del C.N.R. - Via P. Gobetti, 41029 Bologna, Italy

U. KLEIN AND K.-H. MACK

Radioastronomisches Institut der Universität Bonn - Auf dem Hügel 71, 53121 Bonn, Germany

AND

L.G. SIJBRING

Kapteyn Astronomical Institute - Postbus 800, 9700 AV Groningen, The Netherlands

1. New radio maps

We have performed sensitive observations of three classical head-tail radio galaxies at $\lambda 11.1$, 6.3, and 2.8 cm using the Effelsberg 100-m telescope (Zech, 1994). Complete maps of the sources 3C129, NGC1265, and 3C465 were obtained, including the distributions of the linearly polarized intensity. Together with the low-frequency interferometric maps these allow a comprehensive study of their radio spectra and, based on models of particle losses, the derivations of particle ages across these sources. The highest frequency involved allows an unambiguous derivation of the projected magnetic field structure, unimpeded by Faraday effects. Here we focus on NGC1265, which is located in the Perseus Cluster.

The head-tail structure is well visible out to a large distance from the source's head. The twin jets in the vicinity of the head are not resolved with our HPBW. As is to be expected, strong linear polarization is found in the tail. The degree of polarization reaches its maximum towards the tail's end, with values exceeding 50%. This means that the fractional polarization is close to its theoretical value.

347

R. Ekers et al. (eds.), Extragalactic Radio Sources, 347–348.

2. Spectral index and particle ageing

We have computed the distributions of the spectral index across NGC1265, with the aim to investigate the particle ageing across its tail. Tails of radio sources are particularly suited for this purpose due to their extended, relaxed structure. The low-frequency data of NGC1265 were obtained by Sijbring (1993) at $\lambda\lambda92$ and 49 cm with WSRT. All maps were smoothed to a common resolution, i.e. $147''$ when the Effelsberg $\lambda6.3$ cm map was involved, or $69''$, if only the map at $\lambda2.8$ cm was used for the computation. The spectral index computed between $\lambda\lambda92$, 49 and 2.8 cm ranges from $\alpha = 0.5$ in the head to $\alpha = 1.6$ at the tail's end ($S_\nu \sim \nu^{-\alpha}$). The spectral index maps computed between $\lambda11.1$, 6.3, and 2.8 cm exhibit a dramatic steepening to $\alpha \geq 2.0$ towards the tail's end in spite of the low angular resolution of the maps. We have used all maps to calculate break frequencies and corresponding particle ages with a fair angular resolution. We have adopted the method described by Carilli et al. (1991). In the faint tails, inverse Compton losses against the 3K background may also be important so that we use the relation between particle age and break frequency as e.g. given by Alexander and Leahy (1987). The break frequencies have been calculated for the so-called JP (Jaffe and Perola, 1973) model, which accounts for re-isotropization of the pitch angles.

The map of the break frequency exhibits values from 17 GHz in the head of the source to 2.7 GHz at the tails' periphery, corresponding to particle ages of ~10 Myrs and ~45 Myrs, respectively. With a Hubble constant of $H_0 = 75$ km s^{-1} Mpc^{-1} the tail's end is at a projected straight distance of 280 kpc from the host galaxy, and probably a lot more, if we account for projection effects and initial bending of the jets. Assuming that the particles were left behind the host galaxy once they emerged from the inner jet regions, they had to travel at a speed of at least ~6000 km s^{-1} in order to get where they are. This is in flat contradiction with the typical velocities of cluster galaxies, even though the member galaxies of the Perseus Cluster are known to have a relatively high velocity dispersion (1308 km s^{-1}). This discrepancy implies that the particles still have to move at a considerable speed along the tails. It is conceivable that the very uniform magnetic field indicated by our polarization measurements alleviates this rapid propagation.

References

Alexander,P., Leahy,J.P.: 1987, *MNRAS* **225**, 1
Carilli,C.L.,Perley,R.A.,Dreher,J.W.,Leahy,J.P.: 1991, *ApJ* **383**, 554
Jaffe,W.J.,Perola,G.C.: 1973, A&A **26**, 423
Sijbring, L.G.: 1993, *Ph.D. thesis*, Univ. Groningen
Zech, G.: 1994, *Diploma thesis*, Univ. Bonn

OBSERVATIONS OF STRAIGHT-ANGLE TAILED
RADIO GALAXIES IN RICH CLUSTERS OF GALAXIES

AILEEN A. O'DONOGHUE
St. Lawrence University, Canton, New York 13617 USA

JEAN A. EILEK
New Mexico Institute of Mining and Technology
Socorro, New Mexico 87801 USA

AND

FRAZER N. OWEN
NRAO-VLA, Socorro, New Mexico 87801 USA

1. The SAT Multifrequency Study

We have begun VLA observations of straight-angle tailed radio sources (SATs) at 1.5 and 4.8 GHz (L and C band) to achieve one arcsecond resolution at each frequency. This will provide a SAT data set similar to the O'Donoghue, Owen, and Eilek (1990) WAT data set with both total intensity and spectral index information. We will use these data to examine morphological and dynamical properties of straight-tailed radio sources in clusters of galaxies.

The SATs in this study span a morphological range across the FRI/II classification. The current wisdom has it that radio sources are either FR-type I or type II and is supported by Owen's (1991) radio-optical diagram in which FRI and FRII sources occupy different parts of radio power/parent galaxy luminosity space. The SATs all fall into the FRI region of his plot, but some display FRII morphology. Are these the low radio power FRII's unrepresented in Eilek's (1996a) luminosity functions?

SATs have not been studied in a way comparable to NATs (*eg.* O'Dea and Owen 1987) or WATs (*eg.* O'Donoghue *et al* 1990, 1993, Loken *et al* 1995, Eilek 1996b, Katz-Stone *et al* 1996a), although a few individual SATs have been studied in great detail (*eg.* 3C31, Perley *et al* 1996; 3C449, Katz-Stone *et al*, 1996b). This is surprising because (1) they seem closest to the

349

R. Ekers et al. (eds.), Extragalactic Radio Sources, 349–350.
© 1996 IAU.

original picture of a type I source, without the complication of bending, and (2) inspection of the snapshot images from the Owen *et al* (1993) survey show that SATs comprise a greater range of morphology than NATs or WATs. We suspect they are, physically, the bridge between classic FRI and FRII sources.

2. Selection of Sources

The Owen *et al* (1993) cluster sample shows that sources which are type I's based on their radio and optical power, are more complex than was apparent in the 3C images upon which the FR classification was based. Approximately 1/3 of the resolved sources in this sample fall into our SAT category and a variety of detail is found within this source type. The 8 sources we are observing span the morphological range, but are concentrated at the ends of the spectrum and at the transition point.

(1) Nearly FRII Sources: 0124+189 and 0745+521 Two sources that show well-collimated, bright jets with associated "backflow" lobes, although without the highly distinctive outer hot shot characteristic of the original FRII classification.

(2) Twin Tail Sources: 0306-237, 0738+441, and 1233+237 Three sources have less well-collimated jet flows and are reminiscent of the classic FRI sources such as 3C449. They also, however, show hints of broad "cocoons" surrounding the main jet, which suggests they may connect directly to the nearly FRII sources sources.

(3) Cocoon Sources: 0136+185, 1530+282, 2055-079 Three sources show jets embedded in strong cocoons or halos, which extend as far as the end of the jets so that they could be either backflow or a lobe being shed as the source grows. The relative strengths of the jet and cocoon varies from source to source, making them appear transitional between the nearly FRII sources and the twin tail sources sources.

References

Eilek, J. A. (1996a), in preparation (luminosity functions).
Eilek, J. A. (1996b) submitted to *Ap. J.* (WATs).
Katz-Stone *et al* (1996a) in preparation (WATs).
Katz-Stone *et al* (1996b) in preparation (3C449).
Loken C. *et al* (1995) *Ap. J.* **445**, 87.
O'Dea, C.P. & Owen, F.N. (1987) *Ap. J.* **316**, 95.
O'Donoghue, A.A., Owen, F.N. & Eilek, J.A. (1990) *Ap. J. Supp.* **72**, 75.
O'Donoghue, A.A., Eilek, J.A. & Owen, F.N. (1993) *Ap. J.* **408**, 428.
Owen, F. N. (1991), Steps Toward a Radio H-R Diagram, in *Jets in Extragalactic Radio Sources, Proceedings of a Workshop Held at Ringberg Castle, Tegernsee, FRG*, ed. H.-J. Rser and K. Meisenheimer (Berlin:Springer-Verlag).
Owen, F.N. White, R.A. & Ge, J. (1993) *Ap. J. Supp.* **87**, 135.
Perley, R.A. *et al* (1996), in preparation.

THE ENVIRONMENT OF HERCULES A

N.A.B. GIZANI AND J.P. LEAHY
University of Manchester,
NRAL, Jodrell Bank, Cheshire, SK11 9DL, UK.

1. Introduction

The bright radio galaxy Hercules A ($z = 0.154$), has power $P_{178MHz} = 8 \times 10^{26}$ W Hz^{-1}sr^{-1} and although this puts it well above the FR break, its structure is intermediate between Fanaroff and Riley classes I and II. With a linear size of 334 kpc (for $H_0=100$ kms^{-1}Mpc^{-1}), Hercules A possesses an unusual jet-dominated morphology and no compact hotspots. The western jet shows partial or full ring-like features that form a linear sequence heading from the core to the lobes and follow an inner jet.

The host galaxy is a cD at the centre of a cluster with an X-ray luminosity of 1.5 $\times 10^{37}$ W (Sadun *et al* 1993), comparable to typical richness 0 to 1 Abell clusters. By studying the environment in detail we hope to find out whether the peculiar structure is related to the unusually dense environment.

2. The Laing-Garrington effect

Strongly asymmetrical Faraday depolarization is revealed by low resolution VLA polarization data (Garrington & Holmes unpublished) at 6 and 18 cm, the side with the stronger jet being less affected. Hence Her A exhibits a strong Laing-Garrington effect (Garrington *et al* 1988, Laing 1988). The depolarization seems stronger at the centre of the source, especially in the more depolarized western lobe. We are currently observing Hercules A at $\sim 1''$ resolution at several wavelengths. Preliminary results at 6 cm showed the polarization to be disordered in the western lobe, and mostly ordered in the eastern lobe. We interpret this disordering as the result of the Faraday rotation which causes depolarization at lower frequency and resolution. This rotation is apparently external, because at high resolution we see rotation and not depolarization. Hercules A is 28% polarized on average at this

R. Ekers et al. (eds.), Extragalactic Radio Sources, 351–352.
© *1996 IAU.*

resolution, and the scale size of the rotation measure structure, and hence the external magnetic field, is ~ 20 kpc at the end of the western lobe, with smaller scales close to the centre. Our A-configuration observations at L band show a strong depolarization of the western lobe, and a very disordered polarization in the east, rather like the western lobe at 6 cm.

All this suggests the depolarization is caused by a centrally condensed medium in which Her A is embedded at a substantial angle. If the depolarization in the west is external, then the rotation measure is not yet resolved at L-band. Most probably the Faraday-active medium is the 'X-ray' gas, with the asymmetry due to the weak jet and associated lobe being behind the bulk of the gas while the other lobe is in front. Therefore the jet asymmetry must be largely due to relativistic boosting.

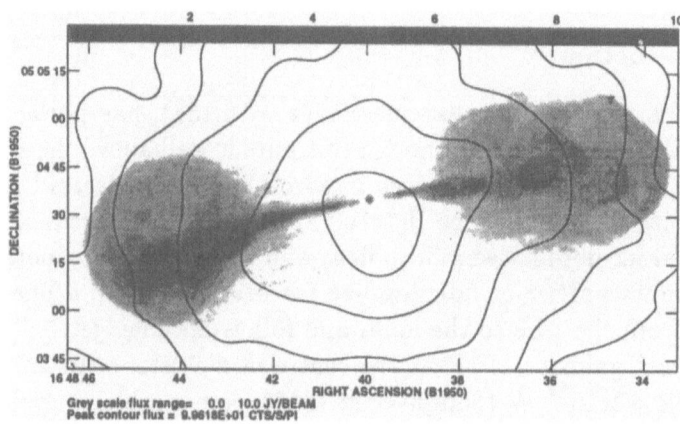

Figure 1. A contour map of the ROSAT image of Hercules A cluster in the 0.5-2 keV band, overlaid on a grey scale of the 6 cm radio map. Contours are separated by a ratio of 2.

3. ROSAT observations

We observed Her A with ROSAT's PSPC. The cluster emission is aligned with the radio axis and extended on roughly the same scale as the lobes. The cluster core is resolved and at most 8% of the total flux can be attributed to a nuclear component. Radial profile plots show that faint X-ray emission extends out to 440 kpc. Global spectral fitting to the Her A cluster source in our PSPC data yields an excellent fit to a Raymond-Smith model with $kT \sim 3$ keV, and $N(H) = 6.2 \times 10^{20}$ cm^{-2}.

References

Garrington, S.T. and Leahy, J.P. Conway, R.G. and Laing, R.A., *Nat.*, 1984, **331**, 147

Laing, R.A., *Nat.*, 1988, **331**, 149

Sadun, A.C. and Hayes, J.J.E., *Astr.Soc.Pac.Publ.*, 1993, **105**, 379

MULTI-WAVELENGTH STUDY OF
ROSAT CLUSTERS OF GALAXIES

A. D. REID AND R. W. HUNSTEAD
School of Physics, University of Sydney
NSW 2006, Australia

AND

M. M. PIERRE
CEA/DSM/DAPNIA CE Saclay, France

We are engaged in a radio–IR–optical–X-ray study of two flux-limited samples of ROSAT clusters of galaxies south of declination $-20°$ (Pierre *et al.*, 1994a). One sample covers an area of 1750 deg^2 in Hydra (Pierre *et al.*, 1994b) and includes some distant ($z \approx 0.3$) and X-ray luminous ($L_X \approx 10^{45}$ erg s^{-1}) clusters. The other sample derives from a volume-limited subset (nominally $z \leq 0.1$) of southern X-ray clusters, which are the focus of an ESO Key Program (Guzzo *et al.*, 1995).

An observing program with the Molonglo Observatory Synthesis Telescope (MOST) at 843 MHz and the Australia Telescope Compact Array (ATCA) at 1.4 and 2.4 GHz has been undertaken to complement existing optical spectroscopy and photometry (ESO and CFHT), pointed X-ray (ROSAT PSPC and HRI) images, and scheduled FIR (ISO) observations.

Our radio observations, together with data at other wavelengths, are helping to give a more comprehensive picture of the cluster environment. As an example, we have chosen a cluster in the Hydra region, ROSAT RXJ 12 54.4 -29 01, part of the A3528 complex at $z = 0.0535$ and located in an important region of the sky, the Shapley 8 supercluster. In Figure 1 we show a preliminary montage of images for this cluster (Pierre *et al.*, 1995, in prep.). There is a strong radio source (A: 0.59 Jy total at 843 MHz) coincident with the X-ray centroid, suggesting that non-thermal emission from an active nucleus may be contaminating the extended thermal X-ray emission from the intracluster medium. MOST detects an additional source (B: 0.52 Jy total) south of the X-ray centroid but still within the X-ray envelope.

R. Ekers et al. (eds.), Extragalactic Radio Sources, 353–354.
© 1996 IAU.

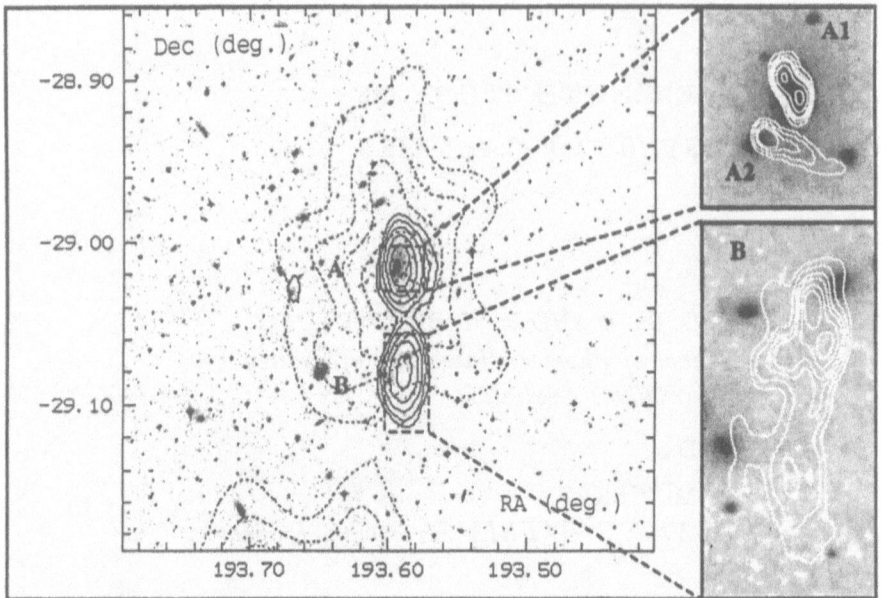

Figure 1. Multi-wavelength overlay in the region of A3528; coordinates are for equinox J2000. *Left:* X-ray contours of RXJ 12 54.4 −29 01 are shown dotted and MOST contours as continuous, against a background of the NASA/STScI digitized sky survey. *Right:* ATCA contour images of source A at 2.4 GHz and B at 1.4 GHz are shown in white, superposed on smoothed greyscale images from the DSS.

The ATCA images reveal considerable structure in both these sources. At 2.4 GHz A is clearly resolved into two separate sources. The northern component A1 is a close double, possibly a wide-angled tail source, coincident with the cD galaxy. The southern component A2 has a head-tail morphology and is most likely identified with the bright galaxy lying close to the radio peak. These identifications are supported by lower-resolution 4.8 GHz VLA images (Gregorini *et al.*, 1994). Source B, which is relatively diffuse, appears to have a narrow-angled tail structure associated with the bright galaxy just to the NW of the two emission peaks. The implied direction of motion is towards the X-ray centre and source A.

References

Gregorini, L., de Ruiter, H.R., Parma, P., Sadler, E.M., Vettolani, G. & Ekers, R.D. (1994), *A&AS*, **106**, 1.

Guzzo, L. and 26 others (1995) in *Wide-Field Spectroscopy and the Distant Universe*, eds. S. J. Maddox and A. Aragón-Salamanca (World Scientific: Singapore), p. 205.

Pierre, M., Hunstead, R., Reid, A. and 10 others (1994a), *ESO Messenger*, **78**, 24.

Pierre, M., Böhringer, H., Ebeling, H., Voges, W., Schuecker, P., Cruddace, R. & MacGillivray, H. (1994b), *A&A*, **290**, 725.

ATCA 13 AND 20 CM OBSERVATIONS OF A3556

S. BARDELLI [1], R. MORGANTI [2,3], T. VENTURI [2],
AND R.W. HUNSTEAD [4]

[1] *Osservatorio Astronomico di Trieste - Italy*
[2] *Istituto di Radioastronomia, CNR- Bologna, Italy*
[3] *CSIRO-ATNF, Epping - Australia*
[4] *University of Sydney - Australia*

1. A3556 and the Radio Observations

We observed the cluster of galaxies A3556 ($< v > = 14300$ km/sec), belonging to the supercluster of galaxies known as the Shapley Concentration (Bardelli et al., 1994, MNRAS 267, 255, and references therein), with the Australia Telescope Compact Array. Our observations took place in the continuum band at 20 cm with the configurations 1.5D and 6C, and at 20 cm and 13 cm in the configuration 6A, for a total of 12×3 hours, with a resolution ranging from $\sim 6''$ to $\sim 40''$. We observed a sky region of $\sim 2° \times 1°$ around the cluster center taking advance of the mosaicing technique developed ATNF.

These observations are part of a larger project, whose main aim is to study the interaction and evolution of clusters of galaxies located in dynamically unrelaxed environments, as is the case for A3556, deriving informations from their radio emission.

2. Radio Emission from the Cluster Center

The radio emission in A3556 is mainly concentrated around its optical center (see Fig. 1), characterised by a cD galaxy with an apparent magnitude $m_V = 14.4$ and by a few other luminous galaxies. Most of the bright members in the cluster have unresolved associated radio emission.

The most prominent feature is the extended radio source J1324-31. The appearance of this radio source is very unusual. Our maps at various resolu-

355

R. Ekers et al. (eds.), Extragalactic Radio Sources, 355–356.

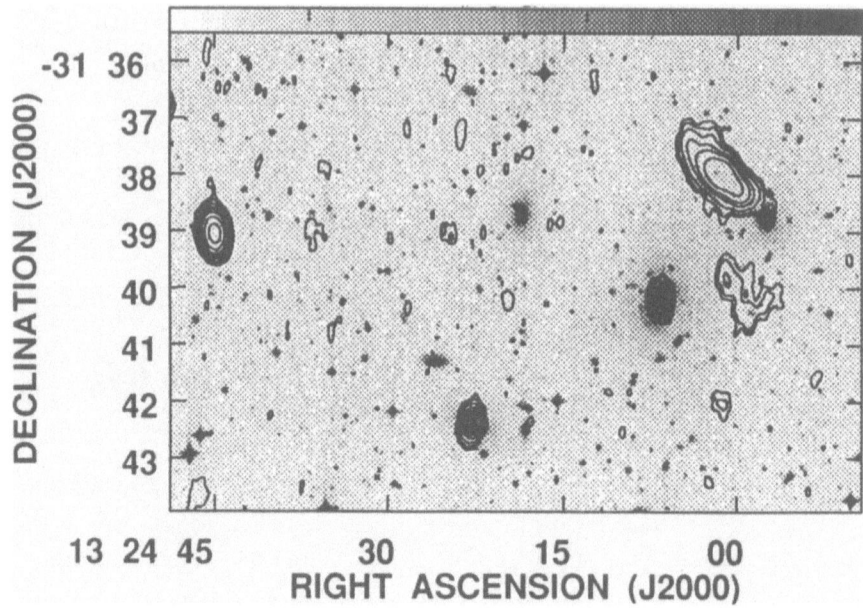

Figure 1.   20 cm image of the central region of A3556 superimposed on the DSS. The restoring beam is is 25.8″ × 15.6″ mas, in p.a. -1.21°. Peak flux is 29.3 mJy/beam, levels -0.3, 0.3, 0.5, 1, 2, 3, 5, 10, 20 mJy/beam.

tions and frequencies show that J1324-31 has a very low surface brightness, and an amorphous morphology, without compact features.

The nature of this source is still unclear and we have been investigating it. J1324-31 could be either a cluster radio source or a background object. Two cluster galaxies are located very close to the radio emission, a 15.6 magnitude elliptical galaxy ($< v > = 15022$ km/s), coincident with the far south end of the radio emission (see Fig. 1), and a 17.9 galaxy ($< v > = 15142$ km/s), located just outside the radio emission. A few other objects, without measured redshift, lie within the radio emission.

Assuming that the source is located at the cluster distance, its projected dimensions on the plane of the sky are $\sim 80 \times 25$ kpc ($H_0 = 100$ km s^{-1}Mpc^{-1}, $q_0 = 0$) and its monochromatic power at 20 cm is logP = 23.05. If we assume that equipartition holds within the source, then our 20 cm and 13 cm observations indicate that the equipartition magnetic field in the source is $H_{eq} \sim 1\mu$G and the minimum energy density is $U_{min} \sim 10^{-13}$ erg cm^{-3}.

A detailed study of J1324-31 as well as of the global radio properties of A3556 is in progress.

THE X-RAY INTRACLUSTER MEDIUM

H. BÖHRINGER

Max-Planck-Institute für extraterrestrische Physik
D 85740 Garching, Germany

1. Introduction

Clusters of galaxies are the largest aggregates of matter that have decoupled from the universal expansion and have approximately evolved to a proper dynamical equilibrium configuration. While in the optical they are just observed as dense concentrations of galaxies, they are seen in X-rays as continuously connected entities through the emission of the X-ray luminous intracluster plasma (e.g. Sarazin 1986). Relativistic particles and magnetic fields are observed in clusters in radio halos, in regions around radio galaxies and through Faraday rotation of the radiation from background radio sources (see contribution by L. Feretti and e.g. Kronberg, 1994). Thus Clusters of galaxies are the largest, well characterized astrophysical laboratories in which plasma physical processes of diffuse media can be studied. In this brief summary the bulk properties of the intracluster medium (ICM) are discussed and two studies of the interaction of galactic radio lobes with the intracluster plasma are presented (related topics are discussed by F. Owen and L. Feretti in this volume).

X-ray observations of galaxy clusters show, that the space between the galaxies in clusters is filled with a hot, tenuous ICM with temperatures between 10 and 100 Million degrees. The X-ray surface brightness distribution reflects the plasma density distribution in the cluster. X-ray spectroscopy provides the means to determine the temperature of the hot intracluster plasma. Assuming that the plasma is approximately in hydrostatic equilibrium in the cluster one can use the density and temperature profile in connection with the hydrostatic equation to derive the mass of the X-ray luminous gas and the total cluster mass. Such studies have been carried out in detail for tens of clusters in particular with the ROSAT Observatory (Trümper 1992). A good general picture of the mass, composition and the

357

R. Ekers et al. (eds.), Extragalactic Radio Sources, 357–360.

structure of galaxy clusters has emerged from these studies and the major results are summarized in the following table.

Components of massive galaxy clusters and their ICM

o **gravitational mass :** $3 \cdot 10^{14} - 3 \cdot 10^{15} M_\odot$

o **galaxies :** $\sim 2 - 7\%$ of the total mass

o **hot intracluster plasma:** $\sim 10 - 30\%$ of the total mass
   X-ray luminosity : $\sim 10^{44} - 5 \cdot 10^{45}$ erg s^{-1}
   energy content : $\sim 10^{62} - 10^{63}$ erg
   energy density (at the core radius) : $\sim 10^{-11} - 10^{-10}$ erg cm^{-3}

o **radio halo:** luminosity $\sim 10^{41} - 10^{42}$ erg s^{-1}
   energy density in relativistic particles: $\sim 10^{-12} - 10^{-11}$ erg cm^{-3}
   energy density in the magnetic field: $\sim 10^{-12}$ erg cm^{-3}
   energy content of the relativistic plasma: $\sim 10^{59} - 10^{60}$ erg
   necessary energy input to power the radio halo: $\sim 10^{42}$ erg s^{-1}

These results imply that the gas mass is much larger than the mass in galaxies and that most of the gravitationally inferred matter is "dark". The energy in the magnetic field and in relativistic particles is generally less than the energy in the thermal ICM and they are therefore unimportant for the gas dynamics and pressure (this conclusion is still under discussion, however).

2. Pressure interaction of radio lobes with the ICM in NGC 1275

The central dominant galaxies of galaxy clusters are found very often to be radio sources displaying radio lobes. The lobes are often bend for which interactions with the cluster ICM are made responsible. A good example is NGC 1275 in the Perseus cluster. Two radio lobes are seen which extend to a radius of almost 1 arcmin from the center (e.g. Pedlar *et al.*, 1990) and are on larger scales embedded in a "mini-radio-halo" (e.g. Noordham and de Bruyn, 1985). In a recent high resolution ROSAT observation a displacement of the hot ICM by the inner radio lobes has clearly been observed now for the first time (Böhringer *et al.*, 1993).

Fig. 1 shows the ROSAT HRI image of NGC 1275. Superposed as contour plot is the radio emission observed by Pedlar *et al.* (1990) with the VLA at 332 MHz. One notes the deficiency of the X-ray emission in the regions of the radio lobes out to a radius of 30 - 40 arcsec (15 - 21 kpc). The pressure in the surrounding ICM and in the lobes is almost equal (within a factor of 3; for a k-value of 100) which supports the picture that the minima in the X-ray surface brightness are due to a displacement by the lobe pressure.

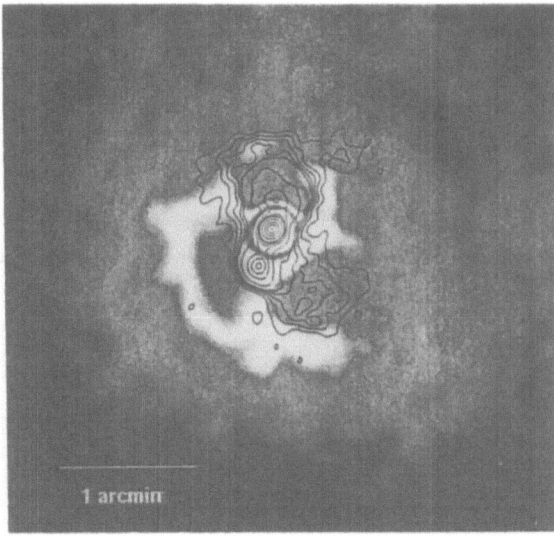

Figure 1. ROSAT HRI image of the central part of the Perseus cluster showing the halo region of the dominant galaxy NGC 1275. A contour map of the radio lobes of this galaxy observed by Pedlar *et al.* (1990) at the VLA at 332 MHz is superimposed onto the X-ray image.

3. The radio and X-ray halo of M87

The opposite effect of radio–X-ray correlation is observed in the halo of M87, the central galaxy of the Virgo cluster (Böhringer *et al.*, 1995). Fig. 2 shows the X-ray halo of the central region of the Virgo cluster from a deep ROSAT PSPC observation, where a spherically symmetric X-ray halo model has been subtracted from the image, and the radio map of Feigelson *et al.* (1987). The positive correlation out to a radius of ∼4 arcmin is clearly visible. A spectroscopic analysis of the X-ray data from the ROSAT observation shows that the excess emission in the radio lobes clearly originates from thermal plasma (Böhringer *et al.* 1995). Thus the emission is not due to the inverse Compton effect caused by the relativistic electrons in the radio lobes as it was speculated earlier (e.g. Andernach *et al.* 1987; Feigelson *et al.* 1987). Why the gas is more luminous, denser and also cooler in the radio lobes is not obvious; it has been speculated (Böhringer *et al.* 1995) that there may be cold gas and magnetic bubbles intermixed with the hot gas causing the observed effects.

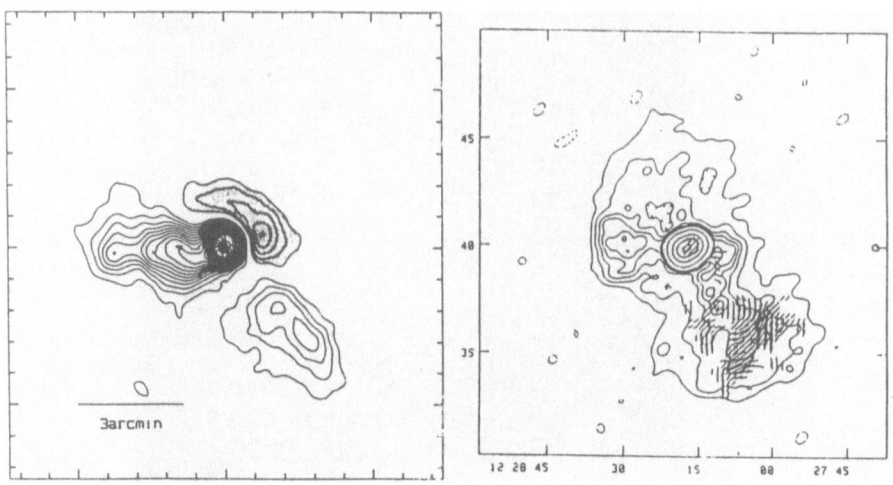

Figure 2. Contour plot (left panel) of the residual image of the X-ray halo of M87. The image was obtained from the cluster image by subtracting a spherically symmetric model halo. The Radio map (right panel) of this region was obtained by Feigelson *et al.* (1978) with the VLA at 1.4 GHz. The nearly perfect correlation of the excess X-ray emission with the radio lobes is well visible.

Acknowledgment

I like to thank the ROSAT team and my colleagues S. Schindler, W. Voges, and D.M. Neumann for their help.

References

Andernach, H., Baker, J.R., von Kap-Herr, A., Wielebinski, R., 1979, *Astron. Astrophys.*, **74**, 93

Böhringer, H., Schwarz, R.A., Briel, U.G., Voges, W., Ebeling, H., Hartner, G., Cruddace, R.G., 1992, in *Clusters and Superclusters of Galaxies*, Fabian, A.C. (ed.), Kluwer, p.71

Böhringer, H., Voges, W., Fabian, A.C., Edge, A.C., Neumann, D.M., 1993, *Mon. Not. R. astr. Soc.*, **264**, L25

Böhringer, H., Nulsen, P.E.J., Braun, R., Fabain, A.C., 1995, *Mon. Not. R. astr. Soc.*, (submitted)

Feigelson, E.D., Wood, P.A.D., Schreier, E.J., Harris, D.E., Reid, M.J., 1987, *Astrophys. J.*, **312**, 101

Kronberg, P., 1994, *Rep. Prog. Phys.*, **57**, 325

Noordham, J.E. and de Bruyn, A.G., 1982, *Nature*, **299** 1

Pedlar, A., Ghataure, H.S., Davies, R.D., Harrison, B.A., Perley, R., Crane, P.C., Unger, S.W., 1990, *Mon. Not. R. astr. Soc.*, **246**, 477

Sarazin, C.L., 1986, *Rev. Mod. Phys.*, **58**, 1

Trümper, J., 1992, *Q. J. R. astr. Soc.*, **33**, 165

A2241: CLUSTERS WITH HEAD-TAILS AT X-RAYS

L. NORCI
Dunsink Observatory
Castleknock, Dublin 15, Ireland

L. FERETTI
Istituto di Radio Astronomia
Via Gobetti 101, I-40129 Bologna, Italy

AND

E.J.A. MEURS
Dunsink Observatory
Castleknock, Dublin 15, Ireland

Abstract. A ROSAT PSPC image of the galaxy cluster Abell 2241 has been obtained, showing X-ray emission from the intracluster medium and from individual objects. The brightness distribution of the cluster gas is used to assess the physical conditions at the location of two tailed radio galaxies (in A2241E and A2241W, at different redshifts). Together with radio and X-ray information on the two galaxies themselves the results are of relevance to the question of energy equipartition in radio sources.

1. Background

A cluster intergalactic medium plays an important role in determining the morphology and evolution of radio sources. The external gas can interact with a radio source in different ways: confining the source, modifying the source morphology via ram-pressure and possibly feeding the active nucleus.

We have obtained ROSAT X-ray data of the region of the cluster Abell 2241, which was previously studied by Bijleveld & Valentijn (1982). A2241 was originally classified as an irregular galaxy cluster, until redshift measurements showed it to consist of two separate clusters projected onto each other: A2241W and A2241E, located at redshifts of 0.0635 and 0.1021, respectively.

R. Ekers et al. (eds.), Extragalactic Radio Sources, 361–362.

According to the radio data obtained with the Westerbork Synthesis Radio Telescope and presented by Bijleveld & Valentijn (1982), both clusters contain some radio emitting galaxies, with most of them showing extended structure. In particular, both clusters contain a radio galaxy characterized by a tailed radio structure: 1657+32 in A2241W and 1658+32 in A2241E. High-resolution VLA radio data for these radio galaxies are reported by Fanti et al. (1987) and references therein. Referring to Einstein data, Bijleveld & Valentijn detected X-ray emission only from A2241E.

2. New X-ray data

We obtained observations of the region of the A2241 clusters both with the PSPC and HRI detectors of the ROSAT satellite, in order to get information both on the large scale X-ray emission and on the pointlike sources. The HRI observation resulted however only in a very short, aborted exposure of ca. 3000 seconds. Maps of the X-ray brightness distribution were produced by binning the photon events in a two dimensional grid and then smoothing with a gaussian filter. The emission around the centre of A2241E covers a large area, whereas a smaller clump is seen at the location of A2241W. Both clusters exhibit extended X-ray emission in this observation.

The central, strongest source coincides with the cD galaxy in A2241E, a blob just East of this source is associated with the A2241E head-tail galaxy. Another extended blob, to the SW, corresponds to the A2241W head-tail galaxy; the cD galaxy in this cluster appears not detected. Two moderately strong point sources, E and W of the central strong source, coincide with a star and a QSO, respectively, and are used for astrometric calibration of the X-ray field.

The HRI exposure detects the central strong source and the QSO to the W. Two other sources are seen in addition.

3. Outlook

After complete analysis of the PSPC field, the combination of X-ray and radio data will be used to investigate the interaction between radio sources and intracluster gas. A comparison of the equipartition pressures determined from the radio observations and the thermal pressures in the X-ray gas is here of particular interest (see Feretti et al. 1992).

References

Bijleveld, W. and Valentijn, E. (1982) A&A 111, 50
Fanti, C., Fanti, R., de Ruiter, H. and Parma, P. (1987) A&AS 69, 57
Feretti, L., Perola, G.C. and Fanti, R. (1992) A&A 265, 9

ASCA OBSERVATIONS OF CLUSTERS OF GALAXIES

KOUJUN YAMASHITA
*Department of Physics, Nagoya University, Furo-cho, Chikusa,
Nagoya 464-01 JAPAN*

1. Introduction

X-ray emissions from clusters are most likely originated from a thin hot plasma in a collisional ionization equilibrium. The optical depth of continuum component is order of 10^{-3}, whereas that of emission lines is around unity. Present emission models used for spectral fitting can not estimate this effect, so that the determination of elemental abundances seems to include large uncertainty. The high resolution spectroscopy with ASCA gives a clue to investigate the physical state of hot intracluster gas and a impact to reconsider the basic atomic processes. This is important issue to deeply understand the structure, formation and evolution of clusters, and the origin of intracluster gas.

The X-ray morphologies are divided into two categories, they are, centrally concentrated and spherically symmetric clusters with small core radii and cooling flows, and largely extended clusters with large core radii and merging substructures. ROSAT all sky survey detected thousands of clusters with the high angular resolution[1]. ASCA makes it possible to derive physical parameters of these clusters from the spatially-resolved spectroscopic observations[2]. The Hubble constant(H_0) was derived by observations of Sunyaev-Zeldovich (hereafter S-Z) effect clusters and by means of resonant scattering effect of emission lines in an intracluster gas.

Here we present the observational results of ASCA and discuss the cooling flow and merging of clusters, and the Hubble constant determination.

2. Observations

ASCA has already observed nearly 150 clusters of galaxies in the energy range of 0.5 - 10 keV and in the redshift range up to 0.5, including group of galaxies, poor clusters, nearby rich clusters ($z<0.1$), distant clusters ($z>0.1$)

R. Ekers et al. (eds.), Extragalactic Radio Sources, 363–366.

and superclusters. Some of S-Z effect clusters and gravitational lensing clusters were also observed. ASCA puts on board four sets of multi-nested thin foil conical mirror X-ray telescope (XRT) incorporated with two X-ray CCDs(SIS) and imaging gas scintillation proportional counters(GIS) for each two sets[3]. The angular resolution and field of view are 3' (HPD), 22'x22' for SIS and 50'(diameter) for GIS, respectively. The energy resolutions are 2% and 8% at 6keV for SIS and GIS, respectively. ASCA instrumentations are described in detail by Serlemitsos et al.[4].

3. Data processing

The spectral analysis of observed data can be done by fitting a model spectra of thermal emissions from a thin hot plasma like Raymond-Smith or MEKA model, taking into account the absorption of neutral hydrogen in the line of sight. Thus the plasma temperature(kT) and abundance of each cluster corrected for the cosmological redshift(z) were derived as shown in Fig.1, where 43 clusters are plotted. It is also important to fit a model of thermal bremsstrahlung and Gaussian lines as free parameters of kT, line intensities and energies. In this case we can derive an electron temperature from the continuum component, an ion temperature from intensity ratio of H-like and He-like ions, abundance from line intensties of each element and the redshift from line energies. This procedure is very useful to qualify observed data. SIS has a capability to resolve emission lines of H-like and He-like ions of heavy elements whereas GIS can distinguish K_α and K_β. Furthermore the optical depth of each line in an intracluster medium can be estimated from the intensity ratio of K_α and K_β. In Fig.2 there is shown an X-ray spectrum of M87(z=0.0038) within the radius of 2.5' observed by SIS and fitted with a thermal bremsstrahlung of kT=1.9keV. Residuals clearly show the existence of several emission lines, such as O, Si, S, Ar, Ca and Fe.

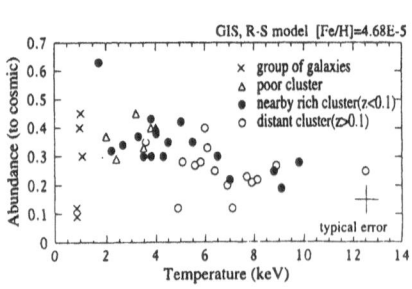

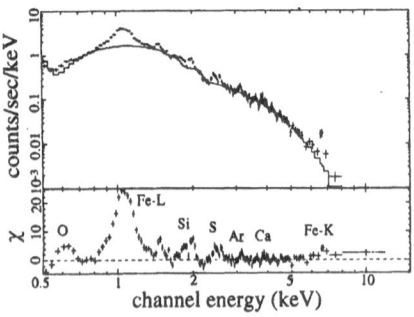

Figure 1. Abundance vs. plasma temperature for 43 clusters

Figure 2. X-ray spectrum of M87(5'ϕ) by SIS

The image processing can be done by folding the response function, which is rather complicated for extended objects. Raw image of A644 observed in 1-10keV for 50 ksec by GIS is shown in Fig.3, which is completely covered in the GIS field of view. The surface brightness distribution against the projected radius is fitted by a standard β-model,

$$I(r)=I(0)(1+(r/r_c)^2)^{-3\beta+1/2},$$

where r_c is core radius, assuming spherical symmetric distribution. r_c and β are obtained to be 3' and 0.68, respectively.

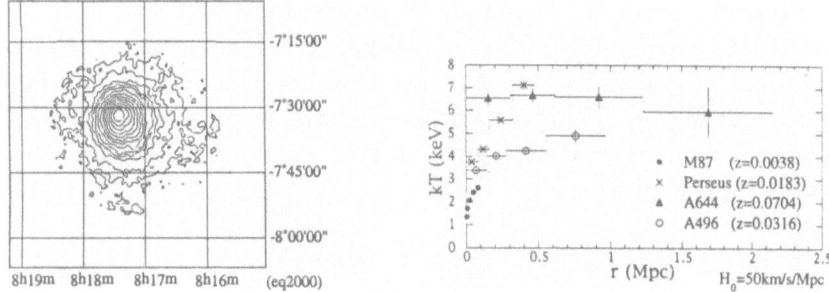

Figure 3. X-ray image of A644 by GIS   *Figure 4.* Plasma temperature(kT) vs. projected radius(r)

4. Cooling flow

Well relaxed clusters with a central dominant galaxy show an evidence of cooling flow which are observationally recognized as excess emissions and absorption in the central core region and radially decreasing temperature distribution to the cluster center. Spectro-imaging observations with ASCA were clearly revealed the radial gradient of temperatures in many clusters as well as the abundance gradient in some cases. In Fig.4 the temperature distribution of M87/Vir, A496, the Per cluster and A644 is shown against the projected radius(Mpc) with H_0=50 km/sec/Mpc. Angular radius of these clusters are divided into annular ring of 2.5', 5', 10' and 15'. Generally speaking, low temperature clusters show steep distribution.

5. Merging

Spherically symmetric clusters without a central dominant galaxy classified as nXD seem to be isothermal and in the hydrostatic equilibrium. These clusters have relatively high temperature, low abundance and less central condensation than cooling flow clusters. The temperature of the Coma cluster averaged over whole cluster was obtained to be 8.09keV by

Ginga. ASCA observed spatially resolved spectra within the radius of 40'. It shows inhomogeneous temperature distribution in the range of 5.7-12keV. The Coma accompanies a subcluster which shows low temperature[5]. This fact indicates that different temperature component corresponds to different substructure going to merge in a cluster potential.

6. Determination of the Hubble constant

Combining the decrement of the brightness temperature of the cosmic microwave background through an intracluster gas (S-Z effect) with plasma temperature obtained by ASCA and the surface brightness profile observed by Einstein, we have derived H_0 to be less than 50 km/s/Mpc for A665 (z=0.1816) and CL0016+16 (z=0.545)[3]. The Hubble space telescope observations of galaxies in the distance to Coma cluster(z=0.0235) gave the value to be 80(+/-17) km/s/Mpc[6]. We have proposed another method by means of resonant scattering effect of Fe-K emission lines in an intracluster gas, as H_0 is expressed as

$$H_0 = 73(km/sec/Mpc)(T/10^8 K)^{-1/2}(\theta/mrad)\frac{<n_e>}{5\times10^{-3}}cm^{-3})\frac{Ab(Fe)}{\tau(K_\alpha)}(z/0.1)$$

where T is plasma temperature, θ angular size, $< n_e >$ mean electron density, Ab(Fe) iron abundance and $\tau(K_\alpha)$ optical depth of Fe-K_α emission lines. This method with ASCA data is applicable to spherically symmetric isothermal clusters in z=0.01-0.1. We expect that it would be possible to derive the z dependence of H_0.

7. Summary

Spatially resolved X-ray spectra of clusters of galaxies have been observed in the energy range of 0.5 - 10 keV and in the redshift range up to 0.5 by ASCA. The cooling flow and merging are directly deduced from the temperature and abundance distribution in an intracluster medium. The determination of the Hubble constant is discussed by means of Sunyaev-Zeldovich effect and resonant scattering effect of Fe-K emission lines.

References

[1] Briel, U.G., and Henry, P.J., 1993, *Astr. Astrophys.*, **278**, 379.
[2] Yamashita, K., 1994, Clusters of Galaxies, eds. F. Durret, A. Mazure and J. Tran Thanh Van, Editions Frontiers, p.153.
[3] Tanka, Y., etal., 1994, *Publ. Astron. Soc. Japan*, **46**, L37.
[4] Serlemitsos, P.J., et al., 1994, *Publ. Astron. Soc. Japan*, **47**, 105.
[5] Briel, U.G., et al., 1992, *Astr. Astrophys.*, **259**, L31.
[6] Freedman, W.L., et al., 1994, *Nature*, **371**,757.

RADIO SOURCES AND THEIR ENVIRONMENT (*DISCUSSION*)

Discussion of the paper presented by <u>OWEN</u> (p. 305)

Macchetto: In the cluster merger calculations that you showed, it was not obvious where the centre of mass (i.e. the position of the CD galaxy) was, and therefore whether the gas would indeed stream pass the galaxy and create the bent radio-lobes.

Owen: The center of mass in the merger simulations evolves as the dark matter in the two cluster merge. The gas which is not important gravitationally tries to follow the changing potential but is limited by its sound speed hydrodynamically. This leads to the long lasting, large scale flows which look very different than the potential and last for many Gigayears.

Rudnick: With the improved statistics, can you tell whether a galaxy has a greater probability of being a radio source if another galaxy in the cluster already is a radio source?

Owen: In nearby clusters there are too few radio galaxies per cluster to sort this out, but this is an exciting prospect in the more distant, very rich clusters.

Ekers: It is very satisfying to see that the idea of large scale motions of the cluster gas are now almost acceptable. An early and very clean argument for large scale motion of cluster gas was made by Toomre in the 1970's when he noted that the large scale curvature of the head tail radio source NGC1265 in the Perseus cluster could not be caused by any possible galaxy orbit; this leads directly to the need for large scale motion of the cluster gas with velocity ≥ 1000 km/sec.

Owen: Yes, the shapes of sources like NGC1265 must be due to a combination of their orbital motion and the velocity field of the cluster gas. The nice thing about the cluster-cluster merger model is that it provides a mechanism for causing the gas motions.

Radhakrishnan: What about using iron's other X-ray lines to measure the motions of cluster gas through Doppler shifts?

Owen: This should be an exciting possibility for future x-ray satellites.

R. Ekers et al. (eds.), Extragalactic Radio Sources, 367–372.

Binette: Supposing the RG wobble within the X-ray gas along the direction of elongation of the X-ray gas, could this be an indication of the ability of the RG to "heat up". the cooling flow gas? This would reduce the mass inflow rate inferred for "cooling flows".

Discussion of the paper presented by <u>DALY</u> (p. 319)

Laing: What do you think is the physical reason why the beam power is anticorrelated with the time for which the source is on?

Daly: The beam power L_j, time the AGN is on t_*, and initial energy available to power the outflow E_i are related: $L_j = E_i/t_*$. Here, it is assumed that the initial energy of a given object is a fixed quantity, as would be the case if it were the spin energy of the central object. If the energy extraction rate L_j were independent of the initial energy, then we would have $L_j \propto t_*^{-1}$, so it is not unreasonable to expect the beam power and time to be anticorrelated. In fact, we were motivated by several puzzling observations that could be understood if we wrote $t_* \propto L_j^{-\beta/3}$ (see Daly 1994, and Guerra and Daly 1995, 1996). The data may be combined in two independent ways to estimate β and both indicate $\beta \simeq 2 \pm 0.5$, and this result is independent of the cosmological model assumed. The physical reason for the anticorrelation is related to the energy extraction mechanism. The data suggest that $L_j \propto E_i^3$ for $\beta \simeq 2$, which in turn may be used to constrain models for the energy extraction.

De Young: If the radio source lifetime decreases with redshift, then the number of sources per unit comoving volume should decrease with increasing redshift. Have you looked for this?

Daly: Not as yet, but this would be a very interesting effect to look for, and could be used to infer something about the evolution of the underlying radio luminosity function for these sources.

Wilson: Do you think that the decline of the linear sizes of the radio sources toward higher redshift is related to the higher inverse Compton losses of the relativistic electrons on the microwave background (the energy density of which increases with redshift)? This effect would tend to "snuff out" large, old radio sources at higher redshift.

Daly: It is true that inverse Compton losses become significant at high redshift as pointed out by Rees and Setti (1968, Nature, 219, No. 5150, 127). However, it is now known that the youngest relativistic electrons are near the radio hotspot, and the older electrons, which would be effected by the inverse Compton cooling, are closer to the center of the source. The effect you mention may make it difficult to observe the radio bridges of high redshift radio sources, but it is unlikely to effect the radio emission near the hotspots, which produce most of the radio emission and are used to define the source size. Thus, we must look to some other physical effect that would cause higher redshift sources to be smaller; we think they are smaller because they have higher beam powers and shorter lifetimes, leading to smaller sources.

Gopal-Krishna: If at higher redshifts radio source engines extinguish faster (followed by a decay in radio luminosity), then wouldn't one expect the radio luminosity function to get increasingly weighted in favor of lower luminosity radio galaxies at earlier cosmological epochs?

Daly: This would be the case if we were able to detect sources with low beam power. It turns out that the beam power and radio power are proportional to one another (see Wan and Daly 1996c). Because the sources studied come from flux limited samples, the radio power increases with redshift. We also see that the beam power increases with redshift, and deduce that this arises because of the correlation between radio power and beam power. Thus, at high redshift we only observe the most powerful sources. In this sense, the decrease of the source size with redshift is a selection effect caused by the fact that at high redshift we observe sources with higher beam power, these sources have shorter lifetimes, and thus are smaller. However, there is another parameter that enters into the source size, which is the lobe propagation velocity. Thus, the decrease of the source sizes with redshift is much weaker than the increase of the beam power with redshift; this follows because the lobe propagation velocity increases with redshift, probably because it depends on the beam power.

Discussion of the paper presented by <u>LACY</u> (p. 321)

De Young: In the $z = 4.41$ object, what is the limiting magnitude in the K band image?

Lacy: K>21.5
(Since the conference ended we have achieved a marginal detection at K$\simeq$22).

Schilizzi: How strong is the radio emission in 6C 0140+326?

Lacy: The radio flux density at 151 MHz is 1.0 Jy.

Discussion of the paper presented by FERETTI (p. 333)

Stocke: Could you please locate the Coma "relic" radio source on the X-ray map of the Coma Cluster.

Feretti: The relic 1253+275 lies beyond the sub-cluster centered on NGC 4839, in the same direction.

Partridge: What is the typical age of the *relic* sources in clusters?

Feretti: It is of the order of 10^8 yrs.

Meisenheimer: Which value of magnetic field did you use for the age estimate and how does that compare with the energy density in the cosmic microwave background?

Feretti: The age of Coma C using the equipartition magnetic field is 10^8 yr. Using the magnetic field derived from rotation measure, the age is much shorter. In the latter case, the energy density of the magnetic field is larger than that in the microwave background, whose equivalent magnetic field is $\sim 4\mu G$.

Dogiel: Where does the energy of relativistic electrons in the Coma cluster halo come from? Is it a part of kinetic energy of the tail galaxy orbiting around the cluster center or is it the energy of CR electrons emitted by this galaxy?

Feretti: The relativistic electrons are likely to be deposited by the tailed radio galaxy NGC4869, and reaccelerated by the energy available from a cluster merger process and possibly galactic wakes.

Eilek: The problem of *sources* of the relativistic electrons is important; they cannot diffuse from the cluster center fast enough to supply the haloes. Do you know if all six of the large-scale haloes have outlying radio galaxies which can supply the particles?

Feretti: The best studied haloes, ie. those in A1656, A2255, A2256 and A2319, have embedded tailed radio galaxies.
The clusters A2163 and A2218 are distant, and less studied. Higher resolution data are necessary to distinguish discrete sources from the broad halo radio emission.

Harris: We expect little or no polarization for haloes because of turbulence making smaller cell sizes and mixing in thermal plasma. Could the quasi-linear polarized structure in A2256 (20% polarization) be the remnant of an old tailed radio galaxy?

Feretti: The halo in A2256 is complex: it actually consists of a very low brightness region at the cluster center and a stronger, polarised feature in the N-W part. This higher brightness region could be the remnant of more tailed radio galaxies.

Discussion of the paper presented by $\underline{BÖHRINGER}$ (p. 357)

Binette: Cooling shocks due to the radio jets might explain the excess emission in X-ray emission related to the radio-structure in M87. Why are shocks not considered any more as a mechanism which compresses the gas and adjusts pressure imbalances?

Böhringer: The main problem is not a problem of pressure imbalance, because we see that the gas is colder in the radio lobes than outside and, with the higher density inferred from the high surface brightness, the lobe gas and the surrounding ICM can still be at pressure equilibrium. The main question is: Why is the gas cooler in the lobes instead of being heated by the disturbances? It may thus be possible that adiabatic waves or weak shocks introduce a temporary clumpiness in the gas speeding up the radiative cooling rather than heating the ICM.

Fraix-Burnet: What is the meaning of the dark blue region West of the nucleus on the M87 X-ray map?

Böhringer: The dark blue region in the X-ray halo image of M87 is a surface brightness deficiency located roughly at the same position angle as the optical and radio jet. Therefore it could in principle also originate from an X-ray gas displacement as in the case of NGC 1275. But this feature is not so significant as in NGC 1275.

Macchetto: I do not understand how you get baryonic densities as high as 30% in clusters such as A2218. Work by Ellis et al, using HST data and ground based redshifts, seem to indicate $M/L \approx 100$ and upper limits to the baryonic density of 10%.

Böhringer: The examples I have shown refer to nearby clusters and a Hubble constant of 50. A baryon mass fraction of 30% is derived if we integrate over the whole cluster out to 3Mpc (in the case of Coma or Perseus). If one looks only at the central part of the cluster and uses

$H_0 = 100 kms^{-1}Mpc^{-1}$ the baryon fraction is lower. The lensing results by Ellis et al may refer to this case. The discrepancy for M/L is more difficult to understand; for $H_0 = 50$ we usually find values from 100-300. for $H_0 = 100$ the values are higher by a factor of two.

Discussion of the paper presented by YAMASHITA (p. 363)

Simkin: How does the temperature asymmetry which you found in the Coma cluster compare with the radio galaxy distribution?

Yamashita: We would like to compare the ASCA result with the radio galaxy distribution. We have not checked any radio data in the Coma region. If the high temperature region coincides to dense distribution of radio galaxies, it seems that the contribution of non-thermal component is expected to be significant.

UNIFIED SCHEMES FOR EXTRAGALACTIC RADIO SOURCES

GOPAL-KRISHNA
NCRA-TIFR, Poona University Campus, PUNE-411007, India

1. Introduction

The most widely discussed class of unified schemes for radio-loud extra-galactic sources attempts to interpret their seemingly disparate types as the same objects seen from different directions. The orientation dependence is attributed to relativistic beaming of the nonthermal jet and, possibly, anisotropic obscuration/re-radiation of the nuclear emission by a circum-nuclear distribution of dusty material with polar openings, possibly a torus [1]. Although alternative approaches have been mooted for unifying radio galaxies (RGs) and quasars (QSRs) by incorporating a strong jet-environment interaction [2], or temporally decaying nuclear prominence [3], the orientation based unified scheme, thanks to its rich predictive potential, has been subjected to a multitude of observational tests and its pros and cons have been discussed extensively in recent reviews [1] [4–7]. Here we briefly address some recent developments, including the claim that the radio size measurements of powerful RGs and QSRs are incompatible with orientation being the primary distinction between them. On balance, it seems that while the basic orientation picture can broadly explain the bulk of the observations, its viability could be much enhanced by taking into consideration the (inevitable) temporal evolution of radio sources.

2. Unification of low–luminosity sources (BL Lacs & FR I RGs)

Due to the statistical similarity in isotropic attributes, such as extended radio emission and the host galaxy, and based on the analyses of radio/X-ray luminosity functions, FR I galaxies have long been favoured as the parent (misaligned) population of BL Lac objects [e.g., 7-9]. The needed evidence for radiation anisotropy in the bases of radio jets of FR I galaxies is furnished by the recent detection/inference of (i) relativistic motion

R. Ekers et al. (eds.), Extragalactic Radio Sources, 373–378.
© 1996 *IAU.*

within the cores of several FR-I galaxies (e.g., [10, 11]), and (ii) polarization asymmetry between their radio lobe pairs [12]. The RG(FR I) – BL Lac unification is further supported by the growing evidence that both lie in moderately clustered environments (Abell richness class 0)[13,14], though BL Lacs appear to avoid very rich clusters [15,16] (see, however, [7, 13]).

Taking a clue from their matching X-ray luminosity but distinctly lower optical polarization, radio/optical luminosities and variability, it has been argued that the X-ray selected BL Lacs (XBLs) are viewed at intermediate orientation between the apparently more active radio-selected BL Lacs (RBLs) and the parent FR I galaxies (e.g., [16–18]). However, the implication that XBLs are much more numerous than RBLs is challenged by the proposal that the XBLs may in fact be the small minority of cases where the peak of the synchrotron spectrum extends up to soft X-ray energies [19, 7]. This controversy remains to be settled, leaving open the question of a 'transitional' population within the FR I unified scheme (Sect.5).

3. Unification of high–luminosity (FR II) radio sources

In this version of the unified scheme, the lobe–dominated QSRs (LDQs) and core-dominated QSRs (CDQs/blazars) are increasingly aligned versions of powerful radio galaxies (PRGs). Recent reviews [5, 7] summarize and update the evidence for this hypothesis, employing orientation independent properties, such as extended radio emission, [O II] 3727 emission, environmental clustering, near-IR (stellar) emission of the host galaxy [20] and far-IR emission (even at $\lambda_{rest} \sim 50~\mu$, the nuclear continuum is either beamed, or re-radiated anisotropically by the torus, though at longer wavelengths it becomes increasingly isotropic [21, 22]). Further support to the unified scheme comes from the recent detection of scattered Mg II broad emission line in the UV spectrum of the nearest PRG Cygnus A [23], though this object may still pose potential concerns to the unified scheme [24].

Additional supporting evidences emerging from recent radio data are: (i) The increasing 'apparent' brightness temperature along the orientation sequence NLRG–LDQ–CDQ, as deduced from radio flux variability [25]; (ii) growing evidence from VLBI for the apparent motion to be usually faster in the nuclei of more core-dominated sources [26]; (iii) peaking of the radio spectra of the *cores* of LDQs at a few times smaller rest-frame frequency, on average, than the spectra of CDQs [27-29]; (iv) the greater lobe depolarization asymmetry observed in QSRs, compared to PRGs [30]; (v) the large-scale radio structure of QSRs appearing more bent than that of PRGs, which is shown to be consistent with the critical misalignment angle $\theta_c \sim 45^o$ being the dividing line between PRGs and QSRs [31, 32]; (this is also consistent with the statistics of jet opening angle in PRGs and

QSRs; [33]). Note that structural asymmetries consistent with such a value of θ_c are noticeable also among compact–steep–spectrum radio sources [34].

4.   The radio size 'anomaly'. Does it spell doom for unification?

An important clue to the orientation based unification came from the result that in the metre-wavelength selected 3CRR sample, where the axes of the radio sources should be randomly oriented, the ratio, R, of the median radio sizes of broad–line objects (QSRs) and narrow–line objects (PRGs) is smaller than unity (R $\sim$ 0.5, for z > 0.5) [35]. However, the lack of such a trend at z < 0.5, bolstered by a similar behaviour reported for some other metre-wavelength samples a few times deeper than the 3CRR sample, has evoked serious doubts about the unified scheme [6, 36-38]. We suggest that even the result R $\sim$ 1 can be explained, despite PRGs being oriented closer to the sky plane than QSRs, provided the following simple, empirically deduced temporal evolution of FR II sources is taken into account [5].

Firstly, recall that a typical powerful radio source during its lifetime T $\simeq 10^7 - 10^8$ yr grows to a size L $\approx 10^2$ kpc such that the expansion velocity V $\propto P^\alpha$, with $\alpha \approx 0.3$ [39], and the (nearly uniform) expansion to L $\approx 10^2$ kpc is accompanied by roughly an order–of–magnitude decrease in the radio luminosity P [40,41]. Secondly, the apparent increase in the QSR–to–PRG number ratio (f_q) with flux density, suggests that intrinsically more powerful radio sources have larger torus openings angles (2 ψ) [5, 6, 42,43].

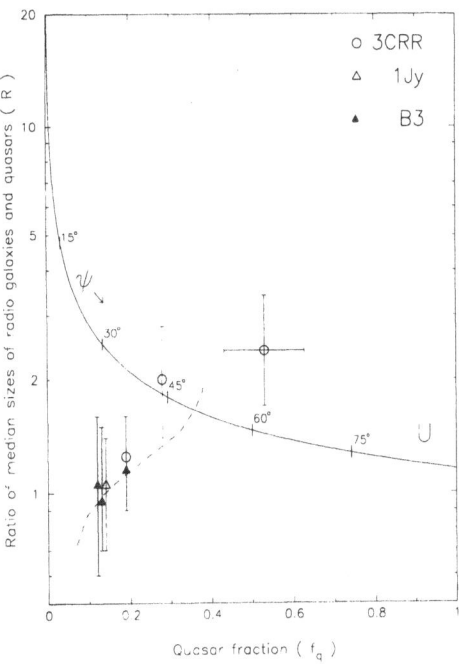

Ratio of median sizes of radio galaxies and quasars (R)

Quasar fraction (f_q)

Fig. 1: A plot of R *versus* f_q for the 3 metre-wavelength samples; the data points are from ref.[6], so also the prediction of the orientation unified scheme for a range of ψ (curve "U"). The dashed curve shows a prediction of our model incorporating temporal evolution into the unified scheme [Sect. 4]. The adopted values for the input parameters, consistent with the observations (Sect. 4, ref.[44]), are: (i) an e-folding time of 10^7 yr for the decay of P, (ii) a radio source 'injection' spectrum: $n(P_0) \propto P_0^{-2}$; (iii) radio size L $\propto P_0^{0.3} \times t^{0.8}$, and (iv) $\psi = 0.4 \log (P_0/10^{26}$ WHz$^{-1})$ where $0.2 \leq \psi \leq \psi_{max} = 1$ rad. The dashed curve spans 2 orders-of-magnitude in P increasing to the right (just as the data points). Details are in ref.[44].

Now, consider the sources **observed** near a radio luminosity P below that of the most powerful born **sources**. At any given time, such sources would include young (hence small) **sources** freshly born with that level of luminosity, as well as ageing **sources that** were born with higher luminosity in the past, but have since then **faded** to P and, concurrently, expanded to larger sizes. Since these older, **expanded** sources with larger sizes should have **a** higher quasar–fraction (f_q) owing to their higher initial luminosities, P_o (and correspondingly larger ψ), **the** median size of the QSRs in a sample taken near the luminosity P may well approach, or even exceed that of the PRGs [5]. A quantitative prediction based on this simple picture matches the radio size data quite well (Fig. 1 & ref. [44]) and, moreover, explains simultaneously the difference reported, e.g., in refs. [6] & [37], between the radio size — luminosity correlations for PRGs and QSRs. Furthermore, in this picture, except in the metre-wavelength samples selected at the highest radio luminosities, QSRs would, on average, be *older* (hence, intrinsically larger) than PRGs. This is opposite to the pattern envisioned in some alternative unification scenarios (e.g., [3]).

5. Towards a single unified scheme for FR I and FR II sources?

Unlike the FR II unification wherein the LDQs appear at intermediate orientation between the CDQs and the (misaligned) PRGs, the FR I unified scheme lacks a well established 'transitional' population between the BL Lacs and the FR I galaxies (Sect.2). To bridge this conceptual gap and thus devise a common framework for the two unified schemes, a few possibilities have been proposed (see refs. [7, 5] for comments), as noted below.

A link between the two schemes is hinted, firstly by the near absence of QSRs among low–luminosity (FR I) sources, plus the evidence for wider torus openings in more powerful sources (Sect. 4). Since, plausibly, the decreasing ψ at lower P could approach near the FR I / FR II break, the typical relativistic beaming angle of the nonthermal jet, a direct view of the nuclear region in FR I sources (a pre–requisite for QSR classification) would be unavoidably accompanied by a Doppler–boosted nuclear continuum jet. Due to this and the obscuration of the nuclear region by the material stripped by the jet from the narrow torus funnels, the aligned FR I sources would mostly be classified as BL Lacs. The lack of a transitional, LDQ type population among FR I sources could thus be understood without invoking a conceptual dichotomy between the FR I and FR II unified schemes [45, 42]. (For evidence for dust in the torus funnels, see ref. [46]).

On the other hand, an analysis of the radio and X–ray luminosity functions has led Maraschi & Rovetti [47] to propose a 'generalized' unified scheme, based on an expanded 'parent' population including all steep-spectrum sources (FR I, FR II & LDQs), whose beamed population would include all flat–spectrum sources (BL Lacs & CDQs). Vagnetti & Spera [48], instead, posit that the canonical BL Lacs are the remnants of (distant) CDQs, by postulating an increase in the jet's Lorenz factor with cosmic time, which leads to an increased beaming of the continuum (The superluminal velocities and jet/counter–jet ratios are predicted in ref. [49]).

6. Potential problems and some outstanding issues

The reported disparity between the asymmetry properties of the CIV λ 1549 line from LDQs and CDQs remain to be understood within the orientation scenario [50, 51]. Another intriguing recent finding is that the broad–line radio galaxies (BLRGs) have flatter mid–IR spectra compared to *both* QSRs and PRGs [22], weakening the general notion that BLRGs are intermediate to PRGs and QSRs in orientation. Other outstanding questions include:

(i) Is an important subset of FR I galaxies devoid of a BL Lac nucleus, as argued in ref. [16] for galaxies in rich clusters (also, [15] see, however [7]).

(ii) Are the low–excitation FR II galaxies parents of some high–luminosity BL Lacs [52,53]? Note that their nuclear 'dullness' could be transitory, making their exclusion from the statistics of FR II sources potentially unsafe.

(iii) Is alignment the key difference between XLBs and RBLs (Sect.2)?

(iv) Does a jet's Lorentz factor correlate with its power (e.g., [54,25,26])?

(v) Do FR I sources, too, possess a broad–line–region? Or, their central engines are basically different from those of the FR II sources (e.g., [55])?

References

[1] Antonucci, R. 1993, *ARAA*, **31**, 473.

[2] Norman, C., and Miley, G. 1984, *A&A*, **141**, 85.

[3] Hutchings, J.B., Price, R., and Gower, A.C. 1988, *ApJ*, **329**, 122.

[4] Barthel, P.D. 1994, in The Physics of Active Galaxies, G. Bicknell, M. Dopita and P.Quinn, eds. (ASP, San Francisco), vol. 54, p.175.

[5] Gopal-Krishna 1995, in Quasars and AGN: High Resolution Imaging, K.I. Kellermann and M.H. Cohen, eds. (Nat. Acad. Sci., Washington), in press.

[6] Singal, A.K. 1995, in Quasars and AGN: High Resolution Imaging, K.I. Kellermann and M.H. Cohen, eds., (Nat. Acad. Sci., Washington), in press.

[7] Urry, C.M., and Padovani, P. 1995, *PASP*, **107**, 803.

[8] Browne, I.W.A., and Jackson, N. 1992, in Physics of AGN, W.J. Duschl and S.J.Wagner, eds. (Springer-Verlag), p.618.

[9] Blandford, R.D., and Rees, M.J. 1978, in Pittsberg Conf. on BL Lacs, A.M. Wolfe, ed., Pittsberg Univ. Press, p.328.

[10] Giovannini, G., et al., *these proceedings*.

[11] Venturi, T. et al. 1995, ApJ, in press (Bologna preprint: BAP06-1995-029-IRA).

[12] Parma, P., *these proceedings*.

[13] Smith, E.P., O'Dea, C.P., and Baum, S.A. 1995, *ApJ*, **441**, 113.

[14] Pesce, J.E., Falomo, R., and Treves, A. 1995, *AJ*, **110**, 1554.

[15] Owen, F.N., Ledlow, M.J., and Keel, W.C. 1995, in preparation.

[16] Wurtz, R., Stocke, J., and Yee, H.K.C., 1995, *ApJ* (Suppl.), in press.

[17] Morris, S.L., Stocke, J.T., Gioia, I.M., Schild, R.E., Wolter, A., Maccacaro, T., and Della Ceca, R. 1991, *ApJ*, **380**, 49.

[18] Urry, C.M., Padovani, P., and Stickel, M. 1991, *ApJ*, **382**, 501.

[19] Padovani, P., and Giommi, P. 1995, *ApJ*, **444**, 567.

[20] Kukula, M.J. et al., *these proceedings*.

[21] Heckman, T.M., O'Dea, C.P., Baum, S.A., and Laurikainen, E. 1994, *ApJ*, **428**, 65.

[22] Hes, R., Barthel, P.D., and Hoekstra, H. 1995, *A&A*, **303**, 8.

[23] Antonucci, R., Hurt, T., and Kinney, A. 1994, *Nature*, **371**, 313.

[24] Harris, D.E., *these proceedings*.

[25] Terasranta, H., and Valtaoja, E. 1994, *A&A*, **283**, 51.

[26] Vermeulen, R., *these proceedings*.

[27] Antonucci, R., Barvainis, R., and Alloin, D. 1990, *ApJ*, **353**, 416.

[28] Gopal-Krishna, and Steppe, H., 1991, in Variability of AGN, H.R. Miller and P.J. Wiita, eds. (CUP), p.194.

[29] Athreya, R., Kapahi, V.K., McCarthy, P.J., and van Breugel, W., *these proceedings*.

[30] Garrington, S.T., Holmes, G.F., and Saikia, D.J., *these proceedings*.

[31] Best, P.N., Bailer, D.M., Longair, M.S., and Riley, J.M. 1995, *MNRAS*, **275**, 1171.

[32] Lister, M.L., Hutchings, J.B., and Gower, A.C. 1994, *ApJ*, **427**, 125.

[33] Oppenheimer, B.R., and Biretta, J.A. 1994, *AJ*, **107**, 892.

[34] Saikia, D.J., Jeyakumar, S., Wiita, P., Sanghera, H.S., and Spencer, R. 1995, *MN-RAS*, **276**, 1215.

[35] Barthel, P.D., 1989, *ApJ*, **336**, 606.

[36] Kapahi, V.K., Athreya, R.M., Subrahmanya, C.R., Hunstead, R.W., Baker, J.C., McCarthy, P.J., and van Breugel, W. 1995, *JAA (Suppl.)*, **16**, 125.

[37] Kapahi, V.K., Athreya,R.M., Subrahmanya, C.R., McCarthy, P.J., van Breugel, W., Baker, J.C., and Hunstead, R.W., *these proceedings*.

[38] Blundell, K. et al., *these proceedings*.

[39] Alexander, P., and Leahy, J.P. 1987, *MNRAS*, **225**, 1.

[40] Fanti, C., Fanti, R., Dallacasa, D., Schilizzi, R.T., and Stanghellini, C. 1995, in Quasars and AGN: High Resolution Imaging, K.I. Kellermann and M.H. Cohen, eds., (Nat. Acad. Sci., Washington), in press.

[41] Readhead, A.C.S., Taylor, G.B., Pearson, T.J., and Wilkinson, P.N. 1995, in Quasars and AGN: High Resolution Imaging, K.I. Kellermann and M.H. Cohen, eds., (Nat. Acad. Sci., Washington), in press.

[42] Falcke, H., Gopal-Krishna, and Biermann, P.L. 1995, *A&A*, **298**, 395.

[43] Lawrence, A. 1991, *MNRAS*, **252**, 586.

[44] Gopal-Krishna, Kulkarni, V.K., and Wiita, P.J. 1995, submitted.

[45] Gopal-Krishna, 1995, *JAA (Suppl.)*, **16**, 153.

[46] Baker, J.C., and Hunstead, R.W. 1995, *ApJ*, **452**, L95.

[47] Maraschi, L., and Rovetti, F. 1994, *ApJ*, **436**, 79.

[48] Vagnetti, F., and Spera, R. 1994, *ApJ*, **436**, 611.

[49] Vagnetti, F., *these proceedings*.

[50] Corbin, M.R., and Francis, P.J. 1994, *AJ*, **108**, 2016.

[51] Wills, B.J. et al. 1995, *ApJ*, **447**, 139.

[52] Laing, R.A., Jenkins, C.R., Wall, J.V., and Unger, S.W., 1994, in The Physics of Active Galaxies, G. Bicknell, M. Dopita and P.Quinn, eds. (ASP, San Francisco), vol. 54, p.227.

[53] Hine, R.G., and Longair, M.S. 1979, *MNRAS*, **188**, 111.

[54] Morganti, R., Osterloo, T., Fosbury, R., and Tadhunter, C. 1995, *MNRAS*, **274**, 393.

[55] Baum, S.A., Zirbel, E.L., and O'Dea, C.P., 1995, *ApJ*, **451**, 88.

THE UNIFICATION OF RADIO-LOUD AGN

C. M. URRY
Space Telescope Science Institute
3700 San Martin Drive, Baltimore, Maryland, 21218

AND

PAOLO PADOVANI
Dipartimento di Fisica, II Università di Roma "Tor Vergata"
Via della Ricerca Scientifica 1, I-00133 Roma

1. Summary

In a recent review paper we summarized the current status of unification of radio-loud AGN (Urry & Padovani 1995 PASP 107, 803), connecting high-luminosity (FR II) radio galaxies with quasars, and low-luminosity (FR I) radio galaxies with BL Lac objects. Unified schemes are motivated by the knowledge that AGN appearance depends strongly on orientation (Fig. 1): optical/UV light from the centers of many AGN is obscured by circumnuclear matter, and in radio-loud AGN, bipolar relativistic jets beam light along the jet axes. Understanding these radiation anisotropies allows us to unify apparently distinct classes of AGN that differ primarily because of orientation.

Our review described the classification and general properties of AGN and summarized the evidence for anisotropic emission caused by circumnuclear obscuration and relativistic beaming. We outlined the evidence, both observed isotropic properties and statistical arguments, for connecting FR IIs with quasars and FR Is with BL Lacs. The population statistics (with beaming) are in accordance with available data and suggest $\gamma \sim 5$ for low-luminosity AGN and $\gamma \sim 10$ for high-luminosity AGN. The distinctions between X-ray-selected and radio-selected BL Lac objects, and between BL Lacs and flat-spectrum variable quasars, still not understood, provide clues to the underlying physics of blazars. Our review discussed several possible problems and complications, and concluded with a list of the ten questions we believe are the most pressing in this field.

R. Ekers et al. (eds.), Extragalactic Radio Sources, 379–380.

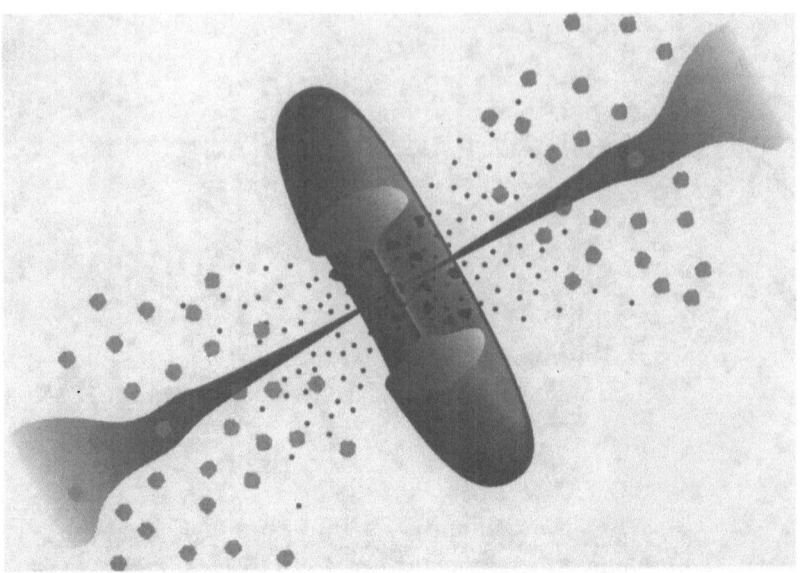

Figure 1.   Current paradigm for radio-loud AGN: central black hole (scales for $10^8 \mathcal{M}_\odot$ black hole, $R \sim 3 \times 10^{13}$ cm), surrounded by accretion disk ($r \sim 1 - 30 \times 10^{14}$ cm), broad-line clouds ($\sim 2 - 20 \times 10^{16}$ cm from BH), thick dusty torus obscuring transverse lines of sight (inner radius $\sim 10^{17}$ cm), and narrow line clouds ($10^{18} - 10^{20}$ cm). Hot electrons scatter continuum and BLR photons into line of sight and may produce hard X-ray continuum. Radio jets (shown as diffuse FR I-type) are relativistic near the BH (10^{17}), produce relativistically beamed optical through γ-ray emission in the inner regions, and can extend to several times 10^{24} cm;

THE TEN MOST IMPORTANT QUESTIONS

1. Are BL Lacs or obscured quasars in all radio galaxies?

2. What is the relation between BL Lacs and FSRQ?

3. Are superluminal velocities, core-to-extended flux ratios, and jet/counter-jet ratios commensurate with beaming?

4. Is the bulk Lorentz factor higher in high-luminosity radio sources (quasars, FR IIs) than in low-luminosity radio sources (BL Lacs, FR Is)?

5. Do FR Is have broad emission line regions?

6. What is the relation between FR Is and FR IIs?

7. How do jets form and propagate?

8. What is the physical cause of the radio loud – radio quiet distinction?

9. Where are the narrow-line (Type 2) radio-quiet quasars?

10. What are the fundamental parameters governing the central engine, and is it powered by a black hole?

ORIENTATION EFFECTS AND SEYFERT CORES

R. P. NORRIS
Australia Telescope National Facility
PO Box 76, Epping, NSW2121, Australia

AND

A. L. ROY
Australia Telescope National Facility, University of Sydney,
and NRAO

1. Introduction

Activity in galaxies takes on a bewildering array of guises. Not all of this diversity comes from fundamentally diverse physics, but if we can understand the relationships between different types of galaxy then we may be able to perceive the underlying physics. This taxonomic approach aims to determine which properties are common to several types of galaxy, and which properties differ. The danger in this process is that it is easy to invent spurious relationships between similar, but quite distinct, types of object.

We may compare the state of extragalactic astronomy to that of stellar astronomy around the turn of the century, when astronomers were trying to understand the relationships between the different types of star. When Hertzsprung & Russell chose to plot colour against luminosity, then the main sequence, and other evolutionary groups, became clear, and subsequently enabled the physics of stars to be determined.

In extragalactic astronomy, we would like to know which axes to plot in order to reveal the underlying simplicity which will guide us to the physics. However, unlike the case of stars, the extragalactic H-R diagram will almost certainly not be two-dimensional - it is easy to list at least six axes which we know from our physical understanding to be independent.

However, we do know that at least one of these axes will represent orientation, since galaxies are not spherically symmetric. Orientation is particularly simple and has enormous predictive power, and so we need to understand the effect of orientation, so that we can disentangle it from

381

R. Ekers et al. (eds.), Extragalactic Radio Sources, 381–384.

other axes which may have deeper physical significance. The principal aim of this paper is to study a few of the relationships between different types of galaxy, and, in particular, the effect that orientation may have on Seyfert galaxies.

2. Do Sy1 and Sy2 differ only in orientation?

Seyfert galaxies can be grouped into two broad classes: Seyfert 1 (Sy1) and Seyfert 2 (Sy2). The observational differences between them are primarily that Sy1s have broader permitted lines and a stronger nuclear continuum than Sy2s. A naive interpretation might be that Sy1s are simply more energetic than Sy2s, but this is inconsistent with properties such as far-infrared (FIR) or radio luminosity, which are similar for both Sy1s and Sy2s.

An alternative explanation is that all Seyfert galaxies contain a nucleus, consisting of a compact continuum source and a broad line region (BLR), surrounded by a dusty torus. When viewed from above the plane of the torus, the broad permitted lines from the BLR and the optical continuum from the nucleus are both visible, and the galaxy is classified as a Sy1. When viewed from within the plane of the torus, the torus obscures our view of the nucleus at optical wavelengths, and the galaxy is classified as a Sy2. An extreme version of this model proposes that starburst galaxies are Seyfert galaxies, the narrow-line region (NLR) of which is also obscured, so that all we observe is the starburst emission in the outer part of the galaxy.

To test this hypothesis, Roy et al. (1994) examined the compact radio cores in carefully matched samples of 157 Sy1 and Sy2 galaxies, using the Parkes-Tidbinbilla Interferometer (a 300-km radio-linked interferometer). Since the dusty torus should be transparent to radio waves, the unified orientation model predicts that an equal number of cores should be seen in Sy1 and Sy2 galaxies. Alternatively, if Sy1s are more energetic versions of Sy2s, or if relativistic beaming were significant, then we should see a greater number of cores in Sy1s than in Sy2s.

The experiment gave a completely unexpected result: significantly more cores were seen in Sy2s (with a 48% detection rate) than in Sy1s (with a 26% detection rate). This firm observational result was at variance with any existing theory, and so demanded an explanation. Roy et al. proposed a variation of the unified orientation model. In this variation, the clouds in the narrow- line region may be optically thick at centimeter wavelengths, and so obscure our view of the radio emission from the compact core. An alternative model proposed by J. Miller (private communication) is that clouds in the broad-line region are optically thick, with the same effect.

3. The role of starburst activity in Seyferts

Normal spiral and starburst galaxies show a tight correlation between their radio and FIR luminosity (e.g. Wunderlich et al., 1987). This correlation is remarkably tight and extends over five orders of magnitude in luminosity. While its cause is still not completely understood, it cannot be accounted for by selection effects or bias. It is almost certainly the result of the star formation process, and can therefore be used as an indicator of star formation activity.

Norris et al. (1988) showed that Seyfert galaxies also roughly obey this correlation, although they have a greater scatter about the correlation than do starbursts or normal galaxies. This suggests that the radio emission from Seyfert galaxies may be dominated by starburst activity. This view is supported by other evidence (e.g. Mouri & Taniguchi 1992) that Seyfert activity is often accompanied by starburst activity.

Baum et al. (1993) have imaged a number of Seyfert galaxies at radio wavelengths. They found that, while the central jets appear to be oriented randomly with respect to the host galaxy, the broad outer radio lobes tend to be oriented perpendicularly to the host disc. This indicates that buoyancy may play a role in determining the direction of the lobes. In this respect, and in their overall morphology, these outer Seyfert lobes resemble the starburst-driven "superwinds" seen in some starburst galaxies (Heckman et al. 1987; Unger et al. 1989).

Baum et al. also showed that, for those Seyferts whose radio emission exceeded that predicted from the radio-FIR correlation, the excess radio emission was largely accounted for by the nuclear (sub-kpc) region. This implies that the outer lobes of these Seyfert galaxies follow the standard radio-FIR correlation. For the ratio of radio-to-FIR luminosity to be the same for these lobes as for starburst galaxies would be very surprising if they were driven by a different mechanism. Therefore, this offers some support for these outer lobes being driven by starburst rather than Seyfert activity. Roy et al. (1995), on the other hand, show that the inner jets do not follow the radio-FIR correlation.

Thus there is some support for the idea that while the inner jets of Seyfert galaxies are indeed tightly coupled to the Seyfert activity, the lobes in the outer parts may be driven by starburst rather than Seyfert activity. An alternative hypothesis is that there is some common underlying mechanism that makes two different energy sources follow the same relationship, and behave in similar ways. However, we should be cautious of naive attempts to unify Seyferts with radio galaxies and radio-loud quasars, as some of the properties of Seyferts may be dominated by starburst activity

4. The radio-loud/radio-quiet connection

Although spiral galaxies and starburst galaxies lie on the well-known radio-FIR correlation, many elliptical radio galaxies and radio-loud quasars have an excess of radio luminosity which places them well off this correlation. None of the unified schemes has yet successfully explained why galaxies fall into these two different groups - the radio-quiet galaxies, most of which are spirals, and the radio-loud galaxies, all of which are ellipticals. Particularly striking is the absence of a single known radio-loud spiral galaxy.

In an attempt to explore the transition from radio-quiet to radio-loud objects, we have cross-correlated the Parkes-MIT-NRAO 6-cm survey (Wright et al. 1995) with the IRAS point source catalogue (Roy & Norris 1995), and have produced a pilot sample of 20 radio-loud, gas-rich objects. These include a BL Lac object, six Seyferts, and four spiral galaxies. Particularly striking is 00182-7112, which is a Sy2 with a 0.1 Jy VLBI core at z=0.3276. It appears that this group includes radio-excess spirals, which may be transition objects between the radio-loud and radio-quiet classes of galaxy.

5. Conclusion

Our studies of the relationships between various types of galaxy show that:

• Sy1/Sy2 core detection rates are inconsistent with the standard orientation unification model (or any other model), but we can reconcile them by invoking a high NLR optical depth;

• Seyferts roughly follow the radio-FIR correlation, and in many cases the radio emission from Seyferts is dominated by starburst activity. Some Seyfert-like features may instead be due to starburst activity;

• there exists a class of intermediate objects between radio-loud and radio-quiet, including some "radio-loud Seyferts", which may represent a transition from radio-quiet to radio- loud objects.

References

Baum, S. A., O'Dea, C. P., Dallacassa, D., de Bruyn, A. G., & Pedlar, A., 1993, *ApJ*, **419**, 553.

Heckman, T. M., Armus, L., & Miley, G. K., 1987, *AJ*, **93**, 276.

Mouri, H., & Taniguchi, Y., 1992, *ApJ*, **386**, 68.

Norris, R. P., Allen, D. A., & Roche, P. F., 1988, *MNRAS*, **234**, 773.

Roy, A. L., Norris, R. P., Kesteven, M. J., Troup, E. R., & Reynolds, J. E., 1994, *ApJ*, **432**, 496.

Roy, A. L., & Norris, R. P., 1995, *MNRAS*, submitted.

Roy, A. L., Norris, R. P., Kesteven, M. J., Troup, E. R., & Reynolds, J. E., 1995, in preparation.

Unger, S. W., Pedlar, A., & Hummel, E., 1989, *A&A*, **208**, 14.

Wright, A. E., Griffith, M. R., Burke, B. F., & Ekers, R. D., 1995, *ApJS*, **97**, 347.

Wunderlich, E., Klein, U., & Wielebinski, R., 1987, *A&AS*, **69**, 487.

NEW OBSERVATIONS OF BL LACERTAE OBJECTS

JOHN T. STOCKE
Center for Astrophysics and Space Astronomy
University of Colorado - Boulder
Boulder, CO 80309-0389 USA

Abstract. This contribution is divided into three parts: **1.** A summary of mostly published work that describes the new type of BL Lac Objects discovered with X-ray satellites, "X-ray Bright BL Lacs (XBLs)", and their relationship to the previously known "radio bright BL Lacs (RBLs)." Most (but not all?) of the differences between XBLs and RBLs are explicable using a "viewing angle" model in which the soft X-ray emission emanates into a much wider cone ($23\text{-}30°$) than the radio emission ($8\text{-}10°$). **2.** New ROSAT PSPC soft X-ray observations are presented for complete samples of XBLs and RBLs which may explain the absence of luminous optical emission lines in BL Lacs. **3.** A deep, subarcsec optical imaging survey of a large (50) sample of both XBLs and RBLs conducted at the Canada-France-Hawaii 3.6m Telescope (CFHT), whose purpose was to characterize the host galaxies and clustering environment of BL Lacs. This work defines more precisely the "parent population" of BL Lacs and identifies a small number of discrepant objects which may not be "beamed FR 1s".

1. X-Ray Loud BL Lacertae Objects (XBLs)

The discovery of numerous new BL Lac Objects with various X-ray satellite telescopes (Einstein, EXOSAT, ROSAT) has revitalized BL Lac research by giving us many new objects to study, new models to create and new disagreements to discuss. With $\sim$200 BL Lacs now known, it is evident that, while XBLs and RBLs share enough common features to both be called BL Lacs, these groups differ systematically in the following ways:

RADIO: While most XBLs and RBLs have the extended radio power levels and morphologies of FR 1s, their core dominance values differ sys-

R. Ekers et al. (eds.), Extragalactic Radio Sources, 385–388.

tematically as would be expected if we are viewing them from different angles relative to their jet axes. Specifically, using data from Perlman & Stocke (1993) as updated by Rector et al. (1996) to constitute nearly complete samples of RBLs and XBLs we find the mean core dominance values for FR 1s:XBLs:RBLs to be 0.1:1.5:20 suggesting (for a single assumed $\Gamma = 5$) mean viewing angles of $60^o : 23^o : 10^o$ and relative volume densities of 130:5:1. These values are similar to those suggested by Urry & Padovani (1995) from other considerations. However, the RBL sample contains a larger number (although still a small percentage) of high-z, high radio power level sources which appear to be FR 2s rather than FR 1s.

OPTICAL: While the optical spectra of both RBLs and XBLs can both be described as nearly featureless (Stickel et al. 1991; Morris et al. 1991), many more RBLs are either very featureless or possess only weak, low luminosity emission lines (Rector et al. 1996). High quality optical spectra of XBLs almost always show weak absorption features of of an underlying host galaxy. The maximum optical polarization percentages of the class of XBLs are significantly less than RBLs and, most interestingly, $\sim 80\%$ of XBLs show "preferred position angles" ($\pm 10^o$) of optical polarization, which RBLs rarely show (Jannuzi et al. 1994). The preferred polarization position angles seem particularly suggestive of viewing angle differences.

X-RAY: The soft (0.1-2.4 keV) X-ray spectra of RBLs and XBLs are quite similar and both have significantly steeper spectra than quasars ($<\alpha>_{ph}$ = -2.2 and -2.4 for RBLs and XBLs compared to -1.5 for quasars; Perlman et al. 1996; Urry et al. 1996). At harder X-ray energies the spectra of RBLs may flatten while XBLs steepen (Sambruna et al. 1994). The γ-ray detections of BL Lacs are dominated by RBLs (von Montigny et al. 1995).

These various properties are explicable if XBLs are viewed further from the jet axis than RBLs, requiring the soft X-ray beam to be broader (but not necessarily at lower Γ) than the radio and optical beams. However, the overall spectral energy distributions (SED) of XBLs and RBLs differ significantly, with the XBL SED peaking in the UVX while the RBL SED peaks in the IR/optical. Such a large difference is difficult to understand in the context of the simplest viewing angle models and may require other differences between XBLs and RBLs (e.g. electron energy distribution; Sambruna et al. 1996; Padovani & Giommi 1995). The large difference in $<V/V_{max}>$ values for RBLs and XBLs (0.62 and 0.34) also remains a problem for the unification of these objects. These new best values reflect the addition of 3 new BL Lacs to both the 1 Jy RBL and EMSS XBL samples, including three low luminosity XBLs discovered recently with the ROSAT HRI (Rector et al. 1996) as suggested by Browne & Marcha (1993).

2. Soft X-Ray Spectra and the Absence of Strong Emission Lines

The viewing angle model may also explain the absence of luminous emission lines in BL Lacs. Guilbert, Fabian & McCray (1983; GFM83) suggested that the emission line regions in BL Lacs could be unstable due to having a very soft X-ray emission spectrum. In contrast, the harder X-ray spectra of quasars create a quasi-stable, two-phase medium, similar to the ISM of the Galaxy, in which the warm clouds imbedded in a hot substrate are the broad (and possibly also the narrow) emission line clouds. By this model the very soft X-ray spectra of BL Lacs are insufficient to create a substrate hot enough to allow quasi-stable warm clouds. While our new ROSAT PSPC survey of BL Lacs supports this hypothesis, the hard X-rays seen in some BL Lacs (primarily RBLs) seem to make the GFM83 model untenable.

However, in the viewing angle model, the soft X-ray emission emanates from the nucleus in a significantly wider cone than the radio (and inverse-Compton X-ray) emission. Thus, the potential emission line regions of BL Lacs are illuminated primarily by the steep, soft X-ray continuum, not by the harder inverse-Compton emission, making warm clouds unstable over much of the near-nuclear region and suppressing luminous line emission.

3. The CFHT Deep Optical Imaging Survey of BL Lac Objects

A deep optical imaging survey of the host galaxy and meta-galactic environment of 50 BL Lac Objects at $z < 0.65$ was conducted at CFHT in Gunn-r band and typically during subarcsec seeing (WSY96; W96). These observations were sufficient to resolve 46 of the 50 BL Lacs observed, 32 well-enough to determine an unambiguous host galaxy morphology. Of these 32, 29 are definite or likely ellipticals and 3 are spirals (PKS 1413+135; MS 0205+351; OQ 530). MS 0205+351 also has the lowest luminosity "host galaxy" in the survey ($M_r = -21.7$ compared to a $<M_r> = -23.2$) and is the only object whose core is definitely decentered (> 0.1 arcsec) with respect to its "host" (Stocke, Wurtz & Perlman 1995). An HST image confirms that the "host" of PKS 1413+135 is an edge-on spiral (McHardy et al. 1994). The VLBI structure of PKS 1413+135 is unique amongst BL Lacs in being two-sided, clearly a problem for relativistic beaming models; see Perlman et al. (this conference) for a stunning VLBA map. These unusual BL Lacs are candidates for background quasars whose properties are, at least partially, produced by microlensing due to stars in a foreground galaxy.

Our analysis of BL Lac clustering environments confirms the Prestage & Peacock (1988) result that BL Lacs avoid rich clusters. W96 finds that $<B_{gg}>$ (BL Lacs) $= 250\,\mathrm{Mpc}^{1.77}$, where B_{gg} is the amplitude of the BL Lac-galaxy covariance function. For reference Abell richness classes 0,1 & 2 have $<B_{gg}> = 360, 645, 945\,\mathrm{Mpc}^{1.77}$, so that a typical BL Lac environment is that of an Abell richness class ≤ 0 cluster. Only five BL Lacs are in richness ≥ 1 clusters and four of those are at $z \geq 1/2$ (the low-z, rich cluster BL Lac is PKS 0548-322). Despite the long-standing belief that BL Lacs are "beamed" FR 1 radio galaxies, the host galaxies and clustering environs of BL Lacs are better matched by FR 2s than by FR 1s in the sense that $\sim 20\%$ of FR 1s are in brighter host galaxies and richer clusters than any BL Lac (these are rich cluster brightest cluster galaxies; BCGs). The evolution in BL Lac cluster environments is also better matched by FR 2s than by FR 1s. But because the extended radio power levels and morphologies of most BL Lacs are those of FR 1s, we do NOT suggest that BL Lacs are beamed FR 2s. Rather we suggest a minor modification of the BL Lac parent population so that the FR 1 radio galaxies in the richest clusters (and specifically BCGs) cannot be BL Lacs. Following Bicknell (1994) this could be due to a more rapid deceleration of the radio jet by the enhanced ICM pressure in a rich cluster or to a greater collimation of the jet in rich cluster FR 1s compared to FR 1s in poor clusters (Stocke & Burns 1978). This also explains the absence of BL Lacs in a large sample of Abell cluster FR 1s (Owen, Ledlow & Keel 1996).

This report is a very brief summary of the PhD dissertations of Drs. Eric Perlman and Ron Wurtz and of Mr. Travis Rector at the University of Colorado. This work was supported by a NASA Long Term Space Astrophysics Grant to JTS.

References

Bicknell, G.V. 1994 *ApJ*, **422**, 542.
Browne, I.W.A. & Marcha, M.J.M. 1993 *MNRAS*, **261**, 795.
Guilbert, P.W., Fabian, A.C. & McCray, R. 1983 *ApJ*, **266**, 466 (GFM83).
Jannuzi, B.T., Smith, P. & Elston, R. 1994 *ApJ*, **428**, 130.
McHardy, I.M. et al. 1994 *MNRAS*, **268**, 681.
Morris, S.L. et al. 1991 *ApJ*, **380**, 49.
Owen, F.N., Ledlow, M.J. & Keel, W.C. 1996, *AJ*, Jan. issue.
Padovani, P. & Giommi, P. 1995 *ApJ*, **444**, 567.
Perlman, E.S. & Stocke, J.T. 1993 *ApJ*, **406**, 430.
Perlman, E.S., Stocke, J.T., Wang, Q.D. & Morris, S.L. 1996 *ApJ* Jan. 10 issue.
Prestage, R.M. & Peacock, J.A. 1988 *MNRAS*, **230**, 131.
Rector, T., Stocke, J.T. & Perlman, E.S. 1996, in prep.
Sambruna, R.M., Urry, C.M. & Maraschi, L. 1996 *ApJ*, in press.
Sambruna, R.M. et al. 1994 *ApJ*, **434**, 468.
Stickel, M., Padovani, P., Urry, C.M. & Fried, J.W. 1991 *ApJ*, **374**,431.
Stocke, J.T. & Burns, J.O. 1978 *ApJ*, **319**, 671.
Stocke, J.T., Wurtz, R. & Perlman, E.S. 1995 *ApJ*, Nov 10.
Urry, C.M. & Padovani, P. 1995 *PASP*, **107**,803.
Urry, C.M. et al. 1996 *ApJ*, in press.
Wurtz, R., Stocke, J.T. & Yee, H.K.C. 1996 *ApJS*, Mar issue (WSY96).
Wurtz, R., Stocke, J.T., Ellingson, E.E. & Yee, H.K.C. 1996, in prep. (W96).
von Montigny, C. et al. 1995 *ApJ*, **440**, 525.

THE UNIFICATION OF QUASARS AND RADIO GALAXIES

SPERELLO DI SEREGO ALIGHIERI
Osservatorio Astrofisico di Arcetri
Largo E. Fermi, 5 - 50125 Firenze - Italy

1. Introduction

The evidence for misdirected quasars in the nuclei of powerful radio galaxies (PRG) is now solid: **1.** strong perpendicular polarization of the UV continuum, produced by the scattering of light from the hidden nucleus, has been observed in 10 (out of 11) PRG with $z > 0.7$ (Cimatti et al. 1993); **2.** polarized broad MgII2800 has been seen in 6 PRG (di Serego Alighieri et al. 1994, Antonucci et al. 1994), showing that the hidden nucleus is indeed a quasar. These findings are receiving authoritative confirmation from observations with the Keck telescope (Cimatti, Cohen and van Breugel in these Proceedings). Here I would like to concentrate on recent developments of our understanding of the unification between quasars and PRG.

2. A dust scattering model for powerful radio galaxies

With Alessandro Manzini we have developed a model to reproduce the spectral energy distribution (SED) and the polarization of the light from the hidden quasar which is scattered in our direction by dust in the ISM of the PRG (Manzini and di Serego Alighieri 1995). This model uses Mie single scattering, the MRN dust size distribution and the geometry suggested by the Unified Model, and deals consistently with extinction and scattering by the same dust. The main results are: **1.** the dust scattering efficiency can be either blue, red or flat, depending on the amount of dust; therefore a flat scattered light does not necessarily imply electron scattering. **2.** The 2175Å feature, which appears in emission in the MRN albedo, but is not generally observed in the SED of PRG at high redshift, can be eliminated including a proper amount of extinction. **3.** The polarization resulting from the model is between 5 and 50%, not uncomfortably higher than the observed values, and varies with the mean scattering angle.

R. Ekers et al. (eds.), Extragalactic Radio Sources, 389–390.

This model, together with improved stellar population synthesis models (D. Villani in these Proceedings) and a model of the nebular continuum (Dickson et al. 1995), have been used to fit the SED and the polarization observed in a number of PRG, giving constraints on the luminosity of the hidden quasar, on the dust content, and on the stellar population ages in distant PRG.

3. [OIII] anisotropies in powerful radio galaxies

Jackson and Browne (1990) have shown that the [OIII] luminosity of quasars is larger than that of PRG for a sample of radio sources at intermediate redshifts. This result has been regarded as a failure of the Unified Model, since the [OIII] line was believed to be radiated isotropically and then it should be observed independently of orientation. However high ionization lines with a large critical density could be emitted close enough to the nucleus to be partially within the obscuring "torus" and therefore escape anisotropically. We have made spectropolarimetric observations of 6 PRG and of 3 quasars to detect possible anisotropies in the [OIII] emission of PRG. We have preliminarily detected polarized [OIII] in 5 PRG, indicating anisotropic emission, while [OIII] is not polarized in the 3 quasars observed (di Serego Alighieri et al. in preparation).

4. Outlook

These recent improvements in our understanding of the active components of PRG are leading to a refinement of the Unified Model of the most powerful AGN, and give us confidence that also stellar populations can be studied in these objects at cosmological distances. Further improvements can be expected in the following areas: studies of how the importance of anisotropic emission in AGN changes with redshift and radio power, estimates of the dilution in the UV polarization in PRG from detailed spectropolarimetry of broad emission lines, studies of the IR polarization and of the 2175Å feature as a dust diagnostic, searches for stellar and interstellar absorptions, and the development of inhomogeneous dust scattering models.

References

Antonucci, R., Hurt, T. & Kinney, A., 1994, *Nature*, **371**, 313.

Cimatti, A., di Serego Alighieri, S., Fosbury, R.A.E., Salvati, M. & Taylor, D., 1993, *M.N.R.A.S.*, **264**, 421.

di Serego Alighieri, S., Cimatti, A. & Fosbury, R.A.E., 1994, *ApJ*, **431**, 123.

Dickson, R., Tadhunter, C., Shaw, M., Clark, N. & Morganti, R., 1995, *M.N.R.A.S.*, **273**, L29.

Jackson, N. & Browne, I.W.A., 1990, *Nature*, **342**, 43.

Manzini, A. & di Serego Alighieri, S., 1995, *A & A*, submitted.

THE EVOLUTIONARY UNIFIED SCHEME, JET ASYMMETRY, AND SUPERLUMINAL MOTION

FAUSTO VAGNETTI

Dipartimento di Fisica, Università di Roma "Tor Vergata"

Via della Ricerca Scientifica, I-00133 Roma, Italy

The evolutionary unified scheme (Vagnetti *et al.*, 1991; Vagnetti and Spera, 1994) is the first attempt to describe the radio source populations accounting at the same time for their cosmic evolution and their anisotropic phenomena. It achieves an interconnection between anisotropy and evolution through a changing balance between two optical components, an isotropic "thermal" one and a relativistic beam. It predicts an evolution of the radio-optical ratio, interpreted as a slow increase of the Lorentz factor with the cosmic time (Vagnetti *et al.*, 1991). It also predicts a link between the high and low-power unified schemes, and in particular between high and low-redshift blazars, and between high and low-redshift radio galaxies.

If this scenario is valid in this simple form, it seems that high redshift sources oriented at intermediate viewing angles, i.e. the steep-spectrum quasars, are destined to become low-power radio galaxies at low redshift. It has therefore been investigated the distribution of quasars and radio galaxies in the viewing angle–redshift plane (Vagnetti and Spera, 1994), based on the changing balance between the nuclear optical components (both isotropic and anisotropic) and the host galaxy luminosity. In fact, without invoking any obscuration mechanism, a substantial fraction of low-redshift sources at intermediate viewing angles are predicted to be seen as radio galaxies.

It is thus interesting to explore the distributions of orientation indicators such as the jet asymmetry $J = (\delta_{jet}/\delta_{counterjet})^{2+\alpha_r} = [(1 + \beta \cos \theta)/(1 - \beta \cos \theta)]^{2+\alpha_r}$ and the apparent transverse velocity $\beta_{app} = \beta \sin \theta/(1 - \beta \cos \theta)$ separately for quasars and radio galaxies. As such parameters are also strong functions of the Lorentz factor, this is also the way to look for the predicted Γ increase.

Preliminary results are here presented for the second of three models proposed in Vagnetti and Spera (1994), which includes a distribution of Γ

R. Ekers et al. (eds.), Extragalactic Radio Sources, 391–392.

(with Γ_{max} increasing with cosmic time), but not convolutions through the luminosity functions of quasars and host galaxies, as in the most accurate model of Vagnetti and Spera (1994). Therefore the results, shown in Figure 1, are to be considered only as trends. However, the dominant effect is clearly the strong change of the shapes of the distributions of both parameters, due to transmutation of steep-spectrum quasars into radio galaxies. Substantial populations of radio galaxies with high asymmetry and superluminal motion are expected at low redshifts. The effect of the Γ increase is instead visible at the high end of the distributions, although modest and with very small numbers, such that very large samples would be required to compare the expectations to reality.

References

Vagnetti, F., Giallongo, E., and Cavaliere, A. 1991, *ApJ*, **368**, 366
Vagnetti, F., and Spera, R. 1994, *ApJ*, **436**, 611

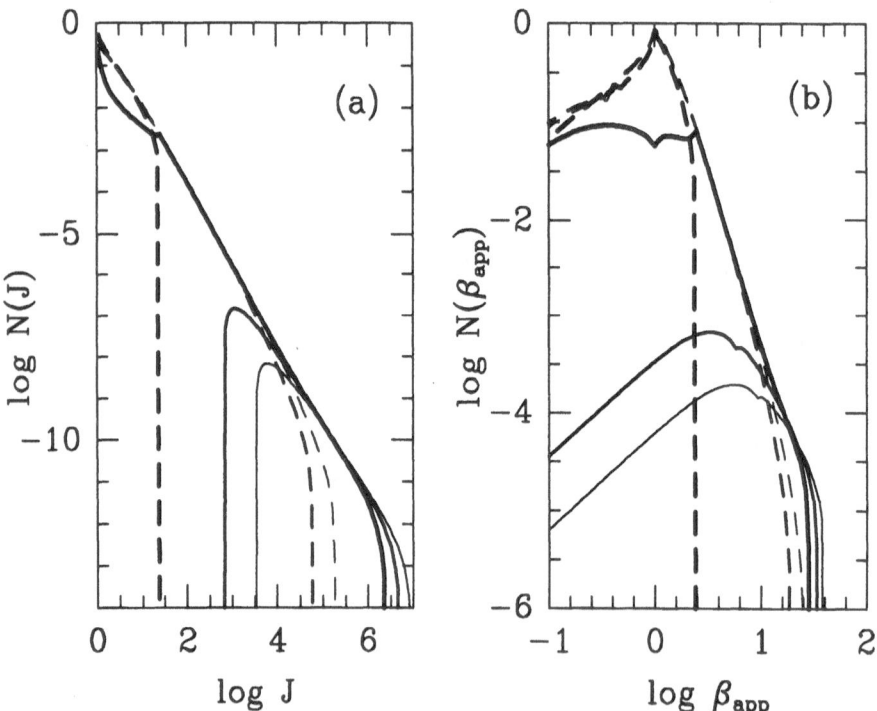

Figure 1.   Predicted distributions of jet asymmetry (a) and of apparent transverse velocity (b). Continuous lines: quasars; dashed lines: radio galaxies. Lines of increasing thickness represent distributions at increasing redshift: 0, 0.2, 0.5. The ordinates represent fractions of sources normalized to the combined population quasars + radio galaxies.

RADIO STRUCTURES OF THE MRC 1-JY SOURCES AND THE UNIFICATION OF RADIO GALAXIES AND QUASARS

V. K. KAPAHI[1], R. M. ATHREYA[1], C. R. SUBRAHMANYA[1]

[1] *NCRA-TIFR, P.Bag 3, Ganeshkhind, Pune-411001, India*

P. J. MCCARTHY[2], W. VAN BREUGEL[3], J. C. BAKER[4]

[2] *OCIW, 813 Santa Barbara St., Pasadena, Ca 91107, USA*
[3] *IGPP, LLNL, Livermore, Ca 94550, USA*
[4] *MRAO, Madingley Road, Cambridge CB3 0HE, UK*

AND

R.W.HUNSTEAD [5]

[5] *Astrophy. Dept., University of Sydney, NSW2006, Australia*

The viewing angle of $\sim 45^\circ$ (between the jet axis and the line of sight) dividing quasars and radio galaxies in the orientation-based unified scheme (Barthel 1989) is based largely on the observation that in the redshift range of $0.5 < z < 1.0$ in the low-frequency 3CRR sample, there are about twice as many radio galaxies than quasars and the median linear size of galaxies is about twice that of the quasars. The relative numbers and sizes of quasars in other redshift ranges in the 3CRR are however in conflict with the simple unified scheme even if the opening angle is considered to evolve with epoch or radio luminosity (Kapahi 1990; Singal 1993). Larger sources samples, preferably selected at low radio frequencies, are clearly important in such statistical studies.

We report here the preliminary results of a comparison of the numbers and sizes of quasars and radio galaxies in a much larger complete sample of MRC radio sources defined by $S_{408} > 0.95$ Jy and $-30^\circ < \delta < -20^\circ$, for which we have made VLA radio maps, near–complete optical identifications and are attempting to obtain spectroscopic redshifts (Kapahi et. al, 1996). Of the 558 sources in our sample we have defined a complete sample of 105 quasars (with 95% spectroscopic redshifts) and 447 radio galaxies (with 60% measured redshifts and 22% estimated from K magnitudes).

R. Ekers et al. (eds.), Extragalactic Radio Sources, 393–394.

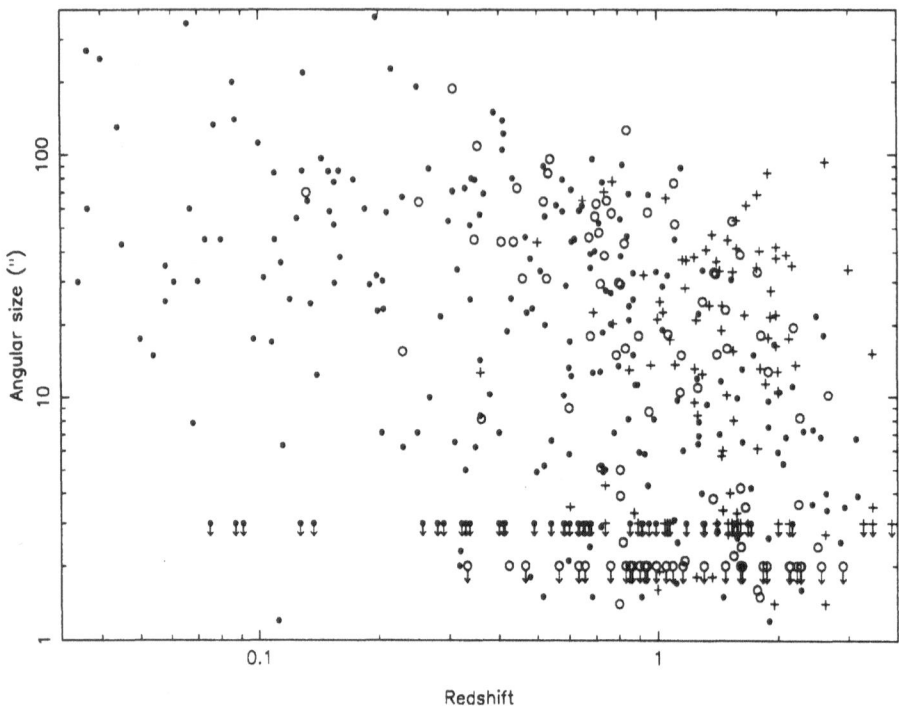

Figure 1. The angular size – redshift plot for radio galaxies ('•' measured z, '+' estimated z) and quasars (○) with known z in the MRC 1–Jy samples.

The angular size – redshift plot for all the quasars and galaxies with known or estimated redshifts is shown in Figure 1. Surprisingly, we see no systematic difference between galaxies and quasars in either the maximum angular sizes or the range of angular sizes, contrary to the prediction of the unified model which requires the projected linear sizes of quasars to be almost a factor of two smaller than for galaxies. The medium linear sizes of galaxies and quasar in the redshift ranges of 0.25 to 0.5, 0.5 to 1.0 and 1.0 to 2.0 do not, in fact, show any significant difference. There is also no systematic change in the quasar fraction as a function of redshift. Our new sample thus provides little evidence for any epoch or luminosity dependence of the torus opening angle and the data are hard to understand in the simple orientation based unification of quasars and radio galaxies.

References

Barthel, P.D. (1989) *Ap.J.*, **Vol.336**, 606.
Kapahi, V.K. (1990) in 'Parsec-scale Radio Jets' (Eds. J.A.Zensus & T.J.Pearson), Cambridge Univ. Press, p304.
Kapahi, V.K. et al. (1990) The MRC 1-Jy sample of Radio Galaxies – *This volume*.
Singal, A.K. (1993) *MNRAS*, **Vol.263**, 139.

SPECTRAL INDEX ASYMMETRIES
IN LOW-Z RADIO GALAXIES

J. DENNETT-THORPE[1], A.H. BRIDLE[2], R.A. LAING[3]
AND P.A.G. SCHEUER[1]

[1] *MRAO, Cavendish Lab., Madingley Rd,Cambridge, UK*
[2] *NRAO, Edgemont Rd, Charlottesville, VA, USA*
[3] *RGO, Madingley Rd, Cambridge, UK*

1. Introduction

In recent years a number of correlations have been observed between the large-scale properties of FRII radio galaxies and quasars. Some correlations find their simplest explanations in terms of environmental effects, others in terms of beaming/orientation effects. The problem, however, is that there is an implied correlation between these different *types* of effects. This is a report on work underway to investigate this directly and exclude the confusion introduced by use of different samples.

The correlations relevant to this work are:

1 counter-jet side ⟷ most rapidly depolarized lobe: explained by beaming theories with the counter-jet side further from us, and viewed through more Faraday medium. (Garrington *et al.*, 1991), 39/47 sources.

2 short side ⟷ most rapidly depolarized lobe: explained as an environmental effect with shorter lobe behind denser Faraday medium. (Liu & Pooley, 1989), 7/10 sources.

3 most rapidly depolarized lobe ⟷ steeper spectrum: not simply explained in terms of orientation and beaming effects but implies an intrinsic source asymmetry. (Liu & Pooley, 1989), 11/12 sources.

4 counter-jet side ⟷ steeper lobe spectrum (Garrington *et al.*, 1991), 36/47 sources.

The correlations with depolarization [1,2 & 3] are compatible only if the *counter-jet side* also correlates with *short side* and *steeper lobe spectrum*. Whilst the Garrington-Laing effect [1] is naturally explained as an

R. Ekers et al. (eds.), Extragalactic Radio Sources, 395–396.

orientation effect, it is hard to see why either lobe length or lobe spectrum should depend strongly on orientation (i.e. jet sidedness).

Observationally it is found that the correlation of jet side with lobe length is very weak (Scheuer, 1995). The correlation of jet side with lobe spectrum was found by Garrington *et al* [4], but they had insufficient resolution to rule out the possibility that this was simply due to a Doppler-boosted hot-spot on the jet side.

The challenge is to disentangle the intrinsic, environmental and orientation effects. To this end we have selected two different samples, one high-powered quasar sample (Bridle *et al.*, 1994, J.Dennett-Thorpe *et al.*, in prep.) and one low-power sample. Beaming models indicate that the first sample should be dominated by orientation effects whilst in the second intrinsic and environmental effects should be dominant.

2. Results

Twelve $z < 0.15$ low-powered sources with detectable jets (from Black, 1991) now have suitable multi-array VLA observations at two frequencies (archival and new data). To date six sources have been analysed. Asymmetries, although small, are apparent in the spectra of these sources, both in the integrated flux and in regions of similar surface brightness.

There is *no* correlation between jet side and spectral index difference. This rules out explanations purely in terms of orientation (e.g. Blundell & Alexander, 1994). This is expected for standard models in a sample dominated by NLRG, but the continued existence of asymmetries indicates that they are not caused solely by differential ages or Doppler effects.

Although the sample size is small there is a strong correlation between *lobe area* and *spectrum* (5/6 sources): lobes with the larger surface area have a flatter spectrum throughout the source. We also note however that there is *not* a strong correlation with lobe length.

The spectral index and surface brightness (dependent differently on geometry, injection spectra and magnetic field history) do not show a one-to-one correspondence so the correlations presented here represent crude averages. The data will be analysed in various other ways and a third frequency added to a number of sources.

References

Black, A.R.S., 1991, PhD thesis, University of Cambridge.

Blundell, K.M., & Alexander,P., 1994, *MNRAS*, **267**, 241

Bridle, A.H., Laing, R.A., Scheuer P.A.G. & Turner, S.T., 1994, in *First Stromlo Symposium*, eds. Bicknell, G.V., Dopita, M.A. & Quinn, P.J.

Garrington, S.T., Conway, R.G. & Leahy, J.P, 1991, *MNRAS*, **250**, 171

Liu, R., Pooley, G., 1991, *MNRAS*, **249**, 343

Scheuer, P.A.G., 1995, *MNRAS*, (in press)

DEPOLARIZATION STUDIES OF RADIO SOURCES AND THE UNIFIED SCHEME

S.T. GARRINGTON AND G.F. HOLMES
The University of Manchester, NRAL Jodrell Bank, UK

AND

D.J. SAIKIA
NCRA, TIFR, Pune, India

1. Introduction

In powerful FRII radio galaxies and quasars the polarization of the two radio lobes is often very asymmetric and strongly correlated with jet sidedness. These sources usually have one-sided jets and the lobe on the jet side depolarizes more slowly than the one on the counter-jet side (Laing 1988; Garrington et al. 1988). This correlation can be interpreted in terms of depolarization of the receding lobe by a halo of hot gas surrounding the radio source and is in the expected sense if the jet asymmetry is due to Doppler boosting.

2. The FRI sources

Low luminosity FRI sources have diffuse outer radio lobes or plumes. Their jets are reasonably symmetric on larger scales but often asymmetric closer to the nucleus. While the outer plumes may be transonic (Bicknell et al. 1990), several of the VLBI scale jets have moderately relativistic velocities (Giovannini, these proceedings). We have observed a sample of low- and intermediate-luminosity sources with the VLA at λ 6 and 18/20 cm at matched resolutions of 5 or 15 arcsec. Larger sources were also observed with the WSRT at λ 49 cm.

Many sources (large and small) show little depolarization between λ 6 and 18/20 cm. Several sources show significant, symmetric depolarization: these sources are predominantly 'fat' or twisted FRI sources. Several sources show strong asymmetric depolarization, with less depolarization on the side

R. Ekers et al. (eds.), Extragalactic Radio Sources, 397–399.
© 1996 IAU.

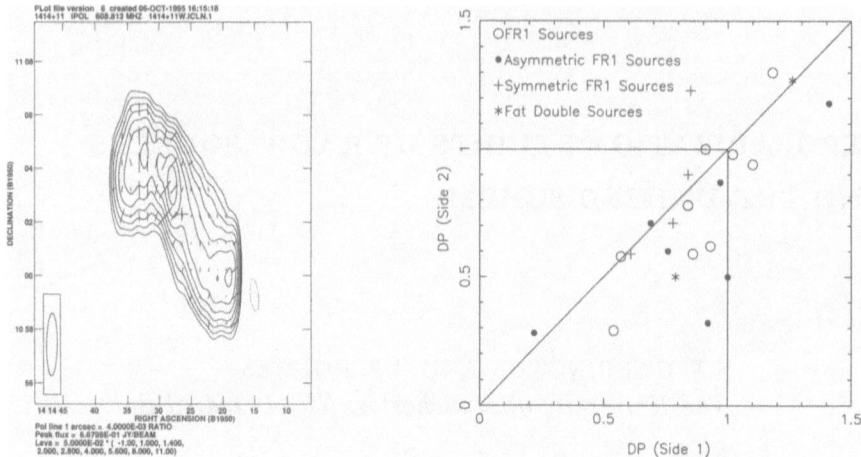

Figure 1. Left: the radio galaxy 3C296 observed with WSRT at λ49cm. The southern jet, which is much fainter in higher resolution maps, depolarizes between 18 and 49 cm.
Figure 2. Right: plot of DP ($= m_{18}/m_6$ or $= m_{49}/m_{18}$) for the two jets or lobes. Side 1 has the brighter jet at high resolution.

of the source with the brighter jet. These sources include the fat doubles and FRI sources with the more asymmetric jets. In 3C296 (Figure 1), the depolarization asymmetry becomes very pronounced in the WSRT 49cm map. Figure 2 compares the measured depolarization ratio (low values of DP correspond to stronger depolarization) for the two sides of each source: The correlation between jet sidedness and depolarization asymmetry is seen in the cluster of sources below the diagonal line and is in the same sense as the more powerful sources.

Our provisional results suggest that FRI sources have low Faraday dispersion, and are generally more symmetric in depolarization than FRII sources. The depolarization asymmetry seen in the sources with the most asymmetric jets is consistent with the view that these jets may be initially relativistic.

3. The FRII radio galaxies and quasars

In the unified model for FRII radio galaxies and quasars (eg Barthel 1989), we would expect the radio galaxies to show the same correlation of depolarization with jet-sidedness, but perhaps at a lower level. We observed a sample of radio galaxies, selected from the 3CR complete sample, with the WSRT at λ49 and 21cm and with the VLA at λ20, 18 and 6cm. Combining these with data from the literature, we find that 19 out of 27 radio galaxies with jets show stronger depolarization on the counter-jet side. The weaker trend seen for radio galaxies appears consistent with the unified scheme.

Since the degree of depolarization asymmetry is stronger at higher redshift and jets are rare in distant radio galaxies, we also constructed a smaller matched sample of 9 radio galaxies and 9 quasars in the redshift range of about 0.3 to 1. The numbers of sources with stronger depolarization on the counter-jet side is 8/9 for the quasars and 5/9 for the radio galaxies (although the most asymmetric cases are in the expected sense). It appears that in radio galaxies the correlation of depolarization with lobe separation is stronger than the relation with jet sidedness, while the reverse is true for quasars (c.f. Laing 1993). While the correlation of depolarization with arm-length may be related to an intrinsic asymmetry in the distribution of gas surrounding the source (McCarthy et al 1991), it would also be a natural consequence of a symmetric, centrally peaked distribution of depolarizing gas.

4. Other sources

A few per cent of radio-loud quasars are highly asymmetric with the bright-ness ratio of the lobes exceeding a few 100:1. This may be due to Doppler boosting (Saikia et al. 1990). We have observed 13 of these asymmetric quasars with the VLA at $\lambda 20$ and 6cm. The new maps reveal weak counter-lobes in some cases, and show that the 'core' is an unrelated source in others. The brighter lobes show little depolarization between $\lambda 20$ and 6cm, consistent with them being the approaching components of double-lobed sources seen through little of the depolarizing medium of the host galaxy or cluster. Some of the smaller sources, however, have low values of polarization at both $\lambda 20$ and 6cm.

From multifrequency studies of the cores of core-dominated quasars, we find that the rotation measures (RM) are generally less than about 30 rad m^{-2} at $\lambda > 6$cm, but extend up to a few hundred rad m^{-2} at shorter wavelengths, perhaps due to contributions from components buried deeper in the core. There is some evidence that the cores of galaxies are much more weakly polarized than quasars (Saikia, Singal & Cornwell 1987), perhaps due to depolarization by the obscuring torus required by unified models.

References

Barthel P.D. 1989, *ApJ*, **336**, 606.
Bicknell G.V. et al., 1990, *ApJ*, **354**, 98.
Garrington S.T, et al., 1988, *Nature*, **331**, 147.
Laing R.A., 1988, *Nature*, **331**, 149.
Laing R.A., 1993, in Astrophysical Jets, eds Burgarella D., et al., CUP, p. 95
McCarthy, P.J., van Breugel, W. and Kapahi,V.K, 1991, *ApJ*, **371**,478.
Saikia D.J., Junor W., Cornwell T.J., Muxlow T.W.B., Shastri P., 1990, *MNRAS*, **245**, 408.
Saikia D.J., Singal A.K., Cornwell T.J., 1987, *MNRAS*, **224**, 379.

THE UV/RADIO CORRELATION OF QUASARS AND ITS IMPLICATIONS FOR UNIFIED SCHEMES.

HEINO FALCKE

University of Maryland, Department of Astronomy, College Park, MD 20742-2421, USA (hfalcke@astro.umd.edu)

1. Introduction

What causes the differences in the central engines of AGN? Are they intrinsically different or are the differences entirely due to environment and orientation? We have postulated that there is basically only one engine, consisting of a black hole and a closely coupled (even symbiotic) jet/disk system with very similar parameters in all AGN [2]. This has lead to interesting results concerning the modelling of the UV/radio correlations [4] of quasars and the difference between FR I and FR II radio galaxies [3].

2. The UV/radio correlation for quasars

If there is a closely coupled jet/disk system in AGN we would expect to find a strong correlation between accretion disk luminosity (UV bump) and radio core luminosity in quasars; and indeed correlations between optical (e.g. O[III] [5]) and radio emission have been demonstrated. We tried to improve upon those correlations by estimating the UV-bump luminosity of quasars directly from optical/UV/x-ray observations of quasars [6] *and* indirectly from emission line luminosities [1]. We combined all estimates for each source into a single estimate for the disk luminosity L_{disk} and plotted this vs. their VLA radio core fluxes at 5 GHz [4]. The UV/radio distribution can then be compared to a simple emission model which takes mass and energy conservation in a coupled jet/disk system into account [2]:
– We find a strong correlation between UV/bump and radio core luminosity in radio weak quasars and a similar trend in radio loud quasars. The former correlation implies that radio emission and UV-bump in *radio weak* quasars have a similar origin (central engine rather than starburst).

R. Ekers et al. (eds.), Extragalactic Radio Sources, 400–401.

– Flat-spectrum compact quasars have brighter radio cores than steep spectrum quasars at the same L_{disk} consistent with the former being relativistically boosted. Width of the distribution and location of the flat-spectrum cores are well modelled by randomly oriented relativistic jets with bulk Lorentz factor $\gamma \sim 6$ at $L_{disk} \sim 10^{46}$ erg/sec.

– The cores of radio loud quasars (with bright extended emission) are brighter than those of radio weak quasars at the same L_{disk}. Hence, the radio-loud/radio-weak dichotomy is already established on the pc scale.

– The total jet power of radio loud quasars must be comparable to their accretion disk luminosity to explain the bright synchrotron emission from their radio cores. This indicates that the jet is produced in the innermost region of the disk where most of the energy is available. Radio loud quasars are very efficient and require that relativistic electrons are in equipartition with the magnetic field (having a minimum Lorentz factor of $\gamma_e \sim 100$). Radio weak quasars can be modelled by basically the same jet if the available electrons are accelerated into a powerlaw starting at $\gamma_e \sim 1$.

– There is a distinct population of a few flat-spectrum quasars which have *total* radio luminosities intermediate between radio loud and radio weak quasars (flat-spectrum intermediate quasars = FIQ). Despite being dominated by a variable, flat-spectrum radio core – which usually is indicative of relativistically boosted jets – their cores are equally bright or even weaker than the cores of lobe-dominated quasars at the same L_{disk}. If the FIQ are boosted quasars their parent population can not be the radio loud quasars as they lack the steep-spectrum lobe emission and their cores are to weak. The only possible parent population therefore are radio weak quasars. Number and offset of the FIQ in the PG sample would then both indicate bulk Lorentz factors of $\gamma \sim 3 - 5$ in *radio weak* quasars!

– In the optically and the radio selected sample there is a void of radio loud *quasars* below a critical L_{disk}. We identify this with the FR I/FR II break and suggest that this might be caused by a torus with power-dependent opening angle [3]. At low powers the torus closes, obscures the central engine for all aspect angles (no quasar signatures) and starts to disrupt the radio jet (causing the FR I morphology). The power-dependent torus will change the length/number statistics for quasars and radio galaxies and the jet torus interaction can modify the jet on the pc scale.

References

[1] Boroson, T.A., Green, R.F. 1992, *ApJS* **80**, 109
[2] Falcke, H., Biermann, P. L. 1995, *A&A* **293**, 665
[3] Falcke, H., Gopal-Krishna, Biermann, P.L. 1995b, *A&A* **298**, 395
[4] Falcke, H., Malkan, M., Biermann, P.L. 1995, *A&A* **298**, 375
[5] Miller, P., Rawlings, S., Saunders, R. 1993, *MNRAS* **263**, 425
[6] Sun, W.H., Malkan, M.A. 1989, *ApJ* **346**, 68

KECK SPECTROPOLARIMETRY OF
HIGH REDSHIFT RADIO GALAXIES

A. CIMATTI*, A. DEY AND W. VAN BREUGEL

Institute of Geophysics and Planetary Physics, LLNL – 7000 East Ave, P.O. Box 808, L-413, Livermore, CA 94550, USA

*On leave from Osservatorio Astrofisico di Arcetri, Italy

R. ANTONUCCI

Physics Dept., University of California at Santa Barbara, CA 93106, USA

AND

H. SPINRAD

Astronomy Dept., University of California at Berkeley – 601 Campbell Hall, Berkeley, CA 94720, USA

1. Introduction

High redshift radio galaxies (HzRGs) are observable up to cosmological distances competitive with the most distant quasars. However, before using them as probes of galaxy evolution, it is crucial to separate the stellar and non-stellar components. One of the most striking properties of HzRGs is the alignment of the UV continuum with the axis of the radio source (*alignment effect*; McCarthy et al. 1987). However, the relative importance of the stellar and non-stellar radiation to the *alignment effect* is still unknown, although a significant fraction is recognized to come from scattering of anisotropic radiation emitted by the obscured nucleus, as expected in the unified model of powerful radio sources (di Serego Alighieri, Cimatti & Fosbury 1994). Spectropolarimetry is the most powerful technique to observe at the same time different radiation components, but the 4m class telescopes can reach a sufficient S/N ratio only on the few brightest objects. Therefore, in order to investigate the origin of the *alignment effect* and to test the validity of the unified model of powerful radio-loud AGN, we have started a pro-

R. Ekers et al. (eds.), Extragalactic Radio Sources, 402–404.
© 1996 *IAU.*

gram of optical spectropolarimetry of HzRGs with the W.M. Keck 10m telescope equipped with the Low Resolution Imaging Spectrometer (LRIS) in polarimetric mode.

2. General results and implications

We observed 6 HzRGs with $0.7< z <1.8$. Table 1 and Figure 1 show respectively the main results and one example of our spectropolarimetry. 3C 368 ($z=1.132$) is not included in Table 1 because the data have been not fully reduced yet. • High linear polarization of the UV continuum is detected in all the observed galaxies (with the exception of 3C 356b). The perpendicularity of $\vec{E}$ to the UV continuum axis and the constancy of $\theta(\lambda)$ suggest that scattering is the dominant polarization mechanism. • The detection of the MgIIλ2800 emission line in polarized flux in at least 3 galaxies suggests that the incident radiation comes from an obscured quasar nucleus and is emitted anisotropically along the radio axis. In particular, the broad and polarized MgIIλ2800 in 3C 324 has velocity and equivalent widths consistent with those observed in radio-loud quasars (Cimatti et al. 1995). On the other hand, the always lower or null polarization of the [OII]λ3727 line implies that this line is emitted isotropically outside the obscuring region. • In the two galaxies analysed in detail so far (3C 256, Dey et al. 1995; 3C 324, Cimatti et al. 1995), we observe *spatially extended* polarization along the UV continuum axis, implying that the scattered flux is spatially extended. This result was possible thanks to the good seeing during the observations and the high S/N ratio of our spectropolarimetry. • These results are in agreement with the requirement of the unified model of powerful radio-loud AGN, where the differences between Type 1 and Type 2 AGN are mainly due to orientation effects and not to intrinsic diversities (Antonucci 1993). Our results also imply that scattered light is a necessary ingredient to explain the *alignment effect*. In particular, we find that scattered light contributes up to 50% of the total UV continuum in 3C 324 (Cimatti et al. 1995). • We detect basically two kinds of $P(\lambda)$: flat and blue. Since the total flux spectra are generally red ($F_\nu \propto \nu^{-1 \div 2}$), the polarized flux spectra $P \times F_\nu$ are either red or flat. This information will be used to investigate in detail the nature of the scattering particles, the incident spectrum and the importance of unpolarized UV radiation.

References

Antonucci, R., 1993, *ARAA*, **31**, 473
Cimatti, A., Dey, A., van Breugel, W., Antonucci, R., Spinrad, H., 1995, *ApJ*, submitted
Dey, A., Cimatti, A., van Breugel, W., Antonucci, R., Spinrad, H., 1995, *ApJ*, submitted
di Serego Alighieri, S., Cimatti, A., Fosbury, R.A.E., 1994, *ApJ*, **431**, 123
McCarthy, P.J., van Breugel, W.J.M., Spinrad, H., Djorgovski, S., 1987, *ApJ*, **321**, L29

TABLE 1. Summary of the results

Galaxy	z	$\Delta\lambda_{rest}$ (Å)	P_{unb} (%)	$\vec{E}$	$P(\lambda)$	$\theta(\lambda)$	$P \times F_{tot}$
3C 441	0.707	2350-5300	2-10	$\perp$	blue	const	const
3C 356a	1.079	1900-4300	7-15	$\perp$	blue	const	const
3C 356b	1.079	1900-4300	2-4	$\perp$	const?	const?	red?
3C 324	1.206	1800-4000	10-13	$\perp$	const	const	red
3C 13	1.351	1700-3800	5-9	$\perp$	const	const	red
3C 256	1.819	1400-3200	10-14	$\perp$	const	const	red

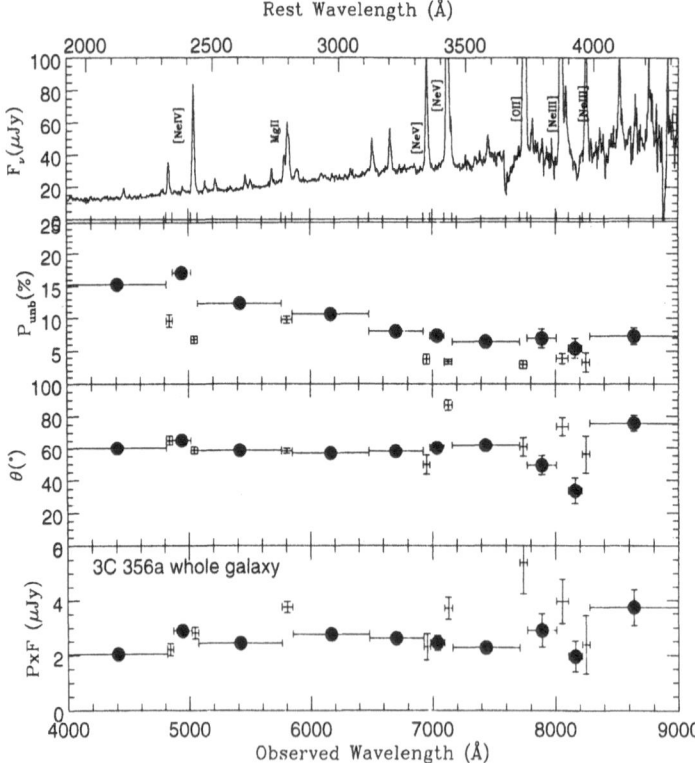

Figure 1. Keck spectropolarimetry of 3C 356a. From top to bottom : the total flux spectrum, the position angle of $\vec{E}$, the degree of polarization and the polarized spectrum.

COMPOSITE OPTICAL SPECTRA OF RADIO QUASARS

JOANNE C. BAKER
Mullard Radio Astronomy Observatory, Cavendish Laboratory, Madingley Rd, Cambridge CB3 0HE, UK.

AND

RICHARD W. HUNSTEAD
Department of Astrophysics, School of Physics, University of Sydney, NSW 2006, Australia.

1. Introduction

Primarily to illustrate the effects of orientation on the optical properties of radio-loud quasars, we have created a set of composite optical spectra (see also Baker & Hunstead 1995). Optical spectra drawn from the 408-MHz Molonglo Quasar Sample (MQS) have been coadded in four sets according to R, the ratio of radio core-to-lobe flux, which is used as an orientation indicator (eg. Orr & Browne 1982). Compact steep-spectrum (CSS) quasars (see review by Fanti, this volume) have been combined separately, revealing for the first time many distinguishing features in their average spectra.

2. The Composite Spectra

Figure 1 reveals clear differences between quasars of different R and the CSS quasars. A detailed comparison is made in Baker & Hunstead (1995). In summary, we find that with decreasing R (i) the optical continuum steepens, (ii) the 3000Å broad emission feature decreases in relative strength, and (iii) the narrow-line equivalent widths, broad line widths and Balmer decrements increase (Baker et al. in prep.). The above trends suggest that the nuclei of many lobe-dominated quasars are viewed through layers of dust, perhaps associated with the hazy outer regions of a torus.

The composite for the CSS quasars in the MQS is especially revealing. The average continuum slope is a very steep power law, with broad features, such as around 3000Å, missing. Enhanced narrow-line emission,

R. Ekers et al. (eds.), Extragalactic Radio Sources, 405–406.
© 1996 *IAU.*

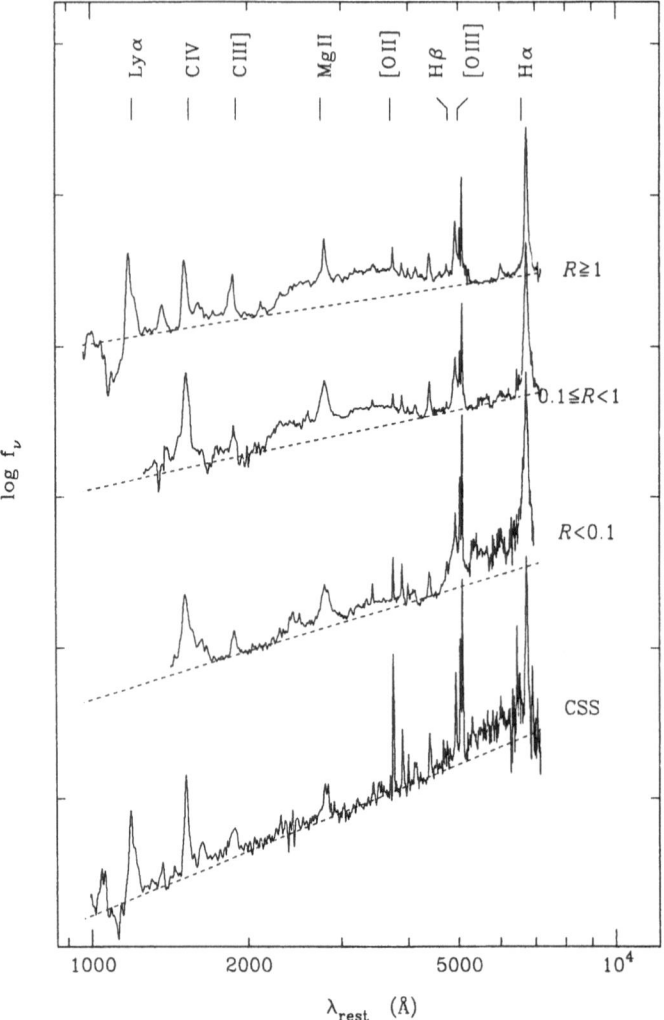

Figure 1. Composite spectra for the MQS, separated according to R (see text). Best-fit power-law spectral slopes are shown as dashed lines (Baker & Hunstead 1995).

particularly low-ionisation species, may indicate strong jet–ISM interactions. Self-absorption of Mg $II\lambda2798$ and the low Ly α/C IV ratio point to heavy absorption in these sources.

References

[1] Antonucci, R.R.J. (1993), *ARAA*, **31**, 473
[2] Baker, J.C. and Hunstead, R.W. (1995), *ApJL*, **452**, L95
[3] Orr, M. and Browne, I.W.A. (1982), *MNRAS*, **200**, 1067

ON THE X-RAY EMISSION FROM THE POWERFUL
RADIO GALAXIES

G. BRUNETTI, G. SETTI, A. COMASTRI
Istituto di Radioastronomia
Via Gobetti, 101, 40129 Bologna, Italy

There is a growing evidence that radio loud quasars and powerful FR II radio galaxies belong to the same population. While X-ray observations of low redshift radio galaxies [1],[2] generally support the unified scheme relating the FR II radio galaxies to the radio quasars, nevertheless detailed studies of the X-ray properties of distant radio galaxies are made difficult due to both the low sensitivity of X-ray-satellites and to the emission of the hot intracluster gas in which they are normally embedded. We point out that significant fluxes of X-rays are produced in the strong radio galaxies by the Inverse Compton (IC) process. In the framework of the unified scheme the radio galaxies are pervaded both by an intense radiation flux from the misdirected hidden quasar and the cosmic microwave background radiation flux (CMB). From the standpoint of IC computation the far and near-IR emissions of the hidden quasar are of particular importance.

Based on the studies of various statistical samples [3],[4] we have adopted a typical quasar continuum spectrum as a combination of several power laws ($F(\nu) \propto \nu^{-\alpha}$): α =0.2, 0.9, 1.34 and 0.6 respectively in the intervals 100 - 50 μm, 50 - 6 μm, 6000 - 650 nm and 650 - 350 nm with an integrated mean luminosity ($L_{<Q>}$) of 6.9 $\times 10^{46}$ erg s^{-1} ($H_0 = 75$ km s^{-1} Mpc^{-1}, $q_0 = 0.5$ are used throughout). We assume that the quasar unabsorbed optical–UV photons are emitted within a cone with half–opening angle = 45° [5] whose axis coincides with the radio axis defined by the hot spots. Since several studies [3],[4] indicate that the mean IR emission of NLRGs is $\sim$ 4 times lower than that of RL quasar, we shall adopt the same geometry for the quasar IR emission cone. We assume that the radio galaxies contain an uniform distribution of relativistic electrons and protons with the same energy density (k=1) and that there is equipartition between the particles and the magnetic fields. Since the IC contribution to the soft X-ray spectrum in our model is mainly due to low energy electrons ($\gamma \sim 50 - 1000$), the strength of

407

R. Ekers et al. (eds.), Extragalactic Radio Sources, 407–408.

the equipartition fields quantities are computed by taking into account the contribution of such particles to the energy density. Ionization losses may become important at low energies and have been taken into account. The spatial distribution of particles and fields has been modelled with a prolate ellipsoid centered on the nucleus and aligned with the radio structure. We have applied our model to the computation of expected soft X−rays fluxes for a sample of very powerful high redshift radio galaxies. In the case of 3C 277.2, 280 and 356 the derived luminosities of the hidden quasars required to match the observed X-ray luminosities of the radio galaxies are close to the average luminosity $L_{<Q>}$. For 3C 277.2 the requested quasar luminosity is $\sim 0.5 \, L_{<Q>}$. The visual magnitude of this quasar would be V$\simeq$16.7, which is very close to the estimate (V=17.2) obtained from spectropolarimetric studies [6]. The ratio between the IC contribution due to the CMB photons and to the hidden quasar radiation field is $\sim$1.2. In the case of 3C 280, by matching the quoted point source flux [7] with our IC model, we require a hidden quasar luminosity $\sim$0.56 $L_{<Q>}$. The ratio between the IC contribution due to the CMB photons and to the hidden quasar would be 0.6. Finally, in the case of 3C 356 the requested quasar luminosity is $\sim$2.5 $L_{<Q>}$, which would entail a ratio between the IC contribution due to CMB and to the hidden quasar of $\sim$0.7. Both the predicted soft X−ray spectrum and brightness distribution are in agreement with the observed ones. This scenario may provide a possible alternative interpretation of the X-ray data to that of the cooling flow hypothesis [1]. We have considered other five high redshift powerful radio galaxies (3C 22, 175.1, 268.1, 352, 441) for which X−ray data are not available yet and have estimated the soft X−ray luminosities predicted by our IC model under the assumption that the hidden quasars have the average luminosity $L_{<Q>}$. The estimated soft X−ray luminosities (in 10^{43} erg/sec) are respectively 2, 0.3, 0.8, 3 and 0.6. As a general remark it should be stressed that the X−ray fluxes (given the radio galaxies volume and the relativistic electrons spectrum) have been computed under the hypothesis of energy equipartition between particles and magnetic fields, so that our model calculations could be considered as a lower limit to the IC contribution.

References

[1] Crawford & Fabian, 1995, *MNRAS*, **273**, 827
[2] Ueno S., Koyama K., Nishida M., 1994, *ApJ*, **431**, L1
[3] Heckman T.M., Chambers K.C., Postman M., 1992, *ApJ*, **391**, 39
[4] Heckman T.M.,et al., 1994, *ApJ*, **428**, 65
[5] Barthel P.D., 1989, *ApJ*, **336**, 606
[6] Di Serego Alighieri S.,et al., 1994, *ApJ*, **431**, 123
[7] Worrall D.M., et al., 1994, *ApJ*, **420L**, 17W

UNIFICATION ISSUES (*DISCUSSION*)

Discussion of the paper presented by <u>GOPAL-KRISHNA</u> (p. 373)

Lacy: Why do you think that the lower QSR fraction in fainter flux radio samples is a luminosity, rather than redshift, effect?

Gopal-Krishna: Narrower torus openings in less luminous sources are expected from theory and also inferred for radio–quiet AGN.

Miley: We have long known that variability and environment play a role in the observed properties of radio sources. You can't explain either the variability of QSRs, or the fact that radio sources don't occur in spirals by orientation alone. As you mentioned, the main attraction of the orientation unification was its simplicity and predictive ability. Now that life is clearly more complicated, wouldn't it be reasonable to start afresh, and re-examine possible scenarios where orientation is not the dominant effect?

Gopal-Krishna: I agree that temporal evolution must be incorporated into any realistic modelling of radio source populations, including their unification. Doing that, I have suggested that the radio size measurements can be reconciled with the basic tenet of the orientation unification, viz., that radio galaxies are oriented closer to the sky plane than quasars.

Laing: I would like to object to George Miley's attack on unification. It is, of course, not necessary for orientation to explain everything. But, unless quasars have side-on counterparts, much of what we believe about radio-source physics must be incorrect, and we have very good evidence at least for the basic relativistic-jet picture.

Urry: While there may be a 'reasonable doubt' about some aspects of unification of radio galaxies and quasars, it is supported by a 'preponderance' of the evidence. A question: Why do you consider the characteristics of BLRGs a problem, why can't they be low-luminosity (local) quasars?

Gopal-Krishna: Apparently, BLRGs differ from *both* radio galaxies and quasars in having distinctly flatter mid–IR spectra.

R. Ekers et al. (eds.), Extragalactic Radio Sources, 409–412.
© 1996 *IAU*.

Discussion of the paper presented by <u>NORRIS</u> (p. 381)

di Serego Alighieri: Have you considered that, since you are working at the low end of the radio luminosity function and your objects are not selected in the radio, your core detections will be largely influenced by small random variations of the intrinsic radio luminosity and therefore cannot be used to test anything?

Norris: It is easy to test statistically whether this result could be produced by random variations. The probability of obtaining this result by chance is less than 1%.

Wilson: To compare the radio properties of Seyfert 1 and 2 cores, one needs to compare the distributions of radio luminosities of the cores, not just the detection rate. When one does that, what is the statistical significance of any difference between the radio core luminosity distributions of Seyfert 1's and 2's.

Norris: The reason for using detection rates, rather than core luminosities, is that it's very difficult to do rigorous statistics on the core luminosities when most are non-detections, as the luminosity distribution is then dominated by upper limits.

de Bruyn: The higher detection rate for Sey 2's reminds me of our own results (de Bruyn and Wilson, 1978, A&A). That (UV-selected) Markarian-Seyfert sample was biased. Your sample could be biased as well (e.g. distance, luminosity?) Have you investigated these effects? What we need is a volume-limited sample constructed from samples selected through various ways (X-ray, UV, IR, radio, optical). Can you now construct such a sample?

Norris: Yes, we are well aware of such potential problems, and we used a FIR-selected sample for that reason, since FIR should be unaffected by orientation. However, we also have optically-selected samples, and 12μm-selected samples, and the results on these are consistent with the FIR- selected sample. It would be nice to have a hard X-ray sample too (soft X-rays are still affected by orientation), but I don't know of any hard X-ray selected sample of sufficient size.

Meier: I am not as worried that there are too few cores in Sy 1's as I am that there are too many cores in Sy 2's. There are two possibilities:
1. these cores are like other flat spectrum cores (in radio quasars, for example). If so, then the fact that they can be seen from the side (in Sy 2's) suggests that cores are not relativistically beamed after all. This

would have important implications for other unified schemes.

or 2. these "cores" are not cores in the true sense, but rather small steep spectrum jets which can be seen from the side. Do you have any spectral information on these cores which would distinguish between these two cases.

Norris: There seems to be little evidence for relativistic beaming in Seyfert cores, and so in this respect they differ from their high-energy cousins. Spectral index is difficult to disentangle from resolution effects, but as far as we can tell the data are consistent with a spectral index of -0.7.

Koekemoer: If you plot the subtracted radio (core) flux vs. the FIR flux, do you still see a weak correlation? If not, can you use this to constrain the amount of FIR emission due to the torus vs. the FIR from the extended galaxy?

Norris: Perhaps surprisingly, we see no significant correlation between the core flux and the FIR flux. I don't know whether this places a significant constraint on the re- processed FIR flux from the torus, but it's an interesting question.

Discussion of the paper presented by di SEREGO ALIGHIERI: (p. 389)

Gelderman: To complete the test of possible anisotropic [OIII] emission, have you observed a sample of quasars, to compare to your quoted radio galaxy results?

di Serego Alighieri: Yes, in addition to the six radio galaxies, we have also observed three quasars, for which [OIII] and [OII] are not detected in the polarised flux. We interpret this as due to the fact that in quasars any anisotropic line flux is pointed towards us and therefore we do not see the scattered component.

Discussion of the paper presented by ATHREYA (p. 393)

Miley: How do you define and isolate the radio cores? Isn't it possible that interaction with the ionized gas could be responsible for the steep spectrum?

Athreya: The earlier steep spectrum cores which turned out to be ionized haloes round the flat spectrum cores were all structures which could be resolved at scales of 2-5 kpc. The cores discussed in this paper are unresolved with a resolution of 0.2″ and likely to be even smaller (at z~2.5, $0.1''$ ~0.7 kpc). We believe that these are more likely to be the parsec scale cores of radio sources. However, higher resolution studies will be able to confirm this.

Laing: Your conclusions about the doppler-shifting of the core break frequency depend on the assumption of a single velocity: if a range of velocity is present, the slower emission will dominate for side-on sources and faster emission for end-on sources.

Athreya: That is correct. However, the dependence on the doppler factor is very weak for the size and the magnetic field calculation (not so for the electron number density). The values presented in the text are only order of magnitude estimates.

Ekers: (commenting on a question by Andrew Wilson). I strongly support Andrew's view that it will be hard to see the small effect of orientation on projected linear size when the distribution of linear sizes is so large. There must be other physical effects which we don't understand which are producing changes 100 times larger than that predicted by orientation.

X–RAY AND GAMMA–RAY EMISSION IN BLAZARS

GABRIELE GHISELLINI
Osservatorio Astronomico di Torino
Strada Osservatorio, 20, I-10025 Pino Torinese, Italy

1. Introduction

More than 50 sources have been detected by EGRET in 4 years of operations (e.g. von Montigny et al. 1995). Almost all of them show the violent characteristics typical of blazars, such as superluminal motions, strong radio emission mainly produced in a flat spectrum core and large amplitude variability at all frequencies. Interestingly, optical polarization does not seem to be required, since more than $1/3$ of the detected sources are less than 3% polarized.

Theoretical activity is obviously hectic in this new field, and I will concentrate in the following only on those models which base the high energy emission on the inverse Compton process.

The overall spectral energy distribution (SED) of γ–loud blazars in a $\nu F(\nu)$ plot shows two broad peaks, the first in the IR–optical–UV band, and the second in the γ–ray band. We can interpret the first as due to synchrotron emission, and the second to the inverse Compton process, which can operate on synchrotron photons or on photons produced outside the γ–ray emitting region and seen amplified (by relativistic effects) in the frame comoving with the blob or jet.

2. Location of the γ–ray emitting region

In the model of Blandford & Levinson (1995) the high energy radiation is produced in a inhomogeneous jet, immersed in a bath of photons produced outside the jet. The jet plasma, flowing relativistically, sees this radiation amplified, and Compton scatters it to high energies. The γ–rays produced in the inner regions of the jet do not survive collisions with the externally produced X–rays, and create $e^{\pm}$ pairs. These pairs, born relativistic, radiatively cool by scattering UV–soft X–ray photons to higher energies. Only

413

R. Ekers et al. (eds.), Extragalactic Radio Sources, 413–416.

further out along the jet, where the X-ray density is lower, the γ-rays of larger energies can escape without being absorbed. The resulting γ-ray spectrum is a superposition of the locally emitted spectra, each with a different high energy cut off due to γ-γ absorption, much like the partially opaque flat radio spectrum of compact radio sources.

The main criticism that can be moved to this model is about the amount of the predicted X-rays. In fact, if the inner regions of the jet produce a γ-ray luminosity L_γ, which is absorbed, one inevitably predicts that this power re-emerges in the X-ray band. This leads to the simple conclusion that the luminosity in the observed 10–100 MeV band should be equal to the X-ray luminosity L_X, contrary to observations (L_X is a factor 10–100 smaller than L_γ, e.g. Dondi & Ghisellini, 1995). Invoking the non-simultaneity of X-ray and EGRET observations does not help, because even if not simultaneously observed in the γ-rays, a γ-loud source should have X-ray flares reaching as much power as seen (at other times) above 100 MeV.

Reversing the argument, we conclude that if dissipation of the primary power occurs too close to the putative black hole and accretion disk, and therefore in a dense environment of X-ray and UV photons, the resulting high energy emission is absorbed, and is reprocessed into too many X-rays. Since this is not observed, the γ-ray emitting region must be thin to γ-γ collisions, and therefore located at some distance R_γ from the black hole, of the same order of the 'γ-ray photosphere': $R_\gamma \geq 10^{16.5}$–10^{17} cm (see Ghisellini & Madau 1995). *Then the dissipation of the primary power into radiation must occur at some distance from the central power-house.*

3. SSC Models

For simplicity, consider a homogenous, one-zone model (which can be a portion of an inhomogeneous jet, as in Maraschi, Ghisellini & Celotti 1992).

In this case the peaks in the SED reflect a break in the electron spectrum $N(\gamma) \propto \gamma^{-n_{1,2}}$ at some energy γ_b, with $n_1 < 3$ below γ_b and $n_2 > 3$ above. Electrons at γ_b produce the first synchrotron peak of luminosity L_S and also scatter this radiation into the self Compton peak (Ghisellini, Maraschi & Dondi 1995).

A change in the normalization of $N(\gamma)$ produces a linear change in the IR-opt-UV, and a quadratic change in the γ-ray band. Other behaviours can exist, especially if γ_b is extremely large, as suggested by the TeV detection of Mkn 421 and Mkn 501. In this case Klein-Nishina effects inhibit the self Compton cooling of the highest energy electrons, which preferentially scatter synchrotron photons of frequencies smaller than the synchrotron peak frequency. *Varying only n_2 results in a change of the synchrotron X-*

rays and the self Compton TeV emission of the same amplitude, while the optical and the GeV emission stays constant.

4. Inverse Compton on external photons

Dermer, Schlickeiser & Mastichiadis (1992) and Sikora, Begelman & Rees (1994) pointed out that if the blob moves relativistically in a photon bath produced outside the blob, then the radiation energy density as seen in the comoving frame is strongly amplified and can overtake the radiation energy density produced by the blob itself.

In these models the Compton peak is produced by scattering off photons produced externally, and the ratio of L_γ to the synchrotron luminosity L_s reflects the ratio of the radiation energy density as seen by the blob and the magnetic energy density U'_{ext}/U_B.

In the model of Sikora et al. (1994), the two peaks in the SED are due to a particle spectrum derived self–consistently, assuming continuous injection and (incomplete) cooling. Since the injected particle spectrum is assumed to be a single power law in the entire energy range, the break in $N(\gamma)$ is due to the particles with $\gamma < \gamma_b$ preferentially escaping instead of cooling. The model is then strongly constrained, because the requirement that the cooling timescale equals the escape time at γ_b fixes the radiation energy density in the comoving frame. This assumption can however be relaxed, if one assumes that the break in $N(\gamma)$ is not due to incomplete cooling, but to the injection mode.

In this model a change in the normalization of $N(\gamma)$ produces a corresponding change of equal amplitude both at synchrotron and inverse Compton energies: the optical and the GeV emissions should therefore vary with similar amplitudes. Variations of different amplitudes occurs if what varies is the bulk Lorentz factor Γ, because the amount of radiation energy density seen in the comoving frame is proportional to Γ^2. For viewing angles $\sim 1/\Gamma$, the observed synchrotron luminosity $L_s \propto \Gamma^4$, while the inverse Compton luminosity $L_C \propto U'_{ext}\Gamma^4 \propto \Gamma^6$. This yields $L_\gamma \propto L_{opt}^{3/2}$.

These considerations have been applied to the simultaneous SED of 3C 279 of June 1991 and beginning of 1993, corresponding to two very different states of the source (Maraschi et al. 1994), and with the γ–rays indeed varying more than the optical–UV emission. Since the two states correspond to two separate events in the life of 3C 279, both the alternatives [change in $N(\gamma)$ or change in Γ] can explain the observations. More secure conclusions could be reached by monitoring simultaneously a γ–loud source in the optical and the γ–rays during a single event, because in that case a change in Γ is unlikely.

5. The 'mirror' model

In the model of Sikora et al. (1994) the BLR and, possibly, some scattering material surrounding the jet are illuminated by the accretion disk. To simplify, assume that the external radiation comes from the reprocessing of the BLR only, located in a spherical shell at distance R_{BLR} from the black hole. As long as the active blob is inside R_{BLR}, it sees a constant U'_{ext}. But the active blob produces some amount of photoinizing radiation (by synchrotron emission) which contributes to the illumination of the BLR clouds. This radiation will be seen greatly amplified by clouds within and in the vicinity of the $1/\Gamma$ emission cone of the blob, and can dominate over the illumination of the accretion disk (Ghisellini & Madau 1995). *In this model the largest γ-ray production occurs when the blob is very close to BLR shell or when it crosses it.*

The crossing time is seen contracted by a factor Γ^2 by the Doppler effect, and therefore a width $\Delta R_{BLR} \sim 3 \times 10^{17}$ cm corresponds, for viewing angles $1/\Gamma$, to a time $t_{var} \sim (10/\Gamma)^2$ days.

The distinguishing feature of the model is, once again, the variability pattern it predicts. In fact assume that the blob becomes active at some time. Optical–UV synchrotron photons will partly ($\sim$90%) cross the BLR without being intercepted by the BLR clouds, and reach the observer together with some amount of γ-rays produced at the same time by SSC or by inverse Compton scattering off the line photons produced by the BLR illuminated by the disk. The remaining 10% of the photoionizing photons are immediately converted into (isotropic) line emission, and part of them will reach the blob after some time. At this point the blob increases its γ-ray production, because of the enhanced radiation field, producing a γ-ray flare. *There is a delay between the optical and γ-ray flares, which should again be of the order of* $t_{delay} \sim \Delta R_{BLR}/(c\Gamma^2) \sim t_{var}$.

Furthermore, for small viewing angles, the observer sees an enhancement of the line emission, due to that part of the BLR illuminated by the blob, simultaneous with the optical flare.

References

Blandford, R.D. & Levinson, A. 1995, *ApJ*, **441**, 79
Dermer, C., Schlickeiser, R. & Mastichiadis, A. 1992, *A&A*, **256**, L27
Dondi, L. & Ghisellini, G. 1995, *MNRAS*, **273**, 583
Ghisellini, G. & Madau, P., 1995, submitted to MNRAS
Ghisellini, G., Maraschi, L. & Dondi, L., 1995, Third Compton Symp., *A&ASS*, in press
Maraschi, L., Ghisellini, G. & Celotti, A. 1992, *ApJ*, **397**, L5
Maraschi, L., et al. 1994, *ApJ*, **435**, L91
Sikora, M., Begelman, M.C. & Rees, M.J. 1994, ApJ, 421, 153
von Montigny, C., et al. 1995, ApJ, 440, 525

COOLING-FLOW MODELS OF THE X-RAY EMISSION AND TEMPERATURE PROFILES OF ELLIPTICAL GALAXIES

GIUSEPPE BERTIN AND THOMAS TONIAZZO
Scuola Normale Superiore, Pisa
piazza dei Cavalieri, 7, I-56126 Pisa, Italy

We compare the results of a spherical, steady-state cooling-flow model with gas loss with the observed X-ray emission profiles and temperatures for five bright elliptical galaxies in Virgo and Fornax. We adopt a "thermal-instability" mass-loss prescription (following Sarazin & Ashe 1989) and refer to galaxy models constrained by radially extended stellar dynamical data (Saglia et al. 1992). Our cooling-flow models are specified by three free parameters: the mass accretion rate $\dot{m}_{ext}$ and the pressure p_{ext} at some external radius r_{ext}, and a dimensionless constant q, which regulates the mass deposition rate along the flow. We find that models with confining pressures of $p_{ext} \sim 4 \div 14 \times 10^3 \mathrm{K}$ cm^{-3} and significant accretion rates of external material, up to $4 \mathrm{M}_\odot/\mathrm{yr}$, provide emission and temperature profiles in good agreement with present-day data. The trend shown in the right frame of the lower panel in the Figure suggests a possible correlation between L_X/L_B and the iron abundance in the gas inside ellipticals. A full account of this work, together with all the relevant references, is given in Bertin & Toniazzo (1995); here below, we add a comparison of the models with some recent ASCA data (Loewenstein et al. 1994; Mushotzky et al. 1994).

This work was partially supported by ASI (under contract 94RS94-202/3 FAE) of Italy.

References

Bertin, G., Toniazzo, T. 1995, *Astrophys. J.* **451**, 111
Loewenstein, M., et al. 1994, *Astrophys. J.* **436**, L75
Mushotzky, R.F., et al. 1994, *Astrophys. J.* **436**, L79
Saglia, R.P., Bertin, G., Stiavelli, M. 1992, *Astrophys. J.* **384**, 433
Sarazin, C.L., Ashe, G.A. 1989, *Astrophys. J.* **345**, 22
Trinchieri, G., Kim, D.-W., Fabbiano, G., Canizares, C.R. 1994, *Astrophys. J.* **428**, 555

R. Ekers et al. (eds.), Extragalactic Radio Sources, 417–418.
© 1996 IAU.

TABLE 1. Properties of the selected models

NGC	q	$\dot{m}_{ext}$ (M$_\odot$/yr)	p_{ext} (10^3 ^{0}K/cm^3)	r_{ext} (kpc)	M_{gas} (10^{10}M$_\odot$)	$\log(L_X)$ (erg/s)	$\langle kT \rangle$ (keV)	$\dot{m}_{ext}/\alpha l$
1399	0.6	4.1	7.6	200	19.55	42.29	1.24	3.65
1404	0.2	0.9	13.7	50	0.88	41.50	0.78	2.13
4374	0.8	0.1	6.3	60	0.51	41.35	0.85	0.07
4472	0.8	0.6	12.6	120	4.55	42.01	1.22	0.19
4636	0.4	3.0	3.8	160	8.20	42.02	0.80	2.02

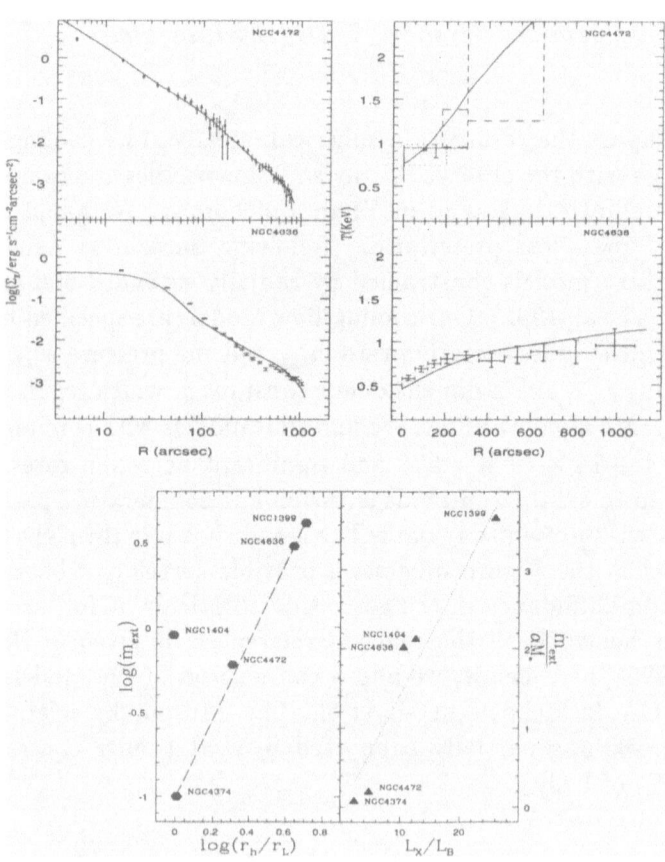

Figure 1.   *Upper panels:* X-ray emission (left) and temperature (right) profiles for the selected models compared with ROSAT (for N4636, see Trinchieri et al. 1994) and Einstein+ASCA+BBXRT (for N4472) data. *Lower panel:* Left: the mass accretion rate $\dot{m}_{ext}$ is found to be larger for more diffuse halos. Here, r_h/r_L is the ratio of the half-mass radius of the *total* mass to that of the luminous mass distribution.   Right: scaling of the dilution parameter $\dot{m}_{ext}/\alpha M_\star$ with the relative X-ray luminosity L_X/L_B.

HIGH-ENERGY GAMMA AND RADIO VARIABILITY OF BLAZARS IN THE MODEL OF NON-STATIONARY JETS

M.M. ROMANOVA
Space Research Institute, Profsoyuznaya 84/32, Moscow, 117810, Russia

AND

R.V.E. LOVELACE
Departments of Astronomy and Applied Physics, Cornell University, Ithaca, NY 14853, USA

A model has been developed for impulsive VLBI jet formation and gamma ray outbursts of Blazars. Propagation of newly expelled matter in the old channel of a jet is calculated supposing that the main driving force is the electromagnetic field. The new outflowing matter overtakes the old matter and forms double, fast or slow magnetosonic shock fronts. In the region of the fronts, the number of particles and their energy increase continuously with propagation time from the central object (Romanova and Lovelace, 1995). Accelerated electrons and positrons in the front interact with a diffuse field of UV photons (inverse Compton scattering), with the magnetic field (synchrotron radiation), and with synchrotron photons (SSC processes), thus creating radiation in a very wide range of bands. The self-consistent relativistic equations for the number of particles, the momentum, energy, and magnetic flux in the front are derived and solved numerically (Lovelace and Romanova, 1995). The time-dependent apparent luminosities in the radio to gamma ray bands are calculated taking into account the Doppler boost of the photons. The model predicts a short outburst of radiation in gamma rays (weeks or so) connected with Compton processes, a sharp (less than a day) outburst in the X-rays with a smooth decrease of the luminosity connected with SSC processes, and synchrotron radiation changing from infrared to radio bands (Fig. 1A). The lepton distribution function was taken as $f_l = K_1/\gamma^2$ in the main energy containing range, $\gamma_1 \leq \gamma \leq \gamma_2$, steeper distribution $f_l = K_2/\gamma^3$ for $\gamma_2 \leq \gamma \leq \gamma_3$, and even steeper for $\gamma \geq \gamma_3$. For $\gamma < \gamma_1$, f_l is assumed negligible as a result of

419

R. Ekers et al. (eds.), Extragalactic Radio Sources, 419–420.

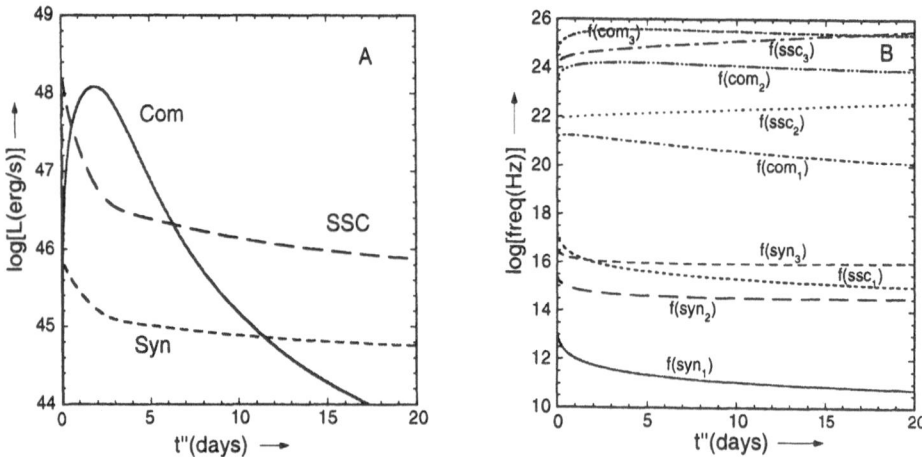

Figure 1. **A.** Apparent luminosities for an observer at an angle $\theta_{obs} = 11.5°$ to the z–axis. **B.** Apparent frequencies of photons radiated by electrons with Lorentz factors γ_1, γ_2, and γ_3 due to different processes

synchrotron self-absorption. The lowest frequency $f(syn_1)$, determined by self-absorption, corresponds initially to the infrared band, and later - to the radio band. From Fig.1B, one can see that radio at 3 mm may start to appear after 2 weeks after outburst. But its maximum may correspond to much later times (months), because $f(syn_1)$ decreases slowly with time. The appearance of the new VLBI component in QSO 0528+134, which approximately coincides with the strong gamma-ray flash and with the beginning of the strong mm radio outburst (Krichbaum, et al. 1995; Pohl, et al. 1995), supports the proposed model.

    Both authors were supported in part by NSF grant AST-9320068. MMR is grateful to RFBR and Organizers of the Symposium for the partial support.

References

Krichbaum, T.P., Britzen, S., Standke, K.J., Witzel, A., Zensus, J.A. (1995) in: *Quasars and AGN: High Resolution Radio Imaging*, ed. M. Cohen and K. Kellerman, Irvine, CA (in press)

Lovelace, R.V.E., Romanova, M.M. (1995) in: Proceedings of the NRAO Workshop *Cygnus A - A study of a Radio Galaxy* , ed. C.L. Carilli & D.E. Harris, Cambridge University Press (in press)

Pohl, M., Reich, W., Krichbaum, T.P., et al. (1995) *Astronomy & Astrophysics*, (in press)

Romanova, M.M., Lovelace, R.V.E. (1995). In: Proceedings of 3rd *Compton Symposium on Gamma-Ray Astronomy and Astrophysics*, Munich, Germany (in press)

COMPTONIZATION AS BLAZAR HIGH ENERGY EMISSION

OLIVER DREISSIGACKER
Landessternwarte Heidelberg, Germany
http://www.lsw.uni-heidelberg.de/~odreissi/

1. Summary

We explain the overall continuous GRAZAR (Gamma Ray Blazar) spectrum from the synchrotron turnover to the EGRET *GeV* detections by means of Comptonization in the parsec scale jet's substructures.

While making use of the constraints on the synchrotron spectrum and other measurable quantities, no exotic particle acceleration is needed to achieve the high energy output.

We show, that the "Lighthouse Model" of blobs of relativistic electrons, travelling with the jet plasma at relativistic speeds, produce both, correct timescales and shapes for the lightcurve, and correct ratios and slopes of the synchrotron, X-ray and γ-ray branches.

2. The Model: Ingredients and Derivations

Babadzhanyants and Belokon (1986) were the first to point out, that the appearance of new VLBI knots in the jet of 3C 345 is strictly correlated to a major outburst in the optical. It is therefore natural, to make the knots responsible for the flare emission.

Quasi periodic peaks during the outburst on timescales much longer than acceleration or cooling times of the involved particles, make it necessary to assume a somewhat periodic modulation process.

The "Lighthouse Model" explains the violent variability in Blazars in terms of a modulation of emission enhancement by a variation of the angle between the line of sight and the emission cone of the relativistically moving source.

Since the matter that later forms the jet plasma was accreted in a disk, it carries angular momentum into the jet. Combining the assumption, that the plasma leaves the disk in the very vicinity of the rapidly rotating black

R. Ekers et al. (eds.), Extragalactic Radio Sources, 421–422.

hole with jet stability calculations from Appl and Camenzind (1993a,b) one obtains the correct timescales for the peak-to-peak periods with a central mass of circa $10^{10} M_\odot$ for the most luminous and therefore massive Blazars.

The form of the lightcurve suggests, that the collimated jet opens at some distance from the core, and the emitted flux drops significantly by adiabatic cooling of the radiating particles. Our results are consistent with the assumption, that the opening occurs when the jet leaves the core of the host galaxy, *i.e.* at a distance of $50 - 100pc$ from the black hole (Kormendy 1994).

The variation of the flux is simultaneous in all regimes originating in the moving component.

Comparing the energy densities of the various radiation fields in the core of the AGN's host galaxy yields that in the rest frame of the blob the synchrotron radiation dominates the lower frequencies, while blue shifted and Doppler enhanced ambient, diffuse radiation dominates the higher energies. Calculating the scattering rates following Blumenthal and Gould 1970 off an evolving electron distribution (Kardashev 1962) and transforming back into the observer's frame shows that during the outburst:

- GHz to optical is dominated by synchrotron emission,
- X-ray is dominated by synchrotron-self-Compton radiation, and
- γ-ray is made by inverse Compton scattered ambient radiation,

all of which originate in the moving component.

A formal treatment on this matter with more details can be found Schramm *et al.* 1993 and Wagner *et al.* 1995, the emission mechanisms and energetics are being addressed in Dreissigacker & Camenzind 1996.

References

Appl, S. and Camenzind, M.: 1993a, *A&A*, **270**, 71
Appl, S. and Camenzind, M.: 1993b, *A&A*, **274**, 699
Babadzhanyants, M. and Belokon, E.T.: 1986, *Astrophysics*, **23**, 639
Bahcall, J.N., et al.: *MNRAS*, in press.
Blumenthal, G.R. and Gould, R.J.: 1970, *Rev. Mod. Phys.*, **42.2**, 237
Dreissigacker, O. and Camenzind, M.: 1996, in prep.
Kardashev, N.S.: 1962, *Sov. Astron.*, **6**, 317.
Schramm, K.J., et al.: 1993, *A&A*, **278**, 391
Steffen, W., et al.: 1995 *A&A*, **302**, 335.
Wagner, S.J. et al.: 1995, *A&A*, **298**, 688-698.
Zensus, J.A., et al.: 1995, *ApJ*, **443**, 35-53.

THE NATURE OF FLAT SPECTRUM SOURCES

D.B. MELROSE

Research Centre for Theoretical Astrophysics
School of Physics, University of Sydney
NSW 2006 Australia

Abstract. Nonthermal radio sources near the Galactic Center with flat or weakly inverted spectra ($S(\omega) \propto \omega^\alpha$ with $\alpha \gtrsim 0$) are attributed to optically thin synchrotron emission from a hard electron energy spectrum, $N(\varepsilon) \propto \varepsilon^{-a}$ with $a = 1 - 2\alpha \lesssim 1$, produced by Fermi acceleration or diffusive shock acceleration at multiple shocks combined with a synchrotron pile up. This basic mechanism is also plausible for flat-spectrum AGN.

1. Introduction

Flat radio spectra, $S(\omega) \propto \omega^\alpha$ with $\alpha \gtrsim 0$, in some classes of AGN have been attributed to self absorption (e.g., Kellermann and Pauliny-Toth 1969). For the self-absorption peak to extend over at least a decade in frequency requires specific geometric properties in the source (e.g., Marscher 1977). This 'cosmic conspiracy' model (Cotton et al. 1980) is not consistent with interferometric data. Moreover, there are nonthermal sources with flat spectra in the Galactic center (GC) region (e.g., Yusef-Zadeh 1989) which are not plausibly self-absorbed. The simplest interpretation of these GC sources is in terms of optically thin synchrotron emission due to a hard electron spectrum, $N(\varepsilon) = K\varepsilon^{-a}$ with $a \lesssim 1$ (e.g., Anantharamaiah et al. 1991). Such an interpretation also needs to be considered for flat-spectrum AGN.

How can such hard electron spectra be produced? Diffusive shock acceleration (DSA) at a single strong shock cannot produce such a hard spectrum even when combined with synchrotron losses. It is suggested here that such hard spectra are produced through multiple DSA (DSA at multiple shocks), or Fermi-like acceleration, combined with synchrotron losses during the acceleration process.

R. Ekers et al. (eds.), Extragalactic Radio Sources, 423–426.

2. Flat spectrum due self absorption

The 'cosmic conspiracy' model for flat spectra requires that $S(\omega)$ depends on the size of the source, described by the radius r in a conical model, only in the combination ωr. The model requires a source with an area $A \propto r^2$, thickness $L \propto r$, a steady outflow implying $K \propto r^{-2}$ and an azimuthal magnetic field $B \propto r^{-1}$. The flat spectrum extends over the frequency range determined by the range of r where the model is valid. The model implies that the linear size of the source is inversely proportional to frequency, which is not supported by interferometric data (Cotton et al. 1980).

3. Flat spectrum sources near the Galactic center

The flat spectrum sources in the GC region include linear filaments $\sim$ 30 pc $\times$ 1 pc oriented nearly perpendicular to the Galactic plane, and threads $\sim$ 30 pc $\times$ 0.5 pc (Yusef-Zadeh 1989). These sources are well removed (by tens of parsec) from the compact source Sgr A*, which has some AGN-like properties. Spectral indices, measured for $\omega/2\pi = 0.3 - 1.5$ GHz are in the range $\alpha \sim 0 - 0.3$. The magnetic field in these sources is relatively strong, $B \sim 10^{-3} - 10^{-2}$ G, implying synchrotron lifetimes $\lesssim 10^4$ yr. These properties favor a model based on in situ acceleration due to shocks or large-scale turbulence (Anantharamaiah et al. 1991).

4. Hardest electron spectra from acceleration alone

Optically thin emission produces a flat spectrum, $\alpha \gtrsim 0$, for $a = 1 - 2\alpha \lesssim 1$. DSA at a single nonrelativistic shock produces a spectrum with $a = (r + 2)/(r - 1) > 2$ for a compression ratio $r < 4$, implying $\alpha < -0.5$. A relativistic shock can produce only a slightly harder ($a > 1.7$) spectrum (Heavens and Drury 1989).

Both Fermi acceleration by MHD turbulence and DSA at many shocks are statistical in character and may be described in terms of diffusion in momentum space. DSA at many shocks implies systematic energy gains at each shock, and adiabatic losses between the shocks (e.g., Achterberg 1990; Schneider 1993). When synchrotron losses are included, the energy spectrum evolves according to

$$\frac{\partial N(\varepsilon)}{\partial t} = \nu_A \frac{\partial}{\partial \varepsilon}\left[\varepsilon^4 \frac{\partial}{\partial \varepsilon}\frac{N(\varepsilon)}{\varepsilon^2}\right] - \frac{\partial}{\partial \varepsilon}\left[\dot{\varepsilon}_{\rm syn}\, N(\varepsilon)\right], \tag{1}$$

where the Fermi acceleration and the synchrotron loss rates are given by

$$\nu_A \sim \bar{\omega}\frac{v_A}{c}\left(\frac{\delta B}{B}\right)^2, \qquad \dot{\varepsilon}_{\rm syn} = -\frac{4}{3}\sigma_T c W_B\left(\frac{\varepsilon}{mc^2}\right)^2, \tag{2}$$

respectively. Here $(\delta B/B)^2 = W_{\mathrm{turb}}/W_B$ is a measure of the level of the turbulence and $\bar{\omega}$ is its mean frequency, or is determined by the frequency with which shocks are encountered, σ_T is the Thomson cross section, v_A is the Alfvén speed and W_B is the magnetic energy density.

In the absence of synchrotron losses, the asymptotic solution for injection at $\varepsilon = \varepsilon_0$ has $N(\varepsilon) \propto \varepsilon^2$ for $\varepsilon < \varepsilon_0$ and $N(\varepsilon) \propto \varepsilon^{-1}$ for $\varepsilon > \varepsilon_0$. Both analytic (Achterberg 1990; Schneider 1993) and numerical (Melrose and Pope 1993) calculations show that the asymptotic spectrum for DSA at multiple shocks is also $a = 1$. Thus either Fermi acceleration or DSA at multiple shocks can account for a nearly flat synchrotron spectrum, $\alpha \lesssim 0$, but not for an inverted spectrum, $\alpha > 0$.

5. Synchrotron blocking

Synchrotron losses combined with acceleration can lead to a pile up at $\varepsilon_{\mathrm{pu}} = |\dot{\varepsilon}_{\mathrm{syn}}|/v_A$, where the average energy gain due to acceleration balances the synchrotron loss. The pile up occurs only if the spectrum is already hard, $a < 2$ (e.g., Melrose 1980, p. 110), and so is not possible for DSA at a single nonrelativistic shock, but is possible for Fermi acceleration or multiple DSA. The pile up hardens the spectrum toward $N(\varepsilon) \propto \varepsilon^2$ below $\varepsilon_{\mathrm{pu}}$. However, in a weak pile up the spectrum should be closer to $a = 1$ than $a = -2$. The resulting synchrotron spectrum has $\alpha = (1-a)/2$ for $a > 2/3$ and $\alpha = 1/3$ for $a < 2/3$ below a peak frequency, ω_{max}, determined by the typical emission frequency of an electron at $\varepsilon_{\mathrm{pu}}$.

6. Idealized Model

Such a hard spectrum forms only when the acceleration is driven hard enough to cause a synchrotron pile up. A simple model for the maximum power input into relativistic particles involves an initial flux of Alfvén waves, e.g., due to a Kelvin-Helmohtz instability, with $\delta B \sim B$. A turbulent cascade transfers this energy to smaller scales where Fermi acceleration becomes effective (Bicknell and Melrose 1982). If synchrotron blocking is important, the power then goes directly into synchrotron emission due to electrons at $\sim \varepsilon_{\mathrm{pu}}$. The following model incorporates these ideas. 1) The power input in turbulence or shocks is $\dot{W}_{\mathrm{turb}} = \eta_1(v_A/L)W_B$ where $\eta_1 = (\delta B/B)^2 = W_{\mathrm{turb}}/W_B$ is of order unity. 2) Synchrotron losses balance this power input, implying $\dot{W}_{\mathrm{turb}} = \frac{4}{3}\sigma_T c\, W_B\, n_{\mathrm{pu}}(\varepsilon_{\mathrm{pu}}/m_e c^2)^2$, where n_{pu} is the number density of particles at $\sim \varepsilon_{\mathrm{pu}}$. 3) The pile up energy is related to the maximum frequency, $\varepsilon_{\mathrm{pu}}/m_e c^2 \approx (\omega_{\mathrm{max}}/\Omega_e)^{1/2}$. 4) There is equipartition of energy, $n_{\mathrm{pu}}\varepsilon_{\mathrm{pu}} = \eta_2 W_B$, with η_2 a constant of order unity.

For the GC filaments relevant input parameters are $\omega_{\mathrm{max}}/2\pi \sim 30\,\mathrm{GHz}$,

$L \sim 1\,\mathrm{pc}$ and $B \sim 10^{-2}\,\mathrm{G}$. The model then requires $v_A/c \sim 10^{-2}$ which is plausible. The emission is optically thin as the model requires. For AGN a pile-up model was applied to infrared/optical emission by Schlickeiser (1984). This requires that $\omega_{\mathrm{max}}/2\pi \sim 10^{13}\,\mathrm{Hz}$ be higher for AGN than for the GC sources. Such an AGN source would then be optically thick at radio frequencies and one needs to modify the model either by including multiple sources or by assuming a smooth r-dependence (cf. Marscher 1977). A smooth model is similar to the 'cosmic conspiracy' model, but differs from it in an important way. Provided that $S(\omega)$ increases with increasing r, optically thin emission dominates the observed emission and there is no implication that the source size should be a strong function of radio frequency.

7. Conclusions

The formation of hard electron spectra due to Fermi acceleration or multiple DSA coupled with synchrotron losses is plausible for flat or weakly inverted synchrotron spectra in the GC region and in AGN. The mechanism needs to be explored in more detail.

References

Achterberg, A. (1990) Particle acceleration by an ensemble of shocks, *A&A* **231**, 251–258

Anantharamaiah, K.R., Pedlar, A., Ekers, R.D., and Goss, W.M. (1991) Radio studies of the Galactic centre – II. The arc, threads and related features at 90 cm (330 MHz), MNRAS, **249**, 263–281

Bicknell, G.V., and Melrose, D.B. (1982) In situ acceleration in extragalactic radio jets, *ApJ* **262**, 511–528

Cotton, W.D. *et al.* (1980) The very flat radio spectrum of 0735+178: A cosmic conspiracy?, *ApJ*, **238**, L123–L128

Heavens, A.F., and Drury, L.O'C. (1989) Relativistic shocks and particle acceleration, *MNRAS*, **235**, 997–1009

Kellermann, K.I., and Pauliny-Toth, I.I.K. (1977) The spectra of opaque radio sources, *ApJ* **155**, L71–L78

Marscher, A.P. (1977) Structure of radio sources with remarkably flat spectra: PKS 0735+178, *AJ*, **82**, 781–784

Melrose, D.B. (1980) *Plasma Astrophysics Volume II Astrophysical Applications*, Gordon & Breach (New York)

Melrose, D.B., and Pope, M.H. (1993) Diffusive shock acceleration by multiple shocks, *Proc. Astron. Soc. Australia*, **10**, 222–224

Schlickeiser, R. (1984) An explanation of abrupt cutoffs in the optical-infrared spectra of non-thermal sources. A new pile-up mechanism for relativistic electron spectra, *A&A*, **136**, 227–236

Schneider, P. (1993) Diffusive particle acceleration by an ensemble of shock waves, *A&A* **278**, 315–327

Yusef-Zadeh, F. (1989) Filamentary structures near the Galactic center, in *The center of the Galaxy IAU Symp. 136*, ed. Morris, M., D. Reidel (Dordrecht), pp. 243–263

ENERGY TRANSPORT IN RADIO GALAXIES AND QUASARS: A WORKSHOP THEORY REVIEW

PHILIP E. HARDEE

University of Alabama
Tuscaloosa, Alabama, 35487, USA

Abstract. A workshop on the dynamics, emission and morphology associated with the highly collimated outflows seen in radio galaxies and quasars was held at The University of Alabama in Tuscaloosa in September of 1995. Complete workshop proceedings will be published in the ASP Conference Series. This workshop review covers the theoretical developments and problems discussed at the workshop.

1. Introduction

It has been over ten years since the Green Bank workshop "Physics of Energy Transport in Extragalactic Radio Sources" was held to address the questions raised by detailed imaging of jets in active galaxies and quasars. In that workshop the focusing topics were (a) Jet Correlations and Observational Constraints, (b) Jet Sidedness, Velocity and Unification, (c) Confinement and Stability, and (d) Particle Acceleration, Entrainment, and Turbulence. Considerable work has occurred in the intervening decade. Much improved data at all wavelengths from radio to γ-rays are now available for these objects, and new families of unified models have been proposed to link classes of objects whose relationships were less clear a decade ago. The VLBA is coming on line and the latest supercomputers make it possible to perform three dimensional numerical experiments. Thus it seemed an auspicious time to review the progress made in the past decade, to evaluate the research being conducted in this area at the present time, and to address the issue of how best to utilize existing observational and computational resources to answer the questions that still remain concerning the highly collimated outflows associated with Radio Galaxies and Quasars.

R. Ekers et al. (eds.), Extragalactic Radio Sources, 427–432.

2. Sub-Parsec and Parsec Scales: Acceleration, Collimation

Energy transport begins with inward transport of fuel from the ISM, conventionally via an accretion disk around a black hole. Andrew Wilson argued that there is some evidence for powerful radio sources being triggered by merger events. The gas must be forced into the vicinity of the nuclear black hole in some special way conducive to the formation of jets, or in a different picture the merger events result in sufficient black hole rotational energy to power the jets. The reason or reasons for the difference in the active nature of spiral and elliptical galaxies, and of radio loud and radio quiet quasars remain an open question. On the other hand, theories of jet formation seem to be converging towards magnetic acceleration and collimation schemes, although Peter Scheuer cautioned that such convergence may be premature.

Magnetic acceleration and collimation schemes where the magnetic field is anchored in a Keplerian accretion disk seem particularly promising at the present time. In these schemes disk rotation twists an initial field which then becomes capable of accelerating a disk corona via magnetic and pressure gradient forces. Collimation is mainly the result of the pinching force associated with the toroidal magnetic field. While numerical experiments give promising results, the experiments are not yet fully self-consistent; e.g., in the work presented by Richard Lovelace the accretion disk is treated as a boundary condition on the electromagnetic fields without consideration of the back reaction on the disk. In addition to back reaction on the disk material, disk instabilities may play an important role in jet production schemes. Dimitris Christodoulou demonstrated the use of an energy variational principle to approach accretion disk instabilities in a way that allows the addition of a toroidal field to the poloidal field configuration assumed in previous stability work (e.g., Balbus & Hawley 1991). This new work suggests an explosive release of an amplified toroidal field which might account for variation in jet speed. Future work will require incorporation of more realistic boundary conditions in the numerical simulations and will also require the investigation of magnetic fields, supported by external currents, that appear to thread a black hole's event horizon. Such schemes extract the rotational energy of a spinning black hole (e.g., Blandford 1992).

Magnetic collimation and acceleration implies strong toroidal magnetic fields and significant jet axial current flows. Thus, current driven instability as opposed to dynamically driven Kelvin-Helmholtz instability (e.g., Birkinshaw 1991) of the jet may be important. Steffan Appl presented results showing that current driven modes have growth rates significantly less than the Kelvin-Helmholtz modes for cases of astrophysical interest. If strong magnetic fields exist in the medium external to the jet then magnetic bending can lead to distorted flow structures. Several numerical simulations presented by Shinji Koide demonstrate that a jet trans-Alfvènic relative to a strong external magnetic field can be redirected. Thus, magnetic bending provides a viable option to jet cloud collisions as a jet deflection mechanism where magnetic fields in the external environment are large.

3. Relativistic Jets: Simulation, Modeling, Emission

A leaning towards relativistic jet production was evident at the workshop and relativistic numerical simulation results from three different groups (P. Hughes & C. Duncan; S. Komissarov & S. Falle; J. Marti & J. Gomez) were presented. At the present time all work remains axisymmetric and an obvious next step is to proceed towards fully three dimensional simulations. Probably the most significant difference between relativistic and non-relativistic jet dynamics lies in the regime where the jet's sound speed is high relative to the flow speed. While non-relativistic transonic jets are very unstable the relativistic jet can be stabilized by a high Lorentz factor. Other differences are more subtle and occur because of the different relationship between energy and momentum fluxes in non-relativistic and relativistic flows. It will be important to quantify the differences between non-relativistic and relativistic flows. One of the important aspects of relativistic jet simulations will be to provide a firm dynamical basis for the calculation of radiation beamed and boosted by the relativistic flow, and from shocks residing within the flow. Both single and multiple shock structures and emission were presented at the workshop and Sergeui Komissarov showed results where the shock structures were moving relativistically. Not surprisingly the appearance of the jets containing such structures proved very dependent on the viewing angle.

The present observational and theoretical situation reviewed by Alan Marscher suggests that theoretical efforts need to focus on the synchrotron processes associated with accelerating, expanding relativistic jets which may contain relativistically moving shocks with Lorentz factor different from that of the relativistically moving outflow. In this regard Markos Georganopoulos presented a detailed calculation of the synchrotron radiation coming from a relativistically moving feature for which amplitudes and variability decrease with frequency and with viewing angle. Such a result is suggestive of the variability difference between X-ray (lesser variability) and radio (greater variability) selected BL Lacs. Future progress in this area depends upon model building and observations involving multi-frequency source monitoring at high frequencies capable of probing the jet, interior to the presently observed VLBI radio core. Model building when combined with such observations ultimately may be capable of determining inner jet dynamics, e.g., the magnetic configuration and the acceleration region, but it will be a challenge to relate the observed appearance and spectrum to the dynamical flow structures and emission processes.

4. Kiloparsec Scales: Entrainment, Structure, FRI/FRII, Ageing

All extragalactic jets may be accelerated up to relativistic speeds, and observed galactic jets provide weak evidence that jets produced by black holes are relativistic (Mirabel & Rodriguez 1995). Evidence for relativistic speeds on the kiloparsec scale is provided by observations of the proper motions of features in the M87 jet reviewed by John Biretta at the workshop. M87 jet features typically move at speeds of 0.5c. That these typical speeds imply relativistic flow is suggested by modeling. For example, a model for knot A presented by Geoffrey Bicknell requires a Lorentz factor of 3 - 5 to move a Kelvin-Helmholtz generated oblique shock at 0.5c along a jet with proper jet density less than the external density by a factor of 10 - 100. Note however that scissors like effects arising from differentially moving multiple mode structures and/or projection of differentially moving near and far side jet structures can produce the observed (Biretta 1993) nearly stationary features and features with speeds up to about 3c. Overly simplistic interpretation of moving features must be avoided in the future.

While subluminal motion does not require sub-relativistic jet speeds, there are compelling reasons to believe that jets must slow down on kiloparsec scales. Slowdown occurs as a result of mixing and entrainment at the jet-external medium interface. As presented by David DeYoung a boundary layer develops in which large scale structures gulp material which is then digested at smaller scales in a non-linear and fully turbulent process. A quasi-steady state exists at the larger scales. Three dimensional MHD simulations presented by Philip Hardee reveal that light jets entrain and mix when the large scale Kelvin-Helmholtz modes develop significant amplitude, and the entrainment of more dense external material leads to relatively rapid slow down and disruption of ordered jet structure. Such interaction should lead to a sheath and spine jet structure such as that espoused by Robert Laing (1993) to explain many of the observed magnetic field and surface brightness structures along kiloparsec scale jets.

Three dimensional MHD simulations performed by David Clarke have revealed that the formation of emission filaments observed in radio lobes is caused by shearing within a turbulent lobe. Michael Norman presented a numerical simulation which showed that jets can remain relatively stable until significant lobe turbulence is encountered if no regular driving frequency such as that provided by jet precession is present. Overall the three dimensional numerical simulations imply that powerful jets protect themselves from serious entrainment effects by inflating a tenuous lobe or cocoon. These jets can remain well collimated until large scale distortions grow to large amplitudes and present results suggest that such jets could remain relativistic to large scales. Less powerful jets not protected by a tenuous lobe or cocoon should slow as they develop an extensive entraining boundary layer and lead to FRI radio sources (e.g., Bicknell 1986). Further evidence for a decelerated relativistic jet FRI radio source model was presented by Geoffrey Bicknell. Relatively simple assumptions relating a galaxy's core radius and pressure to the development of transonic entraining jet flow and to the radio luminosity resulted in an excellent fit to the slope and position of the dividing line in the radio-optical luminosity plane between FRI and FRII type radio sources (e.g., Owen & Ledlow 1994).

The numerical simulations reveal shock structures and magnetic filamentation both within jets and in the lobes. Jean Eilek pointed out that these two features provide in situ acceleration sites and relatively loss free regions for the radiating electron population. These features can serve to explain the discrepancy between dynamically estimated source ages and conventionally determined spectral ageing source ages. However, Larry Rudnick provided compelling reasons to remain sceptical of conventional spectral ageing arguments. In any event, the effects of multiple weak shock particle acceleration sites and particle diffusion through magnetic filaments will need to be included in future modeling efforts.

5. Jets on Cluster Scales

The tenuous plasma in radio jets should prove susceptible to gradients in the external environment and perhaps can be used as probes of the external environment. However, as noted by Mark Birkinshaw, jets create their own environment. Still, the less powerful jets which do not create protective lobes should be more influenced by the environment. In particular, the distorted tailed radio sources must be shaped by interaction with the intracluster medium (ICM). Jagbir Hooda reported results from numerical simulation of the behavior of a three dimensional jet crossing an oblique ISM/ICM interface. Such an oblique interface appears incapable of distorting jets as appears to be necessary to form the Wide Angle Tailed (WAT) structures associated with dominant (D or cD) galaxies that generally lie at the optical or X-ray centroids of clusters. However, Jack Burns noted that numerical simulations of cluster merger events (Röettiger, Burns & Loken 1993) lead to transonic motions of the ICM relative to the optical centroid of the cluster, and that other simulations of jet propagation in crossflows (Loken et al. 1995) show that jets can be bent into the appropriate shapes.

6. Final Remarks

Traditional theory combined with numerical simulations has shown real progress in the past decade. A synergy between theory, numerical simulation, and observation barely envisioned a decade ago has developed. This is particularly apparent in the combined theoretical and numerical studies of jet dynamics, inspired by the radio observations. In part, future development waits on computational and observational hardware. Still, the detailed multi-frequency observations that are now becoming available deserve detailed modeling and interpretive efforts. The major challenge that remains to any modeling and interpretive effort is to tie emission processes and microphysics to the magnetohydrodynamics.

7. Acknowledgements

P. Hardee thanks SOC members G. Bicknell, A. Bridle, J. Burns, R. Laing, A. Marscher, M. Norman, P. Scheuer, & A. Zensus, and also A. Rosen, J. Wardle, & P. Wiita. The NSF provided workshop support through EPSCoR grant OSR-9108761 and research support through grant AST-9318397.

References

Balbus, S.A. & Hawley, J.F. (1991) A Powerful Local Shear Instability in Weakly Magnetized Disks I. Linear Analysis, *Astrophysical Journal*, **376**, pp.214–222.

Bicknell, G.V. (1986) A Model for the Surface Brightness of a Turbulent Low Mach Number Jet. III. Adiabatic Jets of Arbitrary Density Ratio: Application to NGC 315, *Astrophysical Journal*, **305**, pp.109–130.

Biretta, J.A. (1993) The M87 Jet, *Astrophysical Jets*, eds. D. Burgarella, M. Livio, & C.P. O'Dea, Cambridge University Press, Cambridge, pp.263–304.

Birkinshaw, M. (1991) The Stability of Jets, *Beams and Jets in Astrophysics*, ed. P.A. Hughes, Cambridge University Press, Cambridge, pp.278–341.

Blandford, R. (1993) Acceleration and Collimation Mechanisms in Jets, *Astrophysical Jets*, eds. D. Burgarella, M. Livio, & C.P. O'Dea, Cambridge University Press, Cambridge, pp.15–29.

Laing, R.A. (1993) Radio Observations of Jets: Large Scales, *Astrophysical Jets*, eds. D. Burgarella, M. Livio, & C.P. O'Dea, Cambridge University Press, Cambridge, pp.95–119.

Loken, C., Röettiger, K., Burns, J., & Norman, M. (1995) Radio Jet Propagation & Wide-Angle Tailed Radio Sources in Merging Galaxy Cluster Environments, *Astrophysical Journal*, **445**, pp.80–97.

Mirabel, I.F. & Rodriguez, L.F. (1995) Superluminal Motions in our Galaxy, 17th Texas Symposium, *Annals of the New York Academy of Sciences*, in press.

Owen, F.N. & Ledlow, M.J. (1994), The FR I/II Break and the Bivariate Luminosity Function in Abell Clusters of Galaxies, *The First Stromlo Symposium: The Physics of Active Galaxies*, eds. G.V. Bicknell, M.A. Dopita, & J. Quin, ASP Conference Series

Röettiger, K., Burns, J., & Loken, C. (1993) When Clusters Collide - A Numerical Hydro/N-body Simulation of Merging Galaxy Clusters, *Astrophysical Journal Letters*, **407**, pp.53–56.

THE PRODUCTION OF JETS FROM MAGNETIZED ACCRETION DISKS: SIMULATION OF THE BLANDFORD-PAYNE MECHANISM

D.L. MEIER[1], D.G. PAYNE[2], K.R. LIND[3]

[1] *Jet Propulsion Laboratory, Caltech, Pasadena, CA, USA*
[2] *Intel Scalable Systems Division, Pasadena, CA, USA*
[3] *Cray Research Inc., Livermore, CA, USA*

Abstract. We have used massively parallel supercomputer simulations to perform an extensive study of a plausible mechanism for producing the jets seen in extragalactic radio sources – acceleration and collimation by coronal magnetic fields in an accretion disk orbiting a central black hole. We find that such disks can propel jets for a wide range of coronal conditions. The terminal jet velocity is a strong function of the magnetic field equatorial component. Acceleration and collimation are produced by a tight azimuthal field coil generated in the corona, rather than by a stiff poloidal field extending to large distances. The jets are pressure-confined when the external medium pressure is high, but magnetically-confined when it is low.

Blandford and Payne (1982, *M.N.R.A.S.*, **199**, 883) suggested that strong magnetic fields in a differentially-rotating accretion disk might be responsible for generating the jets seen in extragalactic radio sources. In their steady-state, self-similar model magnetic field lines are anchored in a dense, infinitely-thin disk, with the poloidal component of the field making an angle $\theta > 30°$ with the rotation (z) axis. Above the disk is a warm corona, expanding sub-magnetosonically along the field lines. Two jet mechanisms are possible: (1) for sufficient field strength, centrifugal action flings coronal material outward and upward along field lines and differential rotation winds up B_ϕ to virial values, causing magnetic tension ($B_\phi^2/4\pi r$) to squeeze and collimate the outflow; (2) B_ϕ winds up to virial values on a few dynamical times, producing a magnetic pressure explosion ($d(B_\phi^2/8\pi)/dr$) and pushing material upward (Shibata and Uchida 1985, *P.A.S.J.*, **37**, 31).

We have studied magnetized disks under more realistic (non-steady-state, non-self-similar) conditions using our non-relativistic, 2-D axisym-

R. Ekers et al. (eds.), Extragalactic Radio Sources, 433–434.

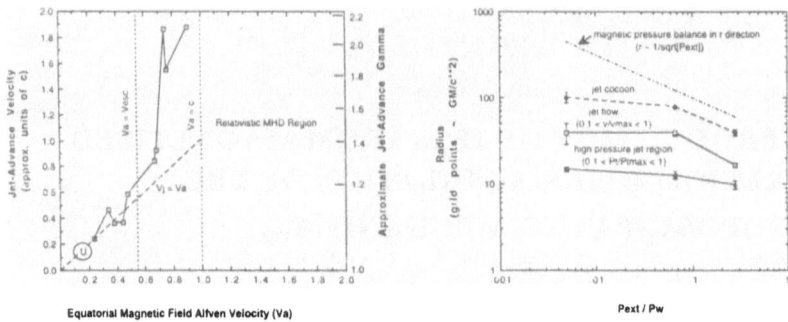

Figure 1. (a) Jet advance speed *vs.* equatorial Alfvén velocity $(V_a = B_{eq}/(4\pi\rho_w)^{1/2})$ for 10 of our models plus that from Ustyugova *et al.* (1995, *Ap. J.*, **439**, L39) (the circle-U point); (b) jet/cocoon radii *vs.* external ambient medium pressure.

metric, magnetohydrodynamic simulation code (Lind, Payne, Meier, and Blandford 1989, *Ap.J.*, **344**, 89). The disk, field, and coronal wind are imposed as initial/boundary conditions along the disk equator, and the flow evolution above the disk is computed. We performed a limited parameter study by varying the disk field $(0.5 \leq V_a \leq 2)$, poloidal field pitch θ, and external pressure $(0.005 \leq P_{ext}/P_w \leq 0.28)$. For all simulations the fixed parameters were: $V_\phi = V_K$ (Keplerian velocity), $c_s = 0.76V_K$, $V_w = 0.07V_K$, and $\rho_{ext} = 0.05\rho_w$. Our 15 simulations had moderate resolution (150 radial $\times$ 300 axial cells), were evolved for ~ 3 inner disk rotation times, and used ~ 12 hours on the Caltech/JPL Intel Touchstone Delta massively parallel supercomputer (the equivalent of ~ 2 weeks of Cray Y-MP time).

We find collimated outflow in *all* our models and that the magnetic explosion mechanism dominates in this region of parameter space. (See Meier, Payne, and Lind 1996, in preparation) For models with the same P_{ext}, but a variety of magnetic field conditions, the jet advance speed appears to be a function of a single parameter – the equatorial magnetic field $(B_{eq} = (B_r^2 + B_\phi^2)^{1/2};$ see figure 1a). For low values of B_{eq} the jet speed remains close to the Alfvén speed, but when V_a exceeds the escape velocity significant jet outflow velocities can occur. While strictly non-relativistic, the model flow velocity can be identified as the spatial component of the 4-velocity (ΓV), if jet thermal and magnetic inertia is small. Our jets reach $\Gamma \sim 2$, but Γ could be much higher for relativistic MHD calculations.

Figure 1b shows the extent of various radial zones in the jet – high pressure core, jet flow, and cocoon – as a function of external pressure, and compares with the trend expected for pure pressure confinement. For external pressures comparable to the coronal pressure the cocoon and jet flow appear to be pressure-confined. But for low external pressures, magnetic stresses provide the confinement. Note that there is always a small magnetically-confined core along the jet axis for all values of P_{ext} studied.

A NUMERICAL STUDY OF RELATIVISTIC JETS

J.A. FONT, J.M. MARTI AND J.M. IBÁÑEZ
Departamento de Astronomía y Astrofísica
Universidad de Valencia, 46100 Burjassot (Valencia), Spain

AND

E. MÜLLER
Max-Planck-Institut für Astrophysik
Karl-Schwarzschild-Str.1, 85740 Garching, Germany

Numerical simulations of supersonic jets are able to explain the structures observed in many VLA images of radio sources. The improvements achieved in classical simulations (see Hardee, these proceedings) are in contrast with the almost complete lack of relativistic simulations the reason being that numerical difficulties arise from the highly relativistic flows typical of extragalactic jets. For our study, we have developed a two-dimensional code which is based on (i) an explicit conservative differencing of the special relativistic hydrodynamics (SRH) equations and (ii) the use of an approximate Riemann solver (see Martí *et al.* 1995a,b and references therein).

The SRH equations have been solved in normalized form by setting the light speed c, the beam radius R_b, and the rest-mass density of the external medium ρ_{em} equal to unity. The flow is then completely characterized by five dimensionless parameters: the beam to external medium rest-mass density ratio, $\eta \equiv \frac{\rho_b}{\rho_{em}}$, the pressure ratio, $K \equiv \frac{p_b}{p_{em}}$, the beam velocity, v_b, the beam Mach number, $M_b \equiv \frac{v_b}{c_s}$ and the adiabatic exponent, γ. In our study these parameters take the following values: $\eta = 0.01, 0.1, 1$, $K = 1$, $v_b = 0.9, 0.99, 0.999$, $M_b = M^{min}, 6.0$ and $\gamma = 4/3, 5/3$. Models with M_b near the minimum value, $M^{min} \equiv v_b/\sqrt{\gamma - 1}$, have large internal energies compared with their rest-mass energy. They are called *hot* models, while highly supersonic jets are referred to as *cold* models.

In all our simulations the jets show the gross morphological features already found in non-relativistic calculations (see, e.g., Norman *et al.* 1982). We find important differences between hot and cold relativistic jets. In hot models the Mach shock at the jet head is permanently present during the

R. Ekers et al. (eds.), Extragalactic Radio Sources, 435–436.

whole simulation. They show naked beams surrounded by lobes instead of cocoons. In hot jets the backflow appearing at the working surface is minimized or, as v_b increases, even non-existing. In addition, they almost completely lack internal structure, because of the pressure equilibrium between the beam and its surroundings. Therefore, the beam/cocoon interface of hot jets is very stable against the growth of pinch instabilities that would evolve into internal shocks. In cold models the Mach configuration is temporarily substituted by a cross-shock during the evolution. These shocks are not as efficient as normal shocks in decelerating the beam flow allowing it to push the contact discontinuity and, eventually, the bow shock. This effect is more important for beams with a lower value of the adiabatic exponent. Backflow is more important in cold models, where stable cocoons can be found in the early stages of the evolution. These cocoons eventually evolve into vortices producing turbulent structures. In cold jets with $\gamma = 5/3$ the cocoon is mainly formed by large vortices while in models with $\gamma = 4/3$ the strong beam collimation causes a large acceleration of the jet. Thus, beam gas is less efficiently redirected into the cocoon, and thinner cocoons with smaller vortices form. Cold models present also a complex structure of internal shocks generated by pressure mismatches between the beam and the overpressured cocoon and by perturbations of the beam boundary by vortices and bulk motions within the cocoon.

For sufficiently large beam flow speeds the estimated propagation velocity of the jet head, obtained by equating the ram pressure of the beam and the external medium in the proper frame of the working surface (see Martí et al. 1994), approaches that of the beam itself. In classical dynamics this only occurs for so-called ballistic jets (i.e. $\eta \gg 1$). Similarly as for classical jets, we define the propagation efficiency δ as the ratio between the mean jet velocity and its corresponding estimate. Our results show (Martí et al. 1995b) that for a wide range of estimated jet propagation speeds ($0.17c - 0.94c$) the efficiencies span the interval $0.76 - 1.24$, i.e. they are significantly larger than their corresponding Newtonian counterparts. We find that hot models have δ very close to one. Highly supersonic (*i.e.* cold) models with $\gamma = 4/3$ have δ greater than one due to the acceleration phase while in models with $\gamma = 5/3$ the efficiency increases with v_b and η, and tends to one for sufficiently dense, highly relativistic models.

References

Hardee, P (1995) IAU Symposium 175: *Extragalactic Radio Sources*
Martí, J. Mª, Müller, E., Ibáñez, J. Mª (1994) *A&A*, **281**, L9
Martí, J. Mª, Müller, Font, J.A., Ibáñez, J. Mª (1995) *ApJ*, **448**, L105
Martí, J. Mª, Müller, Font, J.A., Ibáñez, J. Mª, Marquina, A. (1995) *ApJ* (submitted)
Norman, M. L., Smarr, L., Winkler, K.-H. A., Smith, M. D. (1982) *A&A*, **113**, 285

VIEWING ANGLE AND THE APPEARANCE OF SUPERLUMINAL JETS

H. D. ALLER, M. F. ALLER, P. A. HUGHES, A. MIODUSZEWSKI
University of Michigan – Ann Arbor, MI, 48109-1090

1. Evidence for Changes in Angle of Jet Orientation

The time history of BL Lacertae has shown clear evidence of changes in jet orientation both in the plane of the sky and in the angle to the line of sight (see Figure 1). Models based on transverse shocks in a relativistic flow quantitatively fit the polarization and flux density data well and permit one to determine parameters of the flow such as the bulk Lorentz factor and the angle of the flow to the line of sight (Hughes, Aller and Aller 1989). The orientation of the jet flow to the line of sight changed by approximately 6° between the early 1980 bursts and one in 1991. There have been comparable changes in the orientation of the jet on the plane of the sky. Such changes in jet orientation may be due to a helical flow pattern arising from precession or instability.

2. Simulation Results

The appearance of a jet with relativistic flow speed can be strongly influenced by the viewing angle. We have carried out jet simulations using a hydrodynamical code which admits relativistic velocities. The observed intensity is computed by solving the equation of transfer along the lines of sight using an algorithm which accounts for both the opacity and relativistic effects in the emitting region. Figure 2 shows the case of a collimated flow for which the inflow velocity has been modulated between a Lorentz factor of 1 and 10. When viewed at a large angle, the full lateral extent of the jet is evident, but when observed at an angle of 10° only the inner core of the flow is apparent in the intensity maps. The major cause of this difference is Doppler boosting, and this example underscores the fact that only a small fraction of the true jet flow may be apparent in VLBI maps of

R. Ekers et al. (eds.), Extragalactic Radio Sources, 437–438.

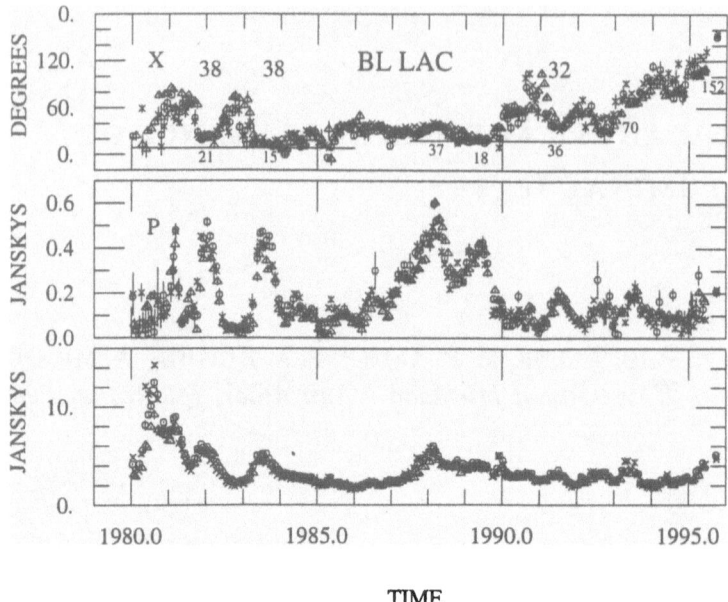

Figure 1. Monthly averages of total flux density and polarization of BL Lac. The derived angles to the line of sight for three bursts are given by the larger numbers. The smaller numbers give the average PA during each burst. The horizontal lines are the observed direction of the jet seen by VLBI (Mutel *et al.* 1990, 1994).

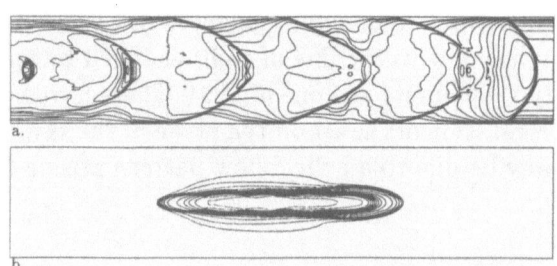

Figure 2. Intensity distribution derived for a relativistic velocity-modulated jet, observed at angles of a. 90°, and b. 10° to the line of sight.

these objects. We give special thanks to Comer Duncan, and we thank the NSF for partial support (AST-9421979).

References

Hughes, P. A., Aller, H. D. and Aller, M. F. 1989, *ApJ*, **341**, 68.

Mutel, R. L., Phillips, R. B., Bumei Su, and Bucciferro, R. R. 1990, *ApJ*, **352**, 81.

Mutel, R. L., Denn, G. R. and Dryer, M. J. 1994, *Proc. of the NRAO Workshop on Compact Extragalactic Radio Sources*, (NRAO: Socorro), ed. Zensus and Kellerman, p. 191.

SIMULATIONS OF SUPERLUMINAL RADIO SOURCES

S.S. KOMISSAROV[1,2], S.A.E.G. FALLE[1]
[1] *Department of Applied Mathematics, The University of Leeds, Leeds LS2 9JT*
[2] *Astrospace Centre, Lebedev Physical Institute, 53, Leninsky Prospect, Moscow B-333, 177924 Russia*

The superluminal knots of VLBI-sources are widely believed to be shocks travelling along relativistic jets. So far only very simple quasi-one dimensional models have been used to study this phenomenon. Recent progress in numerical methods for relativistic fluid dynamics (Falle & Komissarov 1995) make it possible to carry out much more realistic simulations of such flows.

Some superluminal sources (e.g. 4C39.25, Shaffer et al. 1987) have stationary knots as well as moving ones. Daly & Marcher (1988) proposed that these stationary knots could be identified with steady shocks, which would arise quite naturally if the jet is reconfined. Where the relative proximity and brightness of the compact radio sources allow us to observe their parsec scale jets in detail, they often reveal peculiar limb-brightening (e.g. in M87, Reid et al. 1989 and 3C345, Unwin 1993). This may be due to a turbulent boundary layer (Komissarov 1990) but it could also be caused by a reconfinement shock close to the jet boundary.

In this paper we use numerical simulations to study the combined effect of the quasi-stationary reconfinement shock and moving shocks due to variations in the central source on the appearance of a relativistic jet.

We first set up a steady flow in which the jet propagates through an external medium whose pressure decreases like $z^{-1.9}$ where z is the distance from the source. We then allow the jet velocity at the source to vary in such a way that shocks are produced in the gas as soon as it emerges from the source.

The radio emission is determined by assuming that both the magnetic and relativistic pressures are proportional to the thermal pressure. The relativistic version of the radiative transfer equation (Mihalas 1986) is then used to obtain the intensity as seen by a distant observer.

R. Ekers et al. (eds.), Extragalactic Radio Sources, 439–440.

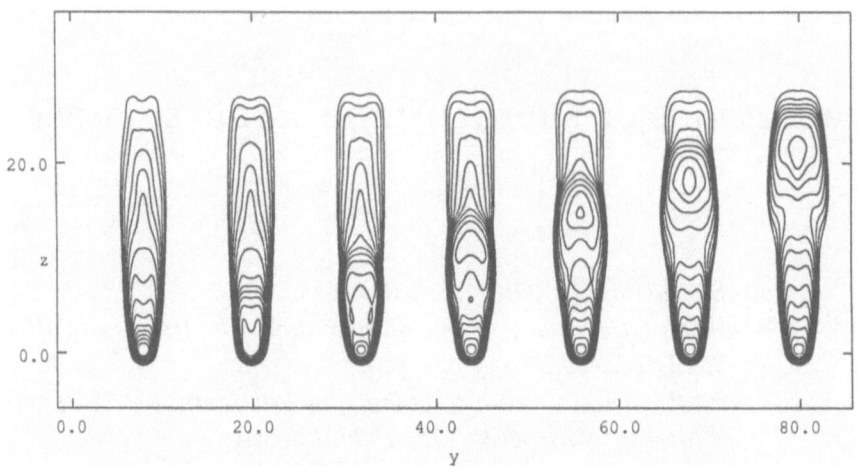

Figure 1. The figures show logarithmic contours of brightness distribution with a dynamic range of 1000:1 at unit time intervals. Note that the speed of light is unity. The images are smoothed with a gaussian beam of size 0.5.

Figure 1 shows the results for an inclination angle of 20°. The most striking property of the images is that they show only one moving knot, whereas there are actually eight travelling shocks in the flow. This is a result of relativistic retardation just like the superluminal motion itself. The apparent speed of the knot is $V_{app} \simeq 4c$, which agrees very well with the analytic expression (Begelman et al. 1984).

Another interesting feature of these maps is the absence of a stationary knot. However, the reconfinement shock is the cause of the limb-brightening in the first and the last thirds of the jet image and the center brightening in the middle. One can also clearly see that the moving knot is limb-brightened when going through the limb-brightened region of the underlying flow. It is interesting that knot N2 in the M87 jet does indeed reveal this kind of structure (Reid et al.,1987).

References

Begelman M.C., Blandford R.D., Rees M.J., 1984, *Rev.Mod.Phys.*, **56**, 255.
Daly R.A., Marscher A.P., 1988, *ApJ.*, **334**, 539.
Falle S.A.E.G., Komissarov S.S., 1995, *Mon.Not.R.astr.Soc.*, in press.
Komissarov S., 1990, Soviet Astron. *Lett.*, **16(4)**, 284.
Mihalas D., 1986, in Winkler K.-H. A. & Norman M. L., eds, Astrophysical Radiational Hydrodynamics, Reidel, Dordrecht, p.61.
Reid M.J., Biretta J.A., Junior W., Maxlow T., Spencer R.E., 1989, *Ap.J.*, **336**, 112.
Shaffer et al., 1987, *Ap.J.*, **314**, L1.
Unwin R.J., 1993, in Davis R.J. & Booth R.S., eds, Sub-arcsecond Radio Astronomy, Cambridge University Press, Cambridge, p.189.

PHYSICS OF THE M87 JET

MITCHELL C. BEGELMAN

JILA, University of Colorado, Boulder, CO 80309-0440, USA

AND

GEOFFREY V. BICKNELL

Astrophysical Theory Centre, Australian National University, Canberra, ACT 0200, Australia

Motivated by the recent determination of proper motions of knots in the M87 jet by Biretta, Zhou, & Owen (1995), we have investigated the relationship between the dynamics of the jet and the kpc-scale radio and optical structure. We interpret the knots as internal shocks in the jet, and find that relativistic effects play an important role in their appearance. In particular, knot A, which appears to be almost transverse to the flow and seems to be viewed nearly edge-on, is in fact a highly oblique shock. The direction of its normal with respect to the flow lies within about 10° of the Mach angle, i.e., the angle of obliquity beyond which no shock is possible. In a frame comoving with the shock front, the incident streamlines make an angle of about 40° − 60° with respect to the shock normal. Knot A has a proper motion corresponding to only 0.51c, and the relativistic effects are modest. Its edge-on appearance is largely fortuitous. We have found, however, that relativistic effects would create a strong bias for shocks with apparent *superluminal* speeds to appear nearly edge-on.

A modest pressure jump of roughly a factor of 5 at knot A is adequate to explain the enhanced synchrotron emissivity, by amplifying the magnetic field and boosting the energies of the upstream population of relativistic electrons. Synchrotron emissivity will be strongly enhanced even if diffusive shock acceleration is inefficient, since the relativistic electrons still undergo shock–drift acceleration (Begelman & Kirk 1990). The required level of compression is consistent with the observed proper motion and small jet deflection, if the jet Lorentz factor is of the order of 3 − 5. Other knots in the M87 jet, as well as the faint striations observed all along the jet (which

441

R. Ekers et al. (eds.), Extragalactic Radio Sources, 441–442.
© *1996 IAU.*

give the impression of helical filaments wound around the jet), could also be oblique shocks.

The observed spectral break frequencies of knots A and B (Biretta, Stern, & Harris 1991) are consistent with particle acceleration at the shock front followed by synchrotron cooling in the post-shock magnetic field, if we assume both that the Lorentz factor exceeds 2 in the post-shock flow and also that the magnetic field is comparable to the equipartition value corrected for relativistic beaming.

We suggest that the nonlinear development of the Kelvin-Helmholtz instability is responsible for triggering the oblique shocks. From the knot proper motions and jet morphology (Owen, Hardee, & Cornwell 1989), we estimate the real and imaginary wavenumbers and group velocity of the most unstable mode. Working backward through the dispersion relation for the instability, we can constrain both the ratio of "cold" to relativistic material inside the jet and the ratio of jet to ambient density. According to this analysis, the M87 jet is composed of a rather relativistic fluid, with an internal sound speed greater than $c/3$ (as compared with an ultrarelativistic fluid, which has sound speed $c/\sqrt{3}$). It propagates through a medium whose density is no more than 10–100 times that of the jet. This implies that the kpc-scale radio lobes are much less dense than the interstellar medium (ISM) in the central regions of the M87 cooling flow. The radio jets in M87 are therefore driving high pressure, low-density bubbles into the surrounding ISM. The radii of the lobes are consistent with this assertion and the inferred energy flux in the jet if the age of the source is $\sim 10^6$ yr, consistent with that estimated by Turland (1975). The resultant expansion of the lobes can comfortably power the excitation of the surrounding optical filaments by driving radiative shocks into preexisting condensations.

Readers desiring more detailed information about this work should consult our recent manuscript (Bicknell & Begelman 1995).

References

Begelman, M. C., & Kirk, J. G. 1990, *ApJ*, **353**, 66
Bicknell, G. V., & Begelman, M. C. 1995, *ApJ*, submitted
Biretta, J. A., Stern, C. P., & Harris, D. E. 1991, *AJ*, **101**, 1632
Biretta, J. A., Zhou, F., & Owen, F. N. 1995, *ApJ*, **447**, 582
Owen, F. N., Hardee, P. E., & Cornwell, T. J. 1989, *ApJ*, **340**, 698
Turland, B. D. 1975, *MNRAS*, 170, 281

RELATIVISTIC ELECTRON-POSITRON CLOUDS IN VLBI JETS

V. DESPRINGRE
Observatoire Midi-Pyrénées
14 Avenue Edouard Belin, 31400 Toulouse, France

AND

D. FRAIX-BURNET
Laboratoire d'Astrophysique de Grenoble
BP53X, 38041 Grenoble Cédex, France

1. Introduction

Extragalactic jets have always had two characteristics : the presence of knots and the requirement for particle acceleration. Shock fronts provide an explanation for both. However, the knotty appearance is less obvious at kp-scale on very high resolution observations from the VLA and the HST. The evidence for shock fronts is therefore weakened. At the pc-scale (or VLBI scale), a lot of these blobs are moving superluminally and they have been interpreted and modelled as shock fronts in a relativistic jet.

The first idea of the present work is to build an alternative model without shock fronts and to propose a different picture for VLBI jets. An exciting outcome of our model is that we will be able to extrapolate back toward the nucleus and synthesize a full spectrum of an AGN with a natural link between the high-energy and radio emissions.

2. Description of the model

We consider a cylindrical and non-relativistic jet. It is characterized by its orientation with respect to the line of sight. A cloud of relativistic e^--e^+ moves on the edges of the jet following a helicoidal trajectory at a relativistic speed. The axis and radius of this helix are assumed to be respectively the axis and the radius of the jet. The shape of the cloud is an ellipsoid. The

443

R. Ekers et al. (eds.), Extragalactic Radio Sources, 443–444.
© 1996 IAU.

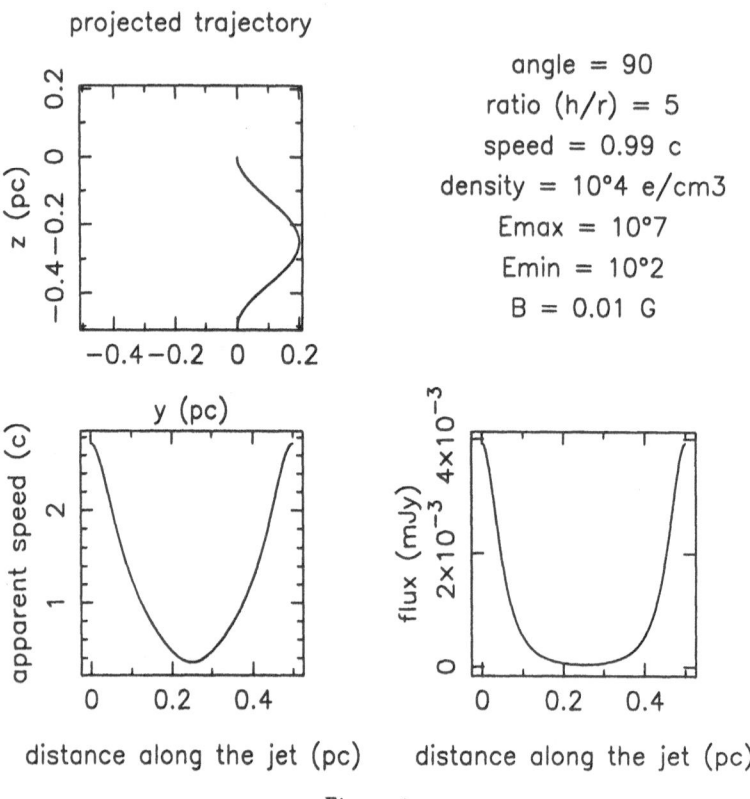

Figure 1.

magnetic field is taken parallel to the trajectory. The particle density inside the cloud is 10^5 cm^{-3} and the energy distribution is a power law of spectral index 2. The full transfer equation with the Stokes parameters is computed.

3. Results

An example of what it is possible to obtain is shown in Fig. 1 with the physical parameters that are used. The evolution of the apparent speed and the intensity vs the distance (projected onto the plane of the sky) along the jet from the core are plotted, as well as the projected trajectory. The preliminary results of this study demonstrate that it is possible to reproduce observed VLBI jets, notably with apparent superluminal speeds even for a jet in the plane of the sky (see Fig. 1).

THE COSMIC NOZZLE

KONJUKOV M.V.
Astrospace Center of P.N.Lebedev Physics Institute of RAS
Moscow Leninskii prospekt 53 Russia

The cosmic nozzle is a channel with flowing hot gas in a system containing the following components: stars, a hydrostatic gaseous corona and a cavity with hot gas in the central part of the system. In the quasi one-dimension approximation, the following solution of the cosmic nozzle problem about cosmic nozzle was obtained

$$
\nu = \nu_0 \pi_g^{3/5}, \qquad w = \sqrt{5\tau_0}\left[1 - \pi_g^{2/5} - \frac{2}{5\tau_0}(\psi_0 - \psi) + \frac{w_0^2}{5\tau_0}\right]^{1/2},
$$

$$
\tau = \tau_0 \pi_g^{2/5}, \qquad \sigma = \sigma_0 \frac{w_0/\sqrt{5\tau_0}}{\pi_g^{3/5}\left[1 - \pi_g^{2/5} - \frac{2}{5\tau_0}(\psi_0 - \psi) + \frac{w_0^2}{5\tau_0}\right]^{1/2}}. \tag{1}
$$

Here (ν, w, τ, σ) are the dimensionless density, hydrodynamical velocity, temperature and cross section of the channel ($\nu_0, w_0, \tau_0, \sigma_0$ are these values at $\xi = 1$), $\phi(\xi)$, $\pi_g(\xi)$ are the gravitational potential and the pressure in the gaseous corona. For a polytropic state equation of the gas in the gaseous corona

$$
\psi_0 - \psi(\xi) = \frac{\kappa_0}{\kappa_0 - 1}\left(1 - \pi_g^{\frac{\kappa_0 - 1}{\kappa_0}}\right) \tag{2}
$$

The investigations of the solutions yield the properties of the flows in cosmic nozzle:

1. The distributions of the density and the temperature of the outflowing hot gas depend only from the pressure in the surrounding gaseous corona $\nu = \nu_0 \pi_g(\xi)^{3/5}$, $\tau = \tau_0 \pi_g(\xi)^{2/5}$. They do not depend on non hydrodynamical forces acting on flowing gas in the channel.

2. For any dimensionless values determining the properties of the cosmic nozzle there is a nonmonotonically changing velocity. It has a maximum at ξ_* determined by the equation $\pi_g(\xi) = \tau_0^{5\kappa_0/(3\kappa_0 - 5)}$. Physically this is

R. Ekers et al. (eds.), Extragalactic Radio Sources, 445–446.

the simple consequence of the transition from $\nu/\nu_g < 1$ to $\nu/\nu_g > 1$. That property of the velocity disappear by the neglecting of gravity forces.

3.In the broad region of the values of the parameters of the cosmic nozzle, there is nonmonotoncally change of the channel cross section, with a minimum at finite distance from the channel origin.

4.The flow in the cosmic nozzle has a narrow layer near the channel origin where the velocity changes sharply.

5.There is the following formula for the velocity in the channel

$$
u = \sqrt{\frac{5kT_0^c}{m}} \left(1 - \tau - \frac{2}{5}\frac{m}{kT_0^c}(\Phi_0 - \Phi) + \frac{mu_0^2}{5kT_0^c} \right)^{1/2}
\tag{3}
$$

where T_0^c is the gas temperature in the cavity. As long as the cavity gas is hot and the gas velocity at the channel origin is subsonic, an asymptotic series for the velocity can be obtained, with main term

$$
u_{as} = \sqrt{\frac{5kT_0^c}{m}} = \sqrt{\frac{5kT_0^g}{3m}}\sqrt{\frac{3kT_0^c}{T_0^g}},
\tag{4}
$$

where $\sqrt{5kT_0^g/3m}$ is the adiabatic sound velocity for corona gas at the boundary with the cavity. The values of u_{as} are nonrelativistic for reasonable temperatures of the cavity and corona.

6. For rough estimations of the energy flow the formula

$$
E_0 = \frac{5}{2}w_0 S_0 n_0^g T_0^g \sqrt{\frac{5kT_0^g}{3m}}
\tag{5}
$$

can be used, where $w_0 = u_0/\sqrt{\frac{5kT_0^c}{3m}}$. Even at the radian measure $S_0 \approx 1$, $w_0 < 1$ and extreme conditions the energy flow is less then $10^{40}\ erg/sec$.

7.There are no physically plausible solutions for the gaseous corona with κ_0 near 1.

8.There are difficulties with the fulfillment of the hydrodynamical approximation in the channel.

9.The set of differential equations for the cosmic nozzle does not have singular points when hydrodynamical velocity equals the local sound velocity.

There are two main consequences from the above obtained results:

1.The cosmic nozzle model cannot explain the main properties of FR-II radio sources.

2.There is no small parameter of the problem for which the set of cosmic nozzle equations is an approximation to the quasi one-dimension gas-dynamic equations.

INTRINSIC KINEMATICS OF MOVING COMPONENTS
IN CURVED RELATIVISTIC JETS

L. LARA[1,2], A. ALBERDI[2,3]

[1] *IRA - CNR, Via Gobetti 101, 40129 Bologna, Italy*
[2] *IAA - CSIC, Apdo. 3004, 18080 Granada, Spain*
[3] *LAEFF, Apdo. 50727, 28080 Madrid, Spain*

1. Description

Superluminal motion in extragalactic jets is successfully explained assuming a relativistic flow of material in a direction close to the observer's line of sight. In rectilinear jets, the apparent velocity measured by the observer (β_{app}) is related with the intrinsic velocity of the moving component (β) and the angle between the jet direction and the observer's line of sight (θ) through the equation (eg. Rees 1966, Nature 211, p.468):

$$\beta_{app} = \frac{\beta \cdot \sin\theta}{1 - \beta \cdot \cos\theta} \tag{1}$$

We generalize Eq.1 for knots travelling along arc of circumferences, given that the trajectory and the line of sight lie in the same plane, i.e. the position angle is constant (Fig.1; Lara & Alberdi, in preparation):

$$\beta_{app} = \frac{\beta \cdot \sin\frac{\Delta\theta_{ij}}{2} \cdot \sin\left(\theta_i + \frac{\Delta\theta_{ij}}{2}\right)}{\frac{\Delta\theta_{ij}}{2} - \beta \cdot \sin\frac{\Delta\theta_{ij}}{2} \cdot \cos\left(\theta_i + \frac{\Delta\theta_{ij}}{2}\right)} \tag{2}$$

For a component travelling along a curved path, we can measure not only its apparent velocity, but also its acceleration or deceleration, the variations of its flux density (S), assumed to be produced by changes in the Doppler factor, and possible variations of the spectral index $(\alpha; S \propto \nu^{\alpha})$. This information, altogether, allows us to obtain the intrinsic velocity of the component and the orientation angle with respect to the observer. Provided

R. Ekers et al. (eds.), Extragalactic Radio Sources, 447–448.

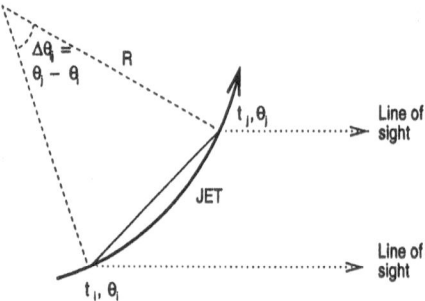

Figure 1. A curved jet observed at two epochs, t_i and t_j

three epochs of observation, t_1, t_2 and t_3, and assuming that β remains constant, we construct a set of equations derived from Eq.2 and from the relation between the observed and the intrinsic flux density through the Doppler factor ($S_{obs} = S_{intr} \cdot D^{3-\alpha}$, for optically thin components):

$$\beta = \frac{\beta_{app} \cdot \frac{\Delta\theta_{12}}{2}}{\sin \frac{\Delta\theta_{12}}{2} \left[\cos\left(\theta + \frac{\Delta\theta_{12}}{2}\right) \cdot \beta_{app} + \sin\left(\theta_1 + \frac{\Delta\theta_{12}}{2}\right)\right]}$$

$$\Delta\beta_{app} = \beta_{app}(\beta, \theta_2, \Delta\theta_{23}) - \beta_{app}(\beta, \theta_1, \Delta\theta_{12})$$

$$\Delta\theta_{ij} = \arccos\left\{\frac{1}{\beta} - \frac{1}{\beta} \cdot (1 - \beta\cos\theta_i)^{\frac{3-\alpha_i}{3-\alpha_j}} \cdot \left[\frac{S_i(\nu_i)\,\nu_j^{\alpha_j}}{S_j(\nu_j)\,\nu_j^{\alpha_j}}\right]^{\frac{1}{3-\alpha_j}} \cdot \gamma^{\frac{\alpha_j-\alpha_i}{3-\alpha_j}}\right\} - \theta_i$$

The spectral index can account for opacity dependent variations in the flux density, although we do not consider the difficulties derived from time-shifts of the turnover frequency which could turn into optically thick the spectrum of the component at a given frequency. After an iterative process we can determine the unknown parameters β, θ_1, θ_2 and θ_3.

2. Application to 4C 39.25

The quasar 4C 39.25 shows a superluminal component, usually called b, travelling with a non-uniform speed and approximately constant position angle between two stationary ones, and gradually increasing its flux density from 1986 on. Applying our analysis to the available experimental data (Alberdi et al. 1993, ApJ 402, p.160; Alberdi et al. 1995, in preparation), we obtain very consistent results for successive epochs, indicating that the orientation of b was $\sim 9°$ in 1984, getting gradually closer to the line of sight. We derive an intrinsic velocity of 0.974 c. These results imply that component b has an intrinsic flux density of ~ 2.4 mJy, and that its deprojected distance to the core is ~ 40 pc ($H_o=100$ km s^{-1} Mpc^{-1}; $q_o=0.5$).

SUPERSONIC JETS AND LOBE MORPHOLOGIES

A. FERRARI
Dipartimento di Fisica Generale, Università di Torino
Via P. Giuria 1, I-10125 Torino, Italy

AND

S. MASSAGLIA, G. BODO
Osservatorio Astronomico di Torino
Strada dell'Osservatorio 20, 10025 Pino Torinese, Italy

1. Introduction

Since the pioneering work by Norman et al. (1982), many numerical studies have been devoted to the analysis of the propagation of a supersonic jet shot into an ambient medium (see Massaglia, Bodo & Ferrari 1995, hereinafter Paper I, and references therein). In spite of these strong efforts many aspects of this problem are still not well understood. This is due to the complexity of the jet-cocoon structure: in fact, the cocoon excites perturbations to the jet flow, which in turn can be amplified by the Kelvin–Helmholtz mechanism and induce a strong activity of the jet's head that affects the cocoon structure. Thus a complex feedback loop mechanism establishes between jet and cocoon which make the dynamics of the interaction very complex.

In addition, when trying a more direct comparison of the results with observations, one must remember that what is observed is an outcome of the distribution of energetic particles and magnetic field and not the bulk of the flowing plasma, and therefore direct comparisons could be misleading. In this paper we focus on the dynamics of the interaction and we describe some properties of the cocoon structure which can be relevant for the observational properties of extragalactic radio sources and have been overlooked in previous studies. We have been able to examine these properties because of the wide exploration of the parameter space especially towards high Mach numbers, typically higher than those discussed in the

R. Ekers et al. (eds.), Extragalactic Radio Sources, 449–452.

present literature, and because our approach allowed us to follow the jet propagation up to very long times and to keep all parts of the cocoon in the computational domain, whereas the usual approach is limited to follow the jet only for one crossing time of the grid and to lose the back part of the cocoon.

2. Results

We study the dynamics of a supersonic, cylindrical, axisymmetric jet continuously injected into a medium initially at rest. We solve numerically the full set of adiabatic, inviscid fluid equations for mass, momentum, and energy conservation.

The numerical scheme, the grid, the code adopted are discussed in Paper I. In the present calculations we take advantage of the particular setup of Paper I, and we perform the calculation in a reference frame where the jet's head is approximately at rest. Therefore, in the initial configuration, the external medium has a uniform velocity $V_h = v_j/(1 + \sqrt{\nu})$ where v_j is the jet velocity in the "laboratory frame" and ν is the ratio of the external to the jet density, and V_h is an approximated advance velocity of the jet's head obtained applying momentum conservation in the front region of the jet (see Paper I). This moving frame is adopted in the computations, but afterwards we will discuss the results obtained, translating them back to the reference frame where the external medium is at rest, i.e. to the "laboratory frame". The system of equations was written in non-dimensional form and the boundary conditions have been set as in Paper I.

The numerical scheme adopted is of PPM (Piecewise Parabolic Method) type and is particularly well suited for studying highly supersonic flows with strong shocks (Woodward & Colella 1984).

The dynamics of the interaction of the jet head with the ambient medium and its dependence on M and ν has been extensively discussed in Paper I, where we have explored the plane (ν, M) with $\nu = 3, 10, 30, 100, 300$ and $M = 3, 10, 30, 100, 300, 1000$. We summarize here the main features of this interaction:

a) Jets with high M (≥ 30) and low ν (≤ 30) have higher head velocities. In this case the backflow compression of the jet behind the head yields the formation of strong biconical shocks that transmit the thrust to the head, increasing the ram pressure on the front region; afterwards the compression reflects at the jet axis and a recurrent process leads to recurrent impulsive accelerations of the head.

b) Jets with low M ($M < 30$), or high M ($M > 30$) and $\nu > 30$, have lower head velocities. In this case the backflow compression is much weaker and the resulting thrust is not sufficient to accelerate the head.

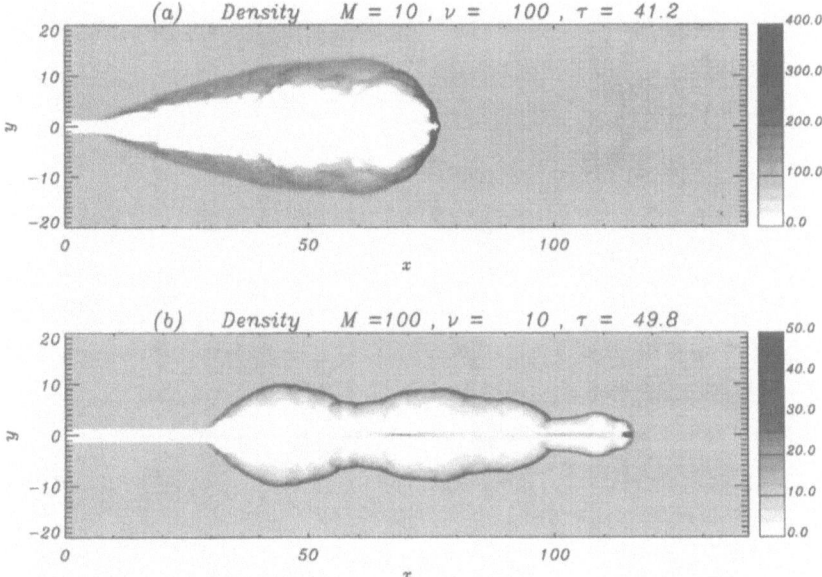

Figure 1. Gray scale image of the density for $M = 10$ and $\nu = 100$ (panel a, 'fat' cocoon), and $M = 100$ and $\nu = 10$ (panel b, 'spearhead' cocoon). This in an example of the two characteristic structures assumed by the cocoon.

The critical parameter is the inclination of biconical shocks: shocks that have a small inclination angle on the axis are successful in accelerating the jet head.

The cocoon morphology reflects the head's evolution. The two classes of jets listed above lead to two different cocoon morphologies: class a) jets form "spearhead" cocoons (Fig. 1b) while class b) jets lead to "fat" cocoons (Fig. 1a).

The application of the results obtained to extragalactic jets can be performed relating the morphologies emerging from the numerical calculations to those observed mainly in the radio band. A basic question that arises at this point is the following: which physical quantity is most suitable to represent the observed radio brightness distribution? If we assume, as usually done in the literature, the density distribution as an actual tracer of the brightness distribution we can compare the former in Fig. 1a) ($M = 10$, $\nu = 100$) and Fig. 1b) ($M = 100$, $\nu = 10$) with the radio maps of 3C296 and 3C223 (Fig. 2), from Leahy & Perley (1991).

From Fig. 1b) we note that the shock that surrounds the cocoon involves matter of the external ambient. Is this shocked region site of particle acceleration? If the answer is positive we can say that indeed the form of the lobe resembles the density distribution of Fig. 1a), with an elongated structure having the front part protruding from the lobe. Similar morpholo-

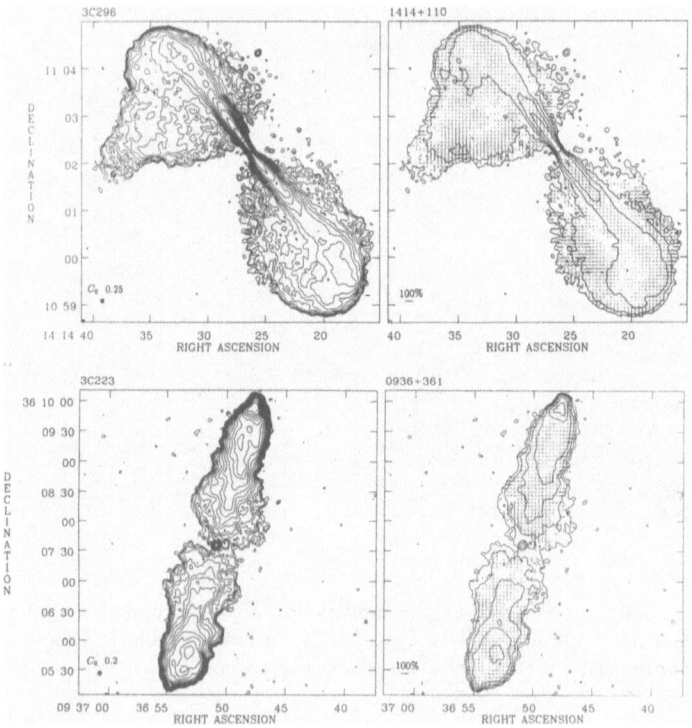

Figure 2.   Brightness distribution and polarization map of 3C296 and 3C223

gies can be found in the sample of high luminosity radio sources by Leahy & Perley (1991); a representative example can be 3C223. In this scheme, the jet would be characterized by a high value of the Mach number accompanied by a moderate value of the density ratio. Moreover, the shock that surrounds the cocoon, according to simulations, must have the effect of enhancing the component of the magnetic field along the shock front, resulting in a high polarization at the source edge, with the polarization vector directed normally to the edge itself. This effect is clearly visible in the polarization map of the source mentioned above.

In the case of slow jet, we note from Fig. 1a) that the shock forms only in the front part of the cocoon. An example of this second kind of morphology can be 3C296.

References

Leahy J. P., Perley R. A., 1991, *AJ*, **102**, 537
S. Massaglia, G. Bodo, A. Ferrari: *A&A* (Paper I) in press (1995)
Norman, M. L., Smarr, L., Winkler, K. H. A., Smith, M. D., 1982, *A&A*, **113**, 285
Woodward, P. R., Colella, P., 1984, J. Comp. Phys. 54, 174

3-D SIMULATIONS OF KELVIN-HELMHOLTZ
INSTABILITIES IN SUPERSONIC JETS

P. ROSSI, G. BODO, S. MASSAGLIA AND A. FERRARI
Osservatorio Astronomico di Torino
Strada dell'Osservatorio 20, 10025 Pino Torinese, Italy

AND

A. MALAGOLI
EFI/LASR, University of Chicago
933 E.56th, Chicago, IL 60637, USA

One of the key processes governing the structure and evolution of astrophysical jets is their interaction with the surrounding medium. A jet can deposit momentum and energy in the ambient medium, and entrain external material. The main physical process responsible for mixing between a jet flow and the ambient medium is the Kelvin-Helmholtz (KH) instability. We have previously analysed the 2D evolution of the axisymmetric modes of a cylindrical jet (Bodo et al 1994) and of the antisymmetric modes of a planar slab jet (Bodo et al 1995). These last are thought to give indications of the 3D evolution of the helical modes of a cylinder, since the linear behavior is very similar. In this contribution we present some preliminary results of fully 3D simulations comparing them with the mentioned 2D results.

The simulations have been performed using a PPM type code; the computational domain is covered by a uniform grid of $128 \times 128 \times 128$ points. In the initial configuration a cylindrical flow, with velocity V_0 directed along the x direction of a cartesian coordinate system, is surrounded by a medium at rest. The jet radius a encompasses 25 grid points and the computational domain extends in the y and z directions from $-2.56a$ to $+2.56a$. Periodic boundary conditions are used at the x boundaries, while free outflow conditions are used at the y and z boundaries. A perturbation to the transverse velocity v_y is superimposed to this equilibrium structure and the evolution is followed by the numerical calculation. Explicit detailed expressions for the equilibrium configuration and the form of the perturbation can be found in Bodo et al (1994, 1995). The system in characterised by two main

R. Ekers et al. (eds.), Extragalactic Radio Sources, 453–454.

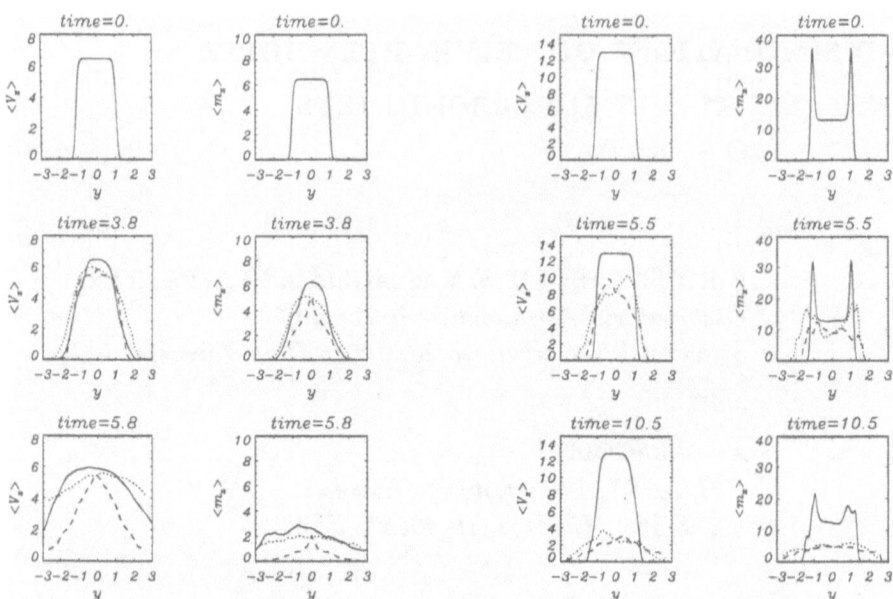

Figure 1. Comparison between the average distributions of the x components of velocity and momentum for the 2D and 3D calculations. The solid line refers to the 2D case, while the dashed and dotted lines refer respectively to a y and z cut in the 3D data. The time is measured in units of the sound crossing time over the jet radius.

parameters: the Mach number M of the flow (with respect to the internal sound speed) and the density ratio ν between external medium and jet. We have performed two simulations with $M = 5$, $\nu = 0.1$ (case A) and $M = 10$, $\nu = 10$ (case B) respectively, i.e. for a jet denser than the ambient medium (case A) and a jet less dense than the ambient medium (case B).

We have compared the distributions of longitudinal velocity and momentum averaged over the periodic x directions with the analogous distributions obtained in the 2D planar slab calculation (Bodo et al 1995). From the figure we can see that, while the distribution are quite similar in case A, the evolution appears to be much faster in 3D for case B.

References

Bodo, G., Massaglia, S., Ferrari, A. and Trussoni, E. (1994), *A&A*, **283**, 655.
Bodo, G., Massaglia, S., Rossi, P., Ferrari, A., Rosner, R. and Malagoli A. (1995), *A&A*, in press.

MHD SELF-SIMILAR SOLUTIONS FOR COLLIMATED JETS

E. TRUSSONI
Osservatorio Astronomico di Torino, Pino T.se, ITALY

C. SAUTY
Observatoire de Paris, DAEC, Meudon, FRANCE

AND

K. TSINGANOS
University of Crete, FORTH, Heraklion, GREECE

The MHD modelling of jets in axisymmetric geometry requires the treatment of the Bernoulli and the transfield equations, that can be treated following a *self-similar* approach. This technique is based on two main assumptions: i) the physical variables are factorized; ii) a suitable scaling law in one direction is prescribed. Solutions self-similar in the r direction (in a spherical frame of reference) have been studied to model collimated winds from disks (Blandford and Payne 1982). Here we present solutions self-similar in the θ direction, suitable to study the collimated wind around the polar axis of a rotating object (Tsinganos and Trussoni 1991, Sauty and Tsinganos 1994). Our basic assumptions are:
- The magnetic flux function, that describes the poloidal components of velocity and magnetic field, is expressed as $A(r, \theta) \propto f(r)\sin^2\theta$.
- The density and the pressure of the plasma are assumed to scale linearly with A: $\rho(r, \theta) \propto 1 + \delta A$ and $P(r, \theta) \propto P_o(r)(1 + KfA)$. Accordingly, the surfaces with equal poloidal Alfvén number M are spherical.

The original MHD equations then reduce to three ordinary differential equations for the four variables M, f, P_o and $P_1(\equiv KfA)$. To close the system we need further assumptions, that define two classes of solutions:

1) P_o and P_1 are related, i.e. $KfA = \kappa = $const; the unknown is f and the shape of the streamlines is deduced. For these solutions a characteristic integral exists (ϵ): it is $\epsilon > 0$ or $\epsilon < 0$ whether the volumetric energy along the polar axis is lower or higher than along the fieldlines, respectively.

2) P_o and P_1 are kept unrelated, then the function f must be prescribed. By assuming $f \propto R^n$ we can choose streamlines radially expanding $(n = 0)$

R. Ekers et al. (eds.), Extragalactic Radio Sources, 455–456.

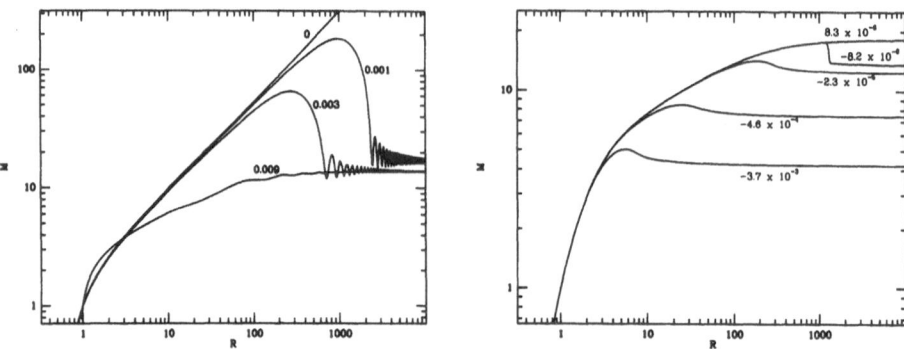

Figure 1. M vs R along the polar axis for solutions of class 1 (left panel), for $\epsilon/\lambda^2 = -0.2$ and some values of κ/λ^2, and class 2 (right panel) for $n = 2$ and some values of q

or cylindrically collimated ($n = 2$). These solutions are characterized by the parameter $q \propto (Q_1/r^2)_{r\to\infty}$.

Typical results are shown in Fig. 1 for $\lambda = 3$ (λ is the ratio of the azimuthal to the poloidal velocity at the Alfvèn point). For both kinds of solutions outflows with superAlfvénic velocities and collimated streamlines can be found. In the former case, for $\kappa = 0$ collimated flows are found when $\epsilon > 0$, while for $\epsilon < 0$ the fieldlines are asymptotically radial and M diverges. For $\kappa > 0$ collimation is found asymptotically also for $\epsilon < 0$: M has a maximum before attaining a constant terminal value. In the collimated region the Alfvèn number has oscillations with decreasing amplitude, related to periodic pinching of the sectional area of the jet. The other class of solutions depend critically on the value of n.

In the case of prescribed streamlines, if they are radially expanding ($n = 0$) the asymptotic velocity slightly increases [$\propto (\log r)^{1/3}$] in the asymptotic region. For collimated fieldlines ($n = 2$, see Fig. 1) superAlfvénic solutions are found with constant asymptotic velocities for several values of q. We notice that again the flow speed can have a maximum before reaching the final velocity, consistently with the behaviour of the other class of solutions.

Our results can be considered complementary with those coming from a r self-similar treatment, not valid on the rotational axis.

References

Blandford R.D. and Payne D.G., 1982, *M.N.R.A.S.*, **199**, 883.
Sauty C. and Tsinganos K., 1994, *Astr. Ap.*, **287**, 893.
Tsinganos K. and Trussoni E., 1991, *Astr. Ap.*, **249**, 156.

MHD EQUILIBRIA OF HELICAL JETS

M. VILLATA AND A. FERRARI
Osservatorio Astronomico di Torino
Strada Osservatorio 20, I-10025 Pino Torinese (TO), Italy

1. Introduction

From recent high angular resolution (~ 0.1 arcsec) observations of some extragalactic jets (in particular we refer to the radio, optical, and ultraviolet images of the M 87 jet, of 3C 66B, and 3C 264) it can be inferred that the synchrotron emission from the jet comes mainly from an external, quasi-cylindrical surface on which bright helical filamentary structures can be clearly recognized.

A fundamental question that may arise in this connection is: how can the observed helical structures be in dynamical equilibrium? The starting point for such an investigation is a magnetohydrodynamic (MHD) description of the underlying physics. The observational evidence of helical morphologies suggesting that the outflow itself presents a helical symmetry urges towards the analytic study of helically symmetric MHD equilibria. The usefulness of this approach has been already stressed in Villata & Tsinganos (1993) and Villata & Ferrari (1994a,b, 1995). In particular, in the last two papers complete solutions of the whole set of time-independent MHD equations in helical symmetry were obtained and detailed models for the interpretation of helical jets (focusing the attention on the M 87 jet) constructed.

In this article we summarize some of the main final results.

2. Main Results and Conclusions

From the results obtained in this study we can conclude that the helical morphologies observed in some astrophysical jets can be interpreted starting from the MHD equilibrium of the underlying magnetized flow. In particular, by considering exact solutions of the time-independent MHD equations, one can construct models for the description of these peculiar structures. Under the assumption that the synchrotron emissivity is proportional to

R. Ekers et al. (eds.), Extragalactic Radio Sources, 457–458.

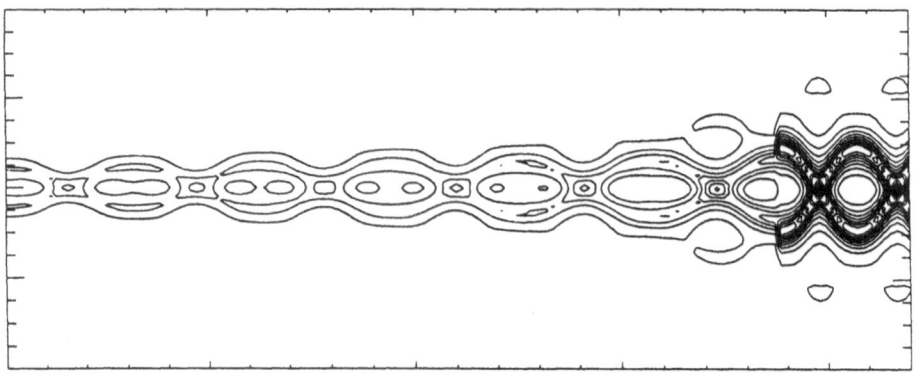

Figure 1. Subalfvénic model for the M 87 jet: equispaced contours of surface brightness

the plasma density ρ and the magnetic field strength squared B^2, we can reproduce the observed high emission helical filaments by the combination of high ρ and high B^2 filaments in conditions of MHD equilibrium.

Moreover, we have constructed specific models for the description of the main features of the M 87 jet: a double-stranded helical filamentary structure having an increasing pitch angle with increasing distance from the galaxy centre and an abrupt structural transition occurring in the outer part of the jet. A subalfvénic model (see Figure 1) seems to be more appropriate for the interpretation of the transition exhibited by the jet: it would be due to the interaction of the flow with the ambient medium at the epoch of the jet birth. The distribution of emission which comes out from this model is in good agreement with the observational one.

In conclusion, this study has provided (i) exact analytical solutions for helically symmetric (or in any other kind of symmetry with one ignorable coordinate) MHD equilibria and (ii) analytical models for the description of helical jets which for the first time allow to derive, by comparison with observations, (iii) the magnitudes of the relevant physical quantities (density, magnetic field, plasma velocity, etc.) by means of exact analytical formulae, thus providing (iv) a complete description of the system.

References

Villata, M. and Ferrari, A. (1994a) Exact Solutions for Helical MHD Equilibria of Astrophysical Jets, *A&A*, Vol. no. **284**, pp. 663–678

Villata, M. and Ferrari, A. (1994b) Exact Solutions for Helical Magnetohydrodynamic Equilibria. II. Nonstatic and Nonbarotropic Solutions, *Phys. Plasmas*, Vol. no. 1, pp. 2200–2206

Villata, M. and Ferrari, A. (1995) Exact Solutions for Helical MHD Equilibria of Astrophysical Jets. II. Non-barotropic Models, *A&A*, Vol. no. **293**, pp. 626–639

Villata, M. and Tsinganos, K. (1993) Exact Solutions for Helical Magnetohydrodynamic Equilibria, *Phys. Fluids B*, Vol. no. **5**, pp. 2153–2164; Erratum (1994), *Phys. Plasmas*, Vol. no. 1, p. 216

A NEW ANALYTICAL JET MODEL

C.R. KAISER AND P. ALEXANDER

MRAO, Cavendish Lab., Madingley Road, Cambridge, UK.

Abstract. We present a self–consistent, self–similar model for classical radio double sources (FRIIs). This model depends only on quantities which can in principle be measured.

1. Motivation and assumptions

Since the standard model for FRIIs was introduced by Scheuer (1974) observations and numerical simulations have provided us with new insights into the properties and mechanisms of these radio sources. In our model we are using some of these results to form our basic assumptions.

First, we note from observations of FRIIs that the ratio of width to length of sources on very different scales are similar (e.g. Black 1992). This suggests that sources are growing in a self–similar way. Therefore we adopt for the volume of the cocoon $V \propto L^3$, where L is the length of the source.

Numerical simulations have shown that the jet itself can be confined by the pressure in the cocoon which is inflated by the shocked jet material. We assume a jet with no opening angle but constant cross section in pressure balance with the cocoon material. This assumption was also made by Begelman et al.(1989) and verified with numerical simulations for their model by Cioffi et al. (1992). Magnetic fields do not play a role in confining the jet and are taken to be dynamically unimportant in the cocoon in the sense that they add to the total pressure, but do not exert a non–isotropic stress, i.e. magnetic field lines are tangled on a small scale.

Assuming a strong, non–radiative shock we find for a relativistic gas in the cocoon ($\Gamma = 4/3$) the sound speed to be about 0.4 times the velocity of the material in the jet. Since this material is relativistic the sound speed in the cocoon is very high and differences in the pressure in different parts of the cocoon will smooth out quickly. Therefore we assume a uniform pressure within the cocoon.

R. Ekers et al. (eds.), Extragalactic Radio Sources, 459–460.

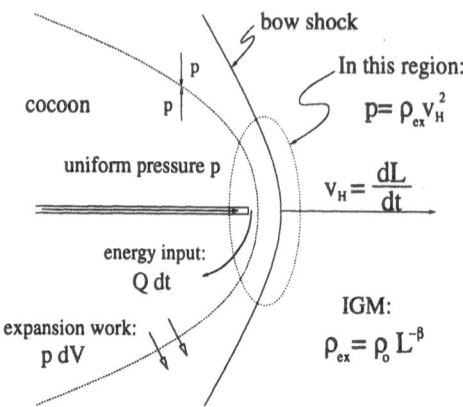

Figure 1. Basic model setup

2. Model dynamics

Fig.1 shows the basic setup of the model. Close to the hotspot we balance the pressure in the cocoon with the ram pressure of the IGM. That means that the whole front of the cocoon has to push through the ambient medium and not just the rather small area where the jet hits the surrounding gas. In order to conserve energy we must have:

$$Q\,dt = \frac{\Gamma}{\Gamma - 1} p\,dV + V\,dp \qquad (1)$$

The jet power Q is assumed to be constant. With this, the expression for the density of the surrounding gas given in fig.1, and the assumption made in section 1 we can derive:

$$L \propto t^{\frac{3}{5-\beta}}, \; p \propto t^{\frac{-4-\beta}{5-\beta}} \qquad (2)$$

The constants of proportionality in both cases depend only on the jet power, the constants ρ_o and β for the external medium, and the ratio of cocoon width to length. All of these can be determined from observations (for Q see Rawlings et al. 1991).

References

Begelmann, M.C. and Cioffi, D.F. 1989, *ApJ*, **345**, L21.
Black, A.R.S. 1992, *Ph.D. thesis*, University of Cambridge.
Cioffi, D.F. and Blondin, J.M. 1992, *ApJ*, **392**, 458.
Rawlings, S. and Saunders, R. 1991, *Nature*, **349**, 138.
Scheuer, P.A.G. 1974, *M.N.R.A.S.*, **166**, 513.

BOUNDARY LAYERS IN EXTRAGALACTIC JETS

DAVID S. DE YOUNG

Kitt Peak National Observatory
Tucson, AZ USA

1. Introduction

The radio morphology of extragalactic radio sources, in particular the FR I objects, has long been recognized to be suggestive of an outflow which is undergoing a strong interaction with the surrounding medium. Two essential elements must be noted in considering this interaction. First, it is inevitable. Second, the interaction will always provide mass and energy transfer from the jet to the ambient medium, and thus it is a loss term in the overall energy balance of the radio source. As will be seen, these losses can be comparable to the overall radio luminosity and hence are a nontrivial constraint on the energy requirements placed on the central engine.

The onset of boundary layer development is driven by the instability of the shear layer between the jet and the surrounding medium. An important experimental result is that such layers are characterized by the growth and persistence of large scale structures within the layer. These structures grow roughly linearly with time in the early stages, and they are the objects which mediate the mass entrainment and energy transfer between the jet and its surroundings. Empirically, the entrainment process appears to be one of "gulping" followed by interior mixing. The experimental data have not given rise to a comprehensive theoretical understanding of this highly nonlinear phenomenon.

2. A New Model for Boundary Layer Development

Consideration of the development of the layer begins in the early nonlinear stage of growth, where small and well separated vortex structures have already formed. The basic questions are the location of the entrainment, the rate of entrainment, and the growth rate of the large structures. Exami-

R. Ekers et al. (eds.), Extragalactic Radio Sources, 461–462.
© 1996 *IAU.*

nation of the empirical data on such boundary layers reveals the following self consistent solution.

On small scales where the vortex size is much less than the jet radius a two dimensional approximation will be appropriate. Between the growing vortices the flow converges from either side to a contact discontinuity between the jet and the surroundings. This flow field places a stagnation point at a location on the contact surface midway between adjacent vortices. Mass flows into the vortices along the contact surface. The initial distance between the small vortices is $2a$, and the initial vortex radius is set at some R_o. The *ansatz* is now to set the azimuthal velocity beyond R_o equal to that created by a point vortex. i.e., far from the vortex the fluid has a velocity field with magnitude decreasing as $1/r$. This velocity field then determines the problem for divergence free flow. Material flows from the stagnation point into the vortex with the mass flow rates given by the flow field. The vortex growth rate is then given by conservation of mass. Thus per unit thickness Δz transverse to the flow the mass inflow is

$$dM/dt = \rho_o \Delta z \int_R^a v_\theta(r)dr. \qquad (1)$$

The corresponding vortex growth rate is given by

$$dR/dt = (1/2\pi)v_\theta(R_o)ln(a/R). \qquad (2)$$

The equation for the vortex radius as a function of time has no closed analytic solution but can be easily integrated numerically. It is obvious that the growth rate will slow as the vortex radius approaches the initial separation $2a$. This approach has been checked by calculating the evolution of a shear layer using kinematic arrays of point vortices. A simulation employing 12000 point vortices shows that stagnation points do arise between the large scale structures as they grow.

3. Results

With these results one can estimate the total mass entrainment rate into a jet by "wrapping" the two dimensional structure onto a jet surface. A minimum entrainment rate results if the process is halted when the vortices grow to fill the separation between them. Using jet velocities of 10,000 km/s, a jet radius of 100 pc, and an ambient number density of 0.01 gives entrainment rates of a few solar masses per year. The kinetic energy lost to the ambient medium by this process is in this case 3×10^{43} erg/s, which is comparable to FR I radio luminosities. Hence the boundary layer interaction can place overall energy demands on the nuclear engine which are comparable to the radio emission and which will decelerate the source and heat its surroundings.

FRACTAL PROPERTIES OF EXTRAGALACTIC JETS: EVIDENCE OF TURBULENCE?

G. BODO, S. MASSAGLIA
Osservatorio Astronomico di Torino
Strada dell'Osservatorio 20, 10025 Pino Torinese, Italy

L. FERETTI
Istituto di Radioastronomia del CNR
Via Gobetti 101, I-40129 Bologna, Italy

AND

A. FERRARI, D. DALL'ANESE
Dipartimento di Fisica Generale, Università di Torino
Via P. Giuria 1, I-10125 Torino, Italy

Several different properties of extragalactic radio sources have been attributed to the effects of turbulence. The morphological appearance of FRI sources has been often interpreted as the result of turbulent entrainment in subsonic or transonic flows (Bicknell 1984, 1986). Moreover, particle acceleration by MHD turbulence via a second order Fermi process is one of the possible ways for accelerating the synchrotron emitting relativistic particles (see Ferrari, Trussoni & Zaninetti 1979). Turbulence appears therefore as an important ingredient in the theoretical modelling of extragalactic radio sources; however, we do not have, unfortunately, any direct evidence of it.

Many recent works on turbulence have analysed the geometrical properties of different kinds of surfaces in turbulent flows. For example, introducing a tracer (a passive scalar) in such a flow, the surface, known as "scalar interface" separating the "coloured" fluid from the "non-coloured" one, acquires fractal properties when advected by turbulent motions. Measures of its fractal dimension D give a value of $D = 2.35 \pm 0.05$ ($D - 1$ for sections), irrespective of the type of flow or of the flow conditions. Similar values have been also obtained for different kinds of surfaces and in different contexts. We recall that the fractal dimension D is defined as $N = r^{-D}$, with N the number of objects with characteristic linear dimension larger than r.

R. Ekers et al. (eds.), Extragalactic Radio Sources, 463–464.
© 1996 IAU.

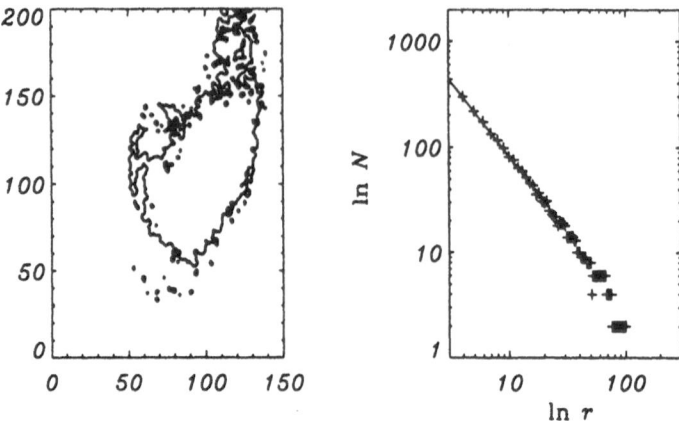

Figure 1. Brightness contour at 3σ level $(1.1 \times 10^{-4}$ Jy/beam) of the southern lobe of 3C449 (left panel); $\ln N$ vs $\ln r$ curve yielding $D = 1.39$ (right panel)

We have analyzed the fractal properties of the boundaries of the brightness distribution (using thresholds corresponding both to 3σ and 5σ levels) in the VLA radio maps of 3 FRI sources: 3C449, 3C465, 4C29.30. We report in Fig. 1 and Table 1 the results obtained.

TABLE 1. Fractal Dimensions

Source	3σ level	5σ level
3C449	1.36	1.21
3C465	1.19	1.28
4C29.30	1.17	1.13

We have also measured the fractal dimension of the scalar interface in the results of numerical simulations of supersonic fluid jets in 2-D (slabs), undergoing Kelvin-Helmholtz instabilities. We can follow there the evolution towards a turbulent state, with D increasing from unity, in the initial state, to $D \simeq 1.35$, in the turbulent state.

References

Bicknell G.V. (1984) *ApJ*, **286**, 68
Bicknell G.V. (1986) *ApJ*, **300**, 591
Ferrari A., Trussoni E., Zaninetti L. (1979) *A&A*, **79**, 190

ISM / RADIO MATERIAL INTERACTION MODELING

P. FERRUIT, L. BINETTE AND E. PECONTAL
CRAL Observatoire de Lyon - UMR142
9, av. Charles André - 69561 Saint-Genis-Laval Cedex, France

With the outcoming of high angular resolution imaging and 3D spectrographic facilities (HST, TIGER-OASIS [1] at the CFH Telescope...), the need of spatially resolved theoretical models of the kinematic interaction of radio material with the ISM (or the IGM for radio-galaxies) is increasing dramatically. In order to fill the gap between pure hydrodynamical models (3D approach, but very poor atomic physic) and photoionisation/shocks ones (planar description, full account for atomic physic processes), we have started to develop 2D or 3D hydrodynamical models including a good description of atomic physic processes (using photoionisation routines from the code MAPPINGS [2]).

We present here the first results of a bowshock model based on that of Taylor et al. [3]. We have improved their description in two main ways :

- by using an adaptive step size when following a cooling particle. This is particularly important in the "catastrophic" cooling zone where the gas temperature drops very rapidly to 10^4 K (see Fig. 1).
- by using MAPPINGS for the computation of the ionization equilibrium and emission line fluxes.

When comparing our results with that of Taylor et al., we have found a large discrepancy between our [OIII] fraction and theirs. This is likely due to charge transfer reactions that they did not take into account (see Fig. 2). It points out the importance of a good atomic physic description.

We have also explored the effect of changing the ionization parameter (by varying the nuclear source flux). This highlights the major role played by photoionization even in a shock based model! As the bulk of the emission originates from the photoionised zone, this puts in question the use of line ratios as shock diagnostics.

Follow up work will aim at introducing out-of-equilibrium processes affecting ionisation states, and at studying in more details the relative contribution of shocks and photoionisation, to the emission.

R. Ekers et al. (eds.), Extragalactic Radio Sources, 465–466.
© 1996 IAU.

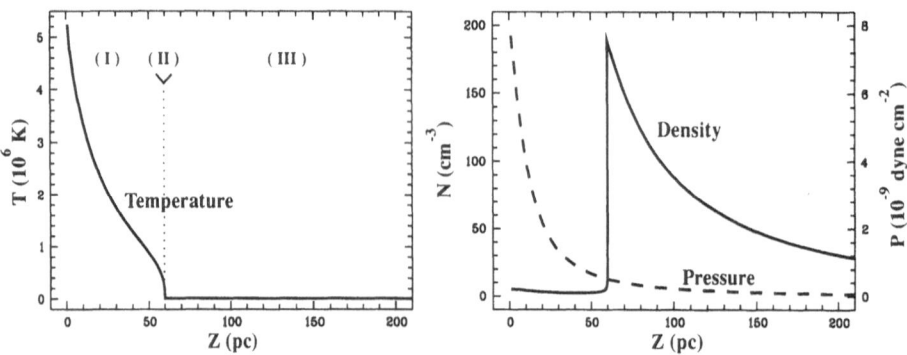

Figure 1. Evolution of a particle that entered the bowshock at Z = 1 pc. Model parameters : shock velocity 700 km s^{-1}; ISM hydrogen number density 1 cm^{-3}; ISM ionization
parameter U = 10^{-3}; power law (spectral index of 1.4) used for the central nuclear source;
geometrical parameters D$_Z$ = 40 pc, B = 0.6 (E model of Taylor et al. with lower ionization parameter). Zone I : the pressure decreases slightly faster than the temperature,
inducing a slow density decrease. Zone II : catastrophic cooling, the temperature goes
to 10^4 K in a few parsec inducing a strong density enhancement (constant pressure).
Zone III : the gas has cooled, the density decreases slowly with the pressure.

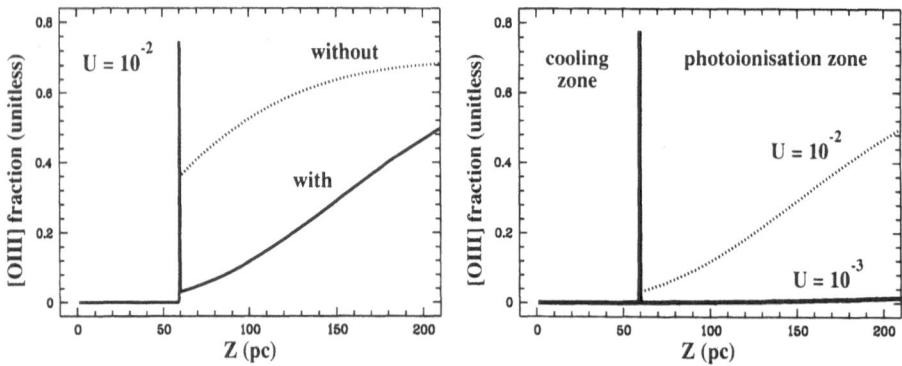

Figure 2. Left : [OIII] fraction as a function of Z with and without including charge
transfer reactions in the modelisation. Right : [OIII] fraction evolution for two different
ISM ionisation parameters. In the cooling zone, the temperature is too high for [OIII] to
exist. In the photoionisation zone, its fraction grows slowly as the ionisation parameter
U increases (the density decreases with Z).

References

[1] R. Bacon, et al., 1995. *A & A Supp.*, **113**, 347.
[2] L. Binette, 1982. Ph.D. thesis, Australian National. University, Canberra
[3] D. Taylor, J.E. Dyson, D.J. Axon, 1992. *MNRAS*, **255**, 351.

NUMERICAL SIMULATIONS OF EXTRAGALACTIC JETS IN AN INHOMOGENEOUS ENVIRONMENT

STEVE HIGGINS, TIM O'BRIEN
Astrophysics Group, Liverpool John Moores University,
Byrom St., Liverpool, L3 3AF, UK

AND

JAMES DUNLOP
Institute for Astronomy, University of Edinburgh,
Blackford Hill, Edinburgh, EH9 3HJ, UK

We have simulated the passage of an extragalactic jet through a medium containing an ensemble of cool, dense clouds. The hydrodynamic code uses the second-order Godunov method of Falle (Falle 1991, van Leer 1979) in three-dimensional, cartesian coordinates. We have estimated the synchrotron emissivity and used this to produce synthetic radio maps. The results are reminiscent of structures seen in many extragalactic radio sources.

There are several lines of evidence to indicate the presence of cool, dense clouds in extragalactic radio sources.

- Polarization maps of radio sources show significant inhomogeneity on scales of less than 5kpc (Pedelty *et al* 1989).
- Line emission at optical wavelengths is observed from regions along the edges of jets, at the outer edges of bends and near knots (Wilson 1993), aligned with the radio axes (Dunlop & Peacock 1993) and associated with the regions of depolarization.
- X-ray observations show that radio galaxies and quasars often lie in the centres of rich clusters with dense, rapidly cooling IGM (Crawford & Fabian 1993), in which cold clouds can condense (Fabian 1993).

The clouds in our simulation have a density contrast of 50 times the ambient medium, and are distributed at random positions in the grid, with a fixed volume filling factor (0.2) and a power law distribution of radius up to a fixed maximum size (7 cells). The computational grid is $90 \times 90 \times 90$ cells, notionally representing a region of $9kpc^3$. The jet has a Mach number

R. Ekers et al. (eds.), Extragalactic Radio Sources, 467–468.

of 10 and a density contrast of 0.01. We assume that jet and clouds are in pressure balance with the ambient material. We have chosen conditions in the ambient medium to be consistent with observations.

The plots show the logarithm of total density in the central plane of the computational grid and radio emission on a linear scale integrated through the grid. As the jet progresses through the grid it can be seen to collide with clouds, producing prominent hotspots in the radio emission. In earlier simulations we have studied the collision of jets with individual clouds (Higgins, O'Brien & Dunlop 1995). These show the deflection of the jet and hotspots giving enhanced emission at the point of deflection.

These spots persist as the jet moves past each cloud and encounters further obstructions. They fade as the clouds are eroded by the passage of the jet. Meanwhile new hotspots form at new encounters, and deflected jet material percolates through the ambient medium, producing filamentary and foamy shock structures. The result is a jet that is made visible by a series of irregular knots, with a crooked ridgeline, multiple hotspots at the head of the jet and filamentary diffuse bridges and lobes.

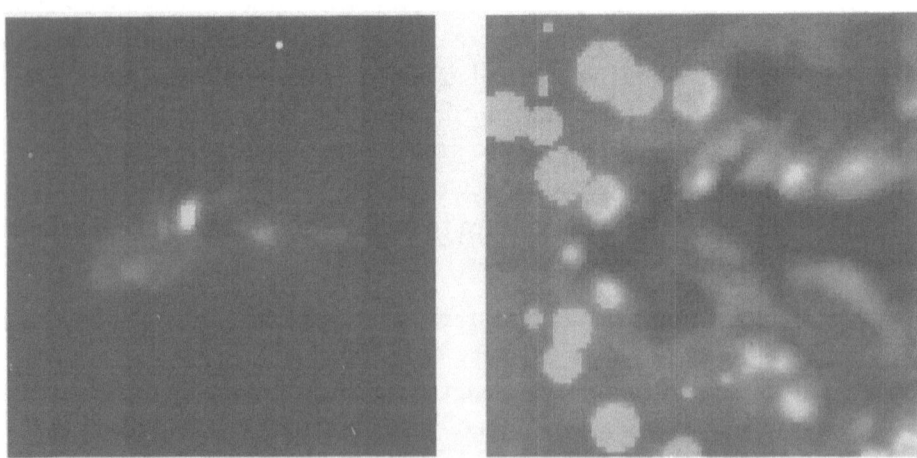

Figure 1.   Density and Synchrotron Emission at 48,000 years

References

Crawford & Fabian (1993) *MNRAS*, **260**, pp.L15

Dunlop, J.S. & Peacock, J.A. (1993) *MNRAS*, **263**, pp.936

Fabian, A.C. (1994) *Ann. Rev. A&A*, **32**, pp.277

Falle, S.A.E.G. (1991) *MNRAS*, **250**, 581

Higgins et al., (1995) In: *'Shocks in Astrophysics'*, eds. Millar, T.J. & Raga A., in press.

van Leer, Bram (1979) *J.Comp.Phys.*, **32**, p101

Pedelty, J. A., Rudnick, L., McCarthy, P.J. & Spinrad, H. (1989) *AJ*, **97**, pp.647

Wilson, A. S. (1993) In: *'Astrophysical Jets'*, STScI Symp.Ser. Vol. 6, eds. Burgarella, Livio & O'Dea, pp.122

SHOCK EXCITATION OF EMISSION LINES
AND THE RELATION TO GPS SOURCES

G.V. BICKNELL[1,2], M.A. DOPITA[2], C.P. O'DEA[3]

[1] *Astrophysical Theory Centre, ANU, Canberra, ACT 0200, Australia*
[2] *Mt. Stromlo & Siding Spring Observatories, Weston PO, ACT 2611, Australia*
[3] *Space Telescope Science Institute, 3700 San Martin Drive, Baltimore MD, 21218, USA*

1. Radio Lobes in Dense Environments

Gigahertz Peak Spectrum (GPS) and Compact Steep Spectrum (CSS) sources have attracted a large amount of attention in recent years since they occupy a surprisingly large fraction of the extragalactic radio source population. In this paper we summarise a theory which attempts to unify the optical emission line and radio properties of these sources. In outline the theory is as follows: The bow shock preceding the radio lobe driven into the ISM by a powerful radio jet ionizes the ISM producing both optical line emission and a free-free absorbing screen. The free-free absorption can explain the relationship between size and turnover frequency (Stanghellini et al., 1995) and the prediction of the line emission is in accord with the observation for a small sample of sources for which optical data are available.

Begelman (1995) has proposed a model of Compact Symmetric Objects (CSOs) in which a "dentist drill jet" (Scheuer, 1982) drives an overpressured lobe. Using Begelman's model, the shock velocity of the expanding lobe is

$$V_{\rm sh} = 2500\,\zeta^{\frac{1}{6}} \left[\frac{F_{E,45}}{(8-n)\,n_0\,(x_0/{\rm kpc})^2} \right]^{\frac{1}{3}} \left(\frac{x}{x_0} \right)^{(n-2)/3} {\rm km\,s^{-1}} \qquad (1)$$

where x is the distance of the hot-spot from the core, n_0 is the number density at $x_0 = 1$ kpc and $\rho \propto r^{-n}$. When $n = 2$, the value favoured

469

R. Ekers et al. (eds.), Extragalactic Radio Sources, 469–470.

by Begelman, the shock velocity is independent of distance. Emission line modelling carried out so far indicates that $500 < V_{sh} < 1000 \, \text{km s}^{-1}$. This is consistent with the typical GPS line widths $\Delta V \sim 750 \, \text{km s}^{-1}$ reported by Gelderman (these proceedings). This constraint implies that for jet energy fluxes $\sim 10^{45} \text{ergs s}^{-1}$ the confining medium is quite dense: $4 \lesssim n_0 \lesssim 30 \text{cm}^{-3}$.

The ionized gas, both post-shock and precursor, contribute to the free-free optical depths of post-shock and precursor regions and these are given by:

$$\tau_{\nu,\text{ps}} \approx 3.9 \times 10^{-4} \, n_0 \, T_{e,4}^{-1.35} \, \nu_9^{-2.1} \left(\frac{V_0}{500 \, \text{km s}^{-1}} \right)^{3.6} \left(\frac{x}{x_0} \right)^{\frac{0.6n-7.2}{3}} \quad (2)$$

$$\tau_{\nu,\text{pc}} \approx 2.4 \times 10^{-4} \, n_0 \, T_{e,4}^{-1.35} \, \nu_9^{-2.1} \left(\frac{V_0}{500 \, \text{km s}^{-1}} \right)^{2.66} \left(\frac{x}{x_0} \right)^{-\frac{0.34n+5.32}{3}} \quad (3)$$

The turnover frequency is given by approximately twice the frequency at which the optical depth becomes unity. For jet energy fluxes $\approx 10^{44} - 10^{46}$ ergs s^{-1} a good fit to the turnover frequency–size data of Stanghellini et al. (1995) is obtained.

2. Emission line luminosity

The software routine MAPINGSII gives for the $H\beta$ luminosity,

$$L(H\beta) = 7.4 \times 10^{-6} \, n_H \left[V_{sh}/100 \, \text{km s}^{-1} \right]^{2.41} A_{sh} \text{ ergs s}^{-1} \quad (4)$$

(Dopita and Sutherland, 1995) where n_H is the ambient Hydrogen number density and A_{sh} is the shock surface area. For $n = 2$ the dynamics of the expanding bubble imply that $\rho_a V_{sh}^3 A_{sh} \approx 0.13 F_E$ and we take $F_E = \kappa^{-1} L_{\text{radio}}$ with $\kappa \approx 0.1$. For a cutoff frequency $\sim 10\times$ the reference frequency, $\nu_0 = 5$ GHz, $L_{\text{radio}} \approx 7.8\nu_0 P_{\nu_0}$ and $L(H\beta)/\nu_0 P_{\nu_0} \approx 0.03 \left(V_{sh}/500 \, \text{km s}^{-1} \right)^{-0.59}$ The geometric mean of the small number of sources we have examined (1345+125, 1934-638, 2352+495, 1718-649) is 0.03, encouraging examination of a larger sample.

The predicted amount of gas could be $\sim 10^{11} - 10^{12} M_{\odot}$ within 10 kpc. The detection of significant amounts of HI in 3 out of 4 GPS sources (J. Conway, these proceedings) is in substantial agreement with this prediction.

References

Begelman, M. C.: 1995, in D.A. Harris (ed.), *Cygnus A.*
Dopita, M. A. and Sutherland, R. S.: 1995, *ApJ*, in press
Scheuer, P. A. G.: 1982, in D. S. Heeschen and C. M. Wade (eds.), *IAU Symposium 97, Extragalactic Radio Sources*, p. 163, Reidel, Dordrecht
Stanghellini, C., O'Dea, C., and Baum, S.: 1995, in preparation
Sutherland, R. S. and Dopita, M. A.: 1993, *ApJS* **88**, 253

DYNAMICAL MODELS OF EMISSION-LINE GAS IN RADIO GALAXIES

Tidal Interactions / Merger Scenarios

A. M. KOEKEMOER

Institut d'Astrophysique de Paris
98bis Bd. Arago, 75014 Paris, France

AND

G. V. BICKNELL

ANU Astrophysical Theory Centre
Australian National University, Canberra, ACT 0200, Australia

1. Introduction

Extended emission-line regions (EELRs) in radio ellipticals are generally thought to trace gas acquired externally, *eg.* through interaction with a gas-rich disk galaxy (Athanassoula and Bosma 1985, Barnes and Hernquist 1992, Hernquist and Mihos 1995). We examine here the dynamical evolution of gas in mergers, focussing on the conditions required for collisions between streams of gas. We find that such collisions can occur over a relatively wide range of encounter geometries, producing large-velocity-amplitude kinematic signatures characteristic of those observed in EELRs. This is relevant to the formation of shocks, which can account for the ionization properties of EELRs (Koekemoer and Bicknell, this conference).

2. Model Results

The evolution of a gas-rich galaxy encountering a large spherical system was modelled using Smoothed Particle Hydrodynamics (SPH, Koekemoer and Bicknell 1995); the results presented here are for a spherical halo mass $M = 10^{12} \, M_\odot$, velocity dispersion $\sigma = 250 \, \mathrm{km \, s^{-1}}$, and core radius $r_c = 1 \, \mathrm{kpc}$. A gas disk of mass $10^9 \, M_\odot$, radius $5 \, \mathrm{kpc}$ and temperature $5000 \, \mathrm{K}$ (warm, partially ionized ISM) was settled in an isothermal halo with $M = 5 \times 10^{10} \, M_\odot$, $\sigma = 100 \, \mathrm{km \, s^{-1}}$, $r_c = 0.5 \, \mathrm{kpc}$. We investigated pericentric distances in the range $r_p = 2 - 10 \, \mathrm{kpc}$, pericentric velocities v_p

471

R. Ekers et al. (eds.), Extragalactic Radio Sources, 471–472.

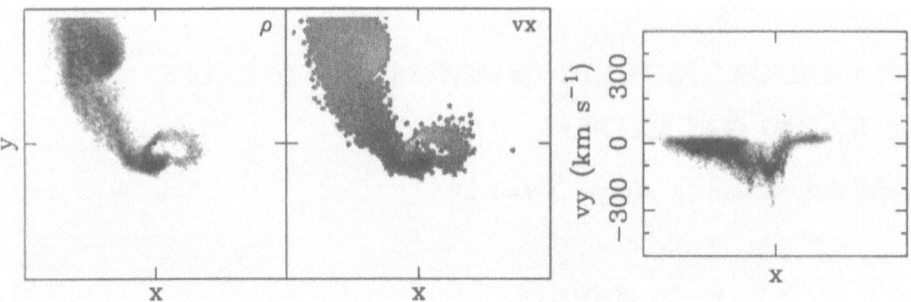

Figure 1. Gas dynamic results as discussed in the text. Plotted are the surface density ρ, the z-velocity component v_x, and the line-of-sight y-velocity v_y projected along the y axis. The v_x greyscale corresponds to $-300 \rightarrow +350\,\mathrm{km\,s^{-1}}$ (white $\rightarrow$ black).

up to unbound orbits (i.e. $990\,\mathrm{km\,s^{-1}}$ for $r_p = 2\,\mathrm{kpc}$), and disk inclinations θ of 0, 45 and 90° to the encounter plane. The volume was gridded into $240 \times 240 \times 80$ smoothing lengths (the smoothing length $h_s = 0.32\,\mathrm{kpc}$).

In Figure 1 we present gas kinematics for a simulation with $r_p = 5\,\mathrm{kpc}$, $v_p = 700\,\mathrm{km\,s^{-1}}$, $\theta = 45°$, at a time $1.5 \times 10^8\,\mathrm{yr}$ after the first pericentric passage. There is substantial interaction between material that has passed through the centre of the potential and that is still infalling; velocity shears of order $200 - 300\,\mathrm{km\,s^{-1}}$ are produced, and persist on timescales of $\sim 1 - 3 \times 10^8\,\mathrm{yr}$ after the initial pericentric passage, depending upon details of the encounter geometry. While the smoothing length is too large to resolve shock structures, shocks with velocities of this order can account for the observed ionization and energetics of the gas.

3. Conclusions

We find that collisions between streams of gas, with relative velocities corresponding to those observed ($\gtrsim 200\,\mathrm{km\,s^{-1}}$), can be produced in a wide range of merger encounter geometries and persist on timescales $\gtrsim 10^8\,\mathrm{yr}$, comparable to the AGN lifetime. The gas morphology changes significantly for different encounter parameters but its large-scale kinematics are affected primarily by the pericentric separation and the depth of the elliptical galaxy potential. Formation of shocks with these velocities can significantly affect the EELR ionization and energetics.

References

Athanassoula, E. and Bosma, A.: 1985, *Ann.Rev.Astron.Ap.* **23**, 147
Barnes, J. E. and Hernquist, L.: 1992, *Ann.Rev.Astron.Ap.* **30**, 705
Hernquist, L. and Mihos, J. C.: 1995, *Ap.J.* **448**, 41
Koekemoer, A. M. and Bicknell, G. V.: 1995, *in preparation*

SHOCK EXCITATION OF EMISSION LINES
IN RADIO GALAXIES

A. M. KOEKEMOER[1], AND G. V. BICKNELL[2]

[1] *Institut d'Astrophysique de Paris*
98bis Bd. Arago, 75014 Paris, France
[2] *ANU Astrophysical Theory Centre*
Australian National University, Canberra, ACT 0200, Australia

1. Introduction

We present evidence for the viability of "auto-ionizing" shocks as the dominant ionization mechanism in extended emission-line regions (EELRs) in two radio galaxies, PKS 0349−27 and PKS 2356−61. The application of this model, rather than the nuclear photoionization hypothesis of unified schemes (Barthel 1989), is motivated by observed EELR properties: large line-of-sight velocity widths (up to $\Delta v \sim 500\,\mathrm{km\,s^{-1}}$ for nearby objects and $\gtrsim 1000\,\mathrm{km\,s^{-1}}$ at higher z); kinematics/excitation relationships (Baum *et al.* 1992); the EELR/radio axis alignment (Chambers *et al.* 1987, McCarthy *et al.* 1987); and the correspondence between the brighter EELR and the shorter radio lobe (McCarthy *et al.* 1991), suggestive of jet/gas interactions. We show that the flux, excitation *and* kinematics across the gas is self-consistently accounted for in terms of shocks as a single physical mechanism, requiring fewer unknown parameters than nuclear photoionization.

2. Shock Model Compared with Observations

The physical basis for auto-ionizing shocks (Sutherland *et al.* 1993, Dopita and Sutherland, 1995) involves cloud-cloud collisions producing strong shocks with temperature $T \propto v^2$. Shocks with $v \gtrsim 200\,\mathrm{km\,s^{-1}}$ produce substantial UV / soft X-rays and photoionize the precursor, which then contributes to the spectrum; the resulting flux and excitation increase with v.

In Figure 1 we plot [OIII]5007/Hβ vs. Δv for all pixels sampling the EELR in PKS 0349−27 (from a "datacube" of longslit spectra taken with

R. Ekers et al. (eds.), Extragalactic Radio Sources, 473–474.

Line / Hβ	Obs	B=0	B=1	B=4
[OII]3726,29	6.27	1.73	4.90	6.97
[NeIII]3869	1.51	0.71	0.92	1.13
[OIII]4363	0.15	0.08	0.07	0.08
HeII4686	0.39	0.28	0.33	0.37
[OIII]5007	11.9	8.26	8.00	9.07
[OI]6300	0.43	2.11	2.36	1.67
[NII]6583	2.54	0.98	2.24	2.40
[SII]6716,31	2.00	2.11	1.89	1.42

TABLE   1.   Line   ratios:
PKS 2356−61.

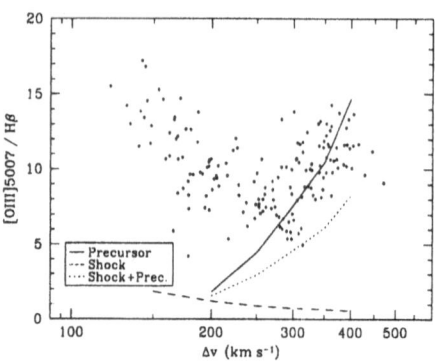

Figure 1.   [OIII]5007/Hβ: PKS 0349−27.

the AAT / RGO spectrograph). The curves are model predictions for the shock, precursor and combined spectrum (low-density-limit photon-bounded precursor, transverse magnetic field parameter $B/n^{1/2} = 0\,\mu\mathrm{G\,cm}^{3/2}$). Gas with $\Delta v \gtrsim 250\,\mathrm{km\,s}^{-1}$ (*ie.* the kinematically disturbed central region) shows a clear trend in excitation, consistent with shock-related precursors. The high ionization of gas with lower Δv implies that it is also photoionized but not physically associated with shocks; its flux and excitation are accounted for by $\sim 5-10\%$ of ionizing flux escaping from shocks in the central region.

In Table 1 we present the low-dispersion line ratios for the central EELR of PKS 2356−61, compared with a single-velocity shock+precursor model (the velocity $\Delta v \sim 400\,\mathrm{km\,s}^{-1}$ corresponds to that observed, while the $B/n^{1/2}$ values represent that of the Galactic ISM, $\sim 3\,\mu\mathrm{G\,cm}^{3/2}$). Almost all the lines (*incl.* [OIII]4363) are accounted for to within $30-50\%$ or better.

3.   Conclusions

Using simple assumptions about shock properties, we have shown that the fluxes, excitation *and* kinematics of gas in PKS 0349−27 and PKS 2356−61 can be self-consistently described by auto-ionizing shocks. This robustness suggests that shocks may play an important role in a wide range of EELRs.

References

Barthel, P. D.: 1989, *Ap.J.* **336**, 606
Baum, S. A., Heckman, T. M. and van Breugel, W. J. M.: 1992, *Ap.J.* **389**, 208
Chambers, K. C., Miley, G. K. and van Breugel, W. J. M.: 1987, *Nature* **329**, 604
Dopita, M. A. and Sutherland, R. S.: 1995, *Ap.J.*, in press
McCarthy, P. J., van Breugel, W. and Kapahi, V. K.: 1991, *Ap.J.* **371**, 478
McCarthy, P. J., van Breugel, W., Spinrad, H. and Djorgovski, S.: 1987, *Ap.J.* **321**, L29
Sutherland, R. S., Bicknell, G. V. and Dopita, M. A.: 1993, *Ap.J.* **414**, 510

COMPARATIVE STUDY OF EXTRAGALACTIC RADIO SOURCES FROM "DOUBLE JET" MODELS

H. DOLE, H. SOL, L. VICENTE
DARC, UPR176 du CNRS, Observatoire de Paris-Meudon,
92195 Meudon – France

It seems now important to consider bulk velocity gradients inside jets (Blandford, 1993; Biretta et al, 1995) and recent results from tomography technique suggest the presence of different components in radio sources (Rudnick, this symposium). A few jet models take explicitly into account two components with different bulk velocities (Smith, Raine, 1985; Baker et al, 1988; Sol et al, 1989; Achatz et al, 1990). A fast beam comes from the vicinity of the black hole while a slower collimated wind is emitted by the accretion disk. When stable, the two components can survive along the jet. If a fast instability develops at a distance D_c from core, the beam is destroyed over some relaxation length. Its energy and momentum are transmitted to the wind. Apparent change in the flow regime such as slowing down, decollimation, local bending or discontinuity in surface brightness is the expected observational signature of such critical zones on radio maps. The exact appearance of the zone depends on the parameters of the two flows and on their interaction regime.

Several plasma microinstabilities can grow in "double jets" and impose specific conditions to ensure stability:

(1) light e^+e^- beams with density ratio $n_b/n_0 \simeq 10^{-3}$ to 0.1.

(2) strong magnetic field along jet axis, $B > B_c = 3.2 \times 10^{-3}n_0^{1/2}$ in CGS units (i.e. plasma and cyclotron frequencies such that $\omega_p < \omega_c$, density $n_0 < 9.7 \times 10^4 B^2$, and Alfven velocity $v_A > 0.02c$) to avoid efficient generation of Langmuir waves.

(3) bulk Lorentz factor $\gamma < \gamma_A = \left(\frac{\omega_c}{\omega_p}\right)\left(\frac{m_p}{m_e}\right)^{1/2} = \left(\frac{m_p}{m_e}\right)\left(\frac{v_A}{c}\right)$ to avoid excitation of Alfven waves: $\gamma_A^{\min} \simeq 43$

(4) $\gamma < \gamma_W = \left(\frac{4v_a}{c}\right)\gamma_A$ to avoid whistler generation: $\gamma_W^{\min} \simeq 4$.

Application to fifty-six jets provides tests of the scenario and comparative studies of different types of sources. To qualitatively describe a source

R. Ekers et al. (eds.), Extragalactic Radio Sources, 475–476.

within two-component models, one needs to interpret some feature along the jet as a critical zone from its apparent morphology. To follow the quantitative approach, one needs some estimates of the magnetic field along the jet and at the critical zone to determine the wind density n_0 from (2).

For WATs, critical zones are identified with the so-called inner hot spots and for FRIs with the start of bright jets after the gaps. In FRIIs inner straight beams come out from nucleus and disappear at brighter knot or decline with small change in direction before entering radio lobes. We tentatively identified the end of these beams with the critical zone. Cross-sectional areas of the two components are deduced from the maps. From (1) to (4), we impose $n_b = 10^{-2}n_0$ and $\gamma = 4$. We first assume $v_0 = 0.02c$ for the wind velocity. Mass flux, net energy flux, and efficiency factor $\varepsilon = L_{\rm rad}/(K_b + K_0)$ for conversion of jet power into radio luminosity are then deduced.

	D_c	B_c	$\dot{M}_b$	$\dot{M}_0$	K_b	K_0	ε
	kpc	μG	$10^{-5}\mathrm{M}_\odot$/yr	$\mathrm{M}_\odot$/yr	10^{42}erg/s		
$\langle$WAT$\rangle$	29.2	14.1	15.5	1.1	105	35.8	0.022
$\langle$FRI$\rangle$	2.9	14.3	0.1	0.007	0.7	0.2	0.12
$\langle$FRII$\rangle$	42.3	42.5	19.9	0.86	135	27.7	0.60
$\langle$FRI$_{CF}\rangle$	1.4	134.0	10.5	0.58	71.5	19.4	0.019

Our study shows that (i) critical zones can be identified in many jets, (ii) the scenario leads to quite realistic values for the $\dot{M}$, K and ε, (iii) different morphological types of sources define specific domains of parameters in $D_c - B_c$ and $K_0/K_f - \varepsilon$ diagrams, together with some continuity from one type of sources to another, (iv) cooling flows harbour smaller but more powerful sources, (v) to keep $\varepsilon < 1$ in all FRII requires $v_0 \simeq 0.06c$ in FRIIs and their inner beam can not lead alone to observed luminosities, (vi) ε is quite different for different types of sources. A Kolmogorov-Smirnov test shows that the distributions of ε are different at the 92% level for WATs and FRIIs. Acceleration of particles appears less efficient in inner hot spots of WATs than in terminal FRII hot spots.

References

Achatz, U., Lesch, H., Schlickeiser, 1990, *AA*, **233**, 391.
Baker, D.N., Borovsky, J.E., Benford, G., Eilek, J.A., 1988, *ApJ*, **326**, 110.
Biretta, J.A., Zhou, F., Owen, F.N., 1995, *ApJ*, **447**, 582.
Blandford, R., 1993, Astrophysical jets, Space Telescope Science Institute Symposium Series, Burgarella, D., Livio, M., O'Dea, C.P., eds, Cambridge University Press, p. 15.
Smith, M.D., Raine, D.J., 1985, MNRAS, **212**, **425**
Sol, H., Pelletier, G., Asséo, E., 1989, MNRAS, **237**, **411**

DISTORTION EFFECTS IN BL LAC RADIO JETS

H. SOL[1], S. APPL[2], L. VICENTE[1]

[1] *DARC, UPR176 du CNRS, Observatoire de Paris-Meudon, 92195 Meudon – France*
[2] *Observatoire Astronomique, 11 rue de l'Université, 67000 Strasbourg – France*

BL Lac objects often show a quite distorted radio morphology. Almost 75% of the BL Lacs for which the information is available show an apparent misalignment angle ΔPA between the VLBI jet and the large scale radio structure larger than 45 degrees. This can be explained by strong enhancement of slight bending due to projection effects, especially if BL Lacs are the most highly beamed sources. However we recently performed a statistical analysis of misalignment angle histograms for 155 extragalactic radio sources of different types and found that the *intrinsic* distortion is significantly more important in BL Lacs than in quasars and even CSS sources. Indeed the best fits of the ΔPA histograms by a simple bend model correspond to $\gamma\psi = 123°$ for BL Lacs, $37°$ for quasars and $36°$ for CSS sources, where ψ and γ are the jet typical intrinsic bend and Lorentz factor within a given class of sources (Appl et al, 1995). If, as currently thought, jets in BL Lacs have smaller Lorentz factors than in quasars, high intrinsic bending and misalignment appear to be the rule in BL Lac sources.

Another property of BL Lac objects which might be characteristic of a class of sources is that their magnetic configuration appears perpendicular to their VLBI jet, while it is commonly longitudinal in other radio sources and quasars (Gabuzda et al, 1992). A simple assumption is that VLBI data mainly reveals the magnetic structure of the external medium in which the nuclear beam is injected. Large scale galactic and intergalactic magnetic fields are not rare in the extragalactic space and likely favour formation of accretion discs in a plane perpendicular to them (Asséo, Sol, 1987). Thus one would expect to observe essentially parallel jets and magnetic fields (as for young stellar objects) if jets are launched perpendicular to discs. However this is no more true if the central black hole has formed much

R. Ekers et al. (eds.), Extragalactic Radio Sources, 477–478.

earlier or from a much smaller region, or has undergone some violent interaction with another galactic nucleus. It has in that case an independent rotation axis and drives the inner part of the accretion disc into its equatorial plane through the Bardeen-Petterson mechanism. An inner nuclear beam can then be injected at high angles to the ambient magnetic field. A wide range of parameters allows propagation and radiation of such beams which present the characteristic VLBI magnetic pattern of BL Lac jets (Sol, Vicente, 1994).

Central engines with highly twisted accretion discs therefore provide an interesting frame for a simple explanation of the magnetic peculiarity of BL Lac jets. Clearly they also lead to high distortion in the global radio morphology when jets are emitted perpendicular to the local plane of the disc. Following the view of "double jet" models presented by Dole et al (this symposium), the VLBI elongation is related to the black hole axis while the large scale radio structure is launched from the outer accretion disc in a direction determined by the large scale properties of the AGN surroundings. High apparent misalignement angles ΔPA and intrinsic distortion are thus directly expected. As a "by-product", this scenario suggests the existence of a population of sources with an intrinsic misalignment close to 90 degrees. Nuclear beams are then injected into the outer part of the discs and perpendicular to the ambient magnetic field, which likely slows them down very efficiently. Such a population provides a straightforward description of the intriguing secondary peak observed at 90 degrees in the histograms of misalignment angles ΔPA (Pearson, Readhead, 1988). Interaction of the nuclear beam with the outer disc and the transverse magnetic field likely favours strong variability, high polarization and enhancement of radiation which can give typical core-dominated properties to such sources (Appl et al, 1995).

Our proposition of highly twisted discs in BL Lac objects has another consequence. Their ionizing cone now intersects an ambient gas distribution with probably different temperature and densities and can induce specific properties of the spectral lines detected in these sources. This illustrates how evolution effects, interaction between galaxies and formation of central engines might be usefully taken into account besides beaming and orientation scheme for unification of active galactic nuclei.

References

Appl, S., Sol, H., Vicente, L., 1995, *A&A*, in press.
Asséo, E., Sol, H., 1987, Physics reports, 148, 307.
Gabusda, D.C., Cawthorne, T.V., Roberts, D.H., Wardle, J.F.C., 1992, *ApJ*, **388**, 40.
Pearson, T.J., Readhead, A.C.S., 1988, *ApJ*, **328**, 114.
Sol, H., Vicente, L., 1994, Multi-wavelength continuum emission of AGN, Courvoisier, T.J.-L., Blecha, A., eds, Kluwer Academic Publishers, p. 473.

DESCRIBING THE ASYMMETRIES IN EXTRAGALACTIC RADIO SOURCES

STANISŁAW RYŚ
Astronomical Observatory of the Jagiellonian University
PL – 30244 KRAKÓW, ul. Orla 171, POLAND,
E-mail: strys@oa.uj.edu.pl

As an ideally symmetrical regime we assume that central engine ejects the same volumes of radio emitting matter – hereafter called 'plasmons' – into two opposite directions. The plasmons have the identical velocities and evolution and are small. We assume that each plasmon on one side has its corresponding twin on the opposite side of the structure. Since in fact observations do not reveal any symmetrical structures we must include into our model some mechanism to reproduce an asymmetry. It is our assumption that such asymmetry is due to the fact that the observer sees the plasmons at their different evolutionary ages and that their velocity vectors have different directions. We are looking for the functions which describe differences between the twins. These functions give us differences in brightness and expected positions of the twins and may be used for interchanging their elements. This exchange leads to obtaining new structure. If this new structure is identical to the original one, then the functions correctly describe the differences between the two sides of the structure (Paper 1).

The *relativistic flip-flop model* described in Paper 2 was tested by applying it to the description of the asymmetry in 33 radio structures selected from the survey by Machalski and Condon (1983). The parameters which represent spatial interval between the switching of activities of the central engine (B) and the relative life-time (τ) are reasonably correlated (correlation coefficient $\rho = 0.59$). Taking into account this value of the correlation coefficient I suggest a linear dependence $B \sim \log(\tau)$ (Fig. 1). If we assume that all sources have the same physical values for twin plasmon delay and characteristic time scale of plasmon shining rate, we obtain simple interpretation of this dependence as follows: *Older structures have smaller parameters B and τ than younger structures.*

R. Ekers et al. (eds.), Extragalactic Radio Sources, 479–480.

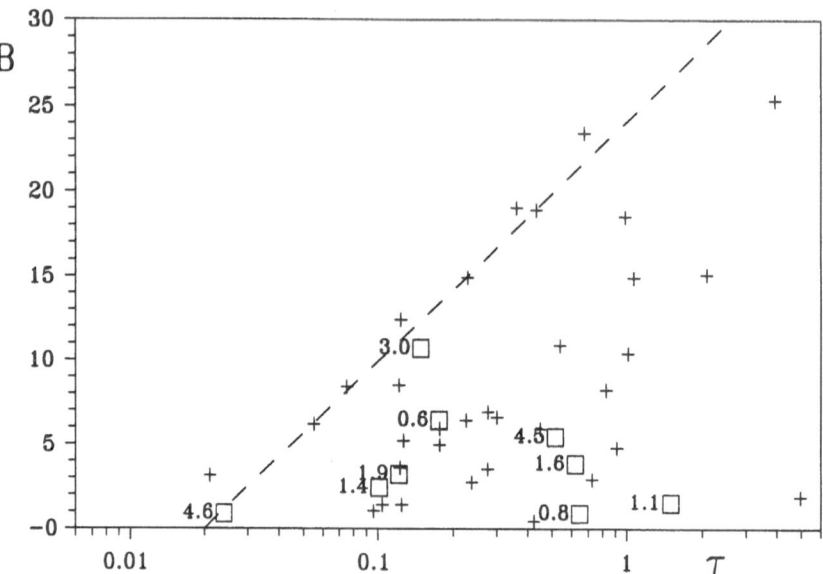

Figure 1. The relationship between B flip-flop parameter and the relative life-time in the model. The dotted line $(B = -0.17 + 8.54 \log(\tau))$ is an example of the evolutionary path of radio structure. $A\&L$ sources are marked by □ and numbers describes their age (in unit of 10^7 years).

In the sample investigated in Paper 2 there are no structures for which the age could be estimated from other independent investigations. Therefore I added nine sources from Alexander & Leahy (1987) sample (hereafter $A\&L$) marked by rectangles in Figure 1. The ages of radio sources determined by $A\&L$ give the sequence without agreement of ages determined by parameters B and τ. This result we can interpret in two ways:
- the simple interpretation is not valid (we need more than one physical model for description of asymmetries – Chyży 1994);
- the dynamical ages of radio structures are different from the values approximated from $A\&L$ (see J. Eilek in this proceedings).

Acknowledgments. I would like to express my gratitude to the IAU Secretary for travel grant. I thank my colleagues K.Maślanka and K. Chyży for fruitful discussion. The work was supported by KBN grant no. 578/P03/95/09.

References

Alexander P., Leahy J.P., (1987) *Astron.Astrophys.*, **225**,1
Chyży K.T., (1994), in "AGN Across the Electromagnetic Spectrum" ed. Courvoisier T.J.-L., Blecha A.R., (Kluwer) p.501
Machalski J., Condon J.J., (1983) *Astron.J.*, **88**,143
Ryś S.: (1994) *Astron.Astrophys.*, **281**,15 (Paper 1)
Ryś S. (1995) submitted to *Astron.Astrophys.* (Paper 2)

ANALYTICAL MODELLING OF LINEAR SIZE EVOLUTION OF POWERFUL RADIO SOURCES

KRZYSZTOF T. CHYŻY

Astronomical Observatory of the Jagiellonian University
ul. Orla 171, 30-244 Kraków, Poland

We assume that two continuous, supersonic jets of plasma advance into a uniform surrounding medium of constant density and pressure. In terms of fluid dynamics their propagation is described by the kinetic model suitable for high Mach number flows, which likely occur in edge-brightened powerful radio sources. In this approximation the jet flowing energy is dominated by the bulk energy of thermal material. The end of the jet enters the undisturbed ambient medium and forms a front shock, where the supplied bulk energy is *in situ* transferred to relativistic particles and magnetic field.

In order to describe the time evolution of a forming cocoon we can divide the source age T into consecutive time intervals Δt that are small enough in comparison with the source age but longer than a characteristic time scale of microphysics processes. During that interval, a piece of a jet material reaches the jet head and shocks. Hydrodynamic simulation and analytical model by Hardee et al. (1992) show that the evolution of such a shocked material is closely approximated by sideways expansion only. Geometrically, the shock region can be approximated by an expanding cylinder.

Following Eilek & Shore (1989) we assume that all kinds of energy in the shock region scale with the net input energy from the jet and that the factors of proportionality are constant during the whole radio source evolution considered. Taking into account the appropriate equations of state, we find the time evolution of all kinds of energy as well as the spectral luminosity of synchrotron emission in an expanding cylinder.

Finally we can evaluate the energy budget of the whole cocoon. A cocoon of the source of age T consists of $n = T/\Delta t$ volume elements. The total energy of the cocoon is simply a sum of energy from all individual elements. By increasing the number of division n, the sum can be replaced by the integral over the time and evaluated.

R. Ekers et al. (eds.), Extragalactic Radio Sources, 481–482.

The model is applied to the cosmological evolution of linear sizes of powerful radio sources. Linear size D of the model structure can be estimated as a product of the object's age and the longitudinal expansion in velocity of the cocoon. According to derived analytical formula the linear size of a radio source depends on the beam characteristics, the interaction of the beam with external medium, the source age T and the cosmological epoch (through redshift z). An important conclusion about this relationship is a power dependence of linear size on spectral radio luminosity P_ν and redshift:

$$D \propto P_\nu^{-0.27}(1+z)^{-2.1}. \tag{1}$$

The model formula can be compared with relations estimated independently for a sample of 152 powerful radio galaxies and 173 extended quasars Chyży & Zięba (1995). For observed radio galaxies, median linear sizes evolves with spectral radio power at 1.4 GHz and redshift as:

$$D \propto P_{1.4}^{-0.35\pm0.03}(1+z)^{-3.4\pm0.4}. \tag{2}$$

The sample of quasars yields the relation:

$$D \propto P_{1.4}^{-0.10\pm0.05}(1+z)^{-1.3\pm0.3}. \tag{3}$$

In that case, the indices in the model dependence locate between those estimated for radio galaxies and quasars. The good correspondence of the model and actual sources confirms in fact the correctness of the model assumptions and approximations.

The decrease of linear sizes with redshift as described by (1) results from the increase of density of external gas in the earlier cosmological epochs. However, two effects contribute to this evolution. The first one arises from a higher in the past ram pressure, exerted on a source by external gas. Carefully examination of (1) reveals that the contribution of this effect is 73% of the total linear size-redshift dependence. The remainder (27%) results from the redshift dependent radiation efficiency.

In the framework of the model, a slightly different relation of linear sizes for radio galaxies and quasars on redshift can be due to different cosmological evolution of the ambient gas density or longer duty cycle for radio galaxy central engines. Differences in luminosity relations can arise from shorter life time for more active objects.

Acknowledgments. This work was supported by the grant from Polish Committee for Scientific Research (KBN), grant no. PB/578/P03/95/09.

References

Chyży K.T., Zięba S. (1995) *A&A*, **303**, 420
Eilek J.A., Shore S.N., 1989 *ApJ* **342**, 187
Hardee P.E., White R.W., Norman M.L, Cooper M.A., Clarke D.A., 1992 *ApJ*, **387**, 460

HOW RADIO SOURCES STAY YOUNG

J.A. EILEK
New Mexico Tech
Socorro, NM, USA

1. The Sources Are Older Than They Look

In studying the dynamical evolution of radio galaxies, I find that the dynamical ages of the sources are older than the ages inferred from spectral breaks, typically by a factor ~ 10. This suggests to me that the spectral ages are misleading.

I model Type I sources as steady flows, described by energy and mass transport integrated across the tails. From this, data on surface brightness and tail width give me the mean velocity field as a function of distance down the tail, and thus the dynamical age of plasma along the tail. Assuming the magnetic field is initially in equipartion, and evolves by flux-freezing along the tail, also gives me the rate of spectral steepening predicted by the standard aging model. When I apply this (in Eilek 1996a) to the two-frequency data of O'Donoghue, Owen & Eilek (1990), I find that the dynamical age at the end of the tail is $\sim 10 - 100$ times longer than the inferred spectral age. That is, the spectrum steepens much more slowly than it should.

I model Type II sources globally, in terms of their linear size and lobe volume. I assume again they are driven by steady jets, supplying energy and momentum to the extended sources. The source size is determined by momentum flux, and the source volume by pressure-driven expansion. Measurements of these two quantities tell us the age and beam power of the source. Assuming the magnetic field in the lobe is supplied by the jet, I can again predict their synchrotron age from standard theory. When I apply these arguments (in Eilek 1996b) to the data of Alexander & Leahy (1986), I find they should be synchrotron old: their integrated spectra should be much steeper than observed. I also apply this analysis to modelling size and luminosity functions of nearby Type II sources, with the same conclusions: the spectra of these sources do not age as fast as they should.

R. Ekers et al. (eds.), Extragalactic Radio Sources, 483–484.

2. Possible Reasons

These contradictions are based on the standard model for synchrotron aging. This model assumes a uniform B field, particles well mixed with this field, and that only radiative (or adiabatic) losses affect the particle spectrum after its initial injection. These are restrictive assumptions: the situation might well be more complex. In particular the magnetic field is very likely not to be uniform throughout the source.

One possibility is *in situ* acceleration. While several types of reacceleration are possible, most share a characteristic result: acceleration in the presence of synchrotron losses leads to a high-energy break in the electron distribution, at some energy E_c where acceleration and loss rates balance. In diffuse radio sources, away from shocks, the acceleration is likely to be from stochastic turbulence. This can create an electron distribution peaked at E_c, rather than a power law (Borovsky & Eilek 1986). Such a distribution can, however, produce a power-law spectrum if the particles sit in an inhomogeneous B field with a power law volume distribution (Eilek & Arendt 1996). The rate of evolution of E_c depends on the local balance of turbulent or shock energy with the B field. Thus, when we measure the spectral steepening we measure the turbulence level, not the age of the plasma.

Another possibility is that the B field is inhomogeneous, and that the electrons move only slowly into high-field regions. Eilek, Melrose & Walker (1996) describe this. They first follow the time evolution of electrons diffusing into a high-field region. They calculate the spectral steepening rate; it is slower than would be the case if the particles were initially well-mixed with the strong field. They also follow the evolution of particles moving between the high and low field regions in a leaky box model. With injection in the low-field region, they find that spectral steepening stops when leakage into the high-field region balances synchrotron losses. At this point, E_c reaches a constant value. Using cross-field diffusion, they find that the diffusion rate in low turbulence levels predicts an E_c which is consistent with observed spectral breaks in the $\sim$ GHz range. Thus, when we measure the spectral steepening we measure the diffusion rate, not the age of the source.

References

Alexander, P, & Leahy, J. P. (1986), *MNRAS*, **225**, 1.
Borovsky, J. O. & Eilek, J. A. (1986), *ApJ*, **308**, 929.
Eilek, J. A. (1996a), submitted to *ApJ*.
Eilek, J. A. (1996b), in preparation.
Eilek, J. A. & Arendt, P. N. (1996), *ApJ*, in press.
Eilek, J. A., Melrose, D. B. & Walker, M. A. W. (1996), submitted to *ApJ*.
O'Donoghue, A. A., Owen, F. N. & Eilek, J. A. (1990), *ApJ Supp*, **72**, 75.

A COCOON TRANSPARENCY AND THE 3C 345 LOW FREQUENCY VARIABILITY

L.I.MATVEENKO
Space Research Institute of RAS, Profsojuznaja 84/32,
117810 Moscow, Russia

Abstract. The structure of the quasar 3C 345 is studied at $\lambda = 49cm$. The core has a low-frequency cut off spectrum and $\alpha \sim 3$. The brightest knot is in the nearest part of the jet. The cocoon wall absorbs low-frequency emission and changes the polarization orientation, RM = 3500 $rad\ m^{-2}$, $B_{||} \sim 100\ \mu$G at the core region.

The studies of the quasar 3C 345 fine structure at $\lambda = 49cm$ with a global VLBI network show that the core emission is weak [1, 2]. The core spectrum has a low frequency cut off and the spectral index in the optically thick part is $\alpha \sim 3$. The compact brightest component corresponds the nearest part of the jet, with a size of $\sim 5 \times 4mas$. In 1983.9-1990.8 the flux density of the component and the solid angle increased by a factor of $\sim$ 2. The brightness temperature was $T_b = 0.5 \cdot 10^{12}K$ and did not change significatively.

According to the black hole model, an accretion disk implies a surrounding medium in the azimuth plane, leaving a relatively free space in the direction of the rotation axis. The relativistic plasma is ejected along the axis, within an angle of $\leq 1str$ [3]. A magnetic field focuses the plasma into thin filaments. The rotation of the black hole (the ejector) twists the filaments around the axis [4, 5] and forms the spiral structure jet [6-8].

The ejector is the source of the synchrotron emission, i.e. what we call the "core". The is core located $\sim 15mas$ E of the brightness peak in the $\lambda = 49cm$ map [1, 2] and has a brightness temperature of $T_b \sim 10^{10}K$. The relativistic plasma flow is surrounded by thermal plasma in the form of a cocoon. The cocoon absorption of the core emission at $\lambda = 49cm$ is equal ~ 100 and the optical depth $\tau \sim 5$. The absorption at 6 cm will be $\leq 20\%$ and the cocoon wall is practically transparent at mm-cm wavelengths.

R. Ekers et al. (eds.), Extragalactic Radio Sources, 485–486.

The time scale of the low frequency variability is $t \sim 1$ yr [9] and the recombination time of the thermal plasma must be ≤ 1 yr. The recombination time is equal $t_r = 10^5 N_e^{-5}$ yr therefore $N_e \sim 10^5$. If the optical depth of the cocoon wall is $\tau \sim 1$, the wall thickness is $l \sim 10^{-3} pc$. The thickness increases with the distance from the core and the total number of thermal electrons in a column is $N_e \sim r^{-2}$. The transparency is determined by the emission measure EM $\sim N_e^2 l$ and varies as $\sim r^{-4}$. The rotation measure $RM \sim N_e B_{||} l$ varies as $\sim r^{-3}$.

The wall is transparent at the centimetre wavelengths, however, the screen can change the orientation of the polarization plane. According to VLBA measurements of 3C 345 with a beam size $\sim 5mas$ [10], the rotation measure in the 18-21 cm band is $RM \sim 28 rad\ m^{-2}$ and the degree of polarization is $\sim 4\%$. The size of polarized region is equal $\sim 5mas$ [11]. The brightest emission region at 18 cm corresponds to the nearest part of the jet [1, 2]. We propose that the polarization emission at 18 cm arises from the same region. The polarization position angle, corrected for the Faraday rotation, gives the position angle of the magnetic field and the jet orientation at the distance of $\sim 5mas$ from the core. The rotation measure, RM $\sim r^{-3}$, will be 3500 $rad\ m^{-2}$ at the core region and the magnetic field is $B_{||} \sim 100\ \mu G$.

The changes in the cocoon wall transparency will change the polarization position angle and the low frequency emission.

This work supported by the Soros's ISF, Grant MFR300.

References

[1] Matveenko, L.I., Graham, D.A., Pauliny-Toth, I.I.K., et al. (1992) *Pis'ma v Astron. Zh.*,Vol. **18**,p 931

[2] Matveenko, L.I., Pauliny-Toth, I.I.K., Baath, L.B., et al. (1996) *Pis'ma v Astron. Zh.*,Vol. **22**,p in press

[3] Begelman, M.C., Blanford, R.D. Rees, M.J. (1981) *Rev.Mod.Phys.*, Vol. **56**,p 255

[4] Shakura, N.I. and Sunyaev, R.A. (1973) *A&A*,Vol. **24**,p 337

[5] Lovelace, R.V. and Berk, H.L. (1991) *Astroph. J.*,Vol. **379**,p 695

[6] Unwin, S.C. and Wehrle, A. (1992) *Astroph.J.*,Vol. **398**,p 74

[7] Krichbaum, T.P., Witzel, A., Graham, D.A.,et al. (1993) *A&A*, Vol. **275**,p 375

[8] Zensus, J.A., Cohen, M.H., and Unwin, S.C. (1995) *Astroph. J.*, Vol. **443**,p 35

[9] Padrielli, L., Eastman, W., Gregorini, L., et al. (1991) *A&A*, Vol. **249**, p 351

[10] Rudnick, L., and Jones, T.W. (1983) *Astron.J.*,Vol. **89**,p 518

[11] Browne, L.F., Roberts, D.H., Wardle, J.F.C. (1994) *Astroph. J.*, Vol. **437**,p 108

SIMULATION OF THE VARIABLE MULTIFREQUENCY RADIO EMISSION AND STRUCTURE OF THE QUASAR 2145+067

Y.Y. KOVALEV
Moscow State University — Vorobjevi gori, 119899 Moscow;
Astro Space Center of the Lebedev Physical Institute —
Profsoyuznaya 84/32, 117810 Moscow, Russia

Here are reported new successful results of analysis for jet in the strong radial magnetic field of an active galactic nuclei, suggested in [1].

15–year observations of the flux density of the quasar 2145+067 at up to 10 frequencies between 0.3 and 230 GHz from [2–4] and other are analyzed. These observations are compared with the calculations from the model, as in [5]. In general, observational curves of flux density and combined spectra versus time are in agreement with fitted model curves (see some of our results in Figure 1a–h).

The structure of this quasar, as it would be observed using a real VLBI beam, is calculated. The calculated structure (Figure 1i) is in qualitative agreement with the observed map for the used epoch from [6].

It is concluded that both the variable emission and the structure of the quasar 2145+067 can be explained by this model. Using fitted model parameters, the luminosity distance to the source has been preliminary estimated as 1200 Mpc. On this reason, we select the source 2145+067 as suitable for radio measurements of the Hubble constant and the deceleration parameter of the Universe by an earlier suggested method [7].

References

[1] Kardashev, N.S. (1969) Epilogue to Russian edition of Burbidge, G.R. and Burbidge, E.M. (1967) Quasars, Freeman. Mir, Moscow.
[2] Mitchell, K.J., et al. (1994) Astrophys. J. Suppl. Ser., Vol. no. 93, p. 441.
[3] Aller, H.D., et al. (1985) Astrophys. J. Suppl. Ser., Vol. no. 59, p. 513.
[4] Teräsranta, H., et al. (1992) Astron. Astrophys. Suppl. Ser., Vol. no. 94, p. 121.
[5] Kovalev, Y.Y. and Larionov, G.M. (1994) Astron. Lett., Vol. no. 20, p. 3.
[6] Wehrle, A.E., et al. (1992) Astrophys. J., Vol. no. 391, p. 589.
[7] Kovalev, Yu.A. (1994) Astron. Astrophys. Tr., Vol. no. 5, p. 67.

R. Ekers et al. (eds.), Extragalactic Radio Sources, 487–488.
© 1996 IAU.

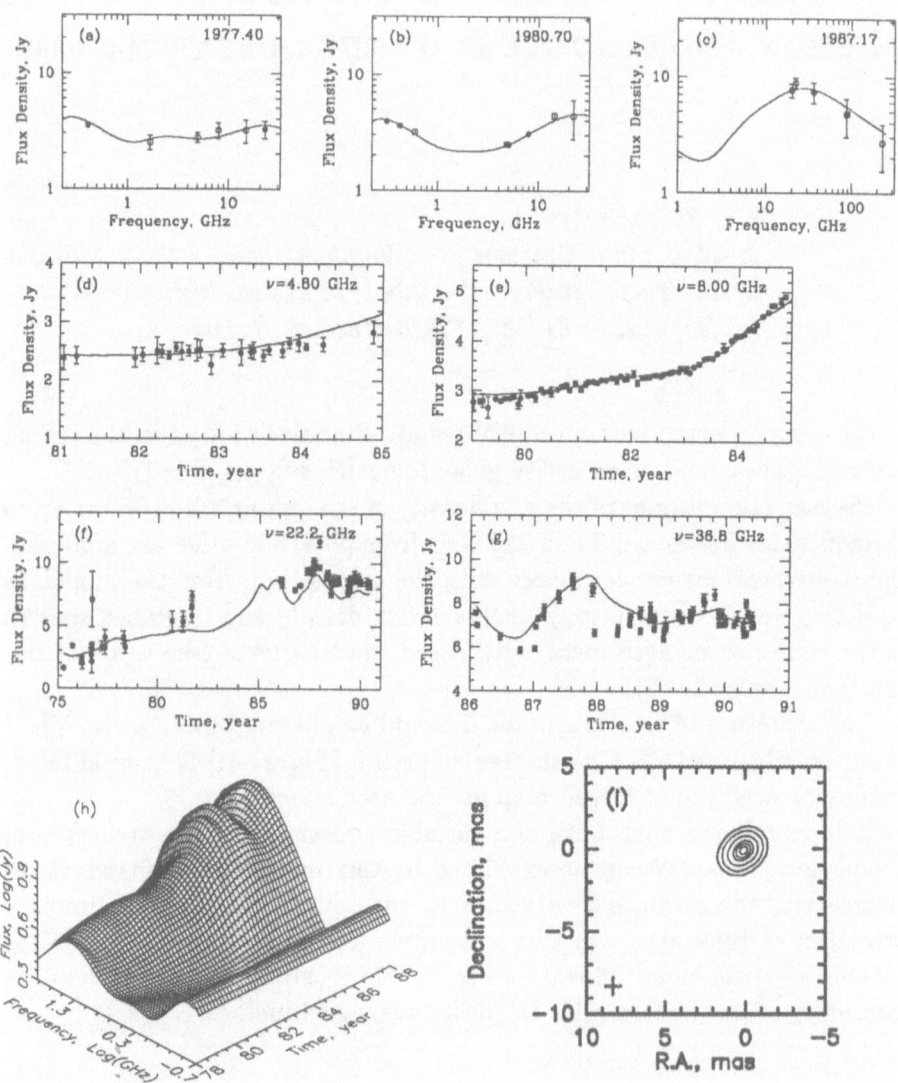

Figure 1. Some results of simulations of the variable multifrequency radio emission and structure for the quasar 2145+067. Quasi–simultaneous spectra for epochs 1977.40 ÷ 1987.17 (a),(b),(c). Dots represent measured or interpolated data. Results of model calculation are shown by solid lines. The evolution of the flux density in time for the frequencies $\nu = 4.8 \div 36.8$ GHz (d),(e),(f),(g), based on the observational data (dots), and model calculations (solid lines). The evolution of model spectra in time (h). Image of the jet in the model for the epoch June 6, 1986 (i) to compare with Figure 2m from [6]. Contours are shown at 1, 5, 20, 70, and 90 % of the peak brightness of 2.03 Jy.

THE SPECTRAL AND SPATIAL DECOMPOSITION
OF EXTRAGALACTIC RADIO SOURCES

L. RUDNICK[1] AND D.M. KATZ-STONE[1,2]

[1] University of Minnesota - Dept. of Astronomy 116 Church St. SE, Minneapolis, MN 55455 USA

[2] U.S. Naval Academy 572 Holloway Road, Annapolis, MD 21402-5026 USA

We have developed a new set of techniques for extracting information from maps of synchrotron emission. Some of these methods are used to determine the shape of the spectrum and then to isolate the contributions of the number and energy of relativistic electrons and the local magnetic field strength to the total intensity images. These are discussed in Katz-Stone, Rudnick & Anderson (1993), Katz-Stone & Rudnick (1994), and Rudnick, Katz-Stone & Anderson, (1994). Here, we describe a method - called spectral tomography - to remove confusion from various features along the line of sight. We find that both structural and spectral confusion are commonplace, and therefore our dynamical and radiative models of extragalactic sources need significant revision.

To conduct a tomography analysis, start with two matched maps $B_{\nu 1}(x, y)$ and $B_{\nu 2}(x, y)$. Then, form a set of tomography maps:

$$B_t(x, y) \equiv B_{\nu 1}(x, y) - t \times B_{\nu 2}(x, y)$$

for a range of values of t. Suppose now that we have a feature, A, with a spectral index α_A. For the tomography map where $\alpha_t \equiv log(t)/log(\frac{\nu 1}{\nu 2}) = \alpha_A$, the feature A will then completely disappear from the map. If another feature is along the same line of sight with A, and has a different spectral index, then it can be seen without confusion (for Cygnus A results, see Rudnick & Katz-Stone, 1996).

The figure below shows part of the spectral tomography analysis for the wide-angle-tail source 1231+674. Each line represents a slice at the same

R. Ekers et al. (eds.), Extragalactic Radio Sources, 489–490.
© 1996 IAU.

declination in the southern tail, for a series of tomography maps B_t. In the top slice, we see a broad peaked feature commonly associated with the flaring jet. In the middle slices, corresponding to a tomography spectral index ≈ 0.5, the central peaked feature disappears, leaving a broader, flat-topped envelope. Below this, the central feature goes negative while the envelope remains positive. The tomography maps show that this behavior is due to a flatter spectrum jet surrounded by a steeper spectrum sheath.

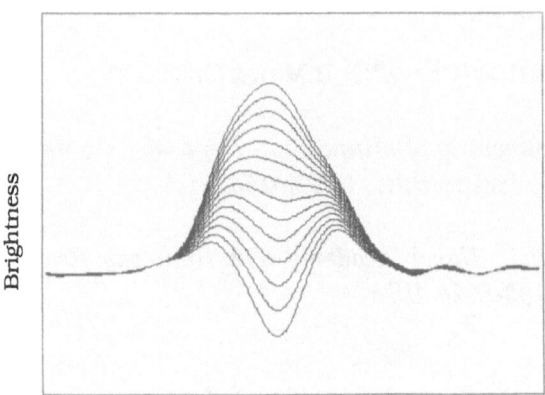

We find similar features in 3C449 and another WAT, 1433+553. After separating the jet from the sheath, there is little or no spectral aging of the jets, even though they *appear* to age when confused with the steep spectrum sheaths. This raises both dynamical questions regarding the origins of the sheaths, (e.g., turbulent layers - Bicknell, 1994 or two-component outflows - Sol, Pelletier & Asseo, 1989), and a need for new aging analyses. Our tentative conclusion is that adiabatically shifted curved electron populations and magnetic fields may dominate the effects we commonly call aging.

ACKNOWLEDGMENTS. This work is supported in part at the University of Minnesota through US NSF grant AST93-18959, using data obtained at the Very Large Array, a facility of the NRAO, operated by AUI under contract with the U.S. NSF. We appreciate the sharing of data by A. O'Donoghue (1231+674, 1433+553), L. Feretti and R. Perley (3C449) and good discussions from too many people to fit here.

References

Bicknell, G.V., 1994, In *ASP Conference Series*, (ed. G. Bicknell, M. Dopita, & P. Quinn) Vol. **54**, 357, ASP.

Katz-Stone, D.M., Rudnick, L. and Anderson, M.C. (1993), *Ap. J.*, **407**, 544

Katz-Stone, D.M. & Rudnick, L., 1994, *Ap. J.*, **426**, 116.

Rudnick, L., Katz-Stone, D.M. & Anderson, M.C., 1994, *Ap. J. Suppl. Ser.*, **90**, 955.

Rudnick, L. & Katz-Stone, D.M., 1996, In *Green Bank Workshop on Cygnus A*, (ed. C. Carilli & D. Harris),58, Cambridge University Press

Sol, H., Pelletier, G., & Asseo, E., 1989, *M.N.R.A.S.*, **237**, 411

RADIO SOURCES MODELLING AND EMISSION MECHANISMS
(*DISCUSSION*)

Discussion of the paper presented by <u>*MELROSE*</u> (p. 423)

Perlman: One of the results from my own observations of PKS 1413+135, one extragalactic flat-spectrum source (which unfortunately is not part of my poster) is that the angular size of the inverted-spectrum core at 13 and 18 cm is in fact proportional to $1/\nu$. I can show you the data later.

Melrose: I am interested to see your data.

Cotton: In Cotton et al. 1980 the flat-spectrum source 0735+178 was observed interferometrically and decomposed into a number of components which were each well represented by a simple, synchrotron model where each component had a peaked spectrum but which added to a flat spectrum. This lead to the term "cosmic conspiracy" as the components "know" about each other.

Melrose: My remarks on the "cosmic conspiracy" refer to the specific model in which the flux density depends only on the product of frequency and radius. I do not believe this mode; can be the correct explanation for flat spectra.

Dogiel: The spectrum generated by multiple shock acceleration strongly depends on the age of an emitting object. As it was shown by Bynol and Toptygin for the case of OB-associations, there is an energy range where the spectrum is $E^{-1}(E < E_0)$ and at higher energies $E > E_0$ the spectrum of electrons is very steep. The position of the break E_0 is a function of time, $E_0 = E_0(t)$.

Melrose: I agree. There is a high energy portion of the energy spectrum whose spectral index is determined by the compression ratio of the shocks. The hard-spectrum with $a \approx 1$ forms at low energy and moves to higher energy with increasing number of encounters with shocks.

Pohl: Inverted non-thermal emission is not only observed from the filaments but also from the diffuse region in the arc, where the B-field is

R. Ekers et al. (eds.), Extragalactic Radio Sources, 491–494.
© 1996 IAU.

much lower. Would your process also work there?

Melrose: I am aware of the observations to which you refer, and my model does not explain such diffuse emission in a natural way. I shall be interested to see whether this observation of diffuse emissions is confirmed.

Pohl: AGN radio outbursts have in the optically thin part of the spectrum an electron spectral index of more than 1 ($\alpha > 0$). On the other hand, in view of the high synchrotron and inverse-Compton losses one would expect the blocking effect and hence the pile-up to be very strong. Isn't this contradictory?

Melrose: I don't see any contradictions. Schlickeiser (1984) has proposed a pile-up model, similar to the one I am suggesting, for optical/IR emission with an inverted spectrum. Note that high synchrotron or inverse Compton losses implies a pile-up only if it occurs at the same time and place as Fermi acceleration. No pile-up occurs for diffusive shock acceleration.

Falcke: I don't see what the GC filaments have to do with the cosmic conspiracy!? Those filaments are associated with large scale magnetic field lines and molecular clouds and are very different from the situation in the AGN. In fact, the only source similar to an AGN, Sgr A* , can well be explained by a "cosmic conspiracy" theory type model. (A&A 278,L1)

Melrose: My view is that the GC sources with flat spectra require hard electron spectra. It seems relevant to ask whether the explanation for flat spectra in the GC might not also be relevant to AGN. I am suggesting that in the absence of clear evidence for self absorption in flat spectral sources, the default explanation should be a hard electron spectrum rather than a "cosmic conspiracy".

Discussion of the paper presented by MEIER *(p. 433)*

Sunyaev: What is the ratio of the magnetic field pressure to the plasma pressure in the disk you assume in your computations as initial value?

Meier: Formally this ratio is infinitesimally small. The *disk* itself in the simulations is infinitely thin and anchors the magnetic field. The material being accelerated is the disk *corona*. The ratio of the magnetic pressure in the corona to the coronal gas pressure is a parameter in our models and varies from about 1/2 to 2. Note that this ratio can be even higher without the field leaving the system because of our assumption of a cold, field-anchoring thin disk below the corona.

Begelman: Do the jets remain well collimated – solely by magnetic tension well beyond the critical surfaces, or is external confinement necessary?

Meier: Yes, the jets remain well collimated, mainly by magnetic tension, as far from the central objects as we have computed them ($\sim$300 gravitational radii), especially in their inner core. For high external pressures, the cocoon appears to be largely pressure confined. However, for very low pressure the cocoon itself, though much larger than in the high pressure case, appears to be magnetically confined. Note that in this mechanism, as opposed to the classical Blandford-Payne process, the critical surfaces are all very close to the disk. The material rises from the disk with trans- or super Alfvenic velocities rather than accelerating slowly through several critical surfaces over a large distance. *Discussion of the paper presented by BEGELMAN* (p. 441)

Romanova: How do you explain the optical and X-ray radiation of M87?

Begelman: The optical and X-ray emission from the knots is readily explained as synchrotron radiation.

Discussion of the paper presented by DE YOUNG (p. 461)

Trussoni: Is the total jump of velocity across the sheer layer subsonic or supersonic?

De Young: In this case it is subsonic. However, this may in fact be the case, since the boundary layer is between the jet and the hot post shock gas behind the bow shock. In these regions the sound speeds are very high.

Eilek: To what extent does the vortex saturation, and thus the asymptotic entrainment rate, depend on the viscosity of the fluid?

De Young: These are very high Reynolds number flows. Empirically, it appears that the complex turbulent dissipation occurs deep inside the large scale structures; hence viscosity is important there. However, the growth and saturation of the large scale structures occurs on a much larger scale and is unaffected by viscosity.

Discussion of the paper presented by FERRUIT (p. 465)

Wilson: Does your model include ionizing radiation from the bow shock or is the ionization completely dominated by photons from the AGN?

Ferruit: No, the only ionization taken into account is from the AGN (no pre-ionization of the ISM by the shock radiation itself.)

Steffen: Do you include thermal mixing of the grid parcels as they move along the bowshock? Do you expect any significant influence of this effect?

Ferruit: No, we did not include any mixing. We can expect a significant effect in the tail where turbulence is likely to occur, but there, the temperature differences between particles are lower. *Discussion of the paper presented by* <u>*EILEK*</u> (p. 483)

Meisenheimer: We do see the spectral break in the lobes of M87 and it is still in the NIR/optical frequency range.

Eilek: Yes, that is an important test case for these questions.

Aller: Do we ever see true breaks in spectra?

Eilek: We do commonly see high-frequency spectral turnovers; but it's not easy to determine the exact shape of the high-frequency spectrum, where the source becomes very faint.

Aller: What about the suggestion that particle energies are redistributed in situ in shocks such that the resulting spectral index is set by the "local" jump conditions.

Eilek: That is certainly an important case of local reacceleration – although I suspect its more relevant to jets and hot spots than to the more diffuse lobes and tails.

Discussion of the paper presented by <u>*RUDNICK*</u> (p. 489)

Ekers: Why call this "tomography"? It is not the usual meaning of the word.

Rudnick: "Tomography" comes from the Greek word "tomos", meaning a section or cut. We use the word deliberately because of its connotation of uncovering hidden features. However, to avoid confusion with the conventional usage, we call the difference maps between two frequencies "spectral tomography", while a related technique, which we introduce elsewhere, we call "polarization tomography".

Ekers: But it still has nothing to do with the process of reconstruction of an object from its "sections" or "cuts".

THE WESTERBORK NORTHERN SKY SURVEY

A.G. DE BRUYN

NFRA, Dwingeloo and

Kapteyn Astronomical Institute, Groningen

Postbus 2, 7990 AA, Dwingeloo, The Netherlands

1. Introduction

The WEsterbork Northern Sky Survey (WENSS) is a large–sky survey being carried out at 92 and 49 cm with the Westerbork Synthesis Radio Telescope (WSRT). At 92 cm WENSS will cover the sky north of declination +30° (an area of 10,000 square degrees) to a limiting (5σ) flux density of about 15-20 mJy. Both linear and circular polarization data is obtained. The polarization images have a noise level of about 2 mJy (1σ). The spatial resolution at 92 cm is about 1′. At 49 cm about one fifth of this area will be covered to somewhat lower flux density and with twice the resolution. The resulting catalogue will contain about 250,000 sources at 92 cm and 30,000 sources at 49 cm.

We expect to finish the observations for the WENSS project in early 1996. The final reductions should be completed at the end of 1996. We expect that the first results of the survey will become available to the as- tronomical community sometime in 1996. The WENSS product will consist of a catalogue of radio sources extracted from the survey and a set of FITS images (each 1024x1024 pixels covering 6°x6° at 92cm). We plan to make them available in both digital (DAT/CD-ROM) and graphical form (atlas). Images will be centered at the locations for the new Palomar Observatory Sky Survey plates. A variety of low resolution images will be made as well to facilitate comparison with other surveys.

2. WENSS data processing

To image large areas of the sky with the WSRT (an east-west synthesis array) within a reasonable amount of time we have to make use of the mosaicing technique whereby the array is repeatedly stepping through a

R. Ekers et al. (eds.), Extragalactic Radio Sources, 495–498.

fixed pattern (a mosaic) on the sky in a relatively short time. At 92 cm the pointing grid has a stepsize of about 1.3°, half the primary beam width, resulting in very uniform sensitivity. In 40 minutes the 14-element array is steered through a mosaic with 80 pointing centres covering a region of about 10°×13° in size. At each pointing we integrate for 20 seconds followed by 10 seconds to move the array to the next field. In 12 hours we thus accumulate 18 observations at each pointing. In order to bring down the sidelobe confusion to acceptable levels each mosaic is observed for 6 periods of 12 hours, with different array configurations, to provide radial sampling in steps of 12 meters. This typically takes six weeks. The starting field during each 12 hour observation is chosen such that we get uniform coverage of the UV-plane leading to very low (1%) sidelobe levels in the dirty images. The net observing time spent on each field is 36 minutes. The sky has been divided into four declination zones. The lower two zones ($+30° < \delta < +43°$, and $+43° < \delta < +56°$) each have mosaics containing 8×10 fields. The high declination zones ($+56° < \delta < +75°$, and $\delta > +75°$) have more complicated pointing patterns resulting in a slight non-uniformity in the sensitivity across the mosaics. The lowest three zones were observed with a bandwidth of 5 MHz split into 7 channels of 0.625 MHz each. The polar cap zone, about 10% of the survey, will be observed in the winter of 1995/96 with a recently commissioned broadband 92cm system which has 8 bands of 5 MHz spanning the range from 307 to 385 MHz. The noise in this part of the survey is expected to be lower by about a factor of 2 (depending on the highly variable RFI interference levels). Due to the large spectral baseline this part of the survey may also be expected to yield useful spectral information for the brighter sources.

The reduction of the data is done in Dwingeloo on a dedicated HP730 workstation with about 3 Gbyte of disk space. Early 1995 a second HP715 workstation with 6 Gbytes of diskspace was added to speed up processing. All data are phase-selfcalibrated to remove ionospheric and instrumental phase errors which dominate the raw images at low frequencies. Fields with strong sources are also selfcalibrated in gain. The positions are calibrated using point sources from the Jodrell-VLA-Astrometric Survey (JVAS) augmented with WSRT 21 cm positions of a few hundred pointlike WENSS-sources. This has enabled us to reduce the systematic position errors to about 1″ across the sky. Subsequent analysis is done at Leiden Observatory. In addition to the large mosaic images we also construct 'frames' of 1024x1024 pixels, separated by 21.1″, yielding images of 6°×6°. After converting them to FITS-format they are sent to Leiden where they are searched for discrete sources. Both the peak flux and integrated flux densities are determined.

3. Scientific drivers and first results

The low frequency, positional accuracy, polarization information, angular resolution and sensitivity to large scale structure make WENSS an important and fundamental database for tackling a wide range of astronomical problems. Below we will describe the science drivers and the projects that we have already started. For many of them first results are becoming available and have led to publications now in press. For lack of space we can not show them here (for a colour display of some of the results we refer to the 1994 Annual Report of the NFRA). Although the dataprocessing for the project is still in full swing, and will remain so for one more year, the scientific exploitation has taken off in a very significant way. Several Ph.D. projects have started in 1995 making use of the mosaiced images and the preliminary catalog.

Below we will describe some of projects for which WENSS can be (and is being) used and which have already led to some exciting discoveries.

radio spectra: WENSS will provide spectral information both internally (325-610 MHz for about 2000 square degrees, 307-385 MHz for the polar cap) and by comparison with radio surveys at other frequencies. In combination with available (6C/7C at 151 MHz, GB6 at 5 GHz) and ongoing (VLA-B/D at 1.4 GHz) surveys this will permit the study of very large numbers of (ultra-)steep, flat and inverted spectrum radio sources:
- Ultra-steep spectra sources with indices between -1.3 and -3. Such spectra are often seen in the most distant radio galaxies, in radio sources which populate rich clusters and in pulsars.
- Flat spectrum sources at low flux levels (25-100 mJy). One of the many uses of such a sample (and one already started in 1994) is the search for radio-loud gravitationally lensed objects (the CLASS project, in collaboration with Jodrell and Caltech astronomers). The flat spectrum sources will also be used to define samples of high-redshift quasars selected in an obscuration free manner.
- Peaked-spectrum sources with maxima in their spectra at a few 100 MHz (CSS peaker) and a few GHz (GPS peaker). This is a little-studied but important class of extragalactic radio source which have typical sizes of ten to a few hundred parsecs.

positions: Apart from reaching fainter sources the WENSS will also yield excellent positional information (from 5-10″ for the faintest sources to better than 2″ for the brighter ones). In a large fraction of the sources this will be sufficient for obtaining optical identifications.

polarization: The sensitive polarization information coupled with the large number of sources give WENSS unique capabilities in searching for radio sources having (anomalously) high linear polarizations at low frequencies. These include pulsars as well as interesting variable extragalactic radio sources. The sensitivity to extended structure (up to one degree, shortest spacing 36 meters) has made it possible to study the large scale distribution of diffuse polarized galactic foreground emission (cf. Wieringa et al. A&A 268, 215, 1993), and will lead to a panoramic view of the Faraday rotation in the magneto-ionic medium within about 1-2 kpc from the sun.

variability: Although not primarily intended to search for variability, the mosaicing technique on which WENSS is based means that information on source variability is available on a variety of timescales ranging from hours to several months. The edges of adjacent mosaics, observed in different years of the survey, will also contain information about very long term source variability. The famous low-frequency variable 4C38.41, recently discovered to be a GRO γ-ray source, decreased its flux density by 20% in just 6 weeks (Peng and de Bruyn, A&A, 301, 25). The bright pulsar B0329+54 revealed itself in the WENSS data through non-cancelling grating rings, caused by 40% variations in its flux density on time scales of weeks.

statistical studies: A combination of WENSS with existing large-sky radio catalogues will produce radio colour-colour diagrams which will enable large numbers of all these sources to be selected to flux-levels fainter by at least an order of magnitude than was previously possible. Using the radio spectral information these various types of sources can be separated. This should provide valuable new data about the evolution of the space density of distant galaxies. In addition, WENSS will allow for the first time studies of large-scale clustering of radio sources to be made which take into account the radio colour discriminant and optical identification information.

giant radio galaxies: The WENSS survey has already detected many tens of radio sources with angular sizes of 10' or more. Most of these will probably turn out to be giant radio galaxies with linear size in excess of 1 Mpc. One of the first giants followed up was identified with a broad-lined Markarian galaxy with a linear size of about 2 Mpc (Röttgering et al, MNRAS, in press).

4. The survey project team

The WENSS project is an NFRA effort requiring the help from a large group of people. In addition to the author the WENSS team consists of George Miley, Yuan Tang, Roeland Rengelink, Martin Bremer, Malcolm Bremer, Huub Röttgering, Klaas Weerstra and Hedy Versteeghe. The calibration and reduction of WENSS data motivated the development of a sophisticated novel software package (NEWSTAR) as well as automated batch-processing procedures. These are constantly being refined with the expert help of the NEWSTAR project team.

FIRST RESULTS FROM THE VLA FIRST SURVEY

R.H. BECKER AND M.D. GREGG
IGPP, LLNL, L-413 Livermore, CA 94551-9900

D.J. HELFAND AND C.M. CRESS
Columbia University, 538 West 120th Street NY, NY 10027

R.L. WHITE
STScI, 3700 San Martin Drive, Baltimore, MD 21218

AND

R. MCMAHON
IoA, Madingley Road, Cambridge, CB3 0HA, U.K.

Abstract. The VLA *FIRST* survey is now in its second year. We have completed mapping over 1500 deg^2 of the North Galactic Cap and present here the catalog of the 138,000 radio sources detected therein. We discuss the statistics of this new catalog including the two-point angular correlation function for all radio emitters, present our optical identification of 24,000 sources using the APM catalog, and report followup studies on radio variability, X-ray source identification, and our bright quasar sample.

1. Introduction

Our survey to collect Faint Images of the Radio Sky at Twenty-cm, *FIRST*, began in April of 1993. Using the VLA in its B configuration, we have completed two seasons of observations and are about to embark on a third. The observations consist of 165 s snapshots on a hexagonal grid of overlapping pointings; 11,000 such grid pointings have been collected to date covering 1,555 deg^2 of the north Galactic cap between 28 deg and 42 deg north. The individual images are weighted and combined to yield the final coadded survey images which are archived at the NRAO and the STScI. The images have a mean rms of 0.13 mJy and pixel sizes of 1.8″. The intent is for the

R. Ekers et al. (eds.), Extragalactic Radio Sources, 499–502.

completed survey to cover the $10,000\, \mathrm{deg}^2$ region of the proposed Sloan Digital Sky Survey.

We have recently published a comprehensive description of the *FIRST* survey (Becker, White, and Helfand 1995a). Today (14 October 1995), we are releasing a source catalog containing over 138,000 sources which have been extracted from these images. The catalog is essentially complete for point sources with peak flux densities > 1 mJy. Extensive astrometric checks have demonstrated systematic position errors of < 0.05″ and 90% confidence error circles of < 1″ for all sources down to the survey limit. Morphological information on sources with sizes > 2″ is also included. The catalog is available from the www site http://sundog.stsci.edu/.

2. The 1994 Catalog

The best way to place this catalog in context is through comparisons with other existing astronomical catalogs. For example, we have made a detailed comparison between the *FIRST* catalog and the APM catalog of optical objects from the POSS I. These comparisons have confirmed the subarcsecond accuracy of the radio positions, and have allowed us to find optical counterparts for ∼ 24,000 radio sources, by far the largest set of radio identifications in existence. For offsets of ≤ 1″, the false matching rate is only 2%. In general, we find that 17% of the radio sources have optical counterparts to the plate limit of the POSS I ($E = 20.0$). The color and morphological information available in the APM catalog allow ready classification for many of the counterparts: blue stellar counterparts have a high probability of being quasars, while red, extended objects are typically radio galaxies at redshifts less than 0.3.

The *FIRST* catalog also contains counterparts to 17% of the X-ray sources in the *ROSAT* WGACAT, providing arcsecond positions for over 1000 serendipitous X-ray emitters which fall in the 190 PSPC pointings covered by the *FIRST* survey to date. This reduction by a factor of 10^2 to 10^3 in the error circle areas for this faint set of X-ray sources will be of great value in optical identification programs. We have also performed comparisons to various stellar catalogs; a description of our successful efforts to find new stellar radio sources is presented in Becker et al. (1995b).

3. The FIRST Bright QSO Survey

A year ago we initiated a program to identify optically bright QSOs detected by the *FIRST* survey. The QSO candidates were identified by comparing the *FIRST* catalog with the APM catalog. Our original selection criteria included all objects brighter than 17.5 mag on the E plate of the POSS I which were classified as stellar in appearance on either plate and which were within 2 arcsec of a *FIRST* radio source. These criteria selected 219 QSO candidates from the 1993 survey area (305 deg^2). To date we have obtained spectra of 151 of the objects; another 25 were readily classified from the literature. Of these 176 objects, there are 69 QSOs, 2 blazars, 32 narrow line galaxies, 41 normal galaxies, and 32 stars. Despite the bright magnitude limit, only 15 of the 69 QSOs found in this 300 deg^2 region had been catalogued previously. Two of the new objects are brighter than 15 (one of which is at a redshift of 0.9), suggesting that the *FIRST* bright quasar survey will be a rich source of targets for HST and FUSE to use in absorption line studies.

An analysis of these initial results indicates that, for the coming observing season, we can define a more stringent set of criteria for the bright QSO search that will be 70% efficient and 95% complete by requiring a 1.1 arcsec match between radio and optical objects, and by excluding objects redder than O-E of 2.0. In the next year the number of QSOs from this project should increase by a factor of 5.

4. Large Scale Structure

The two-point angular correlation function for optical galaxies has been used for twenty years to explore the large scale structure of the Universe. Attempts to calculate this correlation function for radio catalogs have been frustrated by the low source surface density and the large range of redshifts characteristic of a flux-limited radio samples. Peacock and Nicholson (1991) did find a signal for luminous low-redshift radio galaxies, and Kooiman et al. (1995) have found a marginally significant correlation using the 6cm Greenbank survey. The fifty-fold improvement in sensitivity and angular resolution offered by the *FIRST* catalog makes it a logical choice with which to pursue a study of the large-scale structure of the Universe by applying angular correlation function analysis.

We have begun such an effort, and the preliminary results are reported in Cress et al. (1995). Based on the narrow ($\sim 3°$) strip from the 1993 dataset alone, a highly significant correlation function was detected on scales from several arcminutes to several degrees with a slope $\gamma \sim 0.9$, slightly larger than, but marginally consistent with, the value observed for optical galaxy samples. One of the more interesting results of this analysis was the higher correlation amplitude observed for a sample of radio doubles. Whether this result is a consequence of the fact that these resolved sources are, on average, closer, or are, in fact, more highly clustered remains uncertain. We are now extending this work to the new 1500 deg^2 dataset, as well as exploring the use of model radio luminosity functions to derive the 3-D spatial correlation function from these data.

5. Variable Radio Sources

The grid of pointings for the *FIRST* survey is so dense that almost all sources detected by the survey are observed at least twice, allowing us to search for variability in the radio sky. The *FIRST* observations are taken along a single declination strip per observing session with sessions generally separated by 1 to 7 days. Hence the survey samples two time scales, 3 minutes between adjacent fields at the same declination, and several days between fields in neighboring declination strips. Using conservative selection criteria, we have found approximately 400 sources (0.3% of the total) which appear to vary by more than 25%. Strong confirming evidence for the unusual nature of these objects comes from an examination of the optical counterparts of these putative variables. While only 17% of all *FIRST* sources have APM counterparts, the variable sources have counterparts at over twice this rate, and the magnitude distribution is skewed toward bright magnitudes. There is no way that an error in the radio data could produce a positive correlation with the optical sky. Followup observations are planned.

6. Summary

The unique combination of positional precision, depth, and angular resolution provided by FIRST affords us the opportunity to pursue a wide range of scientific programs. In addition to the bright quasar survey, the study of rapid variability, and the analysis of the two-point correlation function described above, a number of other projects are in progress. We are currently obtaining images and spectroscopy of fields containing bent head-tail radio sources in a search for high redshift clusters, examining gravitational lens candidates found in our images, and following up blank *IRAS* fields in a search for high-redshift starburst galaxies. A much larger set of projects can certainly be envisioned, and we hope that our prompt public release of the images and catalogs from *FIRST* will stimulate interest in pursuing them among members of the broader astronomical community.

References

Becker, R.H., White, R.L. and Helfand, D.J. (1995a), *ApJ*, **450**, 559

Becker, R.H. et al. (1995b), in *Radio Emission from Stars and the Sun*, eds. J.M. Paredes and Russ Taylor.

Cress, C.M. et al. (1995), in *Clusters, Lensing, and the Future of the Universe*, ed. V. Trimble, ASP Conf. Ser. (in press).

Kooiman, B.L., Burns, J.O., and Klypin, A.A. (1995), *ApJ*, **448**, 500.

Peacock, J.A. and Nicholson, D. (1991), *MNRAS*, **253**, 307.

THE NRAO VLA D-ARRAY SKY SURVEY (NVSS)

W. D. COTTON, J. J. CONDON AND Q. F. YIN
National Radio Astronomy Observatory [‡]
520 Edgemont Rd, Charlottesville, VA 22903-2475, USA

R. A. PERLEY
National Radio Astronomy Observatory
Post Office Box 0, Socorro, NM 87801-0387, USA

AND

J. J. BRODERICK
Virginia Tech [§]
Physics Department, Virginia Tech, Blacksburg, VA 24061,
USA

1. Introduction

Early large scale radio surveys of the sky were made with instruments with poor imaging quality and were limited to measuring positions and brightnesses of discrete sources. In recent decades radio interferometric arrays have dramatically increased their speed, sensitivity and ability to image the sky and several large scale radio surveys are currently being made with imaging instruments. One of these surveys is discussed in this paper.

2. Survey Design and Goals

The National Radio Astronomy Observatory (NRAO) is using the Very Large Array (VLA) in its most compact configuration, the "D" array, to image the sky north of declination -40° at a frequency of 1.4 GHz (or 20 cm. wavelength). This survey will produce 2326 4° × 4° images in Stokes I, Q, and U (total intensity and linear polarization) with a resolution of 45" FWHM. There will also be corresponding calibrated and edited interferometer data and a list of approximately 2,000,000 discrete sources derived

[‡]The National Radio Astronomy Observatory is a facility of the National Science Foundation operated under cooperative agreement by Associated Universities, Inc.

[§]Research is supported in part by NSF grant AST-93-20547

R. Ekers et al. (eds.), Extragalactic Radio Sources, 503–506.
© 1996 IAU.

from these images. The survey images will have a nearly uniform 5 sigma detection limit of approximately 2.5 mJy/beam which corresponds to 0.75 K for sources approximately the same size as the synthesized beam. The rms position uncertainties range from <1" for S > 10 mJy to approximately 5" at S = 2.5 mJy. The RMS noise in the linearly polarized images is typically 0.3 mJy/beam allowing "1 σ detections" of 12% polarized sources even at the survey limit. Observations in the far north and south where the array is foreshortened in the north–south direction use the the hybrid "DnC" array to maintain the circularity of the synthesized beam. Due to the interferometric nature of the instrument structure on scales larger than about 10' will have a reduced response. It is therefore expected that the majority of sources detected will be extragalactic. The design of this survey is discussed in more detail in [2].

3. Current Status

The observations are being done in three sessions to allow the observations to be done at night. The first of these sessions began in 1993 September, the second in 1995 January and the third will be in the summer of 1996. Approximately two thirds of the sky has been observed although some pointings will have to be repeated due to strong interference. All of the individual pointing data have been imaged from the first session as well as the DnC observations from the second sessions. The final $4° \times 4°$ images from the first session are available as well as a source list derived from them.

4. Availability of Results

The principle data products, images, interferometer data and source list, are being released as they become available. The source list is updated as new images become available. The products of the NVSS are available via anonymous ftp (ftp://nvss.nrao.edu/pub/nvss). Images are available as gzipped FITS files and the source list is kept in the form of a FITS binary table [3]. A browser (http://info.cv.nrao.edu/NVSS/NVSS.html) is available to search the source list. More information is available on the NVSS home page (http://info.cv.nrao.edu/ jcondon/nvss.html).

5. Comparison with UGC galaxies

The UGC catalog of galaxies [5] contains approximately 10^4 galaxies with blue diameter $\theta \geq 1$ arc min and declination $\delta \geq -2.5°$. The positions of the galaxies in the original catalog are only good to $\approx 1'$ which is insufficiently accurate for reliable identification with sources from the NVSS. A subset of the brighter of these galaxies (m < 14.5) have more accurate positions

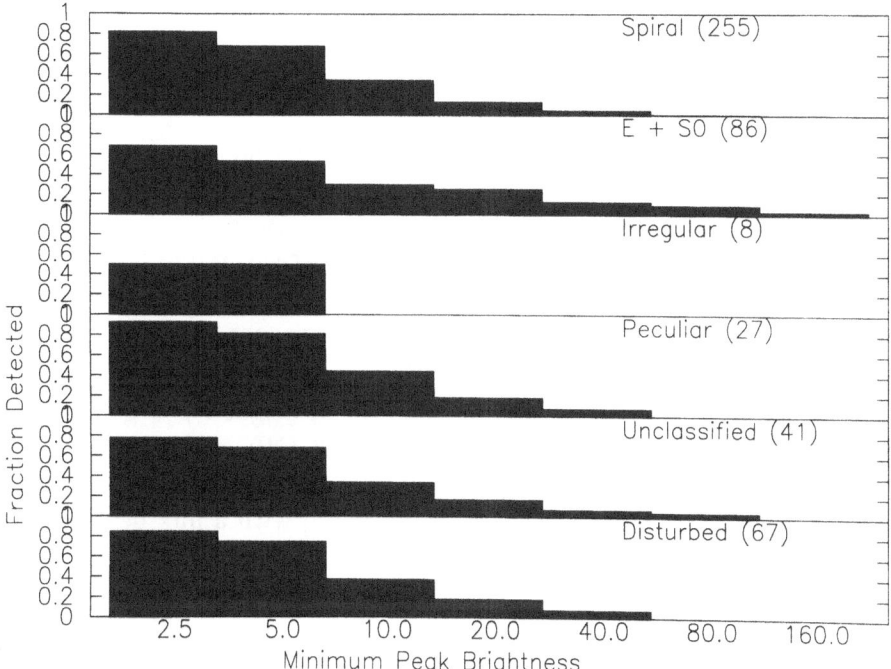

Figure 1. This figure shows the fraction of UGC galaxies with magnitudes between 12.5 and 14.5 detected by the NVSS as a function of limiting peak brightness. Each plot is labeled with the galaxy classification with the number of galaxies in the sample in parentheses. The total number of galaxies is 417, The category "Disturbed" includes those of all morphological types which were indicated as disturbed in some way in the UGC.

($\approx$ 4") measured by [4]. The fields of these galaxies with Right ascensions between 0^h and 10^h and in the magnitude range 12.5 – 14.5 were searched for NVSS sources within 15" of the optical positions. Figure 1 shows the fraction of these galaxies detected by the NVSS as function of limiting peak radio brightness for a variety of morphological categories and for all galaxies marked in the UGC as "disturbed", "distorted", "eruptive", or "peculiar". The relatively small number of objects in Figure 1 is due to the low galactic latitude of the region of overlap between the NVSS and the UGC.

The false detection rate of galaxies within 15" of the radio position should be fairly small. This assertion is supported by the results of a test in which only 3 of 444 galaxies were found "coincident" with radio sources when all the optical positions were given a 5' offset. The high fraction of detections of all categories of galaxies suggests that these relatively nearby galaxies have detectable levels of radio emission – even the ellipticals and lenticulars which are generally gas poor (cf. [7]). The high detection rate of "disturbed" galaxies also suggests that the peculiarities associated with these galaxies may be related to star formation activity. A relatively high fraction ($\approx 60\%$) of these sources are resolved by the NVSS with sizes larger than a few tens of arc seconds. This is to be compared with only 29% of the sources in a sample of 18,000 sources at high galactic latitude appearing resolved.

A comparison of NVSS flux densities with the 60 μ flux densities of IRAS indicate that the majority of spirals and "disturbed" galaxies (72% and 76% respectively) have ratios consistent with star formation. However, only a relatively small fraction (14% – all S0) of the E/S0 galaxies have radio/IR flux ratios consistent with star formation.

The high detection rate of spiral galaxies (210/255 or 82%) is consistent with low–resolution 1.4 GHz VLA images of the brightest spiral and irregular galaxies [1]. The only VLA surveys of bright E/S0 galaxies [6], [8], were made with $\theta \approx 5''$ resolution at $\nu = 4.86$ GHz, so they didn't have the surface brightness sensitivity to detect weak, extended regions. Only 52/198 (25%) E/S0 galaxies were detected by [8] with a flux density limit of 0.5 mJy whereas we detect 59 of 86 (69%) above 2.5 mJy.

The median face on surface brightness of spiral galaxies, $<T> \approx 1K$ at 1.4 GHz, is reachable for the first time by the NVSS. The NVSS will eventually cover the entire region of the UGC allowing a comparison with the nearly 3800 galaxies with accurate optical positions.

References

[1] Condon, J. J. and Cotton, W. D. and Greisen, E. W. and Yin, Q. F. and Perley R. A. and Broderick, J. J., "The NRAO VLA sky survey. I. goals, methods and first results", in preparation, 1996,

[2] Condon, J. J, "A 1.4 GHz atlas of Spiral Galaxies with $B_T \leq +12$ and $\delta \geq -45°$", Astrophys. J. Suppl., 1987, **65**, 485

[3] Cotton, W. D. and Tody, D. and Pence, W. D., "Binary table extension to FITS", Astron. Astrophys. Suppl., 1995, **113**, 159

[4] Dressel, L. L. and Condon, J. J, "Accurate Optical Positions of Bright Galaxies", Astrophys. J. Suppl., 1976, **31**, 187

[5] Nilson, P., Uppsala General Catalog of Galaxies, Uppsala Astronomical Observatory, Uppsala, Sweden, 1973, 456

[6] Sadler, E. M. and Jenkins, C. R. and Kotanyi, C. G., "Low–Luminosity radio sources in early–type galaxies", Mon. Not. Roy, Astron Soc., 1989, **240**, 591

[7] Wrobel, J. M. and Heeschen, D. S., "Current star formation as the origin of kiloparsec-scale radio sources in nearby E/S0 Galaxies", Astrophys. J., 1988, **335**, 677

[8] Wrobel, J. M. and Heeschen, D. S., "Radio Continuum Sources in nearby and Bright E/S0 galaxies: active nuclei versus star formation", Astron. J., 1991, **101**, 148

THE PARKES-MIT-NRAO RADIO SURVEYS

ALAN E. WRIGHT

Parkes Observatory, Australia Telescope National Facility
PO Box 276, Parkes, NSW, 2870, Australia

NIVEN J. TASKER

Macquarie University and the Australia Telescope National Facility
PO Box 276, Parkes, NSW, 2870, Australia

ANN SAVAGE

UK Schmidt Telescope Unit
Private Bag, Coonabarabran, NSW 2357, Australia

AND

ALAN E. VAUGHAN

Macquarie University, Dept Physics & Mathematics,
North Ryde, NSW 2113, Australia

1. Introduction

During 1990, the Parkes radio telescope made a new, deep survey of the southern sky at 4850 MHz (the PMN Survey: see e.g. Griffith and Wright, 1993; Wright *et al.*, 1994). The declination coverage of the survey was from δ = -87° to $+10^o$. The flux limit of the survey was around 30 mJy, although dependent on declination. This survey increased the number of known, southern radio sources by a factor of about 6 to over 65,000.

From 1992 until the present year, we have been re-observing approximately 8,000 of the stronger southern sources ($S_{4850} > 70$ mJy and $\delta <$ -37°) using the ATNF Compact Array at Narrabri. These observations were made in order to obtain accurate (s.e. < 1 arcsec) positions, multi-frequency fluxes, and radio structure.

The third stage of this project was to use these accurate positions, together with the COSMOS digitised sky survey (Yentis *et al.*, 1992), to obtain optical identifications for over 4000 of the objects. We have also

R. Ekers et al. (eds.), Extragalactic Radio Sources, 507–512.

used the Hubble Space Telescope Guide Star Catalogue for the brighter identifications, and the UK Schmidt plates for the fainter identifications.

The principal result of this project has been a large, homogeneous and complete sample of identified extragalactic radio sources. This database is being used to search for southern gravitational lens candidates, to compile an unbiased list of GPS (Gigahertz Peaked Spectrum) radio sources, and to search for very bright QSO candidates that may have previously escaped detection.

In this paper, we describe some of this work and indicate the initial findings from a subsequent programme of optical spectroscopy of a sample of the identified sources.

2. The PMN Surveys and Data

Previous radio surveys of the southern sky have suffered from serious deficiencies. For example, the widely used Parkes 2700 MHz Survey (as contained in the PKSCAT90 database; see Wright and Otrupcek, 1990) does not list sources close to the Galactic Plane and has different flux limits in the different survey zones. This means it can not be easily used for statistical purposes, such as homogeneity tests.

The new PMN survey, taken together with the northern 87GB survey of Jim Condon and colleagues (see e.g. Gregory and Condon, 1991), provides a survey covering the whole sky from $-87° < δ < +75°$ made at the same frequency, with the same receiver and multibeam feed system, and with similar resolutions. And, at a level of about 70 mJy, it is uniform and essentially complete throughout this declination range. Fig. 1 shows a plot of the PMN sources above this flux limit, where the plot is centred on the south celestial pole and extends northwards to $+10°$.

The data from the PMN surveys is available both in the form of point source catalogues (see e.g. Wright *et al*, 1994) or as a set of images. The simplest way to access this data is from the Australia Telescope's anonymous FTP server as:

ftp.atnf.csiro.au and in the area */pub/data/pmn/CA*.

In addition, the Parkes Observatory's Surveys HomePage on the World Wide Web:

http://wwwpks.atnf.csiro.au/databases/surveys/surveys.html

presents a useful and "friendly" interface to this server, as well as providing information about the PMN, and other radio surveys.

Besides presenting a much more extensive and uniform catalogue of radio sources, the PMN surveys have permitted much good science to be done. Correlations have been made with existing radio catalogues at other wavelengths, with optical catalogues and with infrared and X-ray cata-

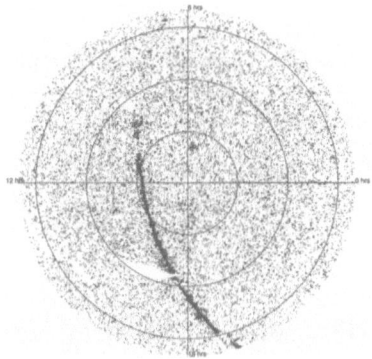

Figure 1. A plot of the stronger sources (S_{4850} greater than 70 mJy) found in the various zones of the PMN Surveys. The plot is centred on the South Celestial Pole and right ascension increases in an anti-clockwise direction starting from the right of the plot. The dark band in the figure marks the position of the Galactic Plane. The Large Magellanic Cloud and the strong radio galaxy Cen A can also be distinguished near (5.5h, −70) and (13.5h, −40) respectively

logues. Furthermore, sensitive tests for anisotropy on the largest scales can be made much more effectively with a deep continuum radio survey than with optical data, which suffers obscuration.

Despite this, however, much more information is revealed about a source once it is optically identified. For example, the redshift is potentially measurable. For this reason, the second stage of our programme attempted to provide optical identifications for as many of the southern ($\delta < -37°$) radio sources as possible. (A similar, and complementary, programme is being undertaken for the more northerly sources by our collaborators at MIT, B. Burke and A. Fletcher, using the Very Large Array.)

3. The ATNF Compact Array Programme

The accuracy of the positions of the sources contained in the point-source catalogues depends on the flux density but is typically around 15 arcsec (standard error) in each coordinate. However, to identify radio sources unambiguously with their optical counterparts, without regard to their optical colour or morphology, positions must be determined with a standard error of less than 2 arcsec — or even more accurately in regions of high star density.

With this in mind, we have undertaken a programme to re-measure the stronger sources contained in the Southern PMN survey. The criteria used to select sources are given in the following table (Table 1).

The observations were made using the Compact Array of the Australia

TABLE 1. The Compact Array Sample zones

Declination Range (J2000)	Flux limits	Galactic Latitude
$-87° < \delta < -73°$	$S_{4850} \geq 50$ mJy	$\|b\| \geq 2°$
$-73° < \delta < -38.5°$	$S_{4850} \geq 70$ mJy	$\|b\| \geq 2°$

Telescope National Facility. Briefly, this array consists of six, 22-m telescopes arranged along an east-west baseline spanning 6 km. A variety of configurations ("6A", "6C" & "6D") was used, with minimum baseline separations ranging from 77 to 337 m and maximum separations of close to 6000 m.

These re-measurements have provided positions with a typical accuracy (standard error) of around 0.6 arcsec in each coordinate for the stronger compact sources, flux densities at both 4800 and 8640 MHz, and structural information for the individual source components which is sufficient to outline the gross morphology of the sources.

Our observing method made brief "cut" observations of the target sources, each lasting about 45 seconds. Every source was observed 3 times: once approximately 4 hrs east of transit, once near transit, and once approximately 4 hrs west of transit. The profile resulting from each cut observation was reduced and displayed in real time, thus permitting data quality to be continuously monitored. When three cuts had been obtained at the three different hour angles, the profiles were reduced and combined to produce positions and fluxes for all source components.

At the present time, over 8,000 sources have been measured and good positional, flux and structural data has been obtained for 7345 components of 6603 sources.

4. Optical Identifications

The accurate positions resulting from our Compact Array observations have been used, together with the COSMOS digitised sky survey (Yentis et al., 1992), to obtain optical identifications for over 4000 of the PMN radio sources (Tasker and Wright, 1993; Tasker, 1996). However, while the COSMOS identifications are useful for the stellar objects, they are unreliable for the faintest or extended sources. We have therefore also used direct inspection of the UK Schmidt SERC J plates for the fainter, complex objects and to calibrate the automatic identification process. In addition, we also cross-correlated the Compact Array positions with the Hubble Space Telescope Guide Star Catalogue (GSC) to identify the brightest objects which may be saturated or unreliable on the COSMOS database.

The fainter $15 < B < 20.5$ optical counterparts to our Compact Array Sample were determined using an automated procedure which interrogated the COSMOS database. The optical identifications were made without regard to optical colour or morphology, and were based solely on coincidence of the radio and optical positions. Identifications were claimed if they were better then 95% reliable, as estimated from the combined error in position and the density of background objects (which varied with position on the sky).

All the COSMOS identifications were confirmed using visual inspection of a set of "postage stamp" images extracted from the Hubble Space Telescope Digital Sky Survey (DSS) CD-ROMs. The reliability of our identification procedure was also calibrated using a control sample of around 500 radio sources which were also identified using original SERC J plate material.

Approximately 40% of all objects in our Compact Array Sample were identified using the COSMOS database alone. And we estimate that a further 5% would be identified if detailed inspection of the SERC J plates were possible for all the radio sources in the sample

Our second identification program used the Hubble GSC to determine the brighter $(9 < B < 15)$ counterparts to the PMN Compact Array sample. We used the GSC to identify all sources within a distance of 10 arcsec from the radio centroid. The reliability of an identification was assessed from a histogram of radio-optical separations. Any candidate identification was also confirmed using the postage stamp images extracted from the DSS images around the relevant PMN survey position. Using this technique, a total of 220 highly-significant, bright identifications were made.

The principal result of the identification work has been an homogeneous and complete sample of over 4000 extragalactic radio sources with reliable optical identifications.

5. Research Projects

Based on the identification work described in the previous section, we are presently engaged in three major astrophysical projects.

The first is using the identifications from the Compact Array radio sample to search for southern gravitational lens candidates. At present, all but a few of the confirmed gravitational lenses lie north of the celestial equator. Since it is very plausible that similar numbers should lie in the south, the dearth of southern confirmations must clearly reflect the lack of reliably identified southern candidates. Our project aims to correct this deficiency.

Secondly, we have compiled an unbiased list of around 30 GPS radio sources, using the fluxes measured in our Compact Array programme together with fluxes from other data sources, where available. We have also obtained optical spectra for almost all of the objects in this sample using the Anglo-Australian 4-m Telescope. These spectra cover the wavelength range 3300–10,000Å and include objects as faint as 21 magnitude (B). The principal goal of this project is to test the conventional "wisdom" that GPS sources are preferentially found at the highest redshifts.

Finally, the aim of our third project was to search for optically very bright — but radio faint — QSO candidates which may have previously escaped detection. To do this, we have obtained optical spectroscopy for the 220 optically-bright identifications mentioned in the previous section. These spectra were obtained using the 1.9-m Mount Stromlo Observatory telescope of the Australian National University. At present, we report that no new, bright QSOs have been identified, although an appreciable number of new ultra-compact galaxies and planetary nebulae have been found.

6. Conclusions

Based on the PMN Southern Survey ($-87^o < \delta < -37^o$) we have described a programme which has obtained accurate positions, fluxes at 4800 and 8600 MHz and structure for over 6600 radio sources using the Australia Telescope Compact Array.

Using these positions we have optically identified over 4000 objects using a mixture of digital and conventional techniques. Furthermore, we have assessed the relative reliability of these two methods.

Finally we are using the identified sources to search for southern gravitational lens candidates; to compile, and obtain optical spectroscopy for, an unbiased list of GPS radio sources; and to search for very bright QSO candidates that may have previously escaped detection.

References

Gregory, P.C. & Condon, J.J. 1991, *ApJS*, **75**, 1011

Griffith, M.R., & Wright A.E. 1993, *AJ*, **105**, 1666

Tasker, N. & Wright, A.E., 1993, Proc. ASA, 10, 4, 320

Tasker, N., Wright, A.E., Griffith, M.R., & Condon, J.J.1994, *AJ*, **107**, 2115

Tasker, N., 1996, PhD thesis, In preparation

Wright, A.E. & Otrupcek, R., eds. 1990, ATNF, "PKSCAT90 - the southern radio database", (Available via "anonymous FTP" from *ftp.atnf.csiro.au* under the subdirectory */pub/data/pkscat90*)

Wright, A.E., Griffith, M.R., Burke, B.F., & Ekers, R.D. 1994, *ApJS*, **91**, 111

Yentis *et al.*, 1992, *"The COSMOS/UKST Catalogue of the Southern Sky"*, in Digitised Optical Sky Surveys, Astrophysics & Space Science Library, 174, 67

THE DISTANT DRAGNS SURVEY

J. D. B. LAW-GREEN
University of Manchester,
NRAL Jodrell Bank, Cheshire SK11 9DH, UK

1. Introduction

DRAGNs (*Double Radio sources Associated with Galactic Nuclei*, Leahy 1991) are the class of powerful extragalactic radio sources thought to be produced by the interaction of a jet with the ambient medium. They exhibit strong cosmological evolution in comoving number density; at $z \sim 2$ the "classical double" FR II DRAGNs were ~ 1000 times as common as they are now (Dunlop & Peacock 1990).

To understand this, systematic studies of complete DRAGN samples at low and high z and differing levels of flux density are required, in order to resolve the $P - z$ ambiguity. The Distant DRAGNs Survey is a long-term project to image with the VLA and MERLIN, matched samples of DRAGNs at high redshift.

2. Observations

The DDS sources are drawn from three optically almost complete low-frequency samples: 18 DRAGNs from 3CR (Spinrad 1985) at $z > 1.5$ and 23 from the overlapping "6C/B2 2-Jy" samples (Allington-Smith 1982, Eales 1985) at $z > 1.7$. Each source is observed at two frequencies from 408 MHz, 1.4 GHz and 5 GHz using MERLIN or VLA. Sub-kpc linear resolution is achieved at 1.4 GHz and higher, probing similar scales to HST imaging. Full polarimetry is obtained at 1.4 GHz and above.

The observations are currently $\sim 60\%$ complete. The project has "long-term" approval status from the MERLIN TAC.

3. Results

Interim analysis of the images obtained so far reveals several trends:-

R. Ekers et al. (eds.), Extragalactic Radio Sources, 513–514.

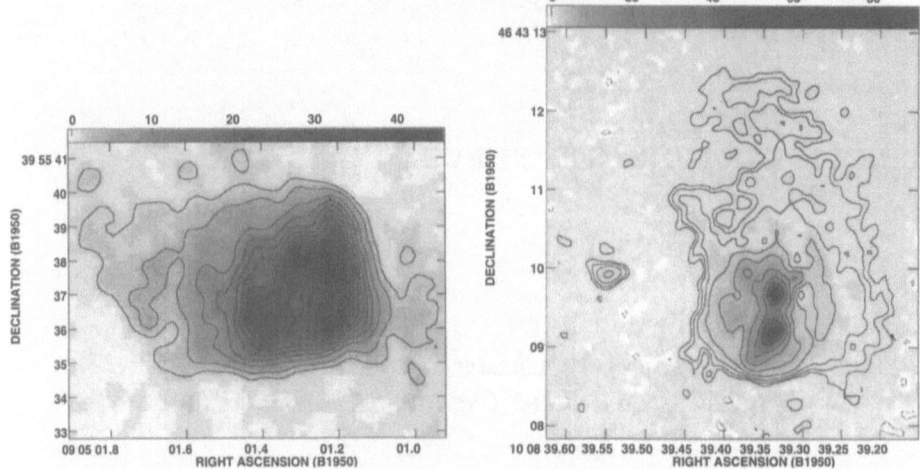

Figure 1. Suspected lensing in distant DRAGNs: (a) "Quad" in western lobe of 4C39.24 (MERLIN 408 MHz; Law-Green et al. 1995). (b) Eastern lobe of 3C 239 with lensed counterimage (MERLIN+VLA 1.4/1.6 GHz MFS; Law-Green et al., in prep.). An $R = 21.8$ foreground galaxy lies very close to the counterimage.

Lensing: Following the method of Kochanek & Lawrence (1990) the expected number of gravitationally lensed sources in our sample is 0.30 in a complete set of MERLIN 1.4 GHz images. We find at least two possible lenses (Figure 1) supporting the idea that distant radio sources are strongly amplified by lensing (Hammer & LeFevre 1989).

Frontflows: Very steep-spectrum ($\alpha > 2$) plumes beyond the radio hotspots seem common in distant DRAGNs, possibly relics of previous outbursts.

Giant: 4C39.24 was found during the DDS survey to be the most distant known giant radio galaxy (Law-Green et al. 1995). The source is $111''$ across at a redshift of $z = 1.883$, equivalent to $D = 690$ kpc ($\Omega_0 = 0.2$).

Compactness: The compactness (fraction of flux in compact hotspots) of distant sources appears greater in less luminous DRAGNs.

References

Allington-Smith J. R., 1982, *MNRAS*, **199**, 611

Dunlop J. S., Peacock J. A., 1990, *MNRAS*, **247**, 19

Eales S. A., 1985, *MNRAS*, **217**, 149

Hammer F., LeFevre O., 1989, *ApJ*, **357**, 88

Kochanek C. S., Lawrence C. R., 1990, *AJ*, **99**, 1700

Law-Green J. D. B., Eales S. A., Leahy J. P., Rawlings S. G., Lacy M., 1995, *MNRAS*, in press

Leahy J. P., 1991, in Röser H.-J., Meisenheimer K., eds. *Jets in Extragalactic Radio Sources*, Springer-Verlag, Berlin, p.1

2-JY SAMPLE OF SOUTHERN RADIO SOURCES

R. MORGANTI[1,2], C.N. TADHUNTER[3], R. DICKSON[3]
M. SHAW[3], T.A. OOSTERLOO[2], AND J.E.REYNOLDS[2]

[1] *Istituto di Radioastronomia, Bologna, Italy*
[2] *Australia Telescope National Facility, Epping, Australia*
[3] *Department of Physics, University of Sheffield, UK*

We have collected multi-waveband (radio, optical and X-ray) data for a complete sample of southern radio sources. The sample includes 88 objects selected from the Wall & Peacock (1985) catalogue that is complete down to $S_{2.7GHz} = 2$ Jy, $\delta < 10°$ and the $z < 0.7$. This database (Tadhunter *et al.* (1993), Morganti *et al.* (1993), and Siebert *et al.* these Proceedings) provides an important tool for investigating the nature of anisotropies and orientation effects in AGN and the physical causes of the correlation between their emission at different frequencies.

1. Radio cores

High-resolution observations of the radio cores have been done using the Parkes-Tidbinbilla real-time interferometer (PTI). We have measured the core flux densities at 2.3 GHz with a resolution of $\sim 0\overset{''}{.}1$ (~ 1 kpc at the redshifts of our sources, $H_o = 50$ km s^{-1} Mpc^{-1} and $q_o = 0$). The radio-core dominance derived using these data ($R = S_{core}/S_{ext}$) shows different distributions for different optical characteristics of the radio galaxies (RG) (Fig. 1). Among the FR II radio galaxies, narrow-line RGs (NLRGs) show the lower value of $\log R$ while broad-line RGs (BLRGs) have the largest $\log R$. There is also a continuity in the distribution of $\log R$ going from the NLRGs to the steep and flat-spectrum quasars (SSQ and FSQ in Fig. 1). This result supports the idea that BLRGs are more beamed toward us (compared to other RGs) either because they are objects intermediate between NLRGs and quasars, or because they represent low-z quasars. This result is also consistent with the higher detection rate and luminosity of

R. Ekers et al. (eds.), Extragalactic Radio Sources, 515–516.

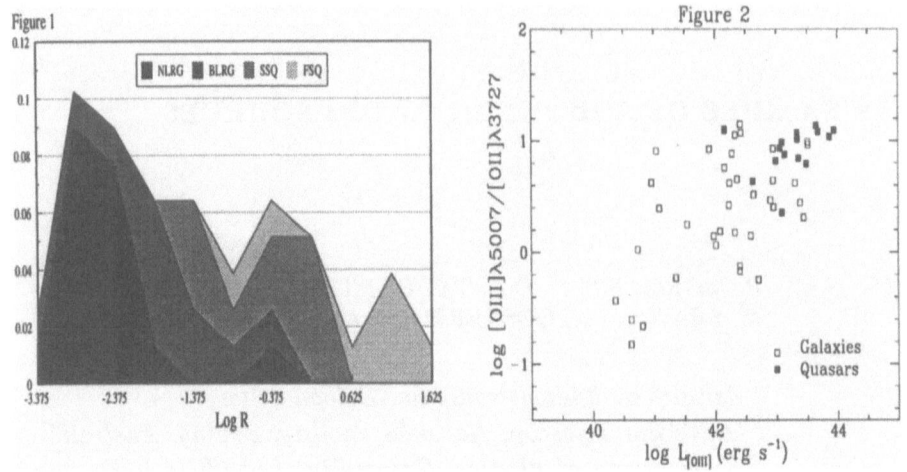

BLRGs in the soft X-rays compared to the other RGs (Siebert *et al.* these Proceedings).

2. Optical properties

Strong emission lines appear only if the total radio power is $\gtrsim 10^{26}$ W Hz^{-1}. However, a significant number of galaxies with only absorption lines or emission lines of small equivalent width in their spectra have FRII radio sources. These objects can be identified with the "low excitation" objects highlighted by Laing (1994). A correlation between [O III]λ5007/[O II]λ3727 v. log $L_{[OIII]}$ is observed at low $L_{[OIII]}$ (Fig. 2) but the correlation flattens for high [O III]λ5007 luminosity (and higher z). This can be explained if the gas properties or the balance between the ionization mechanisms (photoionization or shocks) change with z or radio power.

Optical polarimetry is being carried out for the objects with $0.15 < z < 0.7$ in the B-band (corresponding to the rest-frame UV). Although some of the objects are highly polarized in the UV, for many we have failed to detect significant polarization. In general, the diversity of UV polarization properties we find in powerful radio galaxies at intermediate z suggests that the UV excesses observed in these objects are unlikely to result solely from scattered light.

References

Laing, R.A. 1994, in ASP Conf.Ser. 54, p227;

Morganti, R., Killeen, N.E.B. & Tadhunter, C.N. 1993, *MNRAS*, **263**, 1023;

Tadhunter, C.N., Morganti, R., di Serego Alighieri, S., Fosbury, R.A.E. & Danziger, I.J. 1993, *MNRAS*, **263**, 999;

Wall, J.V. & Peacock, J.A. 1985, *MNRAS*, **216**, 173.

THE MRC 1–JY SAMPLE OF RADIO GALAXIES

V. K. KAPAHI[1], P. J. MCCARTHY[2], R. M. ATHREYA[1]

[1] NCRA–TIFR, P.Bag 3, Ganeshkhind, Pune - 411007, India
[2] OCIW, 813 Santa Barbara St., Pasadena, Ca 91101, USA

AND

W. VAN BREUGEL[3], C. R. SUBRAHMANYA[1]

[3] IGPP, Lawrence Livermore Lab., Livermore, Ca 94550, USA

Most known galaxies at high redshifts have been found by concentrating on ultra steep spectrum ($\alpha \gtrsim 1$) sources in low frequency radio surveys (eg.Rottgering et al., 1994; McCarthy et al., 1990). We have been carrying out a systematic study of a large and complete sample from the 408 MHz Molonglo Reference Catalogue which is not biased by any spectral index criterion. Containing 558 sources, defined by $S_{408} \geq 0.95$ Jy; $-30^\circ < \delta < -20^\circ$ and the RA ranges of $20^h 20^m$ to $06^h 00^m$ and $09^h 20^m$ to $14^h 00^m$, our sample is about 3 times larger in size and 5 times deeper in flux density than the well studied 3CRR sample.

Observations and Current Status : To obtain complete optical identifications we have mapped all the sources using the VLA at 6 cm λ with a resolution of $\sim$ 1" to 5" arc and made CCD images in the r–band using the 1–m and 2.5–m telescopes of the Las Campanas Observatory. We have also imaged a majority of the sources in the K–band. While about 95% of the sources are identified down to r=25, $\sim$ 99% get identified down to K$\sim$19. Spectroscopic redshifts for the sample are being measured at Las Campanas and with the 4–m CTIO telescope at Cerro Tololo.

Apart from the 111 objects forming a complete sample of quasars (including 6 BL Lac objects), we have 447 radio galaxies of which $\sim$ 60% have spectroscopic redshifts. Approximate redshifts for another 22% of the sample can be estimated using the K–z relation. Thus only $\sim$ 18% of the galaxies lack a measured or estimated z and this is largely due to shortage of telescope time at some right ascensions. The available redshifts are therefore likely to be representative of the whole sample. The median redshift of

R. Ekers et al. (eds.), Extragalactic Radio Sources, 517–518.

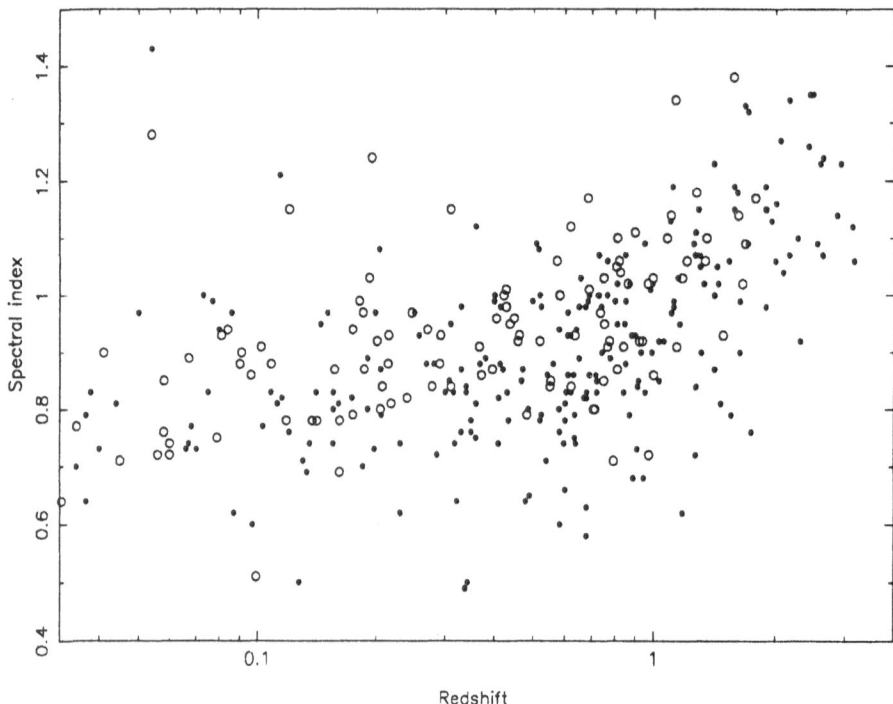

Figure 1. The spectral index – redshift plot for radio galaxies in the MRC (filled circles) and 3CRR (open circles) samples.

the galaxies is ~ 0.8, much larger than the median value of ~ 0.29 for the 3CRR galaxies.

Spectral Index – Redshift Relation : The value of $\alpha_{0.4}^{4.86}$ (between 408 MHz and 4.86 GHz) for all galaxies in the Molonglo Sample with a measured or estimated z is plotted against z in Figure 1 together with the corresponding data for the 3CRR galaxies. A preliminary analysis indicates that α is more likely to correlate with z than with radio luminosity. We also find that most of the α–z correlation could arise from just the cosmological K–correction, as most of the Molonglo galaxies at $z > 2$ have been observed to have spectra that become progressively steeper going from 408 MHz to ~ 8 GHz (Athreya et al. 1996).

References

Athreya et al. (1996) Radio Polarization studies of galaxies at $z> 2$ - *This volume.*
McCarthy, P.J. et al. (1990) *A.J.*, **Vol. 100**, 1014.
Rottgering, A.J.A. et al. (1994) *Astron. Astrophys. Suppl. Ser* ., **Vol.108**, 79.

THE B3-VLA SAMPLE

M. VIGOTTI[1], S.G. DJORGOVSKI[2], L. GREGORINI[1], U. KLEIN[3],
K.H. MACK[3], L. MAXFIELD[2], H.P.REUTER[3], D. THOMPSON[4]

[1] *Istituto di Radioastronomia C.N.R., Bologna, Italy*
[2] *Palomar Observatory, CalTech, Pasadena, CA 91125, USA*
[3] *Radioastronomishes Institut der Universität, Bonn, Germany.*
[4] *Max Planck Institut für Astronomie, Heidelberg, Germany.*

1. Introduction

The use of radio sources to identify the most distant object in the Universe
has been proved to be a very successful approach in observational cosmology. Studies of high flux, powerful 3CR and 1-Jy galaxies show dramatic
evidence for color and luminosity evolution, reaching to look-back times
80% of the Hubble time. In order to disentangle the selection effect, correlation with redshift, and correlation with radio power, it is necessary to
obtain well defined, *complete* samples of radio galaxies at a large range of
redshifts, and with a wide baseline of radio power. We need the identifications of complete samples in the flux range of a factor 10 smaller than 3CR
sample. The B3VLA sample (Vigotti et al. 1989) is a subset of 1050 sources
selected in restricted areas at high galactic latitudes from the B3 survey,
which is complete down to S(408 MHz)=100 mJy. For the B3VLA sample
detailed VLA maps were obtained at 1.4 GHz using A, C and D arrays. We
are conducting a long-term effort to provide optical ID's and redshifts for
well-defined, complete subsamples of the B3VLA survey (Djorgovski et al.
1990, Vigotti et al. 1990, Thompson et al., 1994), a similar effort is being
conducted independently by others. We present here the "status of the art"
for the B3VLA sample: a new low flux sample of 124 QSS selected at meter
wavelenghts, a sample of 194 radio galaxies (77 with measured redshift)
and a sample of 732 Empty Fields (EF : no optical counterpart on POSS I
plate).

R. Ekers et al. (eds.), Extragalactic Radio Sources, 519–521.

2. New Radio observation

The whole sample has been observed in 1994 with the 100m. Effelsberg radiotelescope at 10.6 GHz. As preliminary results we found that the median spectral index in the high-frequency range is shifted by -0.4 from the low frequency one, a figure close to -0.5 as one would expect for a continuous injection model. We observed 64 high-redshift B3VLA radio sources at 230GHz using the bolometer of the IRAM radiotelescope. A total of 13 objects was detected. The radio continuum spectra between 151 MHz and 230 GHz have been compiled for all sample sources. It is found that the observed mm flux densities generally fall on, or below, the extrapolated centimeter radio continuum spectra; this is interpreted in terms of a predominance of nonthermal radiation even at these wavelengths. Excess emission has been detected in only two sources. Attributing this excess to thermal emission from dust, we estimate the mass of the involved dust component to be around a few $\times 10^9$ $M_\odot$.

3. Quasar Sample

172 Quasar candidates were identified on POSS I prints (down to 20.5 red mag) using only the positional coincidence between the optical position and the structural informations of A-configuration (1" resolution) and C-configuration (5" resolution) VLA maps. The sample was divided in 120 blue starlike objects (B) and 52 red or neutral color (B - R $\geq$ 1) starlike objects (N). So any color bias was avoided. For all the candidates (except 35 with literature redshifts available) spectra were obtained at the 3.5m. telescope of Calar Alto (Vigotti et al. 1990, Vigotti et al. 1996 in preparation). The 120 B-QSS candidates splits in 109 QSS - 1 Galaxy - 2 BL Lac - 5 star - 3 Featureless while the 52 N-QSS candidates splits in 15 QSS - 23 Galaxy - 1 BL Lac - 12 star - 1 Featureless. The sample, whose median redshift is 1.25, is almost complete for $S_{408} \geq 0.8Jy$; from a visual inspection to the histogram of the magnitudes we expect $\leq$ 4 QSS with magnitudes fainter than POSS-I limit. In the flux bin from 0.8 Jy to 100 mJy we have 60 QSS, the sample is estimated to be 85% complete and $\sim$ 10 QSS may have magnitudes $r > 20.5$. To test the AGN unified models Quasar samples selected at meter wavelengths are extremely useful, because the selection of sources is largely based on their lobe emission.

4. High Redshift Radio Galaxies

From the complete B3VLA sample, we selected a statistically complete subsample of steep-spectrum EF with $\alpha_{408} \leq -0.9$, small angular size ($\leq$ 20 arcsec), $S_{408} \geq 0.8$ Jy sources. There are 109 Empty Fields (EF) which satisfy these criteria. We have already identified 90% of the sample on deep Gunn r and i CCD images obtained with the Hale 200-inch and with the 60-inch telescopes at Palomar. Up to now we have 60 redshifts ranging from $z = 0.5$ to $z = 3.2$. Only one quasar was discovered in the EF sample, therefore we can safely state that QSS are a small percentage of the EF. The redshifts were obtained at the Palomar 200-inch telescope and at the KECK telescope.

5. Results

The optical study of the EF sample is far to be complete, but we have enough data to make some preliminary comparison. We have combined the high-flux well defined 3C radio galaxies sample with the moderate-flux radio galaxies from B3VLA sample. The Hubble diagrams in the K color show a remarkably smaller scatter than in r band, which for most of these galaxies probes the restframe UV, and is thus more susceptible to the effect of AGN, star formation, or dust. We found a weak correlation between the absolute magnitude in K and r band and the radio power; furthermore a strong correlation is present between the line luminosity and radio power. These two effects are mostly due to an active nucleus, an hidden quasar, which is also responsible for the radio emission. Finally, we examine the behavior of the optical and radio PA alignment. For the B3VLA sources the distribution of the optical-radio alignments, (i.e., absolute PA differences) is nearly uniform. This is in a striking contrast with the situation for the more powerful 3C radio galaxies, where the alignments are quite strong. Combining the B3 and 3C data sets we see that the alignment is prominent in both the high-power and high-redshift sample, as expected, due to the strong correlation between power and redshift in flux limited samples. Then we take the high-redshift ($z > 0.8$) half of the combined B3+3C data set, and split it by median power ($\log P = 27.62$). In this high-redshift sample the high-power subset shows the alignment effect whereas the low-power subset does not. We do a similar test taking the high-power ($\log P > 27.0$) half of the combined B3+3C data set, and split it by median redshift ($z = 0.92$). In this high-power sample the high-redshift subset shows the alignment effect whereas the low-redshift subset does not.

References

Djorgovski S., Thompson D.J., Vigotti M., Grueff G. 1990 *P.A.S.P.* **102**, 113-116.
Thompson D.J.,Djorgovski S., Vigotti M., Grueff G. 1994, *A.J.*, **108**, 828
Vigotti, M., Grueff, G., Perley, R., Clark, B.G., Bridle, A.H.: 1989, *A.J.* **98**, 419
Vigotti M., Merighi R., Vettolani G., Lahulla J.F., Lopez-Arroyo M., 1990, *A&A Suppl. Ser.* **83**, 205-210.

A DEEP 20 CM RADIO MOSAIC OF THE ESO KEY-PROJECT GALAXY REDSHIFT SURVEY

I. PRANDONI[1], L. GREGORINI[1], P. PARMA[1], H.R. DE RUITER[1],
G. VETTOLANI[1], M.H. WIERINGA[2] AND R.D. EKERS[2]

[1] *IRA - CNR, Via Gobetti 101, 40129 Bologna, Italy*
[2] *ATNF, P.O. Box 76, Epping NSW 2121, Australia*

1. Background: The Optical Sample

In two strips of $22° \times 1°$ and $5° \times 1°$ near the SGP Vettolani et al. (1993, *IAU Symposium 161, "Astronomy from Wide Field Imaging"*, H.T. MacGillivray ed., Reidel, in press) have made a deep redshift survey as an ESO Key Project. All the galaxies down to $b_J \sim 19.4$ were observed with the OPTO-PUS multi-fiber spectrograph on the 3.6 m telescope in La Silla, yielding 3348 redshifts. The survey has a typical depth of $z = 0.1$. It fully samples the optical luminosity function down to $B = -15$ and various galaxy populations (e.g. normal galaxies, LSBDs and BCDs) are present. Interestingly, emission lines (OII, Hβ, OIII) have been found in a large fraction of the galaxy spectra ($\sim 40\%$), suggesting strong evolution of the galaxy population in terms of enhanced star formation.

2. ATCA Radio Observations: First Results

We are using the ATCA at 20 cm to image the entire area of the optical survey (mosaic observing mode) with uniform sensitivity (336 pointings), needed for statistical studies. Since our optical sample is rather deep but narrow and 'normal' galaxies are tipically low-power radio sources, deep radio observations are needed. A 3σ radio limit of ~ 0.2 mJy (allowed when sky positions are known) will enable us to detect $P < 10^{21}$ W/Hz for $z < 0.1$.

R. Ekers et al. (eds.), Extragalactic Radio Sources, 523–524.

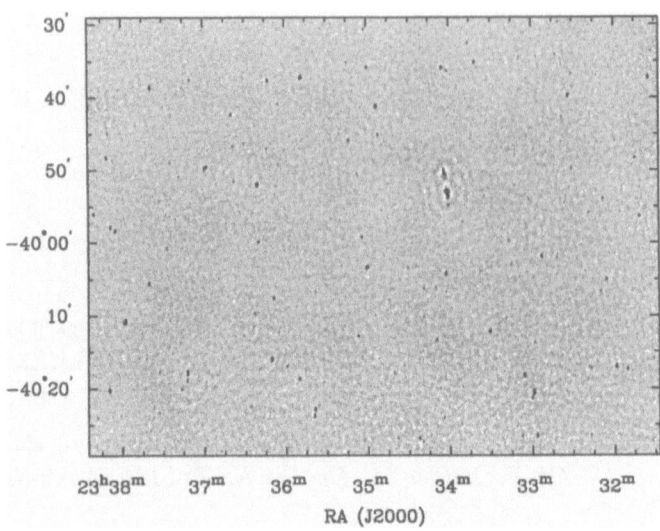

To reach this detection limit we need to observe 1.2 hours per field (with 2×128 MHz bandwidth). All 336 fields can thus be observed in 34 blocks of 12^h. We have already observed $\sim 50\%$ of the entire area and plan to complete the observations by the end of 1996.

Data reduction is almost completed. We get cleaned mosaiced maps of 1.34 sq. degs. with spatial resolution $16'' \times 8''$ (see Figure). The noise level is $\sim 70\,\mu$Jy and is fairly uniform over the entire map. Dynamic range problems cause slightly higher levels of $\sim 80\,\mu$Jy around strong sources ($S_{peak} > 50$ mJy).

In a preliminary analysis of a 4 sq. degs. area we have searched for radio emission associated with the redshift survey galaxies. We used a 3σ-threshold of 0.24 mJy (taking a conservative value of $80\,\mu$Jy for the noise). We found 42 associations. A large fraction ($\sim 50\%$) of them are associated with galaxies showing one or more emission lines. We expect to detect ~ 300 galaxies in the entire area surveyed (27 sq. degs.). The numbers reported above allow us to believe that, when the survey will be completed, a very reliable statistical study of the existing correlations between optical (line activity, colors, morphologies, luminosities, etc.) and radio properties of the various populations represented in the optical sample will be possible.

In the same area we searched for all the radio sources above a 5σ-threshold of 0.4 mJy, getting 360 detections. 40% of them are sub-mJy objects. We expect a total number of ~ 2500 radio sources in the entire area surveyed. This new homogeneous and fairly deep sample of radio sources will allow better studies of e.g., the sub-mJy population, and be useful for the selection of appropriate subsamples aimed at different kinds of analysis.

RADIO OBSERVATIONS OF THE MARANO FIELD: SUB-MILLIJANSKY SOURCE COUNTS AND SPECTRAL INDEX STUDIES

C. GRUPPIONI AND P. PARMA

Istituto di Radioastronomia del CNR, via Gobetti 101, I-40129, Bologna, Italy

AND

H.R. DE RUITER AND G. ZAMORANI

Osservatorio Astronomico di Bologna, via Zamboni 33, I-40126, Bologna, Italy

1. Introduction

The Marano Field (centered at $\alpha(2000)=03^h\ 15^m\ 09^s$, $\delta(2000)=-55°\ 13'\ 57''$) is a deep ROSAT field (flux limit $\sim 4 \times 10^{-15}$ erg cm^{-2} s^{-1}), which has been entirely covered by ESO 3.6 m plates and in the inner part by deep CCD exposures. In order to follow up these data in other wavelength regions, deep radio observations of this field have been carried out with the Australia Telescope Compact Array (ATCA) at 1.370 and 2.378 GHz. The minimum reached rms noise value is $\sim$42 μJy at both frequencies. 80 and 45 sources form complete samples above 5.5 σ_{local} level at 1.370 and 2.378 GHz respectively, in a square area of $\sim$0.34 sq. deg. Almost all of the sources detected at 2.378 GHz have been detected also at 1.370 GHz.

2. The Source Counts

The normalized source counts at 1.370 GHz show a flattening below a few mJy, equivalent to a more rapid increase in the number of faint sources. This change in slope is visible in all survey fields reaching sub-mJy fluxes at 1.4 GHz (*e.g.* Condon & Mitchell 1984; Windhorst et al. 1985). Our normalized differential counts at this frequency are in excellent agreement, both in shape and normalization, with the existing data.

R. Ekers et al. (eds.), Extragalactic Radio Sources, 525–526.

Our 2.378 GHz counts are the deepest at this or similar frequencies (*e.g.* 2.7 GHz). Only predictions based on fluctuation analysis exist at 2.7 GHz down to $\sim$1 mJy (Wall & Cooke 1975). Our counts are in good agreement with these predictions and with the differential 2.7 GHz counts at higher fluxes (Wall & Peacock 1985).

3. Spectral Index Studies

To study the spectral index distribution, sources which are present in one of the two complete samples but not in the other one have been searched for detection down to the $3\sigma_{local}$ level. If no detection was found at this limit, an upper or a lower limit on α ($F_\nu \sim \nu^{-\alpha}$) was established assuming $S_{1.370}$ or $S_{2.378} < 3\sigma_{local}$. The median spectral index for the sources detected at both frequencies is $\alpha_{med} \simeq 0.59$, but, since most of the limits on α are lower limits, the true α_{med} is likely to be larger than this value. Above $S_{1.370} \simeq 0.7$ mJy, where all the sources have been detected at both frequencies, the median spectral index is 0.65 ± 0.09, in good agreement with the results discussed by Windhorst et al. (1993). From the analysis of the spectral index as a function of the 1.370 GHz flux density we find that a significant number of inverted spectrum sources ($\alpha < 0$); appears at fluxes below $\sim$2 mJy. Spectroscopic identifications of the optical counterparts of these objects are in progress, in order to understand their nature and to investigate the recent results of Windhorst et al. (1995) and Hammer et al. (1995). In particular, the last ones found that $\sim$50 % of the μJy population with optical counterpart do have inverted radio spectra. About half of them have been identified with low z, low luminosity blue emission line objects, while the remaining ones are red ellipticals at z>0.75.

References

Condon J.J. and Mitchell K.J. (1984), *AJ*, Vol. no. **89**, p. 610.
Hammer F., Crampton D., Lilly S.J., Le Fèvre O. and Kenet T. (1995), *MNRAS*, Vol. no. 276, p.1085.
Wall J.V. and Cooke D.J. (1975), *MNRAS*, Vol. no. 171, p. 9.
Wall J.V. and Peacock J.A. (1985), *MNRAS*, Vol. no. **216**, p. 173.
Windhorst R.A., Miley G.K., Owen F.N., Kron R.G. and Koo D.C. (1985), *Ap.J.*, Vol. no. **289**, p. 494.
Windhorst R.A., Fomalont E.B., Partridge R.B. and Lowenthal J.D. (1993), *Ap J.*, Vol. no. **405**, p. 498.
Windhorst R.A., Fomalont E.B., Kellermann K.I., Partridge R.B., Richards E., Franklin B.E., Pascarelle S.M. and Griffiths R.E. (1995), *Nature*, Vol. no. **375**, p. 471.

THE VLBA CALIBRATOR SURVEY

A.J. BEASLEY, V. DHAWAN, E.B. FOMALONT, R.C. WALKER
AND J.M. WROBEL
NRAO, Socorro, New Mexico 87801-0387, USA

1. Phase-referencing

Using phase-referencing, the coherent integration time of VLBI observations can be substantially increased, permitting observations of weaker (~mJy) target sources (see e.g. Beasley & Conway 1995). The position of a source can also be accurately measured relative to a reference source, allowing absolute and proper-motion measurements, optical-radio image alignment, and alignment of images made at different frequencies.

To make phase-referencing routinely available to the VLBI community, a suitable grid of phase-reference calibrators is needed. We have begun a VLBA survey of ~3000 flat-spectrum radio sources, selected from the Jodrell Bank–VLA Astrometric Survey (JVAS; Patnaik et al. 1992), in order to derive an "all-sky" catalog of phase-reference calibrators (see Figure 1). When complete, all JVAS sources detected above 100 mJy at 8.4 GHz on VLA A-array scales will be observed with the VLBA S/X dual-frequency system, using a geodetic-style frequency setup of four IFs at 2.3 GHz spanning 100 MHz and four IFs at 8.4 GHz spanning 400 MHz. Each source is observed at three separated hour-angles for 100 s, i.e. a total of 300 s. Approximately 25% of the observing time is spent observing geodetic-grade calibrators. After amplitude and delay calibration in the NRAO AIPS software package, fringe solutions are transferred to GSFC CALC/SOLVE geodetic package to derive source positions. The calibrated data are also automatically imaged using AIPS and the Caltech Difmap package.

2. Scientific Impact

Our sample is statistically complete, and will be the largest set of sources surveyed with VLBI to date. Potential scientific returns from this survey include: (1) the detection of compact gravitational lenses, probing the existence of massive compact lensing objects such as 10^6 $M_\odot$ black holes; (2)

R. Ekers et al. (eds.), Extragalactic Radio Sources, 527–528.

structural information and the brightness temperature distribution at 8.4 and 2.3 GHz for a large sample of AGNs; (3) in combination with optical identifications and redshifts, this structural information will contribute to θ-z and μ-z studies.

3. Current Status

Three of fourteen epochs have been observed to date (11/95), a total of 648 sources in the declination ranges 0–22° and 50–79°. The mean positional accuracy achieved is ~0.7–1 mas rms, sufficient for most phase-referencing purposes. Images from the survey data have typically a 2–3 mJy rms, with a 100:1 dynamic range or better. At present, the plan is to complete all JVAS sources above 200 mJy during 1996, with the entire survey complete by late 1997. The survey positions, images and UV-data will be available online from the NRAO WWW server from early 1996. Special requests to survey selected regions to lower flux densities near important targets will be considered.

VLBA Calibrator Survey Sources -- Observed(crosses), Planned(points)

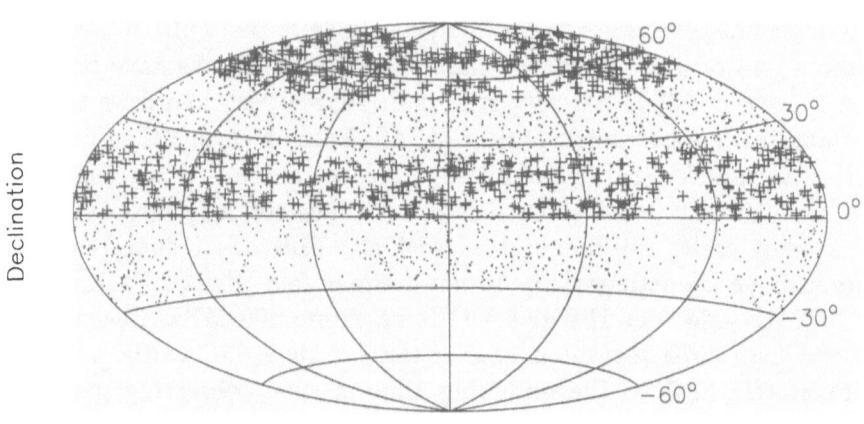

Figure 1.

References

Beasley, A.J. & Conway, J.E. 1995, in *VLBI Phase-Referencing*, Chapter 7, *Proceedings of the NRAO VLBA Summer School*, ASP Conference Series Vol. 82, eds. Zensus, A., Diamond, P.J. and Napier, P.J..
Patnaik, A.R. et al. 1992, *MNRAS*, **254**, 655.

THE VSOP MISSION FOR EXTRAGALACTIC
RADIO SOURCES

HISASHI HIRABAYASHI
The Institute of Space and Astronautical Science
3-1-1 Yoshinodai, Sagamihara, Kanagawa 229 Japan

1. VSOP System Description

The VSOP (VLBI Space Observatory Programme) mission is being developed by the Institute of Space and Astronautical Science (ISAS), in close collaboration with the National Astronomical Observatory (NAO) of Japan. NASA and NRAO of the USA are key collaborating institutions, and most radio-telescopes world-wide will participate in some VSOP observations.

Funding for development of the VSOP satellite, MUSES-B, was first received in 1989. The satellite is scheduled for launch in September 1996. This will be just the tenth anniversary of the first successful space VLBI demonstration, the TDRSS–OVLBI experiment (Tracking and Data Relay Satellite System–Orbiting VLBI Experiment, Levy et al. 1986) at S-band. MUSES-B will place an 8 meter diameter Cassegrain radio-telescope in an elliptical Earth orbit (22,000 km apogee height and 1,000 km perigee height, resulting in an 6.6 hour orbital period) to make the first dedicated space-VLBI observations on baselines up to 30,000 km. Observing frequencies are 1.6, 5 and 22 GHz, with data down-loaded in real-time on a 128 Mbps link supported by a dedicated five station telemetry network (Hirabayashi et al. 1992; Hirosawa and Hirabayashi 1995). The ten-station VSOP correlator will be ready before the spacecraft launch.

2. Observations of Extragalactic Radio Sources with VSOP

On baselines to the same ground antenna, VSOP will have, at 1.6 and 5 GHz, sensitivity a factor of about 6 better than that of TDRSS-OVLBI at 2.3 GHz (including the improvement from the larger VSOP bandwidth).

R. Ekers et al. (eds.), Extragalactic Radio Sources, 529–530.

The sensitivity of a baseline between VSOP and a VLBA antenna will be a factor of ~6 worse than between two VLBA antennas.

Because of the limited sensitivity, like the present-day VLBI, most of the observations will be limited to high brightness sources. Nearly all continuum observations will be of active galactic nuclei (AGNs) with some exceptions for possible flare stars, galactic jet sources, or supernovae events. For baselines between VSOP and VLBA antenna, the estimated number of detectable sources will be 1000~2000 at 5 GHz and 50~100 at 22 GHz.

Mapping periods of 3–4 orbits will be typical for VSOP imaging observations, to build up the UV-coverage for good imaging capability. The high resolution and high dynamic range imaging capability are the most important features of the VSOP mission. Observations with one or two orbits will be used for survey observations or model-fitting type observations. Observations are classified into two categories; peer reviewed Open Program and mission-led Survey Program. The Survey Program will deal with large number of statistical samples, to be observed with a smaller number of ground telescopes.

Monitoring of structure changes will be to some degree limited by the changing UV-coverage. Both the dynamic range and the synthesized beam shape will vary with observing epoch due to the orbital precession. The constraint that sources within 70° of the Sun cannot be observed results in periods when low declination sources are not observable.

3. Science Operation of VSOP

The Open Program is open to proposals from astronomers worldwide, and the proposals will be reviewed based on their scientific merits by an international Science Review Committee, and scheduled taking into account the resource availability. The first Announcement of Opportunity was issued in June 1995 for the first 17 months from the beginning of 1997. For details of VSOP capabilities, readers are advised to consult the VSOP Proposer's Guide, available by anonymous ftp from 133.74.2.131.

The mission's science operations will be led by the VSOP Science Operations Group (VSOG), centered at ISAS with its core members from ISAS and NAO. The scientific policy is determined by VSOP International Science Council (VISC), members of which were selected from world-wide radio-astronomical community.

References

Levy, G.S. et al. (1986) *Science*, **234**, 187
Hirabayashi, H. et al. (1992) *Proc. 18th ISTS (Kagoshima)*, 1813–1820
Hirosawa, H. and Hirabayashi, H. (1995) *IEEE AES Systems Magazine*, June, 17–23

IMAGING OF EXTRAGALACTIC RADIO SOURCES WITH THE VSOP SPACE VLBI MISSION

D.W. MURPHY

Jet Propulsion Laboratory, Caltech, Pasadena, CA, USA

Abstract. In September 1996 the first dedicated VLBI spacecraft, VSOP, will be launched. This Japanese spacecraft operating in conjunction with ground-based VLBI arrays such as the EVN or VLBA will enable observers to routinely undertake VLBI observations with maximum baseline lengths of 2.6 Earth diameters and a resolution of 55 μas at the highest operating frequency of 22 GHz. In this paper we present a brief overview of the imaging capability of the VSOP mission together with an example of an imaging simulation.

Unlike ground-based VLBI experiments, the (u,v) coverage for a space VLBI experiment is a function of both source coordinates and the epoch of observation, since the precession of the orbit plane with a 1.8 year period causes the orientation of the source with respect to the orbit plane to change with time. Furthermore the spacecraft cannot observe sources within 70° of the sun or when the spacecraft is in eclipse. In the nominal 6.6 hour orbit (apogee height = 22000 km and perigee height = 1000 km) with an inclination of 31°, eclipse length will vary with long periods of short (< 40 minutes) or no eclipses to short periods of long (> 80 minutes) eclipses. Sources in or near the orbit plane have linear (u,v) coverages whereas sources that lie along the orbit normal directions have the highest resolution two-dimensional (u,v) coverages. Due to the precession of the orbit plane it will be possible to obtain good two-dimensional (u,v) coverage on any source at least once for a period that lasts several months during the observing period covered by the first VSOP AO (17-months).

In Figure 1 we illustrate the result of a typical imaging simulation with VSOP and the VLBA. Figure 1(a) shows the (u,v) coverage that is obtained for a 4-orbit observation. In this example the (u,v) coverage is intermediate between a source that lies along the orbit normal and one that lies in the

531

R. Ekers et al. (eds.), Extragalactic Radio Sources, 531–532.

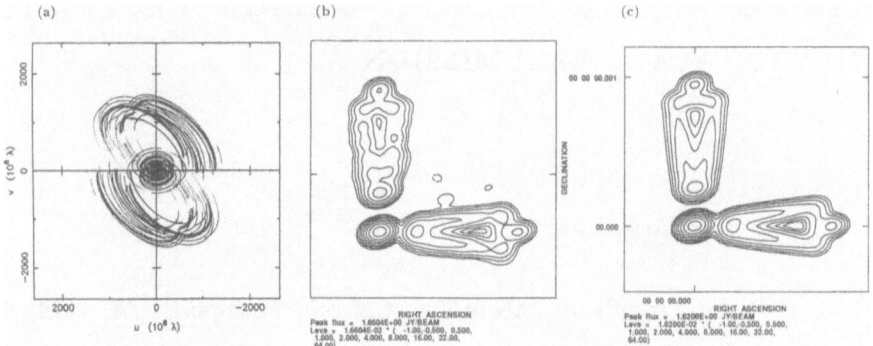

Figure 1. Image Simulation: (a) 4-orbit *(u,v)* coverage; (b) Simulated image ; (c) Source model.

orbit plane. The simulated visibility data have both random antenna-based phase errors and 10% antenna-based amplitude errors. An unbiased 'hands-off' iterative scheme was used to make the image. First of all a low order polynomial is fitted to the upper envelope of the correlated flux versus projected baseline plot. This polynomial is then used to determine what lower *(u,v)* limit to use in the self-calibration stage of the self-calibration/mapping loop. As the image improves (more CLEAN components up to the first negative) the *(u,v)* limit is slowly decreased from a high initial value to finally include all the visibility data. This scheme is found to produce a good image with very few spurious features.

The simulated image is shown in Figure 1(b). This can be directly compared to the source model shown in Figure 1(c), convolved with same beam used for the VSOP+VLBA image. The image has a RMS noise level off-source of 0.7 mJy/beam which compared with a peak flux density of 1.7 Jy/beam gives a dynamic range of 2400:1. However the difference between the image and the model shows that on-source there is RMS error of 12 mJy/beam. Thus the true dynamic range is only 140:1. The main source of error in the image can be traced to the large holes in the *(u,v)* coverage. Higher fidelity images can be made at the expense of loss in angular resolution by choosing to observe sources when the size of the *(u,v)* holes are minimized by projection effects.

Simulations, such as the one shown in Figure 1, have demonstrated that it will be possible to image compact extragalactic radio sources with the VSOP mission. By the time of the next IAU meeting on extragalactic radio sources 'real' space VLBI images not just simulations will be available.

SURVEYS OF RADIO SOURCES (*DISCUSSION*)

General discussion following the group of survey papers presented by
<u>*DE BRUYN, BECKER, COTTON, WRIGHT*</u> (p. 495, 499, 503, 507)

Ekers: How complete are your surveys to extended and multiple component sources.

Wright: I think the best way of ensuring good completeness for extended sources is to survey for them using a large single-dish and get structure for them with an array. This just happens to be the way the PMN work was done and is proceeding!

Becker: Comparison between logN–logS plots for the FIRST catalogue and published data suggest that the FIRST survey is complete at the ~95% level.

de Bruyn: I do not think we could miss sources because they are too big. However, we could fail to properly recognize doubles or multiples (on $3'$–$15'$ scales) and the catalogue could not possibly be complete because of this. There will be a warning in the catalogue if sources are blended. 'Blind' analysis of a catalogue is therefore not without risk.

Cotton: The NVSS should not miss too many sources but associating separated regions of emission with individual sources may be difficult especially in the source catalog. At present association of regions needs to be done by eye from the images. If the source had significant residuals this is marked in the catalog indicating more complex or extended structure.

Condon: Not only does resolution cause sources to disappear form a survey catalogue, it will cause the integrated flux densities of extended catalogued sources to be too low. Users selecting flux-limited samples should watch out for incompleteness at any cut off level, even one well above the nominal catalog limit. This problem is smallest for low-resolution single dish surveys, slightly greater for the WENSS (which has good (u,v)-plane coverage), and worst for the VLA snapshot surveys (which have poor (u,v) - coverage).

R. Ekers et al. (eds.), Extragalactic Radio Sources, 533–534.
© 1996 *IAU.*

Machalsky: Will your observation data be available to international community? If yes, how and when?

Becker: Yes. The images are available by anonymous ftp from NRAO. They are also in the STScI archival system. The catalog is available through WWW at http://sundog.stsci.edu

Jauncey On behalf of VSOP Survey Working Group I would like to note the directed high brightness temperature survey to be undertaken by the VSOP SWG. The aim is to look for high T_b components in a complete all-sky sample of flat-spectrum sources found at 5 Ghz and satisfying $S_5 \geq 1Jy$, $\alpha \leq 0.5$, $|b| \geq 10°$. More details to be found in the VSOP poster and in the VSOP AO.

Hunstead: Comment: The Molonglo Observatory Synthesis Telescope (MOST), operated by the University of Sydney, will be carrying out a survey on the southern sky at 843 MHz south of - 30° beginning $\sim$ mid - 1996 and lasting 5 – 10 years. The MOST has a synthesised beamwidth of $43''$ and a positional accuracy $\sim 1''$. With the new field size of $2°.7$ and upgraded receivers the expected thermal noise level (1σ) is 1mJy in a 12-hour observation. It is anticipated that over 400 000 sources will be detected down to limits $\sim$5mJy. Further information may be obtained from most@physics.usyd.edu.au

Jauncey: Two major advantages of large surveys are: a) High statistical value of large numbers, and b) ability to find rare and unusual new objects. Both of these can be compromised in the data handling etc, by building in biases. An example of this is the colour/morphology selection for quasar indents. There are strong colour-z correlations that show that excluding red stellar idents are often of great interest.

RADIO LUMINOSITY FUNCTIONS

J.J. CONDON

National Radio Astronomy Observatory
520 Edgemont Road, Charlottesville, VA 22903, USA

1. Introduction

Luminosity functions give the space densities of source populations as functions of spectral luminosity. They constrain source physics, clustering, cosmological evolution, and contributions to the radiation background. Extragalactic radio sources span eight decades of luminosity, so dividing them into luminosity ranges effectively separates them by energy generation mechanism and other fundamental properties. Multivariate luminosity functions divide the source population in additional ways (e.g., by optical morphology, optical absolute magnitude, or radio spectral index) selected to emphasize particular source characteristics.

Traditionally, local radio luminosity functions have been based on optically selected galaxies supplemented at high luminosities by optical identifications from general radio surveys. The galaxies were usually grouped by optical morphology: (1) E and S0 galaxies, thought to contain active galactic nuclei (AGN), and (2) "normal" spiral and irregular galaxies whose radio emission is primarily powered by stars and stellar remnants only. Bivariate luminosity functions related the radio and optical luminosities. For example, Hummel's (1981) luminosity functions of spiral galaxies are based on the assumption that the radio and optical luminosities of normal galaxies should be roughly proportional, and distributions of the ratio $\log(R) \equiv \log[S(\mathrm{mJy})] + (B_T - 12.5)/2.5$ were used to show that barred spirals contain more luminous central sources than do ordinary spirals.

New opportunities arise as more data become available (especially in other wavebands) and new statistical questions about source populations are posed. Many optical selection and classification limitations can be overcome by the use of far-infrared (FIR) data from the *IRAS* Faint Source Catalog (Moshir *et al.* 1992). Optical morphology is an imperfect indicator

535

R. Ekers et al. (eds.), Extragalactic Radio Sources, 535–540.

of radio source type: Some spiral galaxies contain radio sources powered by AGN, some low-luminosity sources in E/S0 galaxies appear to be powered by stars (Wrobel & Heeschen 1988), and many radio galaxies have unclassifiable "peculiar" or "disturbed" morphologies. In contrast, the FIR/radio correlation effectively distinguishes nearby normal galaxies (even ultraluminous starbursts are called "normal" galaxies) from genuine AGN with supermassive black holes or other "monsters." An important new application of AGN luminosity functions is testing unified schemes for AGN (Urry & Padovani 1995, and references therein). Relativistic beaming is most important for compact flat-spectrum ($\alpha \equiv -d\ln S/d\ln\nu < 0.5$) sources, so separate luminosity functions for steep- and flat-spectrum sources are needed. Finally, new radio- and FIR-selected samples of low-redshift ($z < 0.1$) galaxies are large enough to reduce the statistical uncertainties limiting radio luminosity functions of normal galaxies.

2. The 1.49 GHz Luminosity Function of Normal Galaxies

The complete sample of 299 spiral and irregular galaxies with $B_T \leq +12$, $|b| \geq 15°$, and $\delta > -45°$ were mapped by the VLA at 1.49 GHz. Following Felten's (1977) "standard" procedure, the local group ($D < 1.7$ Mpc) and galaxies within 10° of the Virgo cluster center were excluded and the 1.49 GHz luminosity function (open circles in Fig. 1) was found by the $(1/V_m)$ method (Condon 1989). The small volume in which spiral galaxies brighter than $B_T = +12$ can be found limits the statistical usefulness of this sample to $L < 10^{22}$ W Hz^{-1} at 1.49 GHz; FIR-selected galaxies are needed to reach higher radio luminosities.

VLA images at 1.49 and 1.425 GHz (Condon et al. 1990, 1995) now cover the *IRAS* Bright Galaxy Sample (BGS) Parts I (Soifer et al. 1989) and II (Sanders et al. 1995) of sources stronger than $S = 5.24$ Jy at $\lambda = 60$ μm in the area $\delta \geq -45°$, $|b| \geq 10°$. Normal galaxies were selected by the requirement $q \equiv \log[(FIR/3.75 \times 10^{12})/(S_{1.49\ \mathrm{GHz}}/10^{26})] \geq +1.85$, where $FIR \equiv 1.26 \times 10^{-14}(2.58S_{60\mu} + S_{100\mu})$. The 465 normal galaxies with $D \geq 1.7$ Mpc and $> 10°$ from Virgo yield the luminosity function indicated by the filled circles in Fig. 1. The solid curve is *not* a fit to the radio data. It is the $\lambda = 60$ μm luminosity function (Saunders et al. 1990, model 17) with the $\lambda = 60$ μm luminosities converted to $\nu = 1.49$ GHz by the mean FIR/radio ratio $\langle S_{60\mu}/S_{1.49\ \mathrm{GHz}}\rangle = 121$ of normal galaxies selected at 1.49 GHz. The close fit is an immediate consequence of the tight FIR/radio correlation. Ultraluminous galaxies should be slightly radio quiet relative to $\lambda = 60$ μm due to free-free absorption in compact starbursts (Condon et al. 1991c). The broken curve showing the Schechter blue luminosity function (Felten 1977) is too steep, so there cannot be a tight blue/radio correlation.

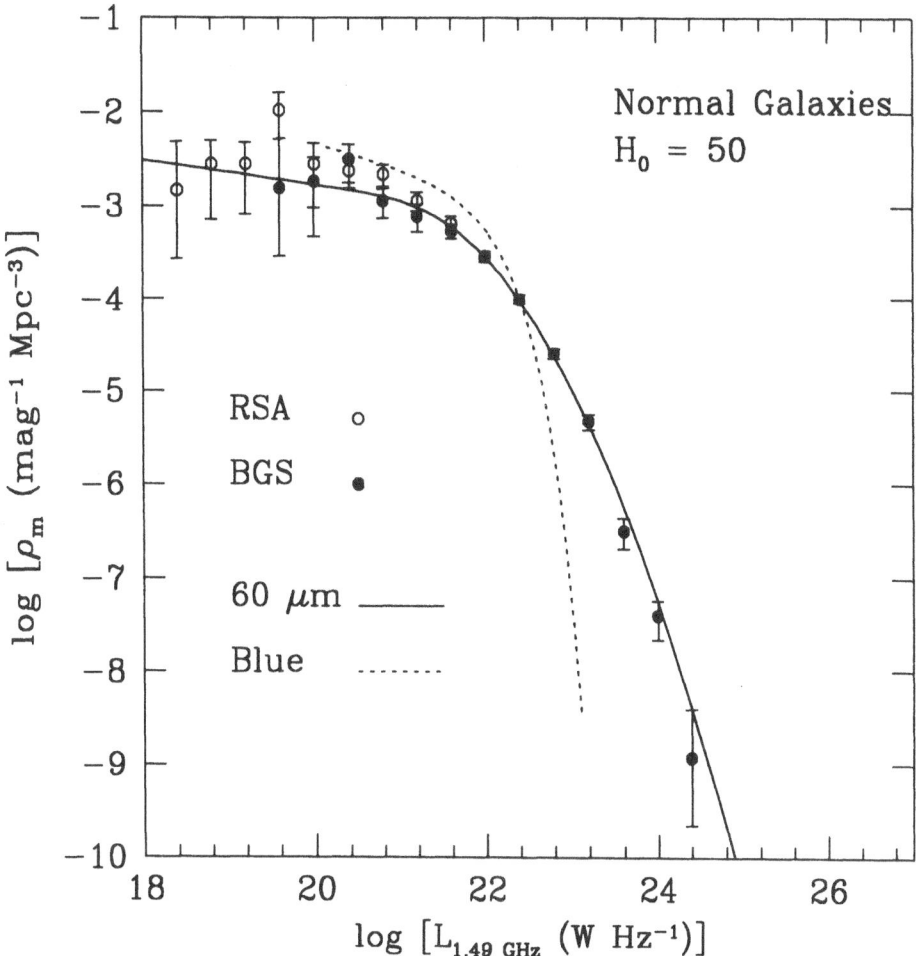

Figure 1. The 1.49 GHz luminosity function of normal galaxies.

This "standard" luminosity function may be high below $\log(L) \approx 21.5$ because the sampled volumes V_m are comparable with the correlation volume of galaxy clustering and are centered on a galaxy—ours. For a two-point correlation function $\xi(r)$, the mean number $\langle N \rangle_P$ of galaxies within a distance r is $\langle N \rangle_P = 4\pi r^3 n/3 + n \int_0^r \xi(x)4\pi x^2 dx$, where n is the density of galaxies averaged over all space (Peebles 1980, Eq. 31.8). If $\xi(r) = (r_0/r)^\gamma$ for $r \leq 2r_0$, $r_0 \approx 10.8$ Mpc, and $\gamma \approx 1.77$ (Davis & Peebles 1983), then the expected density $\langle n \rangle_P$ of galaxies within a distance $r \leq 2r_0$ of our galaxy is $\langle n \rangle_P/n = 1 + 3(r_0/r)^\gamma/(3 - \gamma)$. Thus V_m within $r = 2r_0$ should be multiplied by $\langle n \rangle_P/n$ to yield luminosity functions representative of all space. The open circles and broken curve in Fig. 2 show the radio luminosity func-

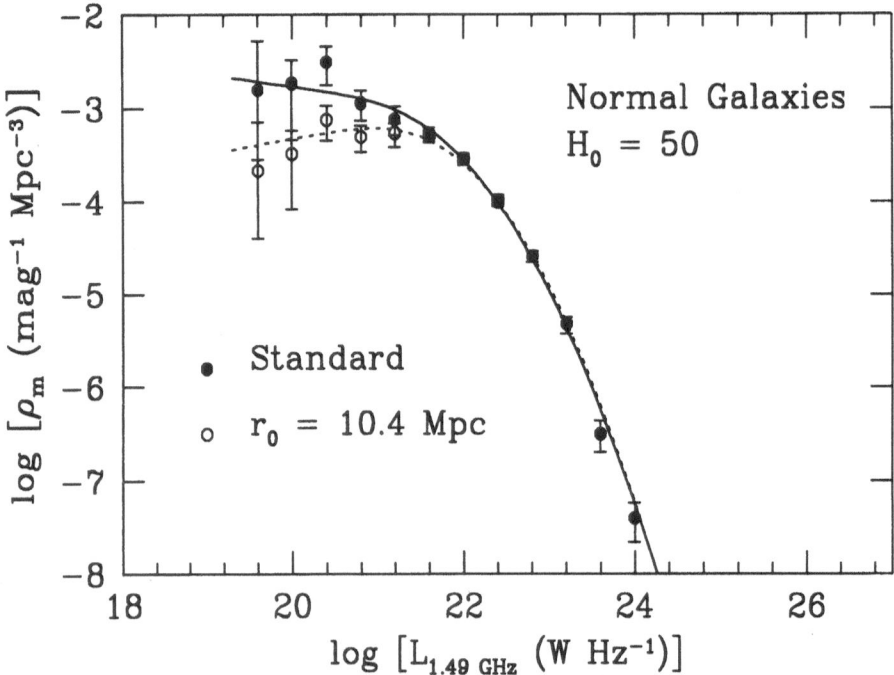

Figure 2.   The 1.49 GHz luminosity function of normal galaxies after correction for clustering (broken curve).

tion of 487 BGS galaxies (including those within $D = 1.7$ Mpc and near Virgo) corrected for the local overdensity. It falls below the $\lambda = 60$ μm luminosity function, in agreement with the nonlinearity of the FIR/radio correlation at low radio luminosities (Condon *et al.* 1991a). The difference between these two curves is a measure of the uncertainty due to clustering, and it can be reduced only by the use of deeper samples with $V_m \gg r_0^3$.

3.   The 4.85 GHz Luminosity Function

A new 4.85 GHz local luminosity function (Fig. 3) was constructed as follows: UGC galaxies with $S \geq 25$ mJy, $5° < \delta < +75°$, and $|b| > 10°$ were identified and classified as normal or AGN with the aid of *IRAS* data (Condon *et al.* 1991b). Luminosity functions based on 128 normal galaxies and 173 AGN were calculated by the "standard" method described in Sec. 2 and are plotted as filled triangles and circles in Fig. 3. Sixteen nearby ($D < 60$ Mpc) early (E/S0) galaxies brighter than $M_P = +14$ and $S = 1.5$ mJy beam^{-1} in the 5″ beam of the VLA (Wrobel & Heeschen 1991) satisfying $u \equiv \log S_{60\mu}/S_{4.85\ \text{GHz}} < 2.0$ extended the AGN luminos-

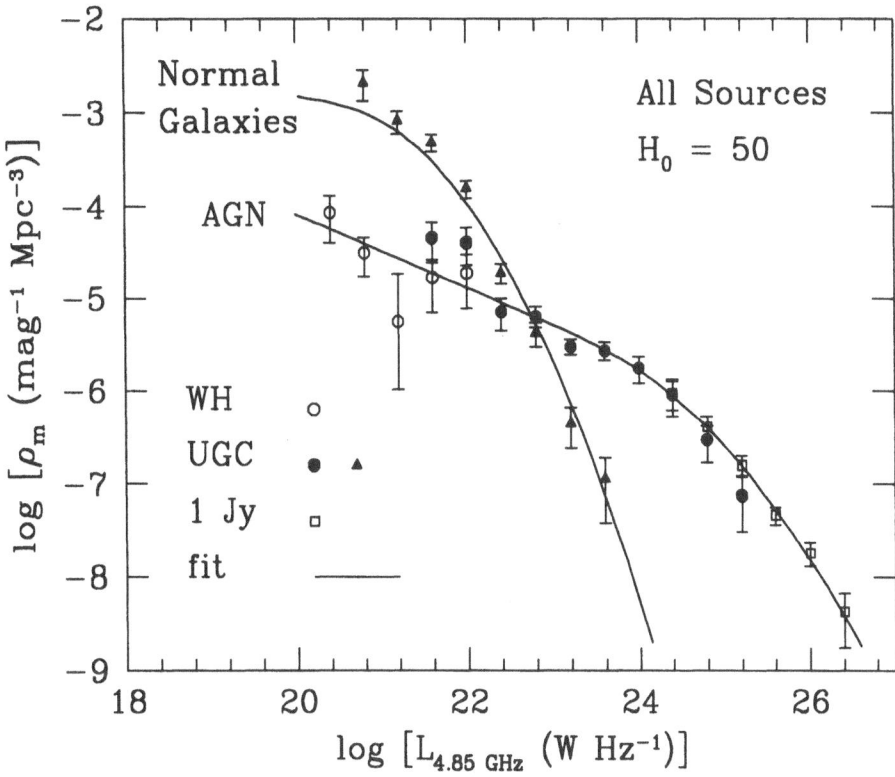

Figure 3. The 4.85 GHz luminosity functions of all radio sources.

ity function as indicated by open circles in Fig. 3. Partial resolution (5″ is only 1.5 kpc at $D = 60$ Mpc) and clustering both increase the uncertainties at low luminosities. Radio identifications of UGC galaxies with the 1.4 GHz NRAO VLA Sky Survey (NVSS) made with 45″ FWHM resolution should ultimately reduce these uncertainties. Optical identifications of the 83 AGN with $z < 0.1$ from the 1 Jy (at 5 GHz) radio source catalog (Stickel *et al.* 1994) were added at high luminosities (open squares in Fig. 3).

Figure 4 shows the luminosity function of flat-spectrum ($\alpha < 0.5$) AGN only, along with the fit to all AGN. The fraction of flat-spectrum sources has a weak minimum near the knee at 10^{24} W Hz^{-1}. The low-luminosity flat-spectrum sources are primarily in S0 galaxies (Condon *et al.* 1991b), suggesting an intrinsic difference not attributable to orientation. Matching this luminosity function is a challenge for relativistic beaming models in which high-luminosity sources are the beamed flat-spectrum cores of low-luminosity sources (cf. Orr & Browne 1982).

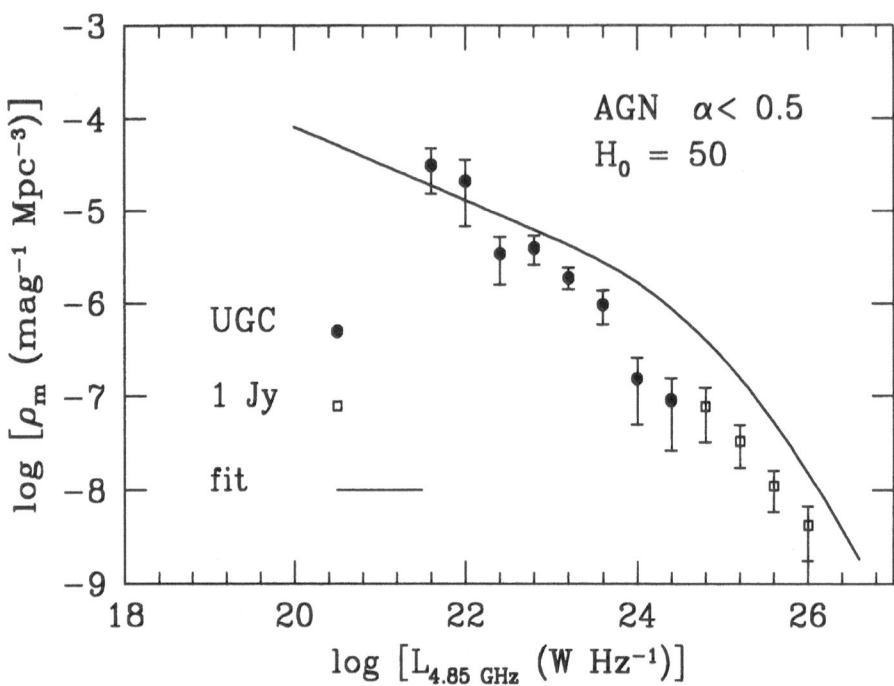

Figure 4. The 4.85 GHz luminosity function of flat-spectrum AGN (data) compared with the that of all AGN (curve).

References

Condon, J.J. 1989, *ApJ*, **338**, 13

Condon, J.J., Anderson, M.L., & Helou, G. 1991a, *ApJ*, **376**, 95

Condon, J.J., Frayer, D.T., & Broderick, J.J. 1991b, *AJ*, **101**, 362

Condon, J.J., Helou, G., Sanders, D.B., & Soifer, B.T. 1990, *ApJS*, **73**, 359

Condon, J.J., Helou, G., Sanders, D.B., & Soifer, B.T. 1995, *ApJS*, in press

Condon, J.J., Huang, Z.-P., Yin, Q.F., & Thuan, T.X.T. 1991c, *ApJ*, **378**, 65

Davis, M., & Peebles, P.J.E. 1983, *ApJ*, **267**, 465

Felten, J.E. 1977, *AJ*, **82**, 861

Hummel, E. 1981, *A&A*, **93**, 93

Moshir, M., *et al.* 1992, Explanatory Supplement to the IRAS Faint Source Survey, Version 2, JPL D-10015 8/92 (Jet Propulsion Laboratory, Pasadena)

Orr, M.J.L., & Browne, I.W.A. 1982, *MNRAS*, **200**, 1067

Peebles, P.J.E. 1980, The Large-Scale Structure of the Universe (Princeton University Press, Princeton)

Sanders, D. B., Egami, E., Lipari, S., Mirabel, I., & Soifer, B. T. 1995, *AJ*, in press

Saunders, W., Rowan-Robinson, M., Lawrence, A., Efstathiou, G., Kaiser, N., Ellis, R.S., & Frenk, C.S. 1990, *MNRAS*, **242**, 318

Soifer, B.T., Boehmer, L., Neugebauer, G., & Sanders, D.B. 1989, *AJ*, **98**, 766

Stickel, M., Meisenheimer, K., & Kühr, H. 1994, *A&AS*, **105**, 211

Urry, C.M., & Padovani, P. 1995, *PASP*, **107**, 803

Wrobel, J.M., & Heeschen, D.S. 1988, *ApJ*, **335**, 677

Wrobel, J.M., & Heeschen, D.S. 1991, *AJ*, **101**, 148

A NUCLEAR LUMINOSITY FUNCTION FOR SEYFERTS

E.J.A. MEURS
Dunsink Observatory
Castleknock, Dublin 15, Ireland

Abstract. Relations between various classes of AGN should show up when examining their LFs. For Seyfert galaxies, such studies have been affected by the prevalence of (essentially) integrated magnitudes. Proper nuclear magnitudes have been determined in an effort of CCD imaging, which enable an appropriate comparison with QSOs.

One way to assess populations of extragalactic objects is offered by their luminosity functions (LFs). Applied to QSOs, radio galaxies, BL Lacs and Seyferts, evolutionary relationships between these categories of AGN, but also with normal galaxies, may be investigated. For an evaluation of the likely connections between Seyferts and QSOs, a potentially serious complication is that most of the magnitudes for Seyfert galaxies refer to rather large apertures. Whereas QSOs are dominated by exceptionally bright cores, the larger aperture magnitudes often only available for Seyferts include substantial contributions from the galaxies themselves. This could notably have influenced discussions of the LF of Markarian Seyfert galaxies such as in Meurs & Wilson (1984).

In order to overcome this contamination from underlying galaxies a dedicated programme of Seyfert photometry was carried out. Short exposure CCD frames in the b band were obtained for some 75 (mostly Markarian) Seyfert galaxies, with the 2.2 m telescope at Calar Alto. Aspects of the data handling have been described in Meurs (1986) and Meurs et al. (1991). The CCD data allow convenient centering on the galaxy nucleus and a suitable extraction area of desired small size is easily selected. Thus, the photometric data will be restricted to the centremost area only. An extraction area of 3 arcsec around the galaxy nuclei was chosen. Detailed photometric data for several Seyferts as available in the literature show that any contribution from the underlying galaxy is not very significant.

R. Ekers et al. (eds.), Extragalactic Radio Sources, 541–542.
© 1996 *IAU*.

The nuclear magnitudes obtained in this way were used to determine an optical nuclear LF (adjusted to B magnitudes) for Markarian Seyfert galaxies, in a fashion similar to the methods employed by Meurs & Wilson (1984). The interesting result is that towards the most luminous Seyfert nuclei the new (nuclear) LF runs close to the former Seyfert LF and the QSO LF. This means that the older (wider aperture) photometric data for the most luminous Seyferts were already dominated by very strong nuclear emission, confirming a conclusion of Meurs & Wilson (1984), and that the continuity of Seyferts with QSOs remains a notable feature also for a LF based on nuclear magnitudes. The latter result will be of importance to evolutionary scenarios linking Seyferts with QSOs.

References

Meurs, E.J.A. (1986) in *The optimization of the use of CCD detectors in astronomy*, Baluteau & D'Odorico (Eds), p. 105

Meurs, E.J.A. and Wilson, A.S. (1984) A&A 136, 206

Meurs, E.J.A., Bonifacio, V. and Lima, N. (1991) *3rd ESO/ST-ECF Data Analysis Workshop*, Grosbol & Warmels (Eds), p. 45

VLA OBSERVATIONS OF THE CAMBRIDGE-CAMBRIDGE ROSAT SURVEY

P. CILIEGI, M. ELVIS, B.J. WILKES
Harvard-Smithsonian Center for Astrophysics, Cambridge, USA

B.J. BOYLE, R.G MCMAHON
Institute of Astronomy, Cambridge, UK

AND

T. MACCACARO
Osservatorio Astronomico di Brera, Milano, Italy

1. Introduction

We report the result of the VLA observations of all the 80 AGN in the Cambridge-Cambridge *ROSAT* Serendipity Survey (CRSS, Boyle et al. 1995), a new well defined sample of 80 X-ray selected AGN with $f_x(0.5\text{-}2.0\text{keV}) \geq 2 \times 10^{-14}$ erg s^{-1} cm^{-2}. Our aim was to obtain a complete classification of the sample members as Radio-loud (RL) or Radio-quiet (RQ) in order to determine well-constrained X-ray luminosity function (XLF) for X-ray selected RQ and RL AGN separately.

Of the 80 AGN in the sample, seven show radio emission at 5 σ level and only two ($2.5^{+4.0}_{-1.7}$ %) qualify as Radio-Loud (RL) objects ($\alpha_{ro} \geq 0.35$, see Ciliegi et al. 1995 for a detailed description of these VLA observations of the CRSS AGN sample). This result, compared with 13% RL in the EMSS sample of AGN (flux limit $f_x(0.3\text{-}3.5\,\text{keV}) \sim 2 \times 10^{-13}$ erg s^{-1} cm^{-2}) confirms the prediction of Della Ceca et al. (1994) that the expected fraction of RL should drops rapidly as the X-ray flux limit is lowered.

2. The X-ray luminosity function

In order to determine well-constrained XLFs for X-ray selected RQ and RL AGN separately, we have combined the CRSS data with the EMSS data. Using the V_e/V_a variable of the $1/V_a$ method of Avni and Bahcall

R. Ekers et al. (eds.), Extragalactic Radio Sources, 543–544.
© 1996 *IAU.*

(1981) we find that both RL and RQ samples exhibit significant cosmological evolution. Following Boyle et al. (1993), we parameterized the XLF with a two-power-law form $\Phi_X(L_X) = \Phi_X^* L_{X_{44}}^{-\gamma_1}$ for $L_X < L_X^*(z = 0)$ and $\Phi_X(L_X) = (\Phi_X^* \times L_{X_{44}}^{-\gamma_2})/L_{X_{44}}^{(\gamma_1 - \gamma_2)}$ for $L_X > L_X^*(z = 0)$ where Φ_X^* is the normalization of the XLF and γ_1 and γ_2 are the faint and bright end slopes respectively. $L_{X_{44}}$ is the 0.3-3.5 keV X-ray luminosity expressed in units of 10^{44} erg s^{-1}.

Using a cosmological model with $q_0=0$ and $H_0=50$, we find that the best-fit parameters are Log $L_X^* = 44.3 \pm 0.2$, $\gamma_1 = 0.80 \pm 0.22$, $\gamma_2 = 3.03 \pm 0.20$ for the RL subsample and Log $L_X^* = 43.9 \pm 0.0.2$, $\gamma_1 = 1.82 \pm 0.15$, $\gamma_2 = 3.70 \pm 0.10$ for the RQ subsample. These data show that the shape of the XLF of the two classes appear to be different both in their low luminosity and high luminosity slopes (parameters γ_1 and γ_2).

We have investigated the possibility of explaining the difference between the XLFs of the two classes of objects in terms of an additional beamed radio-linked component producing X-rays. This component, intrinsically weak, becomes dominant when the direction of the jet with which it is associated is oriented close to the line of sight. In this "X-ray beaming" model, the total X-ray luminosity L_x of AGN can be written as $L_x = L_{xb} + L_{xu}$ where L_{xu} is the unbeamed X-ray luminosity associated with the radio-quiet mechanism which occurs in both RQ and RL and L_{xb} is the beamed X-ray luminosity which is dominant in core-dominated RL due the beaming effect. Using the relation $\mathrm{Log} L_x = 0.13 \times \mathrm{Log}(L_{xb}/L_{xu}) + 27.52$ found by Kembhavi 1993, we obtained the L_{xu} for all the RL AGN in our sample. Using L_{xu} we have re-calculated the XLF for RL AGN. The best-fit parameters for this "unbeamed" XLF are Log $L_X^* = 44.4 \pm 0.2$, $\gamma_1 = 1.65 \pm 0.22$ and $\gamma_2 = 3.66 \pm 0.20$. The XLF for RQ AGN and for unbeamed RL AGN are now consistent (parameters γ_1 and γ_2) within the 1σ errors.

Therefore, we can conclude that the differences in the shape of XLF between RQ and RL AGN can be explained introducing the X-ray beaming model where the "radio-linked" component in RL objects is orientation-dependent, but larger samples of X-ray selected AGN are needed to strengthen this conclusion.

References

Avni Y., Bachall J.N., 1980, *ApJ*, **235**, 694.
Boyle B.J., McMahon R.G., Wilkes B.J. and Elvis M. 1995, *MNRAS*, **272**, 462.
Boyle B.J. et al., 1993, *MNRAS*, **260**, 49.
Ciliegi P. et al., 1995, *MNRAS*, in press.
Della Ceca R. et al., 1994, *ApJ*, **430**, 533.
Kembhavi A.K. 1993, *MNRAS*, **264**, 683.

ON THE COSMOLOGICAL EVOLUTION OF RADIO QUIET
AND RADIO LOUD QUASARS

J. MACHALSKI
Astronomical Observatory, Jagellonian University
ul. Orla 171, PL-30244 Cracow, Poland

Key words: Quasars - cosmological evolution

This contribution gives further insight into the problem of an 'evolution' of the function $G(> R)$ [R is the ratio between radio and optical luminosity of a QSO] suggested by Visnovsky et al.(1992,ApJ,391,560), and recently by Schmidt et al.(1995,AJ,109,473). Using the Optical Luminosity Function for optically selected quasars derived by Boyle et al.(1988,MNRAS, 235,935), and evolving (cf. Schmidt et al. 1995) or non-evolving (cf. Marshall 1987,ApJ,316,84) $G(> R)$ function, an evolutionary function $E(z)$ for the population of both *radio quiet* and *radio loud* quasars is determined by modelling the distributions of redshift $n(z)$, as well as counts of apparent optical blue magnitude $n(B)$ and radio 5 GHz flux density $n(S)$ in samples with *different* radio and optical *limits*, respectively, and fitting them to the available observational data.

Resultant evolutionary function in the form $E(z) \propto (1+z)^{k(z)}$ is shown in Fig.1(left panel). In Fig. 1(right panel), and in Fig.2(left and right panel), the observed distributions of redshift $n(z)$, blue magnitude $n(B)$, and 5 GHz flux density $n(S)$, in different samples of optical and radio QSOs, are compared with the distributions expected from the Model.

The conclusions are as follows:

(1) The resultant function $E(z)$ is strikingly similar to the corresponding one derived from the evolutionary function for the total population of radio sources found by Condon (1984,ApJ,287,461). The $E(z)$ resulting from the present model peaks at $z \approx 2.3$ and slowly decreases for higher redshifts. A similar result of a 'redshift cut-off' was found by Dunlop & Peacock (1990,MNRAS,247,19) for flat- and steep-spectrum radio loud quasars. This is the first (to my knowledge) indication that a similar evolution may apply to luminous but radio quiet QSOs.

R. Ekers et al. (eds.), Extragalactic Radio Sources, 545–546.
© *1996 IAU.*

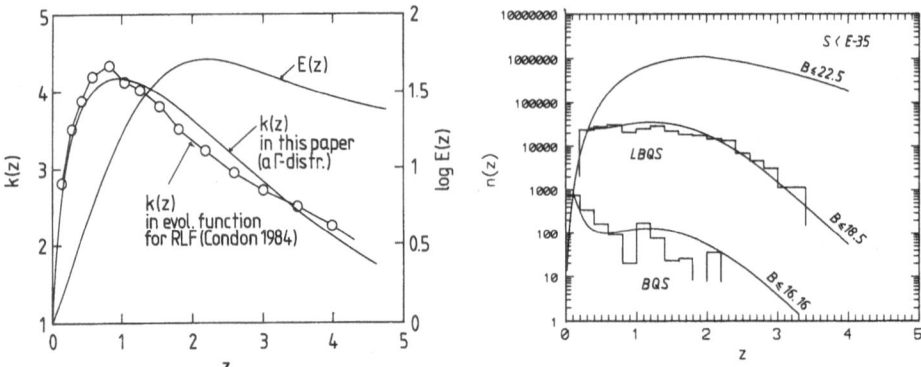

Figure 1. **Left panel:** Fitted power $k(z)$ in the evolutionary function $E(z)$ and the function itself. **Right panel:** Observed redshift distributions for radio quiet QSO, and the model predictions

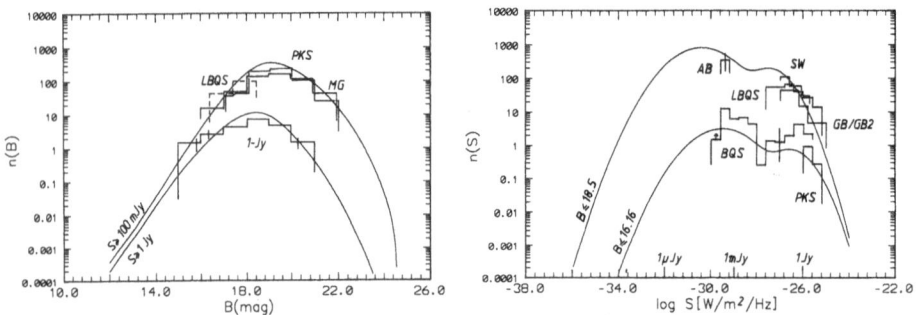

Figure 2. **Left panel:** Observed distributions of B magnitude for radio loud QSO, and the model predictions. **Right panel:** Observed distributions of 5-GHz flux density, and the Model predictions

(2) The same evolution for both populations of QSOs cannot be rejected, although the available data on radio quiet but optically luminous QSOs are still insufficient to prove that.

(3) Existing observational data do not allow yet to constrain predicted distributions $n(z)$, $n(B)$, $n(S)$ for evolving or non-evolving function $G(> R)$. The calculations showed that in order to model these distributions with the function $G(> R, z)$ evolving (decreasing) with redshift [as proposed by Schmidt et al.], a stronger luminosity evolution $E(z)$ would be necessary for QSOs. The fit still can be satisfactory, but evolutionary functions for quasars and all radio sources will be different.

SPACE DISTRIBUTION OF RADIO-SOURCE POPULATIONS

J.V. WALL
Royal Greenwich Observatory
Madingley Road, Cambridge, U.K. CB3 0EZ

This paper sets out the status of data determining the space distribution of extragalactic radio-source populations, describes some recent results from analyses of the data, and indicates why and how the analyses need revision in the light of unified models. It concludes by emphasizing the severity of the effects of large-scale structure on modern survey data.

1. Overview: the basic data N(S) and N(P)

The data required to determine the space distribution for a population of extragalactic sources consist of a) a *source count* N(S), the surface density on the sky determined directly in a survey, and b) at least one *luminosity distribution* N(P), the frequency distribution of radio powers for a complete flux-limited sample, determined by identifying and measuring redshifts for all sample members. The translation of these data into space densities can be described in simple Euclidean terms (Wall, 1983) which make evident the analogies with other analyses such as V/V_m or $1/V_m$ (Schmidt, 1968; Katgert *et al.*, 1979). To determine the spatial distribution, N(P) and N(S) are needed *for each population*. In practice N(P) can be constructed for each object-type by dividing a single N(P) according to the physical properties of its members. However a source-count for each object-type is currently too difficult – the usual approach is to divide N(P) while using a total count to constrain the luminosity-function estimates for the summed populations.

The state of the basic data N(S) and N(P) is summarized in Fig. 1. Definition of source counts at the highest flux densities has reached a fundamental limit: there is no more sky to survey. Counts at the lowest flux densities also approach a fundamental limit: the deepest surveys with the VLA and WSRT reach intensities corresponding to sky surface densities $> 10^7$ per sterad, and the confusion limit is close. All counts show general agreement, regions of overlap using totally different instruments showing

547

R. Ekers et al. (eds.), Extragalactic Radio Sources, 547–552.
© 1996 IAU.

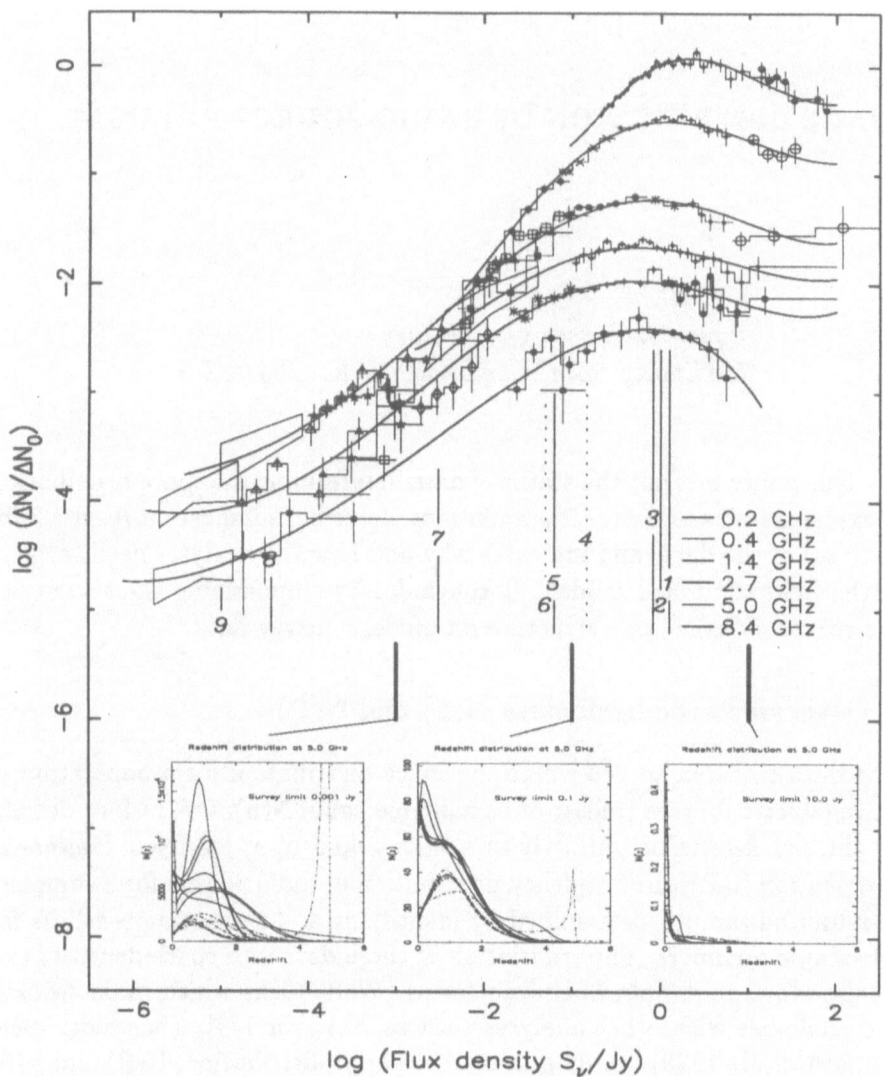

Figure 1. Source counts in relative differential form at 178, 408, 1400, 2700, 5000 and 8440 MHz, descending order. $N_o = K_\nu S_\nu^{-1.5}$, where $K_\nu = 2400, 2730, 3618, 4247, 5677$ and 3738 for the 6 frequencies. References for the surveys are given in Fig. 1 of Wall (1994), with the 10-GHz data replaced here by an 8.44-GHz compilation (Windhorst *et al.* 1993). Polygons represent count estimates from P(D) (background-deflection) analysis. The dashed curves are polynomial least-square fits. Vertical lines indicate equivalent 5-GHz flux-density limits for samples constituting luminosity distributions available for analysis of space distribution. Dotted lines indicate samples-to-be. The limiting flux densities were 'transposed' to 5 GHz with a spectral index of -0.75; actual frequency of compilation is indicated by the row in which each reference is given. These are: 1 Wall & Peacock 1985 and unpublished data; 2 Kühr *et al.* 1981; 3 Laing *et al.* 1983 and unpublished data; 4 R.M. Athreya, this volume; 5 Downes *et al.* (1986); 6 87GB+PMN 5-GHz all-sky survey; 7 Benn *et al.* (1988) and unpublished data; 8 Benn *et al.* (1993); 9 Windhorst *et al.* (1995). Luminosity distributions predicted from the Dunlop-Peacock (1990) space-density models (solid lines for steep-spectrum, dashed for flat-spectrum) are shown under the heavy vertical bars indicating the flux-density limit for each.

few discrepancies (Wall, 1994). Fig. 1 also indicates the equivalent flux-density levels at which luminosity distributions N(P) are complete or approaching completion.

2. Space distribution and the populations

Dunlop & Peacock (1990) analysed the data then available using MEM and free-form epoch-dependent luminosity functions. The uncertainties in space density were mapped by adopting 7 start-point formulations. The analysis was carried out assuming two populations, 'steep-spectrum' and 'flat-spectrum' sources. Principal conclusions were: a) for powerful sources a decline in space density is indicated for epochs corresponding to $z > 3$, b) pure luminosity evolution was permitted by the data, and c) the space distribution for flat and steep-spectrum populations was similar. Predictions of N(P) from the 7 models are shown in Fig. 1. The extreme uncertainties shown at $S_{5GHz} = 0.001$ Jy demonstrate the impact which the new data at these levels will have in improving definition of space distribution.

However, new analyses must do better in terms of the populations: in the face of unified models, retaining 'flat-spectrum' and 'steep-spectrum' as classification is completely erroneous. The common-place populations of the literature are listed in Table 1.

TABLE 1. Extragalactic radio sources

Radio source type	Radio spectrum	Beamed	Membership
Radio galaxies FRI	steep	Y?	1
BL Lac objects	flat	Y	1
Radio-loud QSOs	flat	Y	2
Radio-loud QSOs	steep	Y	2
Radio galaxies FRII	steep	Y?	2,1
GigaHertz-Peak Spectrum sources	'flat'	N?	2
Compact Steep-Spectrum sources	'steep'	N?	2?
Compact Symmetric Objects	flat?	Y	2
High-z radio galaxies	steep	Y	2
Halos, relics	steep	N	3
Starbursters	steep	N	4

Of these 'populations', it is suggested that *memberships 1 (low power) and 2 (high power) encompass virtually all sources catalogued above 1 mJy.* (The relatively few sources of membership 3 (L. Ferretti, this volume) are not considered here, nor is membership 4, included in the tangled populations which appear below 1 mJy; see *e.g.* Windhorst *et al.* 1993, 1995 and F. Hammer, this volume). The considerations are as follows:

1. Unified-model paradigms work (Antonucci, 1993; Urry & Padovani, 1995); in simplest terms, these hold that BL Lac objects are end-on members of FRI radio galaxies (membership 1) while flat-spectrum QSOs are the end-on members of the FRII radio galaxies (membership 2). There are difficulties (*e.g.* Lawrence 1991; Singal 1993) – the picture must be oversimplistic in that (at least) age and environment must play a rôle. In particular uniform optical data for the complete 3CR sample (Laing *et al.*, 1994) show that many *bona fide* FRIIs have very low optical excitation and cannot be part of the QSO paradigm. At lower flux densities, the majority of identifications are with elliptical galaxies showing very weak or no emission lines (Rixon *et al.*, 1991; Wall *et al.*, 1993; Dunlop *et al.*, 1995); many of these have FRII radio structures. These probably represent a major component of the parent population for BL Lac objects. Further evidence is a) the FRII-type structure visible about BL Lac objects (Kollgaard *et al.*, 1992) and b) the inability of FRI galaxies in clusters to provide adequate BL Lac numbers (Owen *et al.*, 1995). FRII radio galaxies thus have membership in both paradigms, as indicated in Table 1.

2. The CSS class of sources (Kapahi, 1981; Peacock & Wall, 1982), comprising $\sim$ 30% of all sources selected at 2.7 GHz, are either young or straight-jacketed versions of the powerful double radio sources (R. Fanti, this volume). GigaHertz-peaked-spectrum (GPS) sources (*e.g.* O'Dea *et al.*, 1991; Snellen *et al.*, 1995) and Compact Symmetric Objects (CSO) (A. Readhead, this volume) are similarly related.

3. But there are questions of taxonomy to be resolved before we understand the relation between the radio-loud classes. Some QSOs typified by 3C48 and 3C119 have distorted radio structures which do not fit comfortably into relativistically-fed double-structure models. Moreover in an investigation to determine the redshift cutoff for radio-beamed QSOs (P.A. Shaver *et al.*, this volume), a 'flat-spectrum' population of radio galaxies occupies the faint-magnitude reaches of the fully-identified sample. It is likely that these faint red galaxies are the objects which Webster *et al.* (1995) considered to be obscured QSOs. The relation of these objects to CSS and GPS sources is uncertain, let alone their rôle in any unified paradigm.

4. A spatial analysis which accounts for the physical relations sketched here is overdue; first steps have been described in the pioneering paper by Orr & Browne (1982), by Urry & Padovani (1995) and by C. Jackson and JVW (this volume). There is already indication that the memberships 1 and 2 of Table 1 are distributed very differently in space. From Longair (1966) on, analyses have found that the powerful objects (membership 2) show strong evolution, while the weaker show little or no change

in co-moving space density. The redshift cutoff of the flat-spectrum QSOs (membership 2) has been detected (Dunlop & Peacock 1990, P.A. Shaver *et al.*, this volume); the lower-power radio galaxies may show a space-density cutoff at a smaller redshift (J.S. Dunlop *et al.*, this volume).

3. Large-scale Structure

The apparent uniformity of the radio sky (*e.g.* Webster 1977) arises quite simply: to detect anisotropies, surveys must reach a level at which large structures each contribute more than one source to the survey. At a redshift of 1, flux-densities of $\lesssim 10$ mJy must be attained before this is the case (Benn & Wall, 1995a). However, surveys now exist which cover most of the sky and reach to (3 - 5) $\times$ 10 mJy, close to this limit. The VLA has embarked on large surveys of the sky (Becker *et al.*, 1995) complete at levels well below that required to see structure.

Statistical investigations of radio-source distribution (Wall *et al.* 1995) have followed two routes: (a) survey analysis with the two-point correlation function, ideally suited to irregularly-shaped areas, and b) prediction of survey-to-survey variation with toy-universe models. The former has been applied to the 87GB and PMN surveys (Wall *et al.*, 1993; Kooiman *et al.*, 1995; Wall *et al.*, 1995), and the signal which is seen at angular scales $< 1°$ is two orders of magnitude stronger than that predicted from galaxy clustering. The second type of analysis modelled the structure by Voronoi tessellation (Benn & Wall, 1995a) to place limits on the scale size of the largest cells. Surveys to 1994 placed a limit of 150 h^{-1} Mpc as the mean distance between cells; analysis of the initial region of the FIRST survey indicates that the limit could be reduced to 50 h^{-1} Mpc. In the context of other constraints on large-scale structure *this limit occupies a critical range between those provided by galaxy surveys and by COBE results.* The imprint of large-scale structure has also been seen directly: some 60 redshifts for sources in the 5C12 survey (Benn & Wall, 1995b) show that the majority of these sources are associated with 1 to 3 other sample members. The projected diameters of these groups range from 10 to 70 h^{-1} Mpc, sizes in accord with the largest structures found in optical/IR surveys. This is the *first direct detection of structure in an unbiassed radio survey.* Multi-object spectrographs on large telescopes open the possibility of tracing structure directly through deep radio surveys out to $z = 1.0$.

'Cosmologically representative' samples from the FIRST and similar new surveys will have to be chosen with care. Moreover it is now no longer adequate to describe radio-source distribution in the simple radial terms of epoch-dependent luminosity functions.

References

Antonucci, R. (1993) *Ann. Rev. Astron. Astrophys.*, **31**, 473

Becker, R., White, R.L. and Helfand, D.J. (1995) *Astrophys. J.*, **450**, 559

Benn, C.R., Wall, J., Grueff, G. and Vigotti, M. (1988) *Mon. Not. R. astr. Soc.*, **230**, 1

Benn, C.R., Rowan-Robinson, M., McMahon, R.G., Broadhurst, T.J. and Lawrence, A. (1993) *Mon. Not. R. astr. Soc.*, **263**, 98

Benn, C.R. and Wall, J.V. (1995a) *Mon. Not. R. astr. Soc.*, **272**, 678

Benn, C.R. and Wall, J.V. (1995b) in *Wide Field Spectroscopy and the Distant Universe*, Proc 35th Herstmonceux Conf., eds Maddox, S.J. and Aragón-Salamanca, A., World Scientific, 184

Downes, A.J.B., Peacock, J.A., Savage, A. and Carrie, D.R. (1986) *Mon. Not. R. astr. Soc.*, **218**, 31

Dunlop, J.S. and Peacock, J.A. (1990) *Mon. Not. R. astr. Soc.*, **247**, 19

Dunlop, J.S. Peacock, J.A. and Windhorst, R.A. (1995) in *Galaxies in the Young Universe*, Proc. Ringberg Conf., eds Hippelein, H. and Meisenheimer, K., in press

Kapahi, V.K. (1981) *Astron. Astrophys. Suppl. Ser.*, **43**, 381

Katgert, P., de Ruiter, H.R. and van der Laan, H. (1979) *Nature*, **280**, 20

Kooiman, B., Burns, J.O. and Klypin, A.A. (1995) *Astrophys. J.*, **448**, 500

Kollgaard, R.I., Wardle, J.F.C., Roberts, D.H. and Gabuzda, D.C. (1992) *Astron. J.*, **104**, 1687

Kühr, H., Witzel, A., Pauliny-Toth, I.I.K. and Nauber, U. (1981) *Astron. Astrophys. Suppl. Ser.*, **45**, 367

Laing, R.A., Riley, J.M. and Longair, M.S. (1983) *Mon. Not. R. astr. Soc.*, **204**, 151

Laing, R.A., Wall, J.V., Jenkins, C.R. and Unger, S.W. (1994) in *The Physics of Active Galaxies*, eds Bicknell, G.V. *et al.*, *ASP Conf. Series*, **54**, 201

Lawrence, A. (1991) *Mon. Not. R. astr. Soc.*, **252**, 586

Longair, M.S. (1966) *Mon. Not. R. astr. Soc.*, **133**, 421

O'Dea, C.P., Baum, S.A. and Stanghellini, C. (1991) *Astrophys. J.*, **380**, 660

Owen, F.N., Ledlow, M.J. and Keel, W.C. (1995) *Astron. J.*, in press

Orr, M.J.L. and Browne, I.W.A. (1982) *Mon. Not. R. astr. Soc.*, **200**, 1067

Peacock, J.A. and Wall, J.V. (1982) *Mon. Not. R. astr. Soc.*, **198**, 843

Rixon, G.T., Wall, J.V. and Benn, C.R. (1991) *Mon. Not. R. astr. Soc.*, **251**, 243

Schmidt, M. (1968) *Ap. J.*, **151**, 393

Singal, A.K. (1993) *Mon. Not. R. astr. Soc.*, **262**, L27

Snellen, I.A.G., Zhang, M., Schilizzi, R.T., Röttgering, H.J.A., de Bruyn, A.G. and Miley, G.K. (1995) *Astron. Astrophys.*, in press

Urry, C.M. and Padovani, P. (1995) *PASP*, **107**, 803

Wall, J.V. (1983) in *The Origin and Evolution of Galaxies*, Proc. NASI, eds Jones, B.J.T. and Jones, J.E., Reidel, 295

Wall, J.V. (1994) *Aust. J. Phys.*, **47**, 625

Wall, J.V. and Peacock, J.A. (1985) *Mon. Not. R. astr. Soc.*, **216**, 173

Wall, J.V., Rixon, G.T. and Benn, C.R. (1993) in *Observational Cosmology*, eds Chincarini, G. *et al.*, *ASP Conf. Ser.*, **51**, 576

Wall, J.V., Benn, C.R. and Loan, A.J. (1995), in *Examining the Big Bang and Diffuse Background Radiations*, Proc. IAU Symp. 168, eds Kafatos, M. *et al.*, Kluwer, in press

Webster, A.S. (1977) in *Radio Astronomy and Cosmology*, Proc. IAU Symp. 74, ed. Jauncey, D.L., Reidel, 75

Webster, R.L., Francis, P.J., Peterson, B.A., Drinkwater, M.J. and Masci, F.J. (1995) *Nature*, **375**, 469

Windhorst, R.A., Fomalont, E.B., Partridge, R.B. and Lowenthal, J.D. (1993) *Astrophys. J.*, **405**, 498

Windhorst, R.A., Fomalont, E.B., Kellermann, K.I., Partridge, R.B., Richards, E., Franklin, B.E., Pascarelle, S.M. and Griffiths, R.E. (1995) *Nature*, **375**, 471

SPACE DENSITIES FOR POWERFUL RADIO SOURCES IN THE LIGHT OF UNIFICATION

C.A. JACKSON
*Institute of Astronomy, University of Cambridge,
Cambridge, CB3 0HA, UK*

AND

J.V. WALL
*Royal Greenwich Observatory, Madingley Road,
Cambridge, CB3 0EZ, UK*

As radio survey frequency is raised the proportion of flat-spectrum sources increases in bright flux-limited samples (*eg* Wall 1994, *Aust J Phys* **47**, 625). Differential source counts show a corresponding broadening of the central maximum due to the increasing proportion of flat-spectrum sources. Orr & Browne (1982, *MNRAS* **200**, 1067) modelled this change in shape of the source count by proposing a unifying scheme which states that the core-dominated, flat-spectrum radio sources are the steep-spectrum sources with their cores Doppler-boosted due to the alignment of the jets with the line of sight.

Investigation of the space densities of radio sources should proceed with populations which are physically delineated; in the face of unified models, the traditional division into 'flat-spectrum' and 'steep-spectrum' populations is incorrect. To this end we are undertaking a new space-density analysis to explore the implications of unified-model schemes, including both the radio-loud QSO – FRII radio-galaxy paradigm and the BL Lac – FRI radio-galaxy paradigm (see Urry and Padovani 1995, *PASP* **107**, 803). To test the formalism, our first stage described here uses (1) complete samples and source-count data over a wide frequency range and (2) optimizing techniques to explore *parameterized* evolution and beaming models.

This initial analysis followed the scheme developed by Wall *et. al* (1980, *MNRAS* **193**, 683). Together with a 151-MHz source-count, the 162 steep-spectrum sources in the 3CR sample (Laing *et. al* 1983, *MNRAS* **204**, 151) were used to define the epoch-dependent luminosity function of the 'parent'

R. Ekers et al. (eds.), Extragalactic Radio Sources, 553–554.

population. The best-fit parameters were determined using the *AMOEBA* downhill simplex method in multidimensions (Press *et. al* 1992, *Numerical Recipes in Fortran* (CUP), 402), evaluating χ^2 between the observed and model source counts. For evolution of the form $exp(M(1 - t/t_0))$ the optimal parameters ($\Omega = 1, h = 0.5$) are M=10.92, z_c=4.075 and transition powers between evolving and non-evolving sources at $\log_{10}(P_1) = 25.33$, $\log_{10}(P_2) = 27.57$. This demonstrates that modern data comprising complete redshifts for the 3CR sources plus a deep source count *require* a redshift cut-off in the space density for steep-spectrum sources.

These parameter values and a single spectral index of -0.75 were used to estimate the 5 GHz count of steep-spectrum sources (Figure 1). Inclusion of the flat-spectrum, beamed population at 5 GHz was achieved with two additional parameters, the Lorentz factor γ and the rest frame core-to-extended flux ratio R_c. The observed core-to-extended flux ratio R_{obs} is given by $R_{obs} = R_c([\gamma(1 - \beta\cos\theta)]^{-2+\alpha_{flat}} + [\gamma(1 + \beta\cos\theta)]^{-2+\alpha_{flat}})$ for a source comprising a pair of continuous relativistic jets with bulk plasma velocity βc whose ejection axis is aligned at a random angle θ ($\geq 0°, \leq 90°$) to the line of sight. We adopted $\alpha_{flat} = 0.0$, and took a source as being 'flat-spectrum' for small enough values of $\theta < \theta_c$ such that $R_{obs} \geq 1.0$ and its observed flux density $S_{enhanced} = R_{obs}.S_{\nu_1}$. For $\gamma = 10.0$ and $R_c = 0.02$ ($\theta_c = 8°$), the count of flat-spectrum sources summed with the steep-spectrum source count closely follows the observed count (Figure 1).

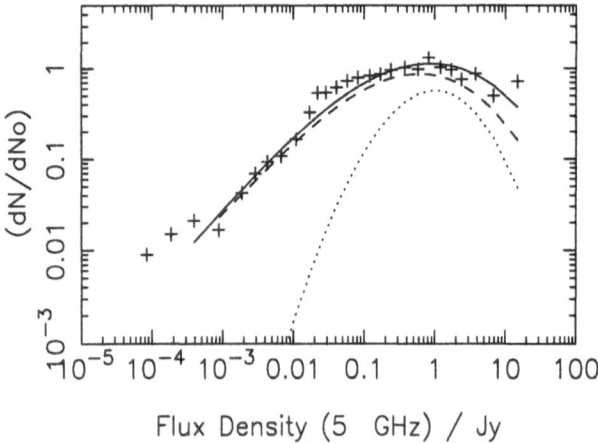

Figure 1. Model and observed source counts at 5 GHz: ++++ observed source count, - - - - model count for steep-spectrum objects, ⋯ model count for flat-spectrum objects, —— total model source count.

This initial analysis demonstrates that (a) a diminution in the space density of 'parent' sources at redshifts above 4 is required, and (b) the FRII – radio-loud QSO unified scheme is consistent with the high-frequency count data for reasonable beaming parameters.

THE RADIO SKY AT MICRO-JY LEVELS

E. B. FOMALONT

National Radio Astronomy Observatory
520 Edgemont Road, Charlottesville, VA 22903, USA

1. Introduction

The properties of the μJy radio sources are described in this paper. The results were obtained from deep VLA observations of four fields, shown in Table 1. Deep optical images have been obtained for the Bootes and Lilly1 fields.

TABLE 1. VLA DEEP OBSERVATIONS

Field	RA	DEC	Freq	Res	S_{lim}	Int	No	Ref
	J2000		GHz	''	μJy	hrs		
Bootes	14 17 59	52 27 12	4.9,1.4	4	16.0	126	62	(1), (2)
Cepheus	03 17 25	80 21 05	8.4	10	22.9	20	14	(3)
Lynx	08 45 04	44 34 05	8.4	10	14.5	62	6	(3)
Lilly1	13 12 17	42 38 06	8.4	6	7.0	159	39	(4)

(1)=Fomalont *et al.* 1991, (2)=Hammer *et al.* 1995, (3)= Windhorst *et al.* 1993, (4)=Windhorst *et al.* 1995, and unpublished

2. The Radio Source Properties

The following radio properties are based on 121 sources in the complete samples. The Euclidean normalized counts of sources at 8.4 GHz, derived from the three 8.4 GHz experiments, are shown in Figure 1. The best fit slope to the 8 GHz counts at μJy levels is $N = 18\ S^{-1.2}$, where N is the number of sources $(\text{arcmin})^{-2}$ with a flux density greater than $S\ \mu$Jy. This corresponds to a differential count of $n = 0.33\ S^{-2.2}(\text{Jy})^{-1}(\text{sr})^{-1}$. The error of the count parameters is about 10%.

R. Ekers et al. (eds.), Extragalactic Radio Sources, 555–558.

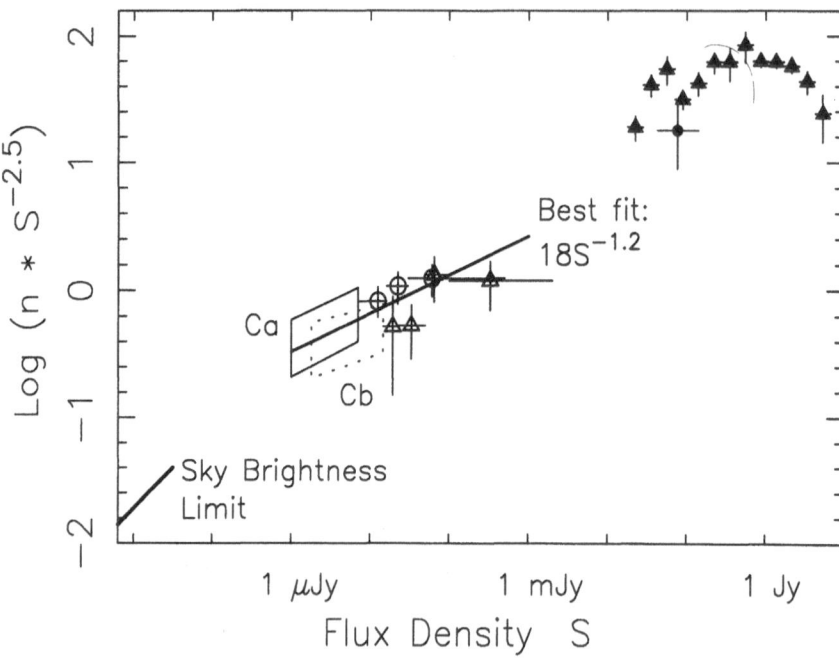

Figure 1. **The Source Counts at 8.4 GHz:** The number, n, of sources $(Jy)^{-1}(sr)^{-1}$ is the ordinate. The plotted points show the count from the detected sources (Lynx and Cepheus fields $\triangle$, the Lilly1 field $\bigcirc$). The trapezoidal boxes are the count derived from the fluctuations in the image caused by weak sources (Ca box for the Lynx and Cepheus fields, Cb box for the Lilly1 field). See Windhorst *et al.* (1993). The best fit slope to the 8 GHz data < 1 mJy is shown by the solid line. The upper limit source count at the nanoJy level which is consistent with less than 0.02 K sky brightness from discrete sources at 8 GHz is shown by the other solid line.

The slope, $\gamma = -2.2$, of the count is surprisingly steep and close to the Euclidean value of -2.5. Since the typical redshift is > 0.5 for the μJy sources, they have evolved with cosmological epoch. However, detailed models cannot be made with the limited sample. An upper limit to the count of sources at the nanoJy levels can be obtained from measurements of the sky brightness. At frequencies between 3 to 50 GHz there is no excess sky brightness contribution, above that of the CBR of 2.74 K, at a level of 0.02 K (Windhorst *et al.*, 1993). The line in the lower left of Figure 1 conforms to this limit.

The estimate of the average angular size of the μJy radio sources have been determined using detailed simulations of the effects of the blending of faint sources (Richard 1996). Of the 38 sources in the Lilly1 field, nine were resolved at 6″ resolution. However, the simulations show that about this number of sources should appear to be resolved, solely from confusion with

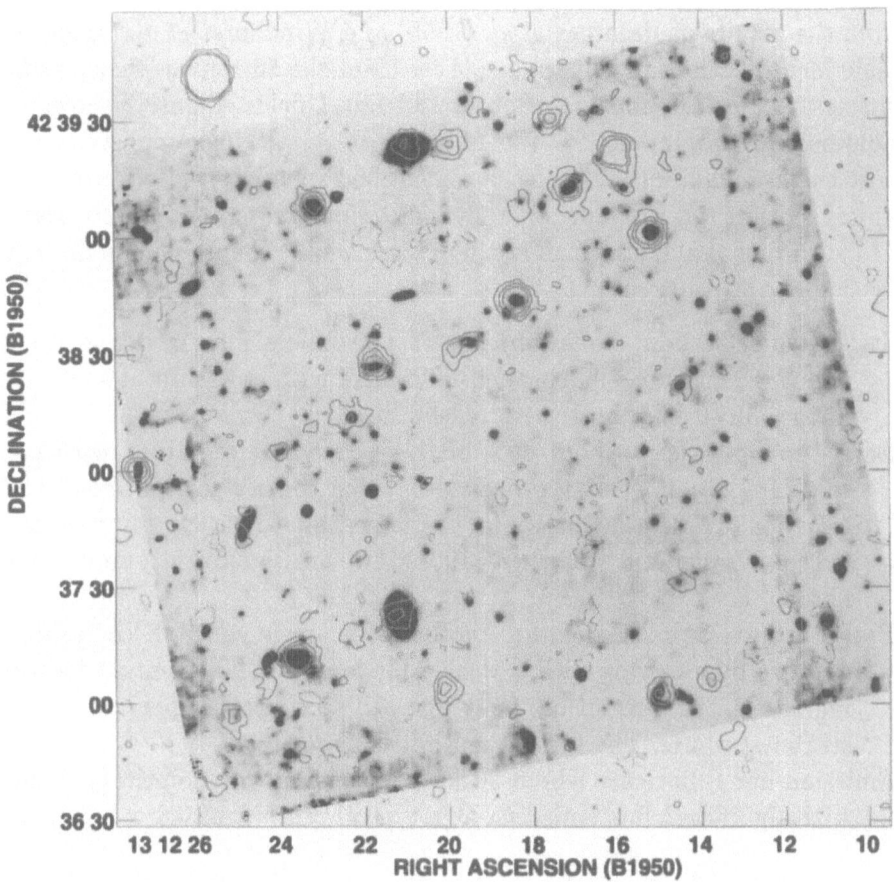

Figure 2. **The Lilly1 Field, Radio and Optical:** The radio image with 6″ resolution is shown by the contour levels at -3.5, 3.5, 7.0, 10.5, 14.0 μJy/beam. The HST WFC I-band image with 0.5″ resolution is shown in grey-scale. Galaxies brighter than 22 m are saturated, with the fainter galaxies at 26 mag. On close inspection of the radio-optical correspondence, only two radio sources are unidentified.

faint background sources. Thus, most of the sources are less than about 2″ in angular size. None of the 39 sources appear larger than 10″; perhaps two sources are about 5″ in size.

Limited spectral information is available for the μJy sources since only the Bootes field was observed at two frequencies. Data in the Lynx and Cepheus field at other frequencies were also available (Windhorst *et al.* 1993). For the μJy sources the average $\alpha = -0.4$ (where $S \sim \nu^{\alpha}$), the same for that of the mJy sources. About 35% of the sources have steep ($\alpha < -0.5$) spectra, 20% have inverted spectra ($\alpha > 0.3$), with the remainder flat spectra. These proportions should be regarded as preliminary.

3. Optical Identifications

From the results of Hammer *et al.* (1995) and Windhorst *et al.* (1995), reliable identifications have been made for 52 of the 56 sources in the radio-optical overlap regions of the Bootes and Lilly1 fields. Figure 2 shows the radio-optical comparison for the Lilly1 field. A striking property is that most bright galaxies (< 22 mag) are identified with μJy radio sources, the typical redshift is 0.5. It appears that radio emission is preferentially associated with those galaxies which are in pairs or small groups. The preliminary division of optical types are:

Stellar Objects: Four stellar objects (7%) have been identified; one M star and three quasars—one in the Bootes field with a redshift of 0.985; two in the Lilly1 field, both with redshift 2.561.

Early-type galaxies: About 40% of the sources are identified with high redshift ($z \geq 1$) early-type galaxies which contain low powered AGN emission. These may be similar to the nearby early-type galaxies studied by Wrobel and Heeschen (1991), but are significantly more luminous.

Post-starburst Spiral Galaxies: Another 30% are so-called 'S+A' galaxies and are blue spirals with a weak AGN and a stellar content indicating recent star formation. Only one or two sources appear to be in a starburst phase. The typical redshift is about 0.5.

Emission-line Ellipticals: About 20% are associated with elliptical galaxies, with significant line emission, at a typical redshift of 0.8.

Observations with many telescopes at several wavelengths will continue in order to determine the properties of these faint radio sources which are an important constituent in the early universe.

Others contributing to these results are K.I.Kellermann (NRAO), R.B. Partridge (Haverford College), E. Richards (University of Virginia), and R. Windhorst (Arizona State University).

References

Fomalont, E.B., Windhorst, R.A., Kristian, J.A., & Kellermann, K.I. 1991, *AJ*, **102**, 1258

Hammer, F., Crampton, D., Lilly, S.J., LeFevre, O., & Kenet, T. 1995 *MNRAS*, **276**, 1085

Richards, E. 1996, in preparation.

Windhorst, R.A., Fomalont, E.B., Kellermann, K.I., Partridge, R.A., Richards, E., Franklin, B.E., Pascarelle, S.M., & Griffiths, R.E. 1995, *Nature*, 471

Windhorst, R.A., Fomalont, E.B., Partridge, R.B., & Lowenthal, J.D. 1993, *ApJ*, 498

Wrobel, J.M., Heeschen, D.S. 1991, *AJ*, **101**, 148

REDSHIFT DISTRIBUTION & NATURE OF μ-JY RADIO SOURCES

F. HAMMER[1], D. CRAMPTON[2], S. LILLY[3], O. LE FÈVRE[1]

[1] *DAEC, Observatoire de Meudon, France*
[2] *DAO, University of Victoria, Canada*
[3] *University of Toronto, Canada*

Abstract. During the preliminary deep imaging phase of our large spectroscopic survey of faint field galaxies (CFRS), one of our fields (10 arcmin × 10 arcmin) was chosen to coincide with the Fomalont et al (1991, A.J. 102,1258, hereafter FWKK) radio source field, including 36 S~ 16μJy radio sources of their complete sample. All sources but two have been identified to V < 25 and/or I_{AB} ≤24, and/or K_{AB} ≤21.

The microJy population is mainly constituted of three distinct populations of galaxies with different redshift regimes: early-type galaxies at z > 0.75 with a low powered AGN in their cores, post-starburst galaxies at intermediate redshifts (z = 0.375 to z = 0.8 or slightly > 1), and emission-line galaxies at z < 0.45 containing AGNs. The fraction of μJy sources with z > 1 could be as high as 30%. Most of the μJy radio sources (> 50%) are likely associated to AGNs, conversely to what is found at mJy levels (mostly starburst galaxies, Benn et al, 1993, MNRAS, 263, 98). Only one galaxy in our sample has a classical starburst spectrum.

The strong decrease of the radio spectral index from sub-mJy to μJy counts appears to be due to a combination of three factors: (1) the emergence of an elliptical population at high redshifts with moderate radio emission (2) an increasing fraction of narrow emission-line AGNs (Seyfert 2 and LINER); (3) a higher contribution of the thermal radiation to the radio emission from spirals, and the almost complete disappearance of starburst galaxies. Details of the results summarized here can be found in Hammer et al (1995, MNRAS, 276, 1085).

R. Ekers et al. (eds.), Extragalactic Radio Sources, 559–560.

1. Spectroscopy of the 25 $I_{AB} \leq 22.5$ radio counterparts.

Among objects enough bright for spectroscopy we find 5 early type galaxies (26%), 6 spirals (32%), 5 emission line galaxies (26%), 1 QSO and 1 M star while we failed to identify 3 objects.

The early type galaxies are rather luminous ($L \approx 2 L^*$), and have redshifts ranging from 0.7 to 1, the latter being the completeness limit for our spectroscopy. Their spectra show a red continuum with a significant 4000Å break and very faint or no [OII] emission. The spirals present disk-like morphologies, are rather luminous galaxies ($L \sim 1.5 \pm 0.4 L^*$) and lie at moderately high redshift ($0.37 < z < 0.81$). They all show moderate [OII] emission (average W =12Å at rest) and relatively strong Balmer continuum and absorption lines (equivalent width ranging from 3 to 5Å), indicating the presence of A and F stars and suggesting that strong star formation occurred in these objects ~ 1 Gyr ago. We classify them as post-starbursting galaxies. The emission line galaxies have relatively moderate luminosities ($L < L^*$), while the emission line ratio of all of them but one are typical of AGNs (from current diagnostic diagrams, see C. Rola, PhD Thesis, 1994).

2. The radio spectral index-colour diagram and the faintest counterparts.

All red ellipticals have inverted radio spectra ($\alpha = -0.4 \pm 0.3$), all the post-starbursts have moderately steep spectra ($\alpha = 0.40 \pm 0.18$), while the bluest emission-line galaxies have inverted spectra, with the notable exception of the most distant one ($\alpha = 0.7$), which could be classified as a starburst. Inverted-spectrum radio emission from ellipticals has been observed in some nearby ellipticals (Wrobel and Heeschen, 1984, Ap. J., 335, 677). This may indicate the presence of a low-power AGN (Rees, 1984, ARA&A, 22, 471), although other alternatives are possible. The post-starburst spirals have radio slopes noticeably flatter than the mJy starburst galaxies, which might indicate an increasing contribution of thermal radiation from star formation (Condon, 1992, A&A Review, 30, 575). Inverted radio spectra are also exhibited by the four very blue galaxies at low and moderate redshift, supporting the hypothesis that AGNs are present in their cores too, producing both the radio emission and the emission lines.

There is considerable evidence that virtually all of the 11 remaining sources (fainter than I_{AB} =22.5) are likely to be at $z > 1$. Indeed all the ones for which we have K photometry have color much redder than any object in the CFRS sample (I-K > 3.2) and are likely early type galaxies at $z > 1$.

THE SPACE DENSITY OF HIGH-REDSHIFT QUASARS

P.A. SHAVER[1], J.V. WALL[2], K.I. KELLERMANN[3], C. JACKSON[4]
AND M.R. HAWKINS[5]

[1] *European Southern Observatory - Karl-Schwarzschild-Str. 2, 85748 Garching, Germany*
[2] *Royal Greenwich Observatory - Madingley Road, Cambridge CB3 0EZ, England*
[3] *National Radio Astronomy Observatory - Edgemont Road, Charlottesville, VA 22903, U.S.A.*
[4] *Institute of Astronomy - University of Cambridge, Madingley Road, CB3 0HA, England*
[5] *Royal Observatory - Blackford Hill, Edinburgh EH9 3HJ, Scotland*

Abstract. An upper limit on the space density of quasars at $z > 5$ is obtained independent of any optical magnitude limit, from the complete identification of a large sample of flat-spectrum radio sources. This upper limit is below the observed space density at $z \sim 2 - 3$, showing that the turnover in space density is real and not merely due to obscuration.

We are completing the optical identification of a large sample of flat-spectrum radio sources, in order to search for high-redshift quasars and determine their space density. The sample is comprised of 896 flat-spectrum sources ($\alpha > -0.4$ between 2.7 and 5 GHz) from the Parkes catalogue with $S_{2.7} \gtrsim 0.25$ Jy in the range $+2.5° > \delta > -80°$, of which 581 were already identified in the catalogue, 83% with quasars. VLA and Australia Telescope positions were obtained for the unidentified sources, accurate to < 1 arcsec, and UKST/COSMOS B_J identifications were then made for 185 of these. CCD imaging observations with EFOSC on the ESO 3.6m telescope in the B, Gunn-i, and Gunn-z bands were made for the remaining sources.

From high-frequency observations of $z \lesssim 2$ flat-spectrum radio QSOs, it is known that the observed spectra between 2.7 and 5 GHz would still be flat for redshifts up to 10. It is also known that intervening Lyman-

561

R. Ekers et al. (eds.), Extragalactic Radio Sources, 561–562.
© 1996 *IAU.*

limit absorption completely obscures $z > 5$ quasars in the optical B-band. Thus, in searching for $z > 5$ quasars we seek flat-spectrum sources which are obscured in the B-band. Tentatively, it appears that all sources can be identified either with galaxies or with stellar objects which are present in the B-band (and therefore not at $z > 5$). Only one very red stellar identification has been found, which was marginally present in the B-band (hence at high redshift, but not at $z > 5$): PKS 1251-407. Its redshift was found to be $z = 4.46$, making it the highest-redshift radio source presently known (Shaver, Wall, Kellermann 1995, MNRAS in press). Objects at higher redshifts could have been found with similar ease. If, as now seems likely, there are no unidentified sources left in the sample which could be associated with quasars at $z > 5$, this gives a firm upper limit on the space density at $z > 5$ *which is independent of any optical magnitude limit.*

Comparison of the upper limit on the space density of radio quasars at high redshift ($5 < z < 7$) with the measured space density for similar objects at lower redshift gives a straightforward measure of the relative space density. A flux density limit of 0.25 Jy was adopted, and a selection was made of similar objects ($\alpha > -0.4$, and radio power corresponding to 0.25 Jy at $z = 7$) at lower redshifts from the Parkes catalogue in the declination range $+10° > \delta > -45°$ where redshift information is 70% complete. This provided *lower* limits on the space density at $z < 5$, to be compared with the *upper* limit at $5 < z < 7$. The latter is about an order of magnitude less than the former, clear evidence of a turnover.

This turnover cannot be due to obscuration, as Ostriker & Heisler (1984 *Ap. J.* **278**, 1) had proposed for optically-selected samples. The radio emission is unaffected by dust, and if all sources are already accounted for, there are none left to be optically obscured quasars at $z > 5$. It also cannot be due to misidentifications - the high positional accuracy and the observed surface density of the identified objects rule out the possibility that all of the $z > 5$ quasars are misidentified with foreground objects. Can it be argued that this turnover applies to *all* QSOs, and not just to radio-loud QSOs? The fact that the radio and optical QSO populations appear to turn over together seems unlikely to be a coincidence. If the optical turnover were entirely due to obscuration, then radio QSOs would also be affected, and it would be unlikely for us to have found PKS 1251-407 rather than a blank field. And the UV background also seems to exhibit a similar peak in redshift, suggestive of an epoch of activity.

We conclude that there is a real turnover in the space density of quasars at high redshift. In the context of this meeting it is interesting to note that, just as radio sources first established the strong *increase* in space density with increasing redshift, they are also required to confirm the *decrease* beyond the "quasar epoch".

COSMOLOGICAL SIZE EVOLUTION OF EXTRAGALACTIC RADIO SOURCES

ASHOK K. SINGAL
Physical Research Laboratory
Astronomy and Astrophysics Division
Navrangpura, Ahmedabad – 380 009, India.
E-mail: asingal@prl.ernet.in

The study of cosmological evolution of the sizes of extragalactic radio sources started about a quarter century back. From the very first angular size-redshift $(\theta$-$z)$ plots (Miley 1968, 1971; Legg 1970) and angular size-flux density $(\theta$-$S)$ plots (Swarup 1975; Kapahi 1975) it became evident that some sort of cosmic epoch-dependent evolution in the size distribution for the population of extragalactic radio source needs to be proposed; the sources at earlier epochs appeared on the average to have smaller physical sizes. However, a suitable luminosity-linear size $(P$-$l)$ correlation among the radio source population could also explain the observations, without invoking a size evolution with redshift. The only reliable way to disentangle these two separate effects is to investigate the size distribution in the luminosity-redshift plane, where one could examine not only the l-z relation for a given luminosity class, but could also check for a P-l correlation in a given redshift bin.

Such an approach was followed by Oort et al. (1987); Singal (1988); Kapahi (1989), who concluded that the physical sizes of powerful radio galaxies (PRGs) appear to evolve rapidly with redshift $(l \propto (1 + z)^{-3})$. In addition it was also pointed out that there is a direct correlation between luminosity and size among PRGs, the more luminous ones are larger in radio sizes. At the same time Singal (1988) cautioned that, unlike in PRGs, quasars do not show a direct luminosity-size correlation. On the contrary, there is a hint of an inverse correlation; the more luminous quasars appear to have smaller physical sizes. Moreover, the size evolution of quasars, if any, appears to be much weaker (see also Barthel and Miley 1988). Later Singal (1993), using a larger sample of sources that included many more galaxies at high redshifts, found the difference between the two distributions

R. Ekers et al. (eds.), Extragalactic Radio Sources, 563–566.

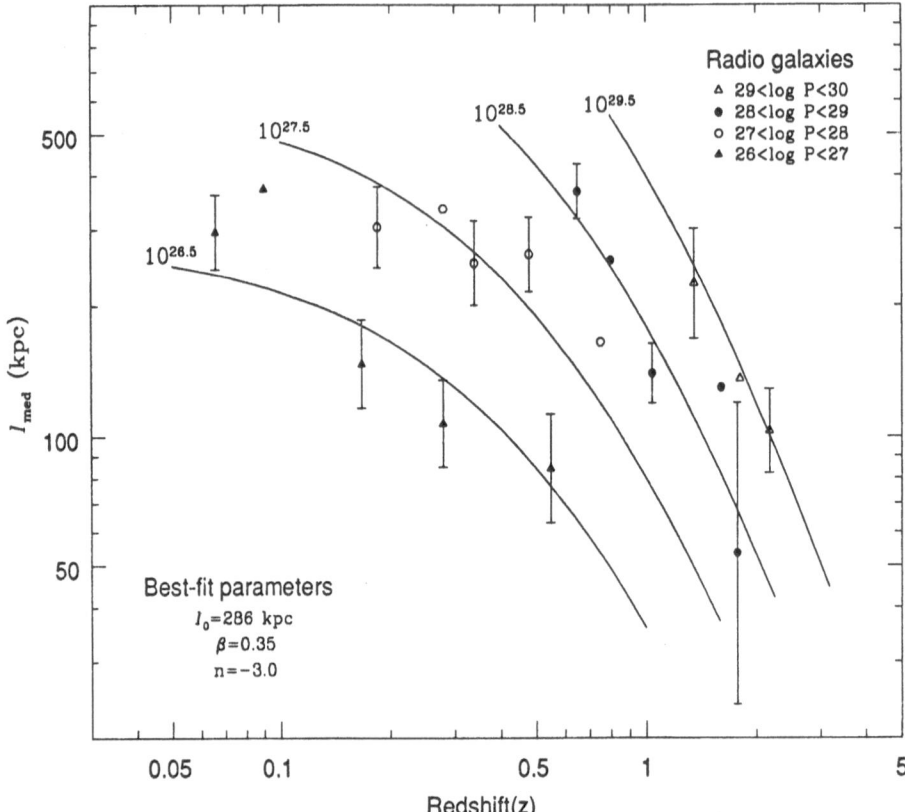

Figure 1.   The change in l_{med} with redshift for radio galaxies in different luminosity bins. The plotted points with error bars are from Singal (1993), while those without error bars represent the 3C and 6C data of Neeser et al. (1995). The family of curves is drawn according to the relation $l = l_0 (P/10^{26.5})^{\beta}(1 + z)^n$, for the best-fit parameter values $l_0 = 286$ kpc, $\beta = 0.35, n = -3$ (Singal 1993).

to be even more statistically significant. This difference in fact, provides a strong evidence against the simple orientation-based unified scheme models (Barthel 1989).

However, there are other publications in the literature with conclusions contrary to the above. For example, Nilsson et al. (1993) have claimed that in their studies they find no significant differences between the radio sizes of quasars and PRGs. They have reported the presence of a *negative* correlation between radio size and power among PRGs, a result in contradiction with almost all other previous studies. Moreover, they find no need for a cosmic evolution of radio source size. On the other hand, Neeser et al. (1995), from their sample containing 3C and 6C sources, have claimed that there is no evidence of a *P-l* correlation among PRGs and that a milder size evolution $((l \propto (1 + z)^{-1.5})$ may be present.

This is a highly confusing situation. Perhaps the first step in resolving

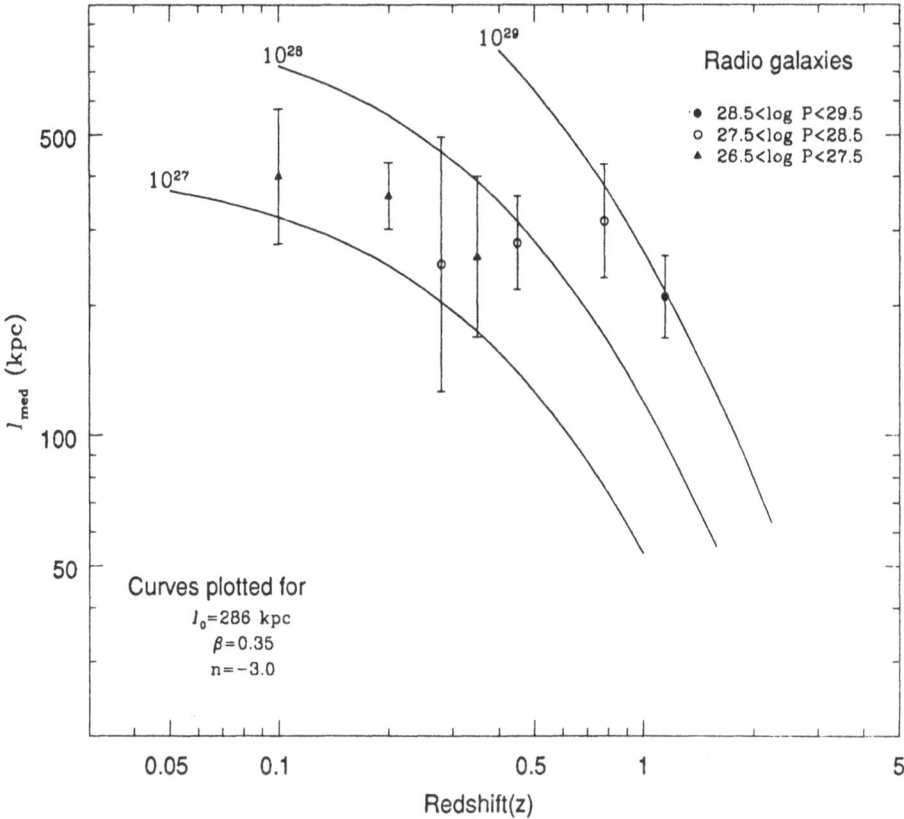

Figure 2.   A plot of l_{med} with redshift for the radio galaxies sample from Nilsson et al. (1993). The family of curves is drawn according to the relation $l = l_0 (P/10^{26.5})^\beta (1+z)^n$, for the parameter values $l_0 = 286$ kpc, $\beta = 0.35, n = -3$ derived by Singal (1993).

these discrepancies would be to check whether the differences lie in the data used in various studies, or are these caused merely by the different statistical methods used by different authors. According to Neeser et al. (1995), the main reason for different estimates does not seem to lie in the different statistical approaches used, and that the differences may lie in the radio size data. We decided to check this here by plotting the median size values estimated from the data used by Nilsson et al. (1993) and Neeser et al. (1995) on Singal's (1993) P-z-l diagram, and to ascertain wherein the discrepancies might lie.

Fig. 1 shows the P-z-l diagram from Singal (1993), depicting the change in l_{med} with redshift for radio galaxies in different luminosity bins. On this diagram we have also plotted l_{med} values in different P-z bins for the sample used by Neeser et al. (1995). Unfortunately it was not possible to give reliable error estimates for these points (especially for the 6C points, mainly because at the moment their tabulated data is not available to us). However, one thing is clear. There is not much discrepancy in the size values in the two data sets. The l_{med} values from Neeser et al. (1995) appear to be quite consistent with the curves drawn using the best-fit parameter estimates from Singal (1993). It also appears that from the 3C and 6C data alone one will be left with a wide range of β and n values, mainly because of a small coverage of the P-z plane.

In Fig. 2 we have plotted the l_{med} values estimated from Nilsson et al. (1993). We have retained the luminosity bins above the FR II limit only. For a comparison we have also plotted the family of curves drawn for the parameter values derived by Singal (1993). There are no significant deviations from the drawn curves. However the plotted points do not, in themselves, put a heavy constraint on the range of β and n, mainly because of the lack of sufficient data especially at higher redshifts in various luminosity bins.

It appears that the differences in results, arrived at by different authors, have arisen mainly because these are based on samples which may be lacking sufficient data in various P-z bins. It appears that the situation is likely to improve only with the availability of larger samples covering larger range in the P-z plane. Samples, of course, have to be selected at metre-wavelengths and with sufficient care so that no radio-size based bias sneaks into them.

In the last IAU Symposium on Extragalactic Radio Sources, Baldwin (1982) had contemplated on the possibility of finding evolutionary tracks of radio sources in the P-D (P-l in the present notation) diagram, the hope was that one may be able to find something akin to an H-R diagram for radio sources. However, the difficulties involved in the case of extragalactic radio sources appear to be of a 'higher dimension'. Because of the cosmological size evolution of these sources we need to consider a 3-dimensional (P-l-z) version of such a diagram if we want to get a hold on the evolutionary tracks of the radio sources.

References

Baldwin, J. E., 1982, in Extragalactic Radio Sources, IAU Symp. 97, eds. Heeschen, D. S., Wade C. M. , Reidel, Dordrecht, p. 21
Barthel, P. D., 1989, ApJ, 336, 606
Barthel, P. D., Miley, G. K., 1988, Nature, 333, 319
Kapahi V. K., 1975, MNRAS, 172, 513
Kapahi, V. K., 1989, AJ, 97,1
Legg, T. H., 1970, Nature, 226, 65
Miley, G. K., 1968, Nature, 218, 933
Miley, G. K., 1971, MNRAS, 152, 477
Neeser, M. J., Eales, S. A., Law-Green, J. D., Leahy, J. P., Rawlings, S., 1995, ApJ, 451, 76
Nilsson, K., Valtonen M. J., Kotilainen J., Jaakkola T., 1993, ApJ, 413, 453
Oort, M. J. A., Katgert, P., Windhorst, R. A., 1987, Nature, 328, 500
Singal, A. K., 1988, MNRAS, 233, 87
Singal, A. K., 1993, MNRAS, 263, 139
Swarup, G., 1975, MNRAS, 172, 501

NEW RESULTS FROM COMPLETE SAMPLES OF FAINT RADIO GALAXIES AND QUASARS

K.M. BLUNDELL[1], S. RAWLINGS[1], S.A. EALES[2], M. LACY[1]
[1] *Oxford University Astrophysics - Keble Road, Oxford, OX1 3RH, U.K.*
[2] *Department of Physics & Astronomy, University of Wales at Cardiff, CF2 3YB, Wales, U.K.*

1. Improved coverage of the luminosity–redshift plane

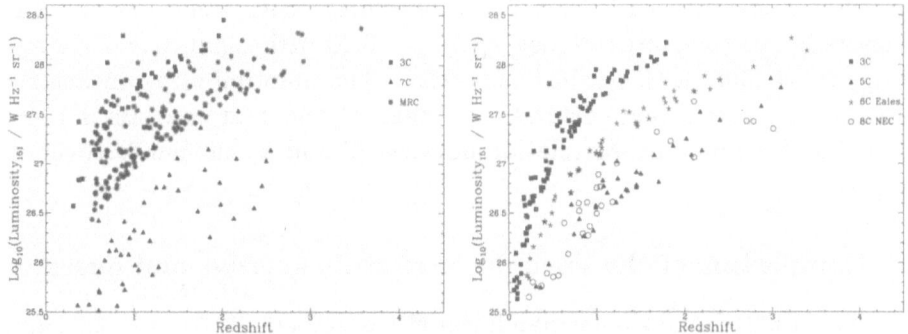

Figure 1. Coverage of the P–z plane with the new flux-limited quasar (left) and RG (right) samples overlaid on the coverage from 3C.

In any flux-limited sample a tight correlation of luminosity *(P)* and red-shift *(z)* is inevitable. It is therefore necessary to obtain complete samples at lower and lower flux-limits in order to have adequate coverage of the P–z plane, essential if we are to decouple the trends in epoch from trends in luminosity. This we have done for a number of flux-limits — giving coverage of the P–z plane seen in Fig. 1. Our redshift information is *spectroscopic*; the results of Eales et al *(in prep.)*, namely the increased scatter in the

R. Ekers et al. (eds.), Extragalactic Radio Sources, 567–568.

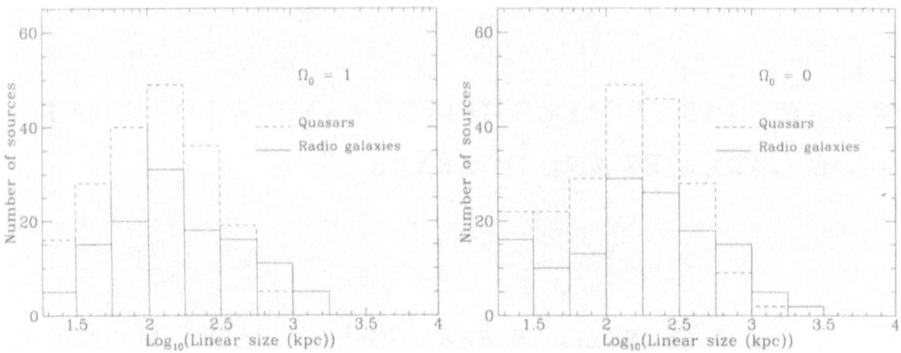

Figure 2. Histograms of the linear sizes of RGs and quasars calculated for two values of Ω.

K-z plot for samples lower in luminosity than 3C, strongly warn us against using redshifts estimated from K-magnitudes.

2. Linear size evolution of radio sources

We calculated the three-way partial rank correlation coefficients (Macklin 1982) for the linear sizes (D) of sources, with their redshifts and luminosities. For a universe with $\Omega = 1$, we obtain for both radio galaxies and quasars in our complete samples, $r_{Dz|P} = -0.43$ with significance 7.6σ and $r_{DP|z} = -0.0067$ with significance 0.11σ. (The notation $r_{Dz|P}$ means the partial rank correlation coefficient between D and z at constant P). We thus find a strong anti-correlation between D and z, but not between D and P.

3. Comparison of the linear sizes of radio galaxies and quasars

Barthel (1989) found the median linear size of RGs in 3C to be ~ 2.2 times that of the quasars in 3C, lending strong support to the unification-by-orientation model of RGs and quasars. For our higher redshift and lower luminosity samples, we find that the ratios of the median lengths in kpc of RGs over quasars for $0 < z < 1$ is 259/157, for $1 < z < 2$ is 119/86, for $z > 1.5$ is 84/77 and for $z > 2$ is actually 56/77. The similarity of the linear size distributions of radio galaxies and quasars can be seen in Fig. 2. We thus conclude that unification without evolution is untenable.

References

Barthel P.D. (1989) *ApJ*, **336**, 606.
Macklin J.T. (1982) *MNRAS*, **199**, 1119.

SPECTRAL INDEX - REDSHIFT RELATION FOR RADIO GALAXIES AND QUASARS

R.D.DAGKESAMANSKII

Astrospace Center of P.N.Lebedev Physics Institute of RAS
Moscow Leniskii prospekt 53 Russia

Cosmological evolution of synchrotron spectra of the powerful extragalactic radio sources was studied by many authors. Some indications of such an evolution had been found firstly by analysis of 'spectral index - flux density' $(\alpha - S)$ relation for the sample of relatively strong radio sources [1,2]. Later Gopal-Krishna and Steppe [3] extended the analysis to weaker sources and found that the slope of $\alpha_{med}(S)$ curve changes dramatically at intermediate flux densities. Gopal-Krishna and Steppe pointed out that the maxima of the $\alpha_{med}(S)$ curve and of differential source counts are at almost the same flux density ranges (see [3], Fig.2). It has to be noticed that the all mentioned results were obtained using the low-frequency spectral indices and on the basis of low frequency samples.

It has been found also that the relation between the mean spectral indices of extragalactic radio sources and their flux densities is mainly due to $\alpha(S)$ relationship for corresponding quasar subsample [1]. For this reason a distribution of quasars on the 'spectral index - redshift' diagram was analysed and it was shown [4] that there were no the relatively flat spectrum $(\alpha > -0.7)$ sources in 3C catalogue which are identified with high redshift quasars $(z > 1.0)$.

Recently, Herbig and Readhead [5] have published some results of their compilation of multi-frequency flux density measurements for three complete samples of strong extragalactic radio sources. The results were presented as tables of the redshifts, luminosities and spectral indices of the sources at three standard rest frequencies: 0.15, 2.5 and 40 GHz. We used this data to study the distribution on (z, α)-plane of the sources from the well-known low-frequency complete sample defined by Laing, Riley and Longair [6]. The sample contains 173 sources and more than 98% of them are identified and have reliable redshifts. So, we can be sure that there is

R. Ekers et al. (eds.), Extragalactic Radio Sources, 569–570.

no any selection effect in the data except for limitations on flux density ($S_{178} \geq 10Jy$) and galactic latitude ($b > 10°$).

Deficit of the distant 'flat' spectrum ($\alpha_{0.15} > -0.7$) quasars in the sample is well seen in their distribution on the ($z, \alpha_{0.154}$)-plane. This confirms the remarkable feature of quasar's ($\alpha - z$) distribution found in [4] from different flux density measurements and for partly different samples of quasars. The same feature peeps out also at ($z - \alpha_{0.15}$) diagram plotted for the subsample of radio galaxies.

Statistical tests applied to the both distributions show that the absence of the 'flat' spectrum objects at high redshifts is significant. A distribution similar to this could be expected if the 'flat' and 'steep' spectrum radio sources have very different luminosity functions. Indeed, if the 'flat' spectrum radio sources are intrinsically much weaker (at low frequencies) compared with the 'steep' spectrum objects then there will be only relatively near 'flat' spectrum objects in flux density limited samples.

However, 'flat' spectrum quasars median luminosity is only 2 times less than the 'steep' spectrum one's. On the other hand, we have found that V/V_{max} median values are very different for the groups of quasars. Indeed, for 12 'flat' spectrum quasars $(V/V_{max})_{med} = 0.33 \pm 0.06$ while for 30 'steep' spectrum quasars $(V/V - max)_{med} = 0.59 \pm 0.04$. The last value is close to the (V/V_{max}) estimates for many other quasar samples and reflects the well-known fact that quasar's spatial density rises when the distance from observer increases. But $(V/V_{max})_{med}$ for the 'flat' spectrum quasars is surprisingly low. Does this mean that their spatial density decrease with the distance from observer ? If it is so, then these objects have to represent some separate group of extragalactic radio sources. This conclusion would have grave consequences and has to be checked carefully.

I would like to thank the Conference organizers for financial help which made me possible to attend this Conference. This work was supported as a part of Russian State Program "Astronomy" (via the SEC "Cosmion").

References

[1] Dagkesamanskii, R.D. (1969) *Astrofizika*,Vol. **5**,pp 297-304

[2] Murdoch, H.S. (1976) *Mon. Not. Roy.Astron. Soc.*,Vol. **177**,p 441

[3] Gopal-Krishna and Steppe, H. (1982) *Astron.Astrophys.*,Vol. **113**,pp 150-154

[4] Dagkesamanskii, R.D. (1970) *Nature*,Vol. **226**,p 432

[5] Herbig, T. and Readhead, A.C.S. (1992) Astroph. J. Suppl.,Vol. **81**,pp 83-124

[6] Laing, R.A., Riley, J.M., and Longair, M.S. (1983) Mon. Not. Roy. Astron. Soc.,Vol. **204**,p 151

HIGH REDSHIFT RADIO GALAXIES

K. MEISENHEIMER, H. HIPPELEIN AND M. NEESER
Max-Planck-Institut für Astronomie
Königstuhl 17, D-69117 Heidelberg

1. Glancing back into the young universe

One hundred years after G. Marconi recorded radio waves over a distance of more than $1000\,m$, the most sensitive radio telescopes are able to detect the radio emission from light travel distances at least 1.4×10^{23} times greater. The electromagnetic waves from these distant objects are red shifted by $\Delta\lambda/\lambda \equiv z > 4$. It is not the mere distance of high redshift objects which is fascinating, but rather the fact that one looks back into the early history of the universe by observing them: Objects at a redshift of 4 shined at a time when the universe had reached only about $1/5$ of its present age.

Celebrating a century of radio transmission it is appropriate to recall briefly the history of how radio sources were identified with more and more distant objects. High redshift astronomy started back in 1960 when Minkowski identified 3C 295 with a galaxy at $z = 0.46$ (at that redshift the age of the universe relative to the present age was $t/t_0 = 0.69$). After the detection of quasars (by M. Schmidt, 1963) and the subsequent discovery that quasars can be found out to a very substantial redshift (3C 9 at $z = 2.01$, i.e. $t/t_0 = 0.32$, Schmidt 1965), the focus of interest shifted to these extremely luminous "beasts". With the development of optical CCD detectors, it also became feasible to identify the less conspicuous radio *galaxies* at high redshift: 3C 13 was identified with a galaxy at $z > 1$ (Spinrad *et al.* 1981) and in 1984 3C 256 was found to be a galaxy at $z = 1.81$ ($t/t_0 = 0.36$, Spinrad & Djorgovski), revealing for the first time the strong Ly-α emission of these galaxies. 1988 brought a break-through in the study of high redshift radio galaxies. Within a few months of Spinrad *et al.* detecting the most distant object in the 3C catalogue (3C 257 at $z = 2.48$), Lilly identified the source 0902+34 from the 2nd Bologna catalogue with a galaxy at $z = 3.39$ and Chambers *et al.* found that 4C 41.17 is located

R. Ekers et al. (eds.), Extragalactic Radio Sources, 571–576.

even at $z = 3.80$ ($t/t_0 = 0.21$). The very extended Ly-α cloud surrounding the latter galaxy emits 10^{38} W, comparable with a moderate quasar. With 4C 41.17 radio galaxies overtook the radio loud quasars in redshift for the first time since 1965 ! During this conference we heard that both the most distant radio galaxy (6C 0140+32 at $z = 4.41$, Lacy *et al.*, these proc.) and the radio loud quasar (P 1251-407 at $z = 4.46$, Shaver, these proc.) allow us to look back into the very distant history of the universe ($t/t_0 < 0.2$). The highest redshift found so far ($z = 4.89$ for the radio *quiet* QSO PC 1247+3406, Schneider *et al.* 1991) is hardly higher than this.

But why expend so much effort in finding these distant radio galaxies when quasars are much easier to detect? Because radio galaxies appear extended in the optical/near-infrared band ! This provides (i) the opportunity to separate the stellar component from the AGN (out-shining everything in QSOs) and (ii) allows one to study the kinematics and ionization structure of the — often very extended — emission line gas. Thus, radio galaxies are our crown witnesses to early galaxy formation and evolution.

2. Radio galaxies as a probe of galaxy evolution

At redshifts $z \gtrsim 1$, any investigation of the stellar spectral energy distribution (SED) has to include observations in the near-infrared ($1 < \lambda < 2.3\,\mu$m). In a seminal piece of work Lilly & Longair (1984) employed JHK-photometry of more than a dozen 3C-galaxies at $z \simeq 1$. They found that none of the galaxies at $z \gtrsim 1$ showed the very red $R - K$ colors expected for a red shifted present-day giant elliptical. In fact, about one half of the galaxies at $z > 1$ were even bluer than passive evolution (i.e. star formation at $z \gg 1$) would predict. The authors concluded that these galaxies contain a substantial fraction of young stars. Many show distorted optical images.

Subsequently it became even more evident that there exists a problem in using radio galaxies as a tracer for galaxy evolution in general. Two groups (Chambers *et al.* and McCarthy *et al.*, 1987) reported a strong correlation between the radio axis and the orientation of the extended emission line regions (EELR) around many of the radio galaxies at $z \gtrsim 1$. In some cases this "alignment effect" even seems to be present in the orientation of the elongated blue continuum, pointing strongly to the existence of a very special radio-related component (RRC) in the SED of these powerful radio sources. As a result the overall SED of powerful radio galaxies cannot be regarded as representative for the stellar SED of galaxies in general.

On the other hand, radio galaxies show a very small scatter in the $K - z$ (Hubble-)diagram (Lilly 1989) and the aligned component becomes less conspicuous with decreasing radio power (Dunlop & Peacock 1993). Thus, we can hope that RRC and stellar SED might eventually be separable.

3. High redshift radio galaxies as a laboratory to study the physics of powerful radio sources

The radio-related component in galaxies at $z \gtrsim 1$ might provide valuable information about the interaction of the radio source with its environment. The kinematics of the EELR may show direct evidence for the protrusion of the *bow-shock* into the surrounding medium or for sideways and backward flow of the radio plasma. The excitation of the EELR could indicate whether the kinetic power of the jet or radiation from a (hidden) AGN dominates the energetics of the RRC. The morphology of both the EELR and the optical continuum should trace the density distribution around the galaxy and thus reveal whether the most powerful radio galaxies are caused by an atypical environment. In addition, the role of galaxy-galaxy interactions for triggering and shaping the radio source can be investigated.

4. The radio-related component at $z \simeq 1$.

We will not attempt to give a general discussion of the alignment effect — an excellent review of this topic has been presented by Clive Tadhunter at this conference. Rather, we will highlight some results of our ongoing project to study a dozen radio galaxies in great detail employing 2D spectroscopy with a Fabry-Pérot-Interferometer (FPI), supplemented by continuum imaging in line-free bands. We have studied ten 3C galaxies in the range $0.47 < z < 1.13$, selected by their very extended emission line regions:

3C 34, 44, 169.1, 265, 337, 352, 356, 368, 435 A, 441.

By discussing three examples we will demonstrate the wide variety encountered in this sample (details are given in the Thesis of Mark Neeser).

3C 368 (z = 1.13) shows a double-lobe structure with a blue-shifted southern and red-shifted northern lobe. There is a perfect correspondence between the extent of the emission line lobes (EELs) and the radio lobes, although the latter protrude about $1''(10$ kpc$)$ further from the source (Meisenheimer & Hippelein 1992). The southern EEL shows a conical structure which, together with the kinematical evidence for an expanding shell, can naturally be explained as external gas which has passed the bow-shocks in front of the radio lobes. Our interpretation of the ground-based FPI-data ($\gtrsim 1''3$ resolution) is confirmed by recent HST results (Longair *et al.* 1995) which — when properly aligned to the radio coordinates[1] show both the 1:1 correspondence of the radio and emission line lobes and the bow-shock morphology of the southern lobe.

---

[1] Based on our astrometry, the brightest peak in the HST image (an M-star) has to be shifted to $18^h 05^m 06^s.429$, $11°01'31''.45$ (J2000), i.e. $\sim 1''$ to NW w.r.t. the radio map.

The ELLs of 3C 368 lag by about 10 kpc behind the present location of the bow-shocks. This behaviour is expected since the shocks should heat the external material to $\gtrsim 10^7$ K and it takes even over-dense parts some 10^7 years to cool down to 20 000 K — the temperature at which O^+ is the most abundant ion. From this observed cooling length and the assumption that ram pressure balances the (minimum) pressure in the radio lobes, we derived a self-consistent model of 3C 368 which is characterized by a rather high external density $n_{ex} = 10^5 \, m^{-3}$ and a moderate expansion velocity $v_{bow} = 1200 \, kms^{-1}$. Thus we conclude that the perfect radio-optical alignment and the high radio power of 3C 368 are caused by a high external density with an atypically smooth distribution. Note, that this model leaves the excitation of the EELR undetermined since the bow-shocks alone cannot provide the necessary energy. Based on the exceptionally low [SIII]λ953.2/[OII]λ372.8 ratio and the absence of anything resembling an radiation cone, we favour the explanation that the excitation is dominated by the interaction between the post-bow-shock gas and the cocoon of the radio source. This interpretation has gained substantial support by the fact that the high extended polarization reported for this source could not be verified with Keck or HST observations (van Breugel, this conf.).

3C 435 A (z = 0.47):   Although this source, with its narrow northern spur of emission line gas extending > 10 kpc beyond the radio lobes, has been called a prime witness for beamed ionizing radiation from a hidden quasar we think that the correct interpretation of the EELR of 3C 435 A is rather different: Most likely the morphology and kinematics of this asymmetric EELR is dominated by a violent interaction between the radio galaxy and a companion galaxy about 5″(30 kpc) towards the north (see Neeser *et al.* 1996). The northern spur of emission line gas much more resembles a tail of tidal debris than a randomly distributed gas ionized by a AGN beam.

3C 265 (z = 0.81):   This is the most extended and most luminous EELR around any radio galaxy with $z < 1$. In spite of its large linear extent the diameter of the EELR is still less than half of the radio source. Both the complicated morphological and kinematical structure of the EELR indicate that it is a superposition of several different processes: Polarimetric evidence (Cohen, this conf.) indicates that a hidden luminous AGN is partly responsible for the ionization of the inner EELR. We derived a similar conclusion for the outermost parts. However, the peculiar velocities observed in the inner region, as well the fact that there is plenty of emission line gas under 90° w.r.t. the radio axis, make it very likely that the presence of several faint gas-rich companions dominates the EELR in the innermost 50 kpc. We regard 3C 265 as an example of powerful radio sources in which galaxy interactions play a major role in supplying the material for the EELR while an AGN (fed by the same process) provides most of the ionizing radiation.

In summary, we conclude from our sample of 3C galaxies: The perfect radio-optical alignment found in the prototype 3C 368 is not typical and is caused by a very fortunate density distribution. Galaxy-galaxy interactions are common in powerful radio galaxies at $z \simeq 1$. Since they should lead to an isotropic distribution of position angles, additional processes (anisotropic ionization from an AGN, brightening of the radio source due to interaction with a companion, etc.) are required to explain the alignment effect. We found several examples in which ionization from a hidden quasar is a likely origin of the alignment. In view of the complicated superposition of the different processes which we found in several sources, we recommend a detailed study of each individual case in order to disentangle the RRC from the typical SED of high redshift galaxies.

5. An example at very high redshift: 4C 41.17.

4C 41.17 ($z = 3.80$, see Chambers *et al.* 1990) is the best studied radio galaxy at very high redshift. Its optical continuum is well aligned with the radio axis (Miley *et al.* 1992). Absolute astrometry indicates that the radio core (Carilli *et al.* 1993) sits exactly in the dark "gap" which is conspicuous on both Ly-α and continuum images taken at $0.6 < \lambda < 2.2\,\mu m$, and is most likely caused by dust absorption. Millimetre and sub-mm observations (Dunlop *et al.*, Chini & Krügel 1994) found a large amount of warm dust, most of which is presumably concentrated near the "gap". We have carried out FPI and long-slit spectroscopy of the Ly-α cloud and narrow band imaging in the [OIII]$\lambda 5007$ line which is red-shifted to $2.40\,\mu m$.

Our main results can be summarized as follows: Although the [OIII] image (low S/N ratio) is consistent with the brightest parts of the Ly-α emission, the "gap" seems deeper in Ly-α than in [OIII] (Hippelein *et al.* 1995). This supports dust absorption as an explanation for the gap. The high-brightness Ly-α/[OIII] region is perfectly aligned with the radio axis and might either be ionized by an UV radiation cone or young stars.

The outer isophotes of the extended Ly-α halo are much rounder than that of the inner high brightness region. At radii $> 2''$ from the core we find substantial Ly-α brightness even at $90°$ w.r.t. the radio axis, which can hardly be explained by UV radiation from an obscured core. Although resonant scattering could broaden the Ly-α emission region, we think that the lack of a sharp boundary between "cone" and halo argues against this interpretation. The red/blue-shift asymmetry in the eastern/western part of the EELR seems to argue for a radio-related kinematics, but the velocity field of the more extended eastern lobe indicates contraction rather than expansion. In addition, a Ly-α absorption at $z = 3.793$, with an equivalent

width of 0.3 nm covers the entire cloud (Hippelein & Meisenheimer 1993,
Hippelein *et al.* 1995).

Associating this absorption with 4C 41.17 (in analogy to the absorption
found in other radio galaxies, see Röttgering this conf.) we propose the
following interpretation: The systemic redshift of 4C 41.17 is traced by the
absorption. The blue-shifted western component (and perhaps the brightest
parts in the east) are caused by the interaction of the radio source with the
surrounding material, most probably ionised by a hidden quasar. The extended Ly-α halo, however, which is red-shifted and contracting (see Fig. 3
in Meisenheimer *et al.* 1994), is due to ongoing accretion of surrounding material (optically thick in Ly-α) onto the density peak at which the source is
located. In this scenario, violent accretion shocks would be responsible for
the extreme Ly-α luminosity and the large velocity width. 4C 41.17 would
therefore be an extremely massive galaxy or cluster core in the process of
formation. The alignment of the radio axis along the elongation of the extended halo is caused by a selection bias; only those radio sources which
happen to be aligned with the density distribution will be sufficiently luminous to exceed the limit of the 4C catalogue. Consequently, there should be
at least $10\times$ more "misaligned 4C 41.17s" with equally bright Ly-α haloes,
but much weaker radio sources.

References

Carilli, C., Owen, F.N., Harris, D.E. 1994: *Astron. J.* **107**, 480
Chambers, K., Miley, G., van Breugel, W. 1987: *Nature* **329**, 604
Chambers, K., Miley, G., van Breugel, W. 1990: *Astrophys. J.* **363**, 21
Chini, R. & Krügel, E. 1994: *Astron. Astrophys.* **288**, L33
Dunlop, J. & Peacock, J. 1993: *Mon. Not. R. astr. Soc.* **263**, 936
Dunlop, J., *et al.* 1994: *Nature* **370**, 347
Hippelein, H. & Meisenheimer, K. 1993: *Nature* **362**, 224
Hippelein, H., Meisenheimer, K. & Röser, H.-J. 1995: in *Galaxies in the Young Universe*
    (eds. H. Hippelein *et al.*), Springer Lecture Notes, p.93
Lilly, S.J. 1988: *Astrophys. J.* **333**, 161
Lilly, S.J. 1989: *Astrophys. J.* **340**, 77
Lilly, S.J. & Longair, M.S. 1984: *Mon. Not. R. astr. Soc.* **211**, 833
Longair, M., Best, P., Röttgering, H. 1995: *Mon. Not. R. astr. Soc.* **275**, L47
McCarthy, P. et. al. 1987: *Astrophys. J. (Lett.)* **321**, LL29
Meisenheimer, K. & Hippelein, H. 1992: *Astron. Astrophys.* **264**, 455
Meisenheimer, K., Hippelein, H., Neeser, M. 1994: in *The Physics of Active Galaxies* (eds.
    G. Bicknell *et al.*), ASP conf. series Vol. 54, p.397
Miley, G., Chambers, K. van Breugel, W., Macchetto, F. 1992: *Astrophys. J.* **401**, L69
Minkowski, R. 1960: *Astrophys. J.* **132**,, 908
Neeser, M., Hippelein, H. & Meisenheimer K. 1996: *in preparation*
Schmidt, M. 1963: *Nature* **197**, 1040
Schmidt, M. 1965: *Astrophys. J.* **141**, 1295
Schneider, D.P., Schmidt, M., Gunn, J.E. 1991: *Astron. J.* **102**, 837
Spinrad, H., Staufer, J., Butcher, H. 1981: *Astrophys. J.* **244**, 382
Spinrad, H. & Djorgovski, S. 1984: *Astrophys. J. (Lett.)* **285**, LL49

HST AND KECK OBSERVATIONS OF HIGH REDSHIFT RADIO GALAXIES

W.J.M. VAN BREUGEL

Institute of Geophysics and Planetary Physics
Lawrence Livermore National Laboratory
P.O. Box 808, L-413, Livermore, CA 94550, USA

1. Introduction

Together with several of my colleagues I have embarked on a comprehensive program to study the radio–aligned restframe UV structures in high redshift radio galaxies (HzRGs) using some of the world's premier optical telescopes: the Hubble Space Telescope for high spatial resolution imaging, and the Keck 10m telescope for high S/N spectropolarimetry. I will discuss some of our latest results from these observations which elucidate, and at the same time obscure, our evolving understanding of HzRGs .

2. HST Imaging of 4C41.17 at z = 3.800

4C41.17 is the highest redshift radio galaxy observed with the refurbished HST to date. Using the WFPC2 with the F702W and F569W filters respectively a deep, line free restframe UV (1500 Å) image and a Ly-α emission line image were obtained.

2.1. ALIGNED UV CONTINUUM AND Ly-α EMISSION

The rest-frame UV continuum of 4C41.17 has a complex, elongated morphology with numerous compact (kpc- sized), bright (~ 0.2 μJy; 'R' = 26.7) components which are aligned, but not coincident, with the inner radio source. The radio core (AGN) is located in a gap between two main regions, both of which are embedded in faint, diffuse emission. This suggests that the AGN is probably obscured, at least in the rest-frame UV.

R. Ekers et al. (eds.), Extragalactic Radio Sources, 577–580.

Imaging and spectropolarimetry of HzRGs , including recent high signal-to-noise spectropolarimetry with Keck (these Proceedings), show strong evidence that a significant fraction of the aligned rest-frame UV continuum in these radio galaxies is scattered light from obscured or 'misdirected' quasar-like AGN. Thus, by analogy, it seems that at least some of the aligned UV continuum in 4C41.17 has a similar origin. The total luminosity of the aligned clumps in 4C41.17 is $log[\nu L_{\nu,aligned}] = 46.9$ (erg s^{-1}) which is comparable to that of radio loud $1 < z < 2$ quasars at this rest wavelength ($46.4 < log[\nu L_\nu] = 47.6$). Thus on purely energetic considerations the aligned rest-frame UV in 4C41.17 may indeed be due to collimated, scattered light from a hidden quasar-like AGN.

The Ly-α emission in 4C41.17 does *not* appear to be directly associated with the UV continuum knots, with a possible exception in the western region of 4C41.17. Instead the line emission seems brightest near the boundaries of the radio components, a feature which is often observed in nearby radio galaxies and which suggests interaction (entrainment, ionization) of the radio source with relatively dense ambient gas. Support for such an interpretation follows also from the high-velocity component seen in 4C41.17, with a total range of 2000 km s^{-1}over the total extent (several arcseconds) of the inner radio source. The dominant source of ionization, AGN, shocks, or starformation is unclear at present, but the arguments above for the probable presence of a hidden quasar in 4C41.17 suggest that photoionization by the AGN is likely. The total Ly-α flux of the aligned component in the HST image is $F_{Lya} = 1.2 \times 10^{-15}$ erg s^{-1} cm^{-2}, implying a Ly-α luminosity of $log[L_{Lya}] = 44.2$ erg s^{-1}, which is comparable to that of steep spectrum quasars.

2.2. COMPANION GROUP

There is an amorphous, 'non-aligned', clumpy group of objects south of 4C41.17 which is embedded in a halo of diffuse continuum emission and which appears connected with the 4C41.17 system. Keck near-IR images have shown that this companion system is very blue ($\alpha_{1500\AA} \sim 0$) and has a luminosity comparable to that of a $M_B = -20.5$ L_* galaxy at this redshift. Thus the companion group may be a galaxy sized system with ongoing star formation. The individual clumps in this system have UV luminosities which are 10 – 100 times higher than those of blue compact dwarf galaxies, in a similar kpc-sized volume, indicating vigorous star formation indeed. The HST observations of 4C41.17 therefore suggest that we may be viewing, for the very first time, the formation of a galaxy through merging of kiloparsec sized, very active starforming regions ('building blocks'), as predicted in dissipative galaxy formation scenarios. These star formation

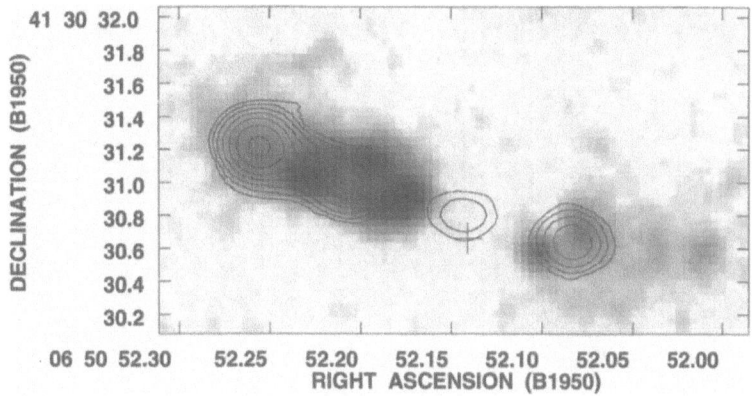

Figure 1. HST WFPC2 image of 4C41.17 with radio contours superimposed

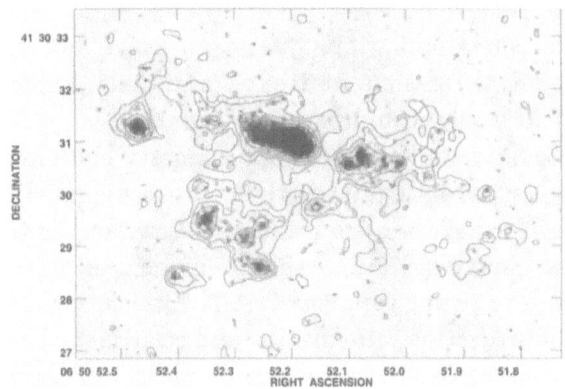

Figure 2. Same data as Fig.1 showing the clumpy companion system

clumps could be sources for dust and dense gas and thus, for example, provide scattering mirrors or ionization material as they are intercepted by the collimated UV-continuum radiation from the 4C41.17 AGN. The AGN itself, one might hypothesize, could have formed through rapid stellar evolution and the formation of a black hole in the center one of these clumps, possibly triggered by the merging process.

3. Keck Spectropolarimetry of HZRGs

During the past few years there has been increasing evidence that aligned restframe UV (< 4000Å) continua in HzRGs may be dominated by scattered light from hidden or mis-directed AGN. The large aperture of the Keck telescope, combined with the often excellent seeing, allows one to obtain vastly superior, high S/N and spatially resolved spectropolarimetry data of HzRGs . Now, for the first time, it has become possible to use the

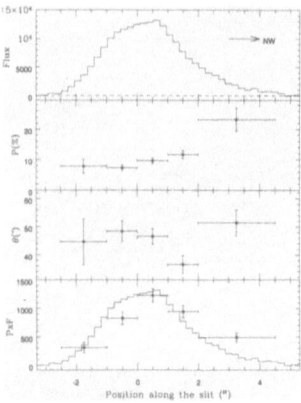

Figure 3. Total flux, percentage polarization, position angle, and polarized flux along the radio axis of 3C256 observed at Keck

full potential of spectropolarimetry as a powerful tool in discovering the nature of the optical continua in HzRGs and their (hidden) AGN, and to probe the ISM in newly forming (active) galaxies.

The first results from our spectropolarimetry observations program at Keck are summarized by Cimatti (these Proceedings). Perhaps the most important overall conclusion from these observations is that there is now little doubt that most HzRGs indeed have hidden, or 'mis-aligned', quasar-like AGN. This is in strong support of AGN 'Unification' models which rely on the effects of projection, obscuration and relativistic beaming to explain many of the observed properties of various classes of active galaxies. Here I would like to emphasize two additional points:

• The spatial extents of the radio/UV–aligned reflection nebulae of HzRGs can be enormous (5″, or 55 kpc in 3C256 at $z = 1.825$ [Dey *et al.* 1995]).

• While a large fraction of the UV continuum is shown to be polarized in HzRGs, a significant fraction is *un*-polarized. Furthermore probably not *all* HzRGs are highly polarized (preliminary analysis of our HST imaging polarimetry and Keck spectropolarimetry shows that the percentage polarization in the proto-type aligned radio galaxy 3C368 may in fact be much less than previously thought). Thus now the interesting question arises as to what the origin is of this unpolarized continuum. Nebular continuum associated with strong emission-line regions appears as an attractive possible source (Tadhunter, these Proceedings).

The work by WvB described here was performed at IGPP/LLNL under the auspices of the U.S. Dept. of Energy under contract W-7405-ENG-48. It is in collaboration with Miley, McCarthy, and Spinrad (4C41.17), and with Dey, Cimatti, Antonucci, and Spinrad (Keck spectropolarimetry), who I thank for allowing me to use these data prior to publication.

HIGH-REDSHIFT MILLI-JANSKY RADIO GALAXIES

J. DUNLOP[1], J. PEACOCK[2], R. WINDHORST[3], H. SPINRAD[4],
A. DEY[4] AND I. WADDINGTON[1]

[1] *Institute for Astronomy, Department of Astronomy, The University
of Edinburgh, Royal Observatory, Edinburgh EH9 3HJ, U.K.*
[2] *Royal Observatory, Edinburgh EH9 3HJ, U.K.*
[3] *Astronomy Department, University of California, Berkeley,
CA 94720, U.S.A.*
[4] *Department of Physics & Astronomy, Arizona State University,
Tempe, AZ 85287-1504, U.S.A.*

The study of radio galaxies selected at mJy flux levels has the potential
to resolve two important issues in observational cosmology provided redshifts can be determined or reliably estimated for complete samples of such
sources. First, the deep flux limit, combined with the shape of the radio
luminosity function means that the redshift distribution of such samples
provides a much more powerful test of the existence of a high-redshift cutoff for radio sources (Dunlop & Peacock 1990) than can be provided by
further studies of brighter radio samples. Second, as a consequence of selection from bright radio surveys, the detailed study of galaxies at $z > 2$ has to
date been confined to objects of extreme radio power (*e.g.* 4C41.17, Chambers *et al.* 1990; B2 0902+34, Eales *et al.* 1993), and it has now become
clear that the ultraviolet-infrared properties of such sources are strongly
contaminated by processes connected to the AGN (Eales & Rawlings 1993;
Dunlop & Peacock 1993). Being 100-1000 times less radio luminous than
these extreme sources, mJy radio galaxies at comparable redshifts should
provide much more representative probes of the formation and evolution of
elliptical galaxies in general.

Accordingly, over the past few years we have been investigating the
properties of radio galaxies with $S_{1.4GHz} > 1$mJy selected from the Leiden-
Berkeley Deep Survey (LBDS) and its extensions (Neuschaefer & Windhorst 1995). This has given a statistically complete sample of 77 galaxies
for which we now possess g, r, i & K photometry (plus J and H for a

R. Ekers et al. (eds.), Extragalactic Radio Sources, 581–582.

subset of sources), enabling us to estimate redshifts from both spectral fitting (Dunlop & Peacock 1993) and from a modified version of the infrared Hubble diagram (Dunlop, Peacock & Windhorst 1995). A new programme of optical spectroscopy with the William Herschell telescope on La Palma, and the Keck telescope in Hawaii has now yielded spectroscopic redshifts for 3 sources along with the detection of a single weak line in a further 2 objects.

These spectroscopic redshifts agree to within 20% of the redshifts estimated from colours and K magnitudes, and give confidence that our estimated redshift distribution for the mJy sample is not likely to be seriously in error. This redshift distribution is consistent with the predictions of a power-independent high-redshift cutoff of the form displayed by luminous sources (Dunlop & Peacock 1990), and is clearly inconsistent with an unchanged radio luminosity function beyond $z \simeq 2$.

Given the ease with which even low-level star-formation activity can mask the properties of an underlying old stellar population, the objects of greatest importance for constraining the epoch of elliptical galaxy formation are the reddest galaxies at $z > 1.5$. We have therefore isolated a subset of 10 objects with $R - K > 5$ and $z_{est} > 1.5$ for intensive study. The initial results from this detailed investigation are extremely interesting. Most excitingly, as a result of 5 hours of integration with the Keck telescope, we have determined an absorption-line redshift of $z = 1.55$ for a mJy radio galaxy which is extremely red ($R - K = 6$), appears to be devoid of emission lines, and has an ultraviolet spectrum very similar to that produced by main-sequence stars of spectral type F/G. A fuller analysis of the rest-frame ultraviolet spectrum of this object indicates that the main-sequence turnoff point in this object must lie near spectral type F2, implying an age of 3.5 Gyr for solar metallicity. The existence of such an old galaxy at $z = 1.55$ sets strong constraints not only on the epoch of elliptical galaxy formation, but also on cosmological models (Dunlop et al. 1995), and so we are currently investigating the robustness of this age estimate as a function of assumed IMF and metallicity.

References

Chambers, K.C., Miley, G.K. & van Breugel, W. (1990) ApJ, **363**, 21.

Dunlop, J.S. & Peacock, J.A. (1990), MNRAS, **247**, 19.

Dunlop, J.S. & Peacock, J.A. (1993), MNRAS, **263**, 936.

Dunlop, J.S., Peacock, J.A. & Windhorst, R.A. (1995) In: 'Galaxies in the Young Universe', Proc. Ringberg Conference, eds. Hippelein, H. & Meisenheimer, K., in press.

Dunlop, J.S., et al. (1995), Nature, submitted.

Eales, S. & Rawlings, S. (1993), ApJ, **411**, 67.

Eales, S., Rawlings, S., Puxley, P., Rocca-Volmerange, B. & Kuntz, K. (1993), Nature **363**, 140.

Neuschaefer, L.W. & Windhorst, R.A. (1995), ApJS, **96**, 371.

DISTANT RADIO GALAXIES: THE STRONG LINK BETWEEN THE RADIO AND OPTICAL EMISSION.

HUUB RÖTTGERING

Leiden Observatory, The Netherlands

Abstract. Recent observations of distant radio galaxies show that there is a strong link between the radio source and the optical continuum and Lyα line emission from the galaxy. This link is discussed in terms of differences in age, orientation and environment between the radio sources.

The double peaked velocity structure of the Lyα emission of the radio source 0943−242 ($z = 2.9$) is likely to be caused by extended regions (e.g. > 13 kpc) of neutral hydrogen with column densities of 10^{19} cm^{-2} (Röttgering et al. 1995).

The Lyα gas associated with the radio galaxy 1243+036 ($z = 3.6$) has three distinct components (van Ojik et al. 1995a): (i) gas with a high velocity dispersion (1550 km s^{-1} FWHM) located inside the radio structure, (ii) enhanced Lyα emission blue-shifted by 1100 km s^{-1} at the location of the strong bend in the radio jet and (iii) Lyα emission extending out well beyond the radio lobes. This emission has a low velocity dispersion (250 km s^{-1} FWHM) and a velocity gradient of 450 km s^{-1} over the extent of the emission, indicative of large scale rotation. We advocate a scenario for the formation of this object in which the outer halo is associated with the accretion of gas during the formation of the galaxy.

The results on these 2 radio galaxies prompted us to study the H I absorption and dynamics of the Lyα gas for a sample of distant radio galaxies (van Ojik, 1995; van Ojik et al. 1995b). Here we briefly summarize the results. (1) In 11 radio galaxies from a sample of 18 we find strong (> 10^{18} cm^{-2}) H I absorption. (2) The radio sources larger than about 50 kpc do not show strong absorption; almost all the smaller ones do. Other clear indications that the radio source size and the Lyα emission are closely linked are that (3) higher velocity dispersions in the Lyα gas are found in the smaller

R. Ekers et al. (eds.), Extragalactic Radio Sources, 583–584.

radio sources and (4) larger radio sources tend to have larger regions of Lyα emission. (5) The amount of the distortion in the Lyα gas strongly correlates with the amount of distortion in the radio sources. (6) It seems that – at least in some sources – there are 2 components in the Lyα halo; an inner halo located within the boundaries of the radio source, that has a high velocity dispersion ($700 - 1600$ km s^{-1}) and an outer halo located outside the radio source that has a low velocity dispersion (~ 300 km s^{-1}).

Finally, (7) the optical morphologies as observed by HST of a complete sample of 3CR radio galaxies in the redshift interval $1 \lesssim z \lesssim 1.3$ are highly dependent upon their radio properties (Best et al. 1995). There is a clear evolution of the optical structures as the size of the radio source increases: small radio sources consist of many bright knots, tightly aligned along the radio axis, whilst more extended sources contain fewer (generally no more than two) bright components and display more diffuse emission.

It seems difficult to understand these trends as reflecting differences in orientation. For example, in such a scenario it is not clear why smaller radio sources show such a pronounced absorption. It is possible to explain most of the trends through a scenario in which the smallest radio sources are in the densest environments. In such a dense environment there is a lot of neutral gas around the small radio sources to absorb the Lyα emission. The radio source heavily interacts with the dense gas leading to disturbed radio morphologies and relatively small radio source sizes. The large differences in radio source size within the 3CR sample has to be at least partly due to ageing of the radio source. We therefore conclude that it is likely that both the environment as well as the age of the radio source determine observed differences in these samples of distant radio sources.

Acknowledgements. I would like to thank my collaborators, Philip Best, Malcolm Bremer, Chris Carilli, Dick Hunstead, Malcolm Longair, George Miley and Rob van Ojik for the numerous discussions.

References

Best P., Longair M. S., Röttgering H. J. A., 1995, Evolution of the aligned structures in $z \sim 1$ radio galaxies, MN: submitted

Röttgering H., Hunstead R., Miley G. K., van Ojik R., Wieringa M. H., 1995, *MNRAS*, **277**, 389

van Ojik R., 1995, *Ph.D. thesis*, University of Leiden

van Ojik R., Röttgering H., Carilli C., Miley G., Bremer M., 1995a, A radio galaxy at $z = 3.6$ in a giant rotating Lyman α halo, *A&A*: in press

van Ojik R., Röttgering H. J. A., Miley G. K., Hunstead R., 1995b, The Gaseous Environment of Radio Galaxies in the Early Universe: Kinematics of the Lyman α Emission and Spatially Resolved HI Absorption, *A&A*: submitted

THE R-BAND HUBBLE DIAGRAM FOR GPS GALAXIES.

IGNAS SNELLEN[1], MALCOLM BREMER[1,2], RICHARD SCHILIZZI[3,1], GEORGE MILEY[1] AND ROB VAN OJIK[1]

[1] *Leiden Observatory, The Netherlands*
[2] *Institute of Astronomy, Cambridge, U.K.*
[3] *Joint Institute for VLBI in Europe, Dwingeloo, The Netherlands*

Abstract. The Hubble diagram of GPS galaxies has a low dispersion and a steep slope compared with that for 3C galaxies. The relative faintness of GPS galaxies at high redshift may be due to the absence of the aligned optical/uv component seen in high redshift 3C galaxies. The GPS Hubble relation is too steep to fit with evolution models for passively evolving ellipticals. This could be caused by the dynamical evolution of the GPS galaxies.

Gigahertz Peaked Spectrum (GPS) sources are a class of compact extragalactic radio source with a dominant peak in their radio spectrum at about 1 GHz in frequency. It is believed that GPS sources are confined within the inner regions of their host galaxies, either because they are young radio sources that will evolve into extended radio galaxies (Readhead et al 1994, Fanti et al 1995), or because they are surrounded by a particular dense interstellar medium (O'Dea et al 1991).

We are carrying out a project to investigate the nature of these objects by studying a sample of *faint* GPS sources from the Westerbork Northern Sky Survey (WENSS), and comparing them with brighter samples (e.g. O'Dea et al 1991, Stanghellini et al 1994).

While making an inventory of the optical properties of *bright* GPS sources we noted that the Hubble diagram of GPS galaxies has a low dispersion and a steep slope compared with that for 3C galaxies. A detailed description of this work can be found in Snellen et al 1995.

Figure 1 shows the R-band Hubble diagram for a sample of 22 GPS galaxies found in the literature. Also shown are the Hubble diagram for 3C

R. Ekers et al. (eds.), Extragalactic Radio Sources, 585–587.

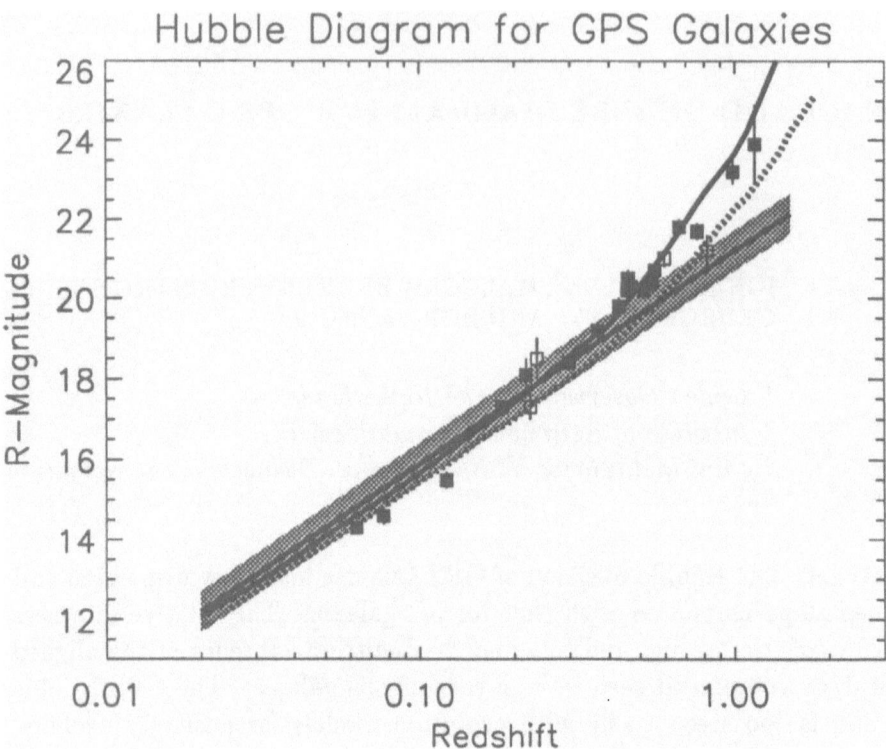

Figure 1. The R-band Hubble diagram for GPS galaxies. The open squares indicate the GPS galaxies at low galactic latitude ($b < 15°$). Note the low dispersion and steep slope of the GPS Hubble relation. The grey band indicates the Hubble relation for 3C galaxies. The 3C galaxies are clearly brighter at high redshift. The solid line is the expected curve for non-evolving ellipticals. The dotted line is the expected curve for passively evolving ellipticals (Bruzual, C-model).

galaxies and models for passively evolving and non-evolving ellipticals. The dispersion around a least squares fit is 0.37 magnitudes, but decreases to 0.28 magnitudes when the four galaxies with the lowest galactic latitude are excluded due to the uncertainty in galactic extinction.

The dispersion is low and the slope is steep in comparison with the Hubble diagram for 3C galaxies (Eales et al 1985). The dispersion and slope are comparable with the Hubble diagram for first ranked cluster galaxies (Hoessel et al 1980), although that is only determined at low redshift. The relation is too steep to fit with a passively evolving elliptical model (Bruzual 1983), but it can be fitted with a model of an elliptical galaxy with an old stellar population which is not evolving (Coleman et al 1980)

Discussion

It appears that the Hubble diagram for GPS galaxies is consistent with them having an old stellar population that undergoes no cosmological evolution. That the galaxies are in a special phase of their evolution (the "GPS" phase) could explain this. However, nearby GPS galaxies show a close companion or double nucleus, which indicates that galactic cannibalism (dynamical evolution) may play an important role in these objects. It is possible that the dynamical and stellar evolution cancel each other out more or less, which makes the Hubble diagram consistent with a non-evolution scenario. This means that the mass of a GPS galaxy would need to increase typically by a factor 2-3 between a redshift of 1 and 0.1, assuming M/L is constant.

We suggest that the GPS galaxies are fainter at high redshift than 3C galaxies, because the former do not have the extra component seen in high redshift 3C galaxies responsible for the optical/radio alignment effect. This aligned light usually completely dominates the rest-frame UV emission of 3C galaxies at $z > 0.6$, although there is wide variation from source to source. If GPS sources are missing this extra aligned component of emission, this would explain why the R-band magnitude of the GPS galaxies are similar to 3C galaxies at low redshift, but fainter at higher redshifts. This would also explain the larger scatter in the 3C Hubble relation, due to source-to-source variation in the aligned component.

The comparable magnitudes and dispersions of the Hubble relations for GPS galaxies and first ranked cluster galaxies suggest that GPS galaxies are located in the centres of clusters. However there is, as yet, no observational evidence for this.

References

Bruzual G.A., 1983, *Rev. Mex. Astr. Astrofiz.*, 8, 63
Coleman G.D., Chi-Chao Wu, Weedman D.W., 1980, *Astrophys. J. Suppl.*, 43, 393
Eales S.A., 1985, *Mon. Not. R. Astr. Soc.*, 213, 899
Fanti C., Fanti R., Dallacasa D., Schilizzi R.T., Spencer R.E., Stanghellini C., *Astr. Astrophys.*, in press
Hoessel J.G., Gunn J.E. and Thuan T.X., 1980, *Astrophys. J.*, 241, 486
O'Dea C.P., Baum S.A., Stanghellini C., 1991, *Astrophys. J.*, 66, 380
Readhead A.C.S., Xu W., Pearson T.J., 1994, *in Compact Extragalactic Radio Sources*, ed. J.A. Zensus and K.I. Kellermann, p17
Snellen I., Bremer M., Schilizzi R., Miley G., van Ojik R., 1995, *Mon. Not. R. Astr. Soc.*, in press
Stanghellini C., O'Dea C.P., Baum S.A., Laurikanen E., 1993, *Astrophys. J. Suppl.*, 88,1

STELLAR POPULATION MODELS OF DISTANT RADIO GALAXIE$

D. VILLANI
Dipartimento di Astronomia, Università di Firenze
Largo E.Fermi 5, Firenze, Italy

AND

S. DI SEREGO ALIGHIERI
Osservatorio Astrofisico di Arcetri
Largo E.Fermi 5, Firenze, Italy

1. Introduction

Stellar populations of high redshift radio galaxies (**HzRG**) (z up to 4.2) are the oldest stellar systems known, that is the ones formed at the earliest cosmological epochs. Therefore they are the best objects for providing us with information about the epoch of galaxy formation. The information on the stellar populations in HzRG are obtained from the study of their Integrated Spectral Energy Distribution (**ISED**) which are gathered both from spectra and integrated magnitudes. The most common approach for the interpretation of colors and spectral features of the energy distribution of galaxies is the Evolutionary Population Synthesis (**EPS**), which has been introduced for the first time by Tinsley in 1972. EPS models have often been used in the past to interpret the ISED of HzRG (Chambers & Charlot 1990; Lilly & Longair 1984; di Serego Alighieri et al. 1994) in order to draw conclusions on the age of the stellar populations and therefore on the epoch of galaxy formation. The results are sometimes conflicting and a number of very recent EPS models have become available (Bressan et al. 1995; Bruzual & Charlot 1993; Buzzoni 1989; Guiderdoni & Rocca-Volmerange 1987): we are therefore analysing the differences between the various EPS models with the aim of assessing their suitability to study the stellar population at early epochs. The EPS models assume for stars a given Initial Mass Function (**IMF**) as well as a Star Formation Rate (**SFR**). Then one can compute the number of stars with given mass present in the galaxy as a function of time. The position of each star in the HR diagram is determined by means of

R. Ekers et al. (eds.), Extragalactic Radio Sources, 588–590.

the isochrones, which are calculated from stellar evolutionary models. The ISED of a galaxy is obtained from the superposition of the spectra of single stars obtained from a stellar spectral library. Thus these models describe the galaxy ISED as a function of the time, giving a complete evolutionary picture.

2. Discussion

The key ingredients of the EPS models available at present are: 1) the assumption for the IMF, the SFR and chemical composition; 2) the library of evolutionary tracks used to calculate isochrones in the HR diagram; 3) the library of stellar spectra adopted to derive the ISED.

We want to stress that the quality of an EPS model depends primarily on the quality and completeness of the library of evolutionary tracks derived from stellar models and the technique adopted to calculate isochrones in the HR diagram.

Analysing the various EPS models, we verify that everyone uses different libraries mainly because these do not extend over the desired range of mass, chemical parameters, and evolutionary phases. Since the libraries of evolutionary tracks differ in the basic input physics, this way of proceeding can be dangerous because it may alter the relative number of stars present in different evolutionary stages, and hence give rise to spurious effects on the final ISED. Moreover it must be emphasized that a crucial point of these models is the choice of the stellar birthrate ($dN = \Psi(t, Z)\Phi(M)dtdM$), for which the theoretical argument is until now a debated question.

We compare, here, the models of Bruzual & Charlot (1993), used until now to model the stellar component of HzRG, with the most recent models of Bressan et al. (1995). We find that: 1) Bressan's models have a larger contribution from red stars in the near-IR by a factor of 2 - 3, decreasing with age; 2) the UV-rising branch is higher in Bressan's models by a factor of 4 at 10 Gyr (figure 1).

To understand these differences between the two models we have to take into account primarily that they were built using different libraries of evolutionary tracks.

The larger contribution from red stars present in the Bressan's models may come from the particular care with which the evolutionary sequences are explicitly calculated up to the late stages, namely the thermally pulsing regime of the asymptotic giant branch phase (TP-AGB) or the central C-ignition (as appropriate for the initial mass of the stars), with respect the semiempirical calculations to which Bruzual et al. were forced by the incompleteness of the tracks of Maeder & Meynet (1991) for the late evolutionary stages. The reason for the higher UV-excess in the Bressan's

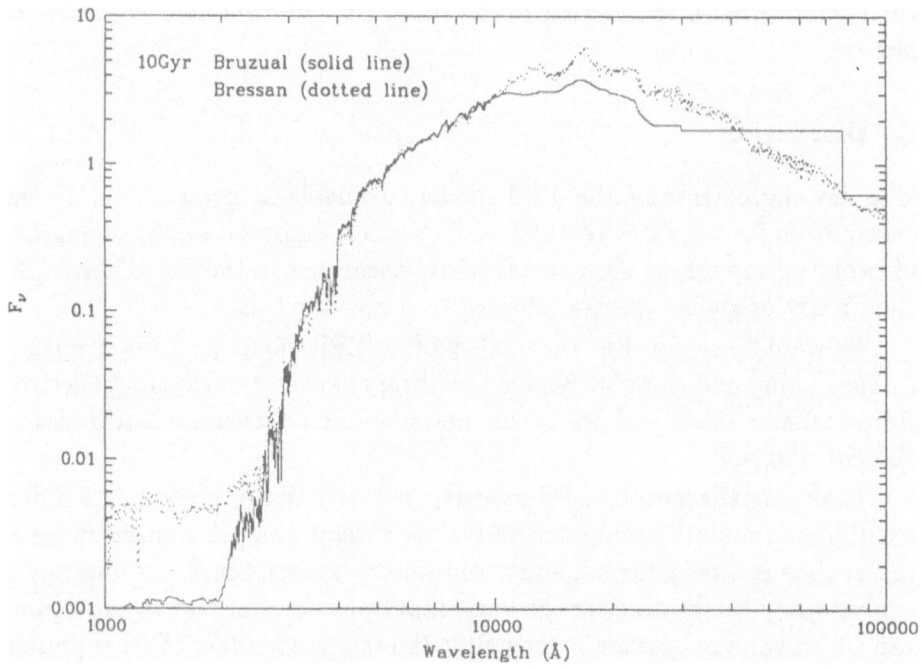

Figure 1. Comparison of the synthesis models of Bruzual et al.(1993) and of Bressan et al. (1995) for an age of 10 Gyr, with a Salpeter IMF and an Instantaneous Burst SFR. Notice the difference in the near-IR and UV fluxes discussed in the text.

models may be due to the different metallicity content, which is <u>crucial</u> for the UV emission, as discussed by Bressan et al. (1994). While Bressan et al. consider the chemical enrichment as a result of galactic evolution and the average value of the final metallicity is above solar (they also take into account the helium enrichment law $\frac{\Delta Y}{\Delta Z}$), Bruzual et al. consider only solar values both for the metallicity and helium content.

References

Bressan, A., Chiosi, C., Fagotto, F., 1994, *ApJS*, **94**, 63.
Bressan, A., Chiosi, C. & Tantalo, R., 1995, *A&A*, submitted.
Bruzual, A.G. & Charlot, S., 1993, *ApJ*, **405**, 538.
Buzzoni, A., 1989, *ApJS*, **71**, 817.
Chambers, K.C.& Charlot, S., 1990, *ApJ*, **348**, L1.
di Serego Alighieri, S., Cimatti, A. & Fosbury, R., 1994, *ApJ*, **431**, 123.
Guiderdoni, B. & Rocca-Volmerange, B., 1987, *A&A*, **186**, 1.
Lilly, S.J. & Longair, M., 1984, *MNRAS*, **211**, 833.
Maeder, A. & Meynet, G., 1991, *A&AS*, **89**, 451.

THE RATAN-600 - VLA - 6M RUSSIAN TELESCOPE: EARLY UNIVERSE PROJECT

Y.N. PARIJSKIJ[1], N.S.SOBOLEVA[1], W.M. GOSS[2], A.I. KOPYLOV[1], O.V. VERKHODANOV[1], A.V. TEMIROVA[1] AND O.P. ZHELENKOVA[1]

[1] *Special Astrophysical Observatory of the Russian Academy of Sci - 357147 Nizhnij Arkhyz Karachaj-Cherkessia, RUSSIA*
[2] *National Radio Astronomocal Observatory - Edgemont Road, Charlottesville VA 22903 U.S.A.*

We present the preliminary results of the "BIG TRIO" project of penetration into the "Dark Age" of the Universe, between the recombination epoch and the epoch of the first QSO using RATAN-600 "Cold Experiment" deep (few mJy) multi-frequency strip survey (Parijskij, Korolkov, 1986; Parijskij et.al,1991,1992). Our general approach is close to the classical SS FRII RG selection rules of very distant galaxies with old stellar population (McCarthy, 1993) with small improvements. Details may be found in (Kopylov et al, 1995). Mean estimated z of our SS FRII RG's is about 1.5, 20% of all objects have z larger than 3. Before the direct spectroscopy of all SS FRII objects we have estimated their redshifts by different methods: using the updated Hubble diagram, θ-z and flux density-z relations and now begun to use multi-color measuring all our objects to estimate "color z" and a stellar age of the parent galaxies using available models of stellar population evolution in such objects (gE class). The first subgroup of 16 objects (not the weakest ones) measured in BVRI gave us the following preliminary results using Bruzual-Charlot model (Bruzual et al., 1993). Looking at this Table, we can make the following statements: mean color z occurred to be very close (within 15%) to the photometric z, but individual redshifts may differ. A mean age of the stellar population is about 1 Gyr, and, at least in some objects, star formation began when the Universe was only 1 Gyr old ($\Omega_0 = 1, H_0 = 50$), that is older than the most distant QSOs. One object, RC0934+0505, happened to be the oldest object found up to now. We estimate, that there are about 10000 objects

R. Ekers et al. (eds.), Extragalactic Radio Sources, 591–592.

TABLE 1. STARS FORMATION EPOCH FROM "BIG TRIO" PROJECT

N	RC-name	B-V	V-R	R-I	R	Z-rg	Star age, Gyr	Z-star formation
1	0837+0446	0.01	0.79	-0.04	22.17	1.5	0.33	1.7
2	0908+0451	0.95	0.90	0.81	19.96	0.5	1.1	0.7
3	0934+0505	0.87	-0.13	1.38	24.07	3.4	1.1	9.5
4	1031+0443	1.17	0.72	1.27	22.18	1.0	1.1	1.4
5	1152+0449	0.22	1.28	1.41	22.45	1.0	1.1	1.4
6	1155+0444	1.55	0.94	0.71	18.95	0.5	1.1	0.7
7	1219+0446	0.61	0.84	0.30	22.25	2.3	1.1	4.1
8	1333+0452	0.46	0.89	1.12	23.64	1.1	0.9	1.4
9	1357+0453	1.16	0.76	1.01	21.17	0.5	1.1	0.7
10	1436+0501	0.07	0.50	0.71	23.47	2.0	0.6	2.6
11	1510+0438	1.30	1.19	1.35	22.69	0.6	1.7	0.9
12	1626+0448	-0.69	0.37	0.14	22.88	2.6	0.35	3.1
13	1646+0501	1.66	1.53	1.28	21.21	0.7	1.7	1.1
14	1703+0502	0.96	0.35	0.96	23.47	3.4	0.6	5.2
15	1740+0502	1.83	0.55	0.64	22.63	0.6	0.9	0.8
16	2013+0508	0.76	0.29	0.23	21.12	2.9	0.4	3.6

with such an age on the sky which can be easily selected (especially with new generation of radio and optical catalogs now in preparation) and fully studied spectroscopically even by present day facilities. The only evolution effect we have noticed is that more distant objects have smaller ages of stellar populations. There is weak z-dependence of the ratio of radio to optical luminosities in the objects of our list, which disappears when we take into account the K- correction to the SED in gE.

This work was done with partial support by ISF Grant 96300, Russian "Cosmion" and RFFI grants 93-02-1738, 95-02-03783.

References

Bruzual G, Charlot, S. (1993) *Ap.J.*, **405**, 538.
McCarthy, P. (1993) *Ann.Rev. of A & Ap.*, **31**, 639.
Kopylov A., Goss W., Parijskij Y., Soboleva N., Zhelenkova O., Temirova A., Vitkovskij Val., Naugolnaya M., Verkhodanov O. (1995) it A.J.,Russian, **72**, 437.
Parijskij, Y., Korolkov, D. (1986) *Ap& Space Phys.Rev.*, **5**,40.
Parijskij, Y., Bursov, N., Lipovka, N., Temirova, A. (1991) *A &A Suppl.Ser.*, **87**, 1.
Parijskij, Y., Bursov, N., Lipovka, N., Soboleva, N., Temirova, A., Chepurnov, A. (1992) *A &A Suppl.Ser.*, **96**, 583.

FAINT RADIO SOURCES AND THE COSMIC MICROWAVE BACKGROUND

ERIC A. RICHARDS

University of Virginia
Charlottesville, VA

1. Introduction & Observations

We have mapped a single field with the VLA to an unprecedented rms sensitivity of 1.5 μJy. Our observations reveal that the excess μJy population (see Windhorst et al (1993), Fomalont et al. (1993)) is continuous down to 1 μJy. In addition, we measure a microwave sky temperature of $\Delta T/T = (1.4 \pm 1.2) \times 10^{-5}$, consistent with microwave decrements we discovered near the center of our map.

We imaged a single field at (J 2000) RA = 13^h12^m and DEC = $+42°38'$ from October 1993 through January 1995 with the VLA, giving us 159 hr of good quality data. Observing with both the C and D configurations gave us a combined synthesized beam of FWHM $\approx$ 6". Our field of view as defined by the FWHM of the primary beam was 312". After proper editing and weighting of the data, we obtained a point source sensitivity of 7 μJy (5 σ). As the brightest source in our field of view was S = 273 μJy, our observations were *not* dynamic range limited.

2. Removing the Foreground Sources

The radio sources observed above our completeness limit in our field are discussed by Fomalont et al. elsewhere in these proceedings. Here we will focus on the subliminal radio sources. In order to gain information on the sky density of sources $\leq$ 7 μJy, we attempted to model the faint radio population. We used the following scheme.

1. We fit an integral source count to *all* sources $\geq$ 7 μJy within our field of view to empirically determine a source count $N(\geq S) = (17\pm2)S^{-1.2\pm0.2}$ per arcmin^2.

R. Ekers et al. (eds.), Extragalactic Radio Sources, 593–594.

2. We randomly populated an area 10 arcmin^2 in size between 0.2 - 300 μJy with point sources according to the above power law.

3. We calculated the visibility function for this simulated sky at the *identical* (u,v) point sampled by the observations and imaged the subsequent simulated data, including receiver noise.

4. We subtracted out all sources ≥ 7 μJy from the simulated map by isolating the (u,v) data associated with each source.

5. We preformed a statistical analysis on the resultant map, comparing it to our *observed* map with all sources ≥ 7 μJy subtracted out.

We find that the excess signal found in the center of our observed residual map is consistent with the contribution expected from faint radio sources between 7 - 1 μJy.

3. Isolating the Cosmic Microwave Background

Our observations are also well suited for small scale ($\theta \leq 1$') CMB anisotropy searches. If we can accurately measure or estimate the relative contribution of the faint radio sources and the receiver noise, then we can measure the smoothness of the microwave sky. Our analysis included the following steps.

1. We preformed a variance analysis in 200" concentric rings about the phase center on our *simulated* residual maps (all sources ≥ 7 μJy removed), smoothed to 60".

2. We used our *observed* residual map, smoothed to 60", for a likewise variance comparison.

3. By dividing our master (u,v) data set into two equal halves, mapping each part identically, and then subtracting the two, we are left with an excellent measurement of the instrumental noise during our observations.

After accounting for receiver noise and faint radio sources, we find an excess signal which we attribute to CMB fluctuations.

4. Conclusion

Our CMB measurement is equivalent to $\Delta T/T = (1.4 \pm 1.2) \times 10^{-5}$. Although experimentally marginal, this is consistent with 'cool spots' we found at higher resolution ($\theta \approx 18$"). The most negative source has a microwave surface brightness of $\Delta T = -0.4$ mK. The most plausible physical mechanism is a SZ cluster, for which we derive a $M_{gas} \approx 10^{13}$ M_{solar} assuming $T_{ICM} \approx 1$ kev. This perspective cluster probably lies at $z = 2.56$, as we have detected two QSO's and three Lyα emitters at this distance in subsequent ground based narrow band imaging.

FSRQ AND THE GAMMA-RAY BACKGROUND

T. DI GIROLAMO, A. COMASTRI, G. SETTI
Istituto di Radioastronomia del CNR
Via Gobetti 101, I–40129 Bologna, Italy

The contribution of Flat Spectrum Radio Quasars (FSRQ) to the γ-ray background (GRB) is modeled. FSRQ are known to make a substantial contribution to the hard (E>100 MeV) background. The so called "MeV bump", however, cannot be accounted for in terms of this class of sources even taking into account the newly discovered class of FSRQ with a broad band spectrum sharply peaked in the MeV range ("MeV blazars"). This is in agreement with the recent results obtained by Kappadath et al. (1995) using COMPTEL data.

The most important discovery of the CGRO EGRET in the field of the extragalactic astronomy is the detection of high energy γ-rays (E>100 MeV) from active galaxies. At present (Thompson et al. 1995) some 50 sources have been identified, the majority with FSRQ while about ten are classified as BL Lac objects. The purpose of this work is to estimate the contribution from FSRQ to the GRB using the available γ-ray properties recently discovered by CGRO COMPTEL and EGRET observations coupled with the available informations at radio and X-ray wavelenghts. The values H_0=50 km s^{-1} Mpc^{-1} and q_0=0 have been used.

Our model for the synthesis of the overall GRB is anchored to the X-ray emission properties of FSRQ at 1 keV. The parameters for the broad band X- and γ-ray spectrum and cosmological evolution of FSRQ have been chosen in order to obtain a good fit to the overall set of observational constraints. All the assumed parameters are consistent, within the errors, with those suggested by the present available observations.

The adopted local emissivity at 1 keV is of 9.7×10^{19} W Hz^{-1} Gpc^{-3}. For a typical FSRQ the spectrum can be represented by a broken power law with α_x=0.5, α_γ=1.2 and a break energy at 7 MeV, while for the "MeV blazars" the mean spectrum has α_x=0, α_γ=2 and a break energy at 2.5 MeV. The evolution is parameterised as a power law such that the luminosity L(z)=L(0)$\times$(1+z)$^\beta$. We have adopted β=3, an evolution cut-off

R. Ekers et al. (eds.), Extragalactic Radio Sources, 595–596.

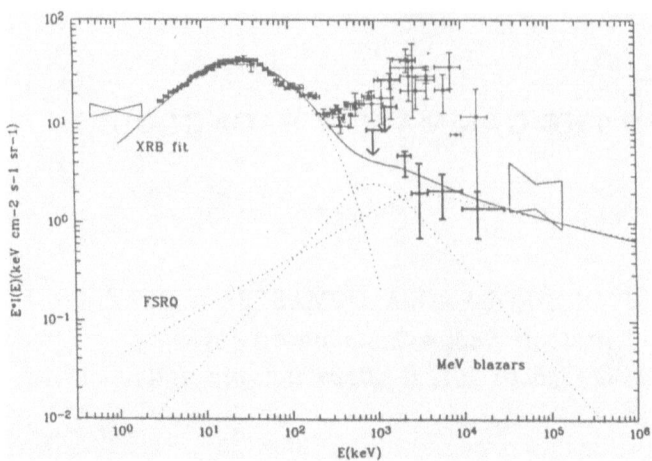

at a redshift $z_{cut}=2.5$ and then a constant emissivity up to $z_{max}=5$. The predicted background intensity in the energy range 1 keV – 1 GeV has been computed assuming that 95% of the local emissivity at 1 keV is due to the FSRQ and the remaining 5% to the "MeV blazars". This ratio reflects the fact that only a few "MeV blazars" have been discovered by COMPTEL compared with the roughly 40 FSRQ in the EGRET band. Our results are shown in the figure. Compared with a selection of data on the X- and γ-ray backgrounds, the solid line is the sum of three distinct contributions (dashed lines): the fit to the X-ray background obtained with the AGN model of Comastri et al. (1995),the FSRQ and the "MeV blazars" contributions to the GRB derived in the present model. The thick error bars in the Mev region represents the recent results obtained by COMPTEL (Kappadath et al. 1995). The contribution of FSRQ to the GRB (>100 MeV) is $\simeq 70\%$.

In order to make full use of the available data we have made an attempt to predict a Log N – Log S relationship and a redshift distribution and compare them with those obtained with EGRET. We have found consistency with the observations by adopting a γ-ray luminosity function directly derived from the radio one via the strict correlation between radio and γ-ray luminosities of EGRET FSRQ. Source variability and our lack of knowledge about the EGRET sky-coverage makes a strict comparison difficult.

As a general conclusion it appears that various classes of AGN are able to account for most of the extragalactic background radiations observed over the very wide energy interval from about 1 keV to tens of GeV.

References

Comastri A., Setti G., Zamorani G., Hasinger G., 1995, *A&A*, **296**, 1
Kappadath S.C., et al., 1995, in *Proceedings of the 3rd COMPTON Symposium*, München, June 1995, in press
Thompson D.J., et al., 1995, *ApJS*, in press

COSMOLOGICAL IMPLICATIONS (*DISCUSSION*)

Discussion of the paper presented by <u>CONDON</u> (p. 535)

Stocke: Since real limits are surface brightness limits not flux, please comment on the possibility of sources missed because they have very low surface brightness.

Condon: Magellanic irregular galaxies have radio surface brightness comparable with the rms brightness fluctuation (≈ 0.1 K at 1.4 GHz) of the radio source background, so they cannot be detected reliably even by the most sensitive radio telescope. The Wrobel & Heeschen (1991) sample of E/S0 AGNs appears to be brightness limited by the VLA observations with 5″ resolution. This limitation can be corrected by the NVSS, which reaches to within an order-of-magnitude of the confusion limit, and Bill Cotton's preliminary results on the UGC galaxies (this volume) reveals a significant new population of low-brightness radio sources in elliptical galaxies.

Ekers: Your simple model to fit the source counts implies that the "normal galaxy" population has the same strong evolution seen in the powerful radio source population.

Condon: That's right. Even though the energy sources (stars and AGN) are quite different, the amount of evolution is about the same. This suggests a common evolutionary mechanism. For example, galaxy collisions are known to trigger starbursts, and they may also increase the fueling rate of central "monsters".

Lari: I wonder if the different estimates of radio luminosity function don't reflect that different type of galaxies have different (optical?) luminosity function. Different luminosity function would reflect the local inhomogeneity.

Condon: If different types of Galaxies (eg., spirals and ellipticals) all had the same radio/optical ratio R, then their radio luminosity functions would mimic their optical luminosity functions, just as the radio and FIR luminosity functions of "normal" galaxies are alike. In fact, the R-distributions of spirals and ellipticals are quite different. This broad

597

R. Ekers et al. (eds.), Extragalactic Radio Sources, 597–602.
© 1996 *IAU.*

range of radio properties is more important than the optical luminosity function differences in accounting for the differing radio luminosity functions of different galaxy types.

Lari: Counts are not always corrected for statistical bias.

Condon: I don't understand; does "counts" refer to source counts (n(s)) or numbers of galaxies in luminosity function bins? What statistical bias?

Discussion of the paper presented by WALL (p. 547)

Padovani: Your "beamed" objects include only the flat-spectrum sources. There are various indicators that steep-spectrum radio quasars are also beamed, although at somewhat larger angles. Did you try to include them in your calculations?

Wall: Our "beamed" objects were in fact drawn from the steep-spectrum luminosity function, using the high-power (evolving) end only. As yet we have not indulged in any refinements of the type you mention; but further work will incorporate them.

Discussion of the paper presented by FOMALONT (p. 555)

de Bruyn: The count convergence at 100 nJy depends on the integrated sky temperature value of 0.02 K (excluding MWB). What is the error on this number and how would you determine it?

Fomalont: The sky brightness temperature has been measured by COBE above 19 GHz and from the ground above 5 GHz. All measurements are consistent with a constant CMB of 2.7K at all these frequencies to an error of no more than 0.02K. Hence, any residual sky brightness caused by radio sources at 8.1 GHz must be less than about 0.02K.

Discussion of the paper presented by HAMMER (p. 559)

Becker: Have you calculated the chance coincidence rate between radio and optical sources?

Hammer: Not exactly. But the coincidence between radio and optical was much better than 1″ for all sources but two.

Wilson: You said that 25% of the radio sources are identified with low redshift, low luminosity AGN's and referred to them as Seyfert 2 or Seyfert 1. You also said these objects have inverted radio spectra. Such radio spectra are extremely unusual among normal Seyfert galaxies. Do you have any suggestions about this difference in the radio spectra?

Hammer: These objects have also other differences with "normal" Seyferts.
- their radio power might be higher ($P \gtrsim 10^{23} W/Hz$)
- their emission line ratios (OIII/H$_\beta$, SII/H$_\alpha$) are often intermediate between LINERS and Seyfert
- their optical luminosities are rather small (substantially smaller than L^*).

Wall: A very large proportion of radio sources in the 10-100 mJy range turn out to be red ellipticals which obey a tight Hubble relation and which have no evidence of emission lines. These are 'proper' double-structure 'AGNs', real radio sources. It may be that your faint passive ellipticals are from the same population.

Discussion of the paper presented by <u>SHAVER</u> (p. 561)

Jauncey: I am concerned that your selection of "no B-band" is based on spectra of optically selected QSOs, which were chosen on the basis of optical colours, in particular for having no B-band emission. So your selection for spectroscopy of "no B-band" is not necessarily a property of the quasars, but of the optical survey. It is known that several of the radio QSOs show emission shortward of the 912 Å limit at the emission line redshift.

Shaver: This criterion is based on a property of the intervening medium, not of the QSOs themselves. At $z > 4 - 5$ the increasing density of intervening Lyman-limit absorbers at high redshift blocks out the flux from the QSO right up to the Lyman limit in the QSO rest frame (5470 Å at $z = 5$). Spectra of the highest-redshift radio-selected quasars do indeed exhibit this phenomenon (*cf.* Hook *et al.* 1995, *MNRAS* **271**, L63; Shaver, Wall, Kellermann 1995, MNRAS in press).

Meisenheimer: How did you account for the radio luminosity function? If it is steep you will only see the most luminous (and rare) objects at high redshift. This has to be folded into your $\Phi(z)$ diagram.

Shaver: We are just concerned about the space density above a given luminosity. The computed $\Phi(z)$ corresponds to all luminosities above the minimum which could be detected at the highest redshift considered, $z = 7$; the space density of quasars above that luminosity decreases at high redshift.

Bicknell: How do you know that high redshift QSOs are not free-free absorbed by surrounding ionized gas?

Shaver: Our main conclusion concerns *intervening* absorption along the line of sight; absorption by gas and dust *intrinsic* to the QSO does not affect that conclusion. We are talking about a particular observed phenomenon *i.e.* quasars like those seen at $z \sim 1 - 2$, and the space density of such objects decreases at high redshift. Of course there must be precursors of "normal" quasars (and galaxies) at still higher redshifts, at least in the form of mass concentrations, but that is a different issue. And even if there are "hidden" quasars enshrouded in dust and ionized gas, if quasars are young objects we would expect such "hidden" quasars to exist at *all* redshifts, so the redshift turnover would be unaffected. Any such objects may be found by conducting searches in the mm/IR or X-ray bands.

Wright: Could you repeat the argument as to why the turnover in space density applies to *all* QSOs (rather than just *radio-loud* QSOs)?

Shaver: Potentially the most compelling argument concerns the UV background intensity measured *in situ* as a function of redshift using the "proximity effect" observed in the statistics of the Lyα forest in QSO spectra (see fig. 2 Bechtold, 1995 *QSO Absorption Lines*, ed. G. Meylan, ESO Astrophysics Symposia, Springer, p. 299). The UV field also seems to exhibit a pronounced maximum as a function of redshift. This coincides roughly with the redshift where the QSO space density peaks, consistent with the UV field having been produced by the QSOs (and other AGN), and suggestive of a universal turnover.

Vermeulen: What is known about the properties (redshifts, luminosities) of the radio galaxies in your sample?

Shaver: We have just begun a study of these optically-faint, flat-spectrum radio galaxies - it is too early to summarize their properties at the moment.

Discussion of the paper presented by <u>SINGAL</u> *(p. 563)*

Gurvits: Similar studies but on the 10^3 times smaller angular scale (i.e. VLBI) indicate a clear difference in the behaviour of $(\theta - z)$ diagrams on the kpc and pc scales, certainly for quasars. Any comments on this difference?

Singal: If quasars differ from radio galaxies on pc or kpc scales then that is again an evidence against the simple unified scheme. However at those scales the size evolution may not be playing any part, at least not in the same way as for more extended structures. In fact the absence of a size evolution at that scale is assumed in interpreting the $(\theta - z)$ plots for determining q_0.

Leahy: As a member of Neeser et al., I'd like to make two comments:
1) We showed Oort et al suffered a major bias due to underestimating the size of distant FR1's through K-dimming.
2) You need complete redshift because the K-z relation breaks down at $z \geq 2$. This is only available for 3C and our sample so far:- it is very important to complete the optical on big samples like Molonglo and B3.

Kapahi: I would like to make two comments:
1. One can make a fairly strong and simple argument (*Kapahi 1989, A.J.* **97**, 1) for cosmological evolution of linear sizes that may not have been appreciated in the literature. FR II galaxies in the 3CR sample have a median linear size between $\sim$250 and 300 kpc almost independent of redshift up to $z \sim 1$. The minimum angular size subtended by a 250 kpc source (for $q_0 = 0.5$) at any redshift is $\sim 30''$. If linear sizes were independent of redshift and luminosity, median angular sizes (θ_m) in complete flux limited samples of radio galaxies can never be $< 30''$. Whereas it is well known that θ_m approaches a value near $10''$ in complete surveys at flux levels between $\sim$1 and 0.1 Jy at 408 MHz. This can only be explained by invoking evolution in linear sizes as the deeper samples have higher median redshift. It is clearly important to check that models for the z and P dependence of linear sizes are consistent with the observed $\theta - S$ relation as well.
2. The fraction of compact steep-spectrum sources in source samples is a function of both survey frequency as well as the flux density limit. This can affect the comparison of median sizes in different samples and must be kept in mind while interpreting the comparisons.

Discussion of the paper presented by MEISENHEIMER (p. 571)

di Serego Alighieri: In your introduction you said that high redshift radio galaxies are particularly useful to study stellar systems in the early universe; you mentioned young stars to explain the blue colours.

Can you say anything about the stellar content in the galaxies you have studied?

Meisenheimer: Not yet. So far we have concentrated on studying the emission line regions. A detailed investigation of the continuum colors (after subtracting the contribution of the emission lines) has still to be done.

Koekemoer: In your K-band data of 4C 41.17, do you detect any companions which might be interacting – and if so, can you constrain the dynamics of Ly-α in the primary galaxy.

Meisenheimer: In fact, we found several neighbours on our K-band image of 4C 41.17. But none of those show significant Ly-α emission within about $\pm 1500 kms^{-1}$ of the systemic redshift of 4C 41.17. So we do not regard them as physical companions.

Bicknell: Do you find any correlation between excitation and velocity dispersion.

Meisenheimer: We can't tell since we only have Fabry- Pérot data for a single line (either [OII] or Ly- α. For redshifts $z < 0.8$ we could in principle determine an [OIII]/[OII] ratio, but we have not tried this up to now.

CONCLUSIONS

L. WOLTJER

Observatoire de Haute Provence,
F-04870 Saint Michel l'Observatoire, France

At the Albuquerque symposium in 1981, Jan Oort presented a general overview in which he listed a number of questions and tentative answers. The first five items in this summary are from his list.

The source of energy

Probably massive $(GM_\odot)$ black holes provide much of the energy, either through accretion or from their "rotational" energy. However, bursts of star formation appear also important in many cases and may contribute significantly. The possibility of extracting energy from binary black holes was also mentioned here; for the mechanism to be effective near relativistic binaries may have to be considered in which case gravitational radiation may pose a problem.

Much of the energy frequently emerges within a narrow cone

Originally focusing by pressure gradients was believed to be important, but now magnetic focusing is frequently invoked and may well be more effective. It is interesting that the Broad Absorption Line quasars (which may, in fact, be representative of the majority of quasars) show that a less focused subrelativistic flow is also possible. The BAL quasars generally have no significant radio emission which could support the notion of magnetic focusing for radio quasars, but orientation dependent effects may also play a role.

R. Ekers et al. (eds.), Extragalactic Radio Sources, 603–608.

Jets are formed on many scales

At least close to the center these jets appear to be relativistic as evidenced by superluminal motions. This supports the belief that they originate in the environment of a black hole. In the less powerful (FRI) radio galaxies the jets appear to slow down to modest velocities (0.01 c) at larger distances (kpc) from the center. The composition of the jets is still under debate with e^+e^- and pe^- plasmas being preferred under different circumstances. Shocks appear to be important in jets, which is confirmed by polarization observations which indicate that transverse magnetic fields occur frequently.

Numerical simulations have been rather successful in modelling the slowing down of jets by entrainment of surrounding matter. Filamentary structure and composite jets with a slower outer sheath appear in the simulations. Instabilities play an important role. Also jets with a continuous input of energy behave rather differently from those with an episodic input, because of the formation of multiple shocks. In this connection also the disappearance of some optical knots in the M87 jet during only a year seen by HST is of interest. In general there is a remarkable similarity in images of Herbig Haro jets (100 Km s^{-1} in interstellar matter), of jets around some galactic (few $M_\odot$) black holes and of extragalactic jets.

Jets have a rather stable direction during millions of years

Frequently this is ascribed to the stable rotation vector of the black hole; small changes in direction may be due to precessional motions. Some jets, especially at lower energies, are bent and this appears to be due to interaction with the gas in clusters of galaxies; this gas has become visible in X-ray observations.

Superluminal motions

At the Albuquerque meeting 7 cases of motions with apparent velocities in excess of that of light were reported. Now the sample includes 100 objects. The most plausible explanation is still relativistic motion at small angles to the line of sight, but the quantitative relation between bulk velocity of matter and pattern velocity is still unclear. One sided or very asymmetrical jets are then understood as due to Doppler effects. Support for such models come from observation of the Faraday rotation of some sources which indicates which side is directed toward us. The assumption is usually made that jets are intrinsically bisymmetrical, but it is still uncertain how much intrinsic asymmetry there is.

In some cases when optical synchrotron radiation is seen in radio lobes the lifetime of the relativistic electrons is no more than 100–1000 years. (Re)acceleration in situ is therefore needed and a Fermi-type shock acceleration is the most likely explanation. The resulting particle energy spectra are beginning to be understood.

X-rays

With Rosat some 10^5 extragalactic X-ray sources have been detected, mainly AGN. Synchro-Compton models still appear to fit the gross characteristics. Of particular importance are X-ray observations of the density and temperature of the gas in clusters, since from this the pressure in the radio lobes may be inferred. The non uniform temperature (ASCA) and the existence of important velocity fields show the clusters to be far more dynamic than previously thought. In some cases inverse Compton radiation from the lobes has been measured yielding values of the magnetic field, which sometimes are close to equipartition values. The fact that in the clusters the mass determined from the X-ray data is about the same as that inferred from gravitational lensing gives confidence that the parameters of the gas can be believed.

Gamma-rays

Of the gamma-ray sources (with $|b| \geq 30°$) observed with GRO/EGRET more than 2/3 have been identified with BL Lacs and OVV quasars with a very broad luminosity function. Since the gamma-ray background is rather weak more abundant objects like Seyferts and radioquiet quasars must be on average intrinsically faint in gamma-rays.

Synchro-Compton models with photons from the Broad Line Region have been presented; closer to the nucleus the $\gamma\gamma$ opacity would be too high for gammas to emerge. To quantify such models simultaneous optical, X and gamma data are needed with a resolution of less than a day. Such data may be correlated still in 1996 with GRO and later with "Spectrum X, gamma" (1998) and Integral (2002).

Unification with Doppler boosting

If relativistic motions occur radio galaxies with jets pointing close to the line of sight could present the appearance of BL Lacs and it is therefore possible to "these two classes of objects. The situation is less clear with the OVV quasars which have emission lines with equivalent width only slightly smaller than ordinary quasars, while their v/c values seem to be even larger than for BL Lacs.

Unification with an absorbing torus

In some Sy 2 and RG polarized broad emission lines (or heavily absorbed X-rays) are seen which unambiguously indicate that Sy 1 or quasar-like nucleus is hidden behind absorbing matter. Statistical arguments in favour of unification on the basis of absorption effects are weaker. The observational results presented here on the diameters of quasars and radio galaxies appear still to be somewhat contradictory. In any case with models in which the opening angle of the torus is a function of power, age and redshift, there is so much freedom that definite conclusions about unification are not easy to come by.

A more direct way to study the effects of an absorbing torus is to observe in the IR where at some wavelength the torus should become transparent and the BLR visible. With ISO, the Infrared Space Observatory to be launched by ESA later this year, many AGN will be observed and the absorption effects quantitatively elucidated.

Radiative ionization - Shock ionization

In most cases it remains difficult to identify the ionizing mechanism. If radiative models do not work, optical depth effects and variability of the ionizing source may yield a fit to the emission line data. If simple shock models do not work multiple shocks, inhomogeneities and precursors can be invoked.

Magnetic field structure

The main distinction is between small scale tangled fields and larger scale flux carrying fields, the latter being needed to explain Faraday effects. Evidence was presented that Faraday effects originate in many cases outside the radio lobes in the cluster gas. The origin of the large scale fields there is ascribed to dynamo processes, but the details are far from clear. Accurate observations of polarization also in the visible with the new large telescopes should yield more definite information on the nature of the magnetic fields.

High-z radio galaxies

Intriguing optical structures in these objects appear to be correlated with radio structures. The roles of AGN ionizing cones, shocks and star formation remain to be elucidated.

Some fraction of the numerous μJy sources appears to have $z \geq 1$, but (post)starburst galaxies and AGN at lower redshifts also contribute to this population. With an estimated 20 sources per $arcmin^2$ stronger than $1 \mu Jy$ there would be 3×10^9 such sources in the Universe, a significant fraction (10% ?) of all galaxies.

Gravitational Lenses

These give the potentially important information on (H_0, q_0), on intergalactic matter and on masses of galaxies and clusters, but accurate modeling is essential. With the lensing effects by clusters, galaxies and stars all superposed this is far from trivial. It is still uncertain how much the lensing affects the images of high-z galaxies.

Evolution

Most quasars are optically variable by more than 20% on time scales of 1 - 10 years, while global quasar lifetimes of $10^7 - 10^8$ years have been estimated. Nothing is known at intermediate time scales. Obviously, the simplest unified model would involve quasars switching on and off. Important variations in the BLR have been observed with PKS 0521-36 and BL Lac itself developing broad lines. Related variations have also been observed in quasars.

Systematic evolutionary effects (e.g. dust removal) have been suggested to relate hyperluminous infrared galaxies and quasars, while (with appropriate luminosity evolution included) the compact steep spectrum doubles could evolve into large radio galaxies. A certain gross evolutionary similarity is seen in the redshift distribution of radio galaxies, radio quasars and radio quiet quasars with a large increase in numbers towards $z = 2$ and a steep decline for $z \geq 3 - 4$. For the radio objects this decline was shown not to be caused by absorption in the universe. A similar study in X-rays is needed to confirm the same conclusion for radio quiet quasars.

Mergers

There is a certain belief that AGN are generally related to galaxy mergers and in many cases the evidence looks very strong. At the same time there is much uncertainty if most gE galaxies are really the result of mergers, in particular because of arguments involving globular cluster frequencies.

Why ellipticals?

As at Albuquerque, we are not yet able to answer the most conspicuous question: what is so special about gE galaxies that all powerful radio galaxies in our neighborhood originate from these and never from spirals?

The future

VSOP and Radio Astron (1996/98) will improve the resolution of VLBI. Image quality may benefit, too. This is also important for the reliable measurement of the proper motions of features with changeable shapes.

ISO (1995) will tell us how frequent BLR's are in radio galaxies and Sy 2's. It also will contribute to the calorimetry of AGN by allowing a better evaluation of the energy radiated by dust.

HST (1997) will acquire a near infrared capability ($\leq 2.5\mu m$) which should allow intermediate and high-z radio galaxies to be observed with less trouble due to dust.

Large samples of radio and X-ray sources ($10^5 - 10^6$) are becoming available, the former from NRAO, Parkes and Westerbork, the latter from ROSAT. Optical identifications and redshifts are lagging behind. With 10% of the 4-m telescope time available in the word a sizeable part could be dealt with in the coming decade. Cooperative arrangements would be desirable.

Simulations of the complex systems involved in the extragalactic radio sources will remain very necessary and hopefully will lead to a more global understanding.

Radio galaxies were first identified in 1949 and quasar redshits in 1963. By the time of the Albuquerque meeting in 1981, a general knowledge of identifications, evolution and energetics existed. Jets and superluminal motions had been detected and the first gravitational lens identified. In the meantime, we have obtained many new data: instead of mJy sources we now discuss μJy sources, instead of a few hundred X-ray sources we now have more than 10^5. Instead of one uncertain gamma-ray AGN we now have several dozens. However, we still do not have a satisfactory answer to the simple question as to what makes an extragalactic radio source.

AUTHOR INDEX
(references are to the first page of papers)

OBJECT INDEX
(references are to the first page of papers)